Metallurgy: Processes, Technologies and Design

Metallurgy: Processes, Technologies and Design

Edited by
Darren Wang

STATES
ACADEMIC PRESS
www.statesacademicpress.com

Published by States Academic Press,
109 South 5th Street,
Brooklyn, NY 11249, USA

ISBN: 978-1-63989-352-2

Cataloging-in-Publication Data

Metallurgy : processes, technologies and design / edited by Darren Wang.
p. cm.
Includes bibliographical references and index.
ISBN 978-1-63989-352-2
1. Metallurgy. 2. Metallurgy--Technique. 3. Metals--Design. I. Wang, Darren.
TN665 .M48 2022
669--dc23

For information on all States Academic Press publications
visit our website at www.statesacademicpress.com

STATES
ACADEMIC PRESS

Contents

Preface

The branch of materials science and engineering concerned with the study of metallic elements is known as metallurgy. It focuses on the chemical and physical behavior of metallic elements, their alloys and inter-metallic compounds. It studies science and technology involved in the production of metals and the engineering aspects of metal components which are used for consumer as well as manufacturing products. Metallurgy can be classified into physical and chemical metallurgy. Physical metallurgy delves into the physical and mechanical properties of metals. Material characterization, physical performance of metals, phase transformations, mechanical metallurgy and failure mechanisms are some of the focus areas of this domain. The reduction and oxidation of metals along with their chemical performance is studied under chemical metallurgy. It also focuses on metal extraction, corrosion and mineral processing. This book traces the progress of this field and highlights some of its key concepts and applications. It includes some of the vital pieces of work being conducted across the world, on various topics related to metallurgy. Students, researchers, experts and all associated with this discipline will benefit alike from this book.

Various studies have approached the subject by analyzing it with a single perspective, but the present book provides diverse methodologies and techniques to address this field. This book contains theories and applications needed for understanding the subject from different perspectives. The aim is to keep the readers informed about the progresses in the field; therefore, the contributions were carefully examined to compile novel researches by specialists from across the globe.

Indeed, the job of the editor is the most crucial and challenging in compiling all chapters into a single book. In the end, I would extend my sincere thanks to the chapter authors for their profound work. I am also thankful for the support provided by my family and colleagues during the compilation of this book.

Editor

Adsorption Equilibrium for Heavy Metal Divalent Ions (Cu^{2+}, Zn^{2+}, and Cd^{2+}) into Zirconium-Based Ferromagnetic Sorbent

Agnes Yung Weng Lee, Soh Fong Lim, S. N. David Chua, Khairuddin Sanaullah, Rubiyah Baini, and Mohammad Omar Abdullah

Faculty of Engineering, Universiti Malaysia Sarawak, 94300 Kota Samarahan, Malaysia

Correspondence should be addressed to Soh Fong Lim; sflim@unimas.my

Academic Editor: Peter Majewski

Zirconium-based ferromagnetic sorbent was fabricated by coprecipitation of Fe^{2+}/Fe^{3+} salts in a zirconium solution and explored as a potential sorbent for removing the Cu^{2+}, Zn^{2+}, and Cd^{2+} from aqueous solution. The sorbent could easily be separated from aqueous solution under the influence of external magnetic field due to the ferromagnetism property. A trimodal distribution was obtained for the sorbent with average particle size of 22.74 μm. The –OH functional groups played an important role for efficient removal of divalent ions. The surface of the sorbent was rough with abundant protuberance while the existence of divalent ions on the sorbent surface after the sorption process was demonstrated. Decontamination of the heavy metal ions was studied as a function of initial metal ions concentration and solution pH. Uptake of the heavy metal ions showed a pH-dependent profile with maximum sorption at around pH 5. The presence of the ferromagnetic sorbent in solution at different initial pH has shown a buffering effect. Equilibrium isotherms were analyzed using Langmuir, Freundlich, Dubinin-Radushkevich, and Temkin isotherm models. Adequacy of fit for the isotherm models based on evaluation of R^2 and ARE has revealed that heavy metal ions decontamination was fitted well with the Freundlich model.

1. Introduction

Heavy metals are natural constituents of earth's crust, but indiscriminate human activities have drastically transformed their geochemical cycles and biochemical balance [1]. At present, pollution of aquatic ecosystems by the heavy metals is a significant problem as the heavy metals constitute certain most hazardous substances that can bioaccumulate in environment and ecosystem. Toxic metal compounds on the earth not only reach the earth's waters (seas, lakes, ponds, and reservoirs), but also can contaminate underground water in trace amounts by leaking from the soil after rain and snow [2]. The toxic metal ions dissolved can ultimately reach top of food chain and therefore developed a risk factor for human health besides causing ecological damage [3].

The risk of heavy metal pollution to public health and wildlife has led to an enlarged interest in development of effective technologies for water purification. Adsorption process has evolved into one of the prominent methods for removing organic and inorganic pollutants in waterway systems [4]. This effective and economic process has been found superior to other techniques such as ion exchange and precipitation for treating aqueous effluents in terms of initial cost, simplicity of design, ease of operation, and insensitivity to toxic substances [5].

In recent years, the application of magnetic materials in solid phase extraction has received considerable attention by taking into account many advantages arising from the inherent characteristics of magnetic particles [6, 7]. The ferromagnetic sorbent can be easily recovered by magnetic separation technology and served as a satisfactory resolution for separation difficulty of ordinary powdered adsorbents. The ferromagnetic sorbent can simply be collected by an external magnetic field and such facile separation is essential to improving the operation efficiency. Additionally, the outstanding separation efficiency allows evacuation of the sorbent saturated with metal ions by means of magnetic captures, indicating a relatively simple technology without generation of secondary effluents [8]. The magnetic separation technique combined with adsorption makes the wastewater treatment process simpler than applying a centrifugation or filtration.

Ferromagnetic iron oxide particles are the most favourable magnetic sorbents used in removal of pollutants due to their better magnetic properties, lower toxicity, and lower price [9, 10]. Besides, hexagonally ordered zirconia was shown to have high affinity towards Co^{2+}, Ni^{2+}, Cu^{2+}, and Zn^{2+} [11]. The findings herein are investigated for possibility of using the zirconium-based ferromagnetic sorbent for adsorptive removal of Cu^{2+}, Zn^{2+}, and Cd^{2+} from aqueous solution. The dynamic behaviour of the metal ions sorption onto the ferromagnetic sorbent was studied with the effects of solution pH and initial metal ion concentration. The linear isotherm data analysis using two-parameter isotherm models was performed to validate the affinity between metal ions and aforementioned sorbent. In order to establish the most appropriate sorption equilibrium correlation, adequacy and accuracy in parameters prediction of the isotherm models were evaluated using error functions, regression coefficient (R^2) and average relative error (ARE). The potential benefit of this study is to provide a certain theoretical basis for operational design and applicable practice of the sorption systems of heavy metals removal, which has drawn increasing attention in the area of water environmental protection.

2. Materials and Methods

2.1. Materials. All chemicals used were of analytical reagent grade and they were used without further purification. The raw materials for sorbent fabrication, zirconium(IV) sulfate ($Zr(SO_4)_2$), iron(II) chloride ($FeCl_2{\cdot}4H_2O$), and iron(III) chloride ($FeCl_3{\cdot}6H_2O$), were obtained from Sigma-Aldrich Co. LLC whereas sodium hydroxide (NaOH) was obtained from Merck. The adsorbates, copper(II) sulfate ($CuSO_4{\cdot}5H_2O$), zinc sulfate ($ZnSO_4{\cdot}7H_2O$), and cadmium nitrate ($Cd(NO_3)_2{\cdot}4H_2O$) were purchased from Fisher Scientific UK Ltd. The stock solution of Cu^{2+}, Zn^{2+}, and Cd^{2+} was prepared by dissolving $CuSO_4{\cdot}5H_2O$, $ZnSO_4{\cdot}7H_2O$, and $Cd(NO_3)_2{\cdot}4H_2O$ in appropriate amounts of ultrapure water produced using Sartorius arium® pro ultrapure water system. All working solutions were freshly prepared by diluting the stock solutions to the desired concentrations, and initial pH value was adjusted to the designed values according to the following experimental design with hydrochloric acid (HCl) and sodium hydroxide (NaOH) solutions. The pH measurements were made using an Accumet AB15 pH meter equipped with a combine electrode (Fisher Scientific, USA).

2.2. Fabrication of the Zirconium-Based Ferromagnetic Sorbent. The zirconium-based ferromagnetic sorbent was fabricated according to the Lim [18] and Zheng et al. [19] chemical coprecipitation method with some modifications. $FeCl_2{\cdot}4H_2O$ and $FeCl_3{\cdot}6H_2O$ (1 : 2) were dissolved in ultrapure water. Zirconium sulfate solution was then injected into the mixed solution. The mixture was mechanically stirred with a Memmert water bath (Model WNE 14) and heated to 70°C. When the temperature was stabilized, 10 M of NaOH was added dropwise until the pH was above 8. Fe-Zr oxides were formed as the precipitate. The precipitate was washed until the pH of supernatant reached approximately 7. The precipitate was separated by magnetic decantation and dried in an oven (BINDER ED 53, Germany) at 70°C for 24 h. Finally, the obtained Fe-Zr sorbent was cooled at room temperature. The product was then ground in an agate mortar, appearing in the form of fine powders, subsequently sieved with a 500 μm stainless steel sieve, and kept at desiccator for further use.

2.3. Instrumental Analyses of the Zirconium-Based Ferromagnetic Sorbent. The particle size distribution of the zirconium-iron oxide particles was measured by Particle Size Analyzer (PSA) (CILAS 1090, French). The particle size analysis by laser diffraction gives measurement range from 0.02 μm to 500 μm in wet mode.

The surface functional groups of the zirconium-based ferromagnetic sorbent before and after metal ion adsorption were detected by an attenuated total reflectance (ATR) technique using a Fourier Transform Infrared (FT-IR) Spectroscopy (Shimadzu IRAffinity-1, Japan). Samples were measured by pressing them against a diamond prism that is positioned horizontally for the infrared microscope analysis. The infrared spectra were recorded in the IR wavelength between 4000 and 400 cm^{-1}. The spectra data was collected with twenty scans at a resolution of 4 cm^{-1} and the apodization function was of the Happ-Genzel type. All the spectra were recorded and plotted in the same scale on the transmittance axis.

Surface morphology of the ferromagnetic sorbent was studied by JEOL JSM-6390LA scanning electron microscope (SEM) with an embedded energy dispersive X-ray analyzer (EDS) for the elemental analysis of the ferromagnetic sorbent. The dry sample was mounted on aluminium plates using double-sided conductive carbon tape and sputter-coated with an Au film in vacuum using a JEOL JFC-1600 Auto Fine Coater prior to measurement. The coated sample was then visualized and examined with the scanning electron microscopy analysis operated at 10 kV. On the other hand, the qualitative EDS spectrum analyses on the specimen surface were conducted at 15 kV accelerating voltage and 35° take-off angle of X-ray at the analytical working distance of 10 mm. A probe current of 1 nA was raster-scanned across the specimens to minimize the risk of sample damage during imaging. All elements were readily observed and mapped using Smiling program.

Potentiometric titration was applied to study the acid-base properties of the ferromagnetic sorbent and the ionic exchange properties with regard to H^+ and OH^- ions. The amount of 0.1 g of sorbent was added to 100 mL ultrapure water. A recorded volume of 0.1 M HCl was added to the mixture. The pH of each addition of acid was recorded in a time interval of 10 min. During the titration, the solution was gently shaken on the water bath and the experiment was carried out at 25 ± 1°C. The experiment was repeated using 0.1 M NaOH to replace the 0.1 M HCl. The acid and base titration experiments were conducted separately.

(a) (b) (c)

FIGURE 1: The zirconium-based ferromagnetic sorbent (a) under presence (wet) and absence (dry) of aqueous solution, (b) its magnetic responses under normal condition when being in contact with a permanent magnet, and (c) magnetic separation after sorption process.

3. Adsorption Experiments

3.1. Equilibrium Study. The heavy metal ions sorption isotherm was determined using batch experiments at varied initial metal ion concentration ranging from 5 to 350 mg/L. In each test, the sorbent at solid/solution ratio of 1 g/L was loaded in the 100 mL Erlenmeyer flasks with solution containing different concentration of heavy metal ion. The flasks were kept sealed in order to minimize losses to atmosphere. Next, the flasks were shaken in a refrigerated incubating shaker (Thermo Scientific 491, Japan) at 200 rpm for 24 h at 25±1°C. After the reaction period, all samples were filtered by 0.45 μm membrane filter and analyzed for heavy metal ions using the atomic absorption spectrophotometer (AAS) (Shimadzu, AA-7000, Japan) equipped with D_2 (deuterium lamp) method as a background correction, an air-acetylene burner, and controlled by the AA software, namely, WizAArd. The hollow cathode lamp was operated according to the elements of Cu, Zn, and Cd at 10 mA, 7 mA, and 5 mA whereas the analytical wavelength was set at 324.8 nm, 213.9 nm, and 228.8 nm, respectively. The amount of heavy metal ions adsorbed onto the sorbent (Q_e) was calculated by mass balance equation of heavy metal ions before and after the sorption as expressed by

$$Q_e = \frac{(C_0 - C_e)V}{W}, \tag{1}$$

where C_0 (mg/L) and C_e (mg/L) are the initial and equilibrated heavy metal ions concentrations, respectively, V (L) is the solution volume, and W (g) is the weight of the sorbent used. Correction for any sorption of heavy metal ions on the Erlenmeyer flask surface was made using the control experiment carried out under identical conditions in the absence of the zirconium-based ferromagnetic sorbent. These experiments indicated no detectable sorption by the Erlenmeyer flask surface and other factors.

3.2. Effect of Solution pH Study. Studies have been performed to identify the effect of solution pH on heavy metal ions sorption and uptake by the zirconium-based ferromagnetic sorbent. The effect of solution pH was investigated in batch mode by varying the initial pH of the solution over the range of 2–7 with initial metal ion concentration, 50 mg/L, while keeping the other experimental conditions similar to equilibrium study. The experiments were carried out in triplicate in order to better understand the system.

3.3. Estimation of Best Fitting Isotherm Models. Over the past few decades, linear regression has been established as a major option in designing liquid-phase adsorption systems. Determination of the best isotherm model through analysis of regression coefficient (R^2) is efficient but this indicator is limited to solving isotherm models that present linear forms [20]. Hence, average relative error (ARE) was employed in order to evaluate the fit of the isotherm equations to the experimental data and it is expressed as in

$$\text{ARE} = \frac{100}{n}\sum_{i=1}^{n}\left|\frac{q_{e,\text{exp}} - q_{e,\text{calc}}}{q_{e,\text{exp}}}\right|_i, \tag{2}$$

where $q_{e,\text{exp}}$ and $q_{e,\text{calc}}$ are the experimental and theoretically calculated adsorption capacity at equilibrium, respectively, while n is the number of data points.

4. Results and Discussion

4.1. Characterization of the Zirconium-Based Ferromagnetic Sorbent. The zirconium-based ferromagnetic sorbent is validated to have a good ferromagnetic property. As demonstrated in Figure 1, the ferromagnetic sorbent is attracted to a permanent magnet which is placed at outer wall of a glass vial that contains the sorbent. The magnetic attraction phenomenon occurs in both presence (wet) and absence (dry) of aqueous solution. Complementary to this, the magnetic sorbent also is able to be separated from dispersions with the application of a magnetic field after sorption process. Figure 1 illustrates that the sorbent is attracted and agglomerated with the aid of a permanent magnet and the supernatant is then discarded. The solution was observed to become limpid when the vial was placed near a permanent magnet

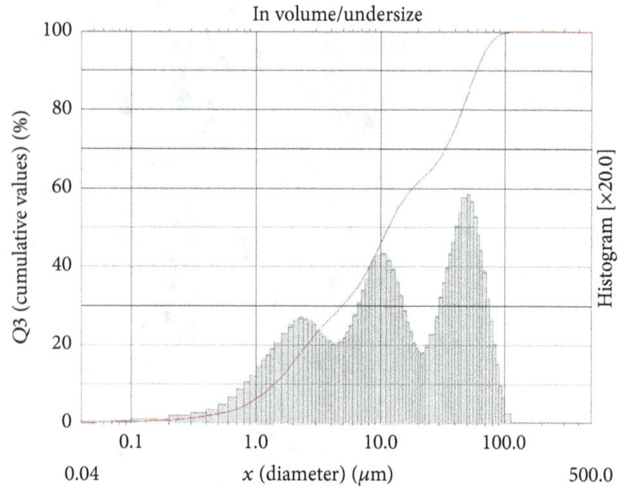

FIGURE 2: Particle size distribution of the zirconium-based ferromagnetic sorbent.

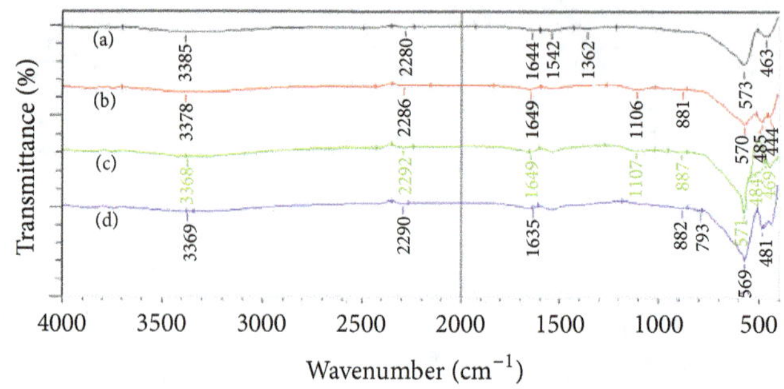

FIGURE 3: FT-IR spectra of the zirconium-based ferromagnetic sorbent (a) before and after sorption of (b) Cu^{2+}, (c) Zn^{2+}, and (d) Cd^{2+} [experimental conditions: sorbent dosage = 1 g/L, initial concentration = 200 mg/L, and contact time = 24 h].

FIGURE 4: SEM micrograph of the zirconium-based ferromagnetic sorbent.

for a few seconds. This sorbent is used to adsorb heavy metal ions and subsequently isolated from the medium by the technique of magnetic separation. Once magnetized, the particles behave like small permanent magnets, thus forming lattice or aggregates due to magnetic interaction [7]. The zirconium-based ferromagnetic sorbent functionalized with the magnetic elements such as iron or oxides of iron and alloys enable the sorbent to have ferromagnetic properties that help rapid and easy separation of the new solid phase from the solution; thus avoiding the time consuming conventional column passing and filtration operations appeared to be promising especially in the case of environmental applications.

The particle size distribution is shown in Figure 2. It was found that this volume-based distribution of the ferromagnetic particles was in the range of 0.06 μm to 110 μm and the average particle size was 22.74 μm. The analysis in the particles led to a trimodal distribution of the dispersion in the micron range. A larger fraction of the powdered particles was within the size range of 20 μm to 100 μm and the median particle size of the zirconium-iron oxide powders was 11.52 μm.

Figure 3 displays the FT-IR spectra obtained for the ferromagnetic sorbent before and after sorption process. The analysis indicated that hydroxyl groups play an important role in the uptake. The stretching vibration of functional group –OH usually forms a broad brand region within the range of 3100–3700 cm^{-1} for the presence of hydroxyl of coordinate water molecules. It was observed that those peaks have altered significantly after metal ion sorption, indicating the involvement in binding of the metal ion and ferromagnetic sorbent. Meanwhile, the prominent peaks at low frequency zone in the range of 500–700 cm^{-1} observed are due to metal oxygen stretching vibration which could be attributed to ZrO and Fe–O stretching vibration bond in the ferromagnetic sorbent. Besides, there were obvious additional peaks in the spectra after sorption (Samples (b)–(d)), showing evidences of Fe–O–H stretching (~882 cm^{-1}) and bending (793 cm^{-1}) vibrations for adsorbed ion species. These imply that the –OH groups are involved during the metal ions sorption. Additionally, there is a peak on the multispectra around 2300 cm^{-1} that corresponds to the asymmetrical stretch of carbon dioxide (CO_2), indicating CO_2 in atmosphere is dissolved into the samples [21].

On the other hand, Figure 4 shows the SEM images of the ferromagnetic sorbent taken at a magnification capacity of ×10000 and a scale bar of 1 μm. The sorbent particles with varying sizes agglomerated together appearing with irregular surface morphology. The surface of sorbent which was rough with abundant protuberance favoured the diffusion of metal ions into the ferromagnetic sorbent. The micrograph revealed irregularly shaped amorphous particles of the ferromagnetic sorbent similar to the characteristic of zirconia [22]. The existence of the very small sized particles along with the large agglomerates could be the explanation for the results obtained from particle size analysis where multimodal distribution was determined (Figure 2). The qualitative EDS spectrum analyses on the specimen surface were demonstrated in Figure 5. Based on Figure 5(a), the EDS spectrum revealed the energy spectrum of the X ray character emitted from the element of iron (Fe), zirconium (Zr), and oxygen (O). Furthermore, no impurities were detected that would be attributable to contamination of chemical precursors. The

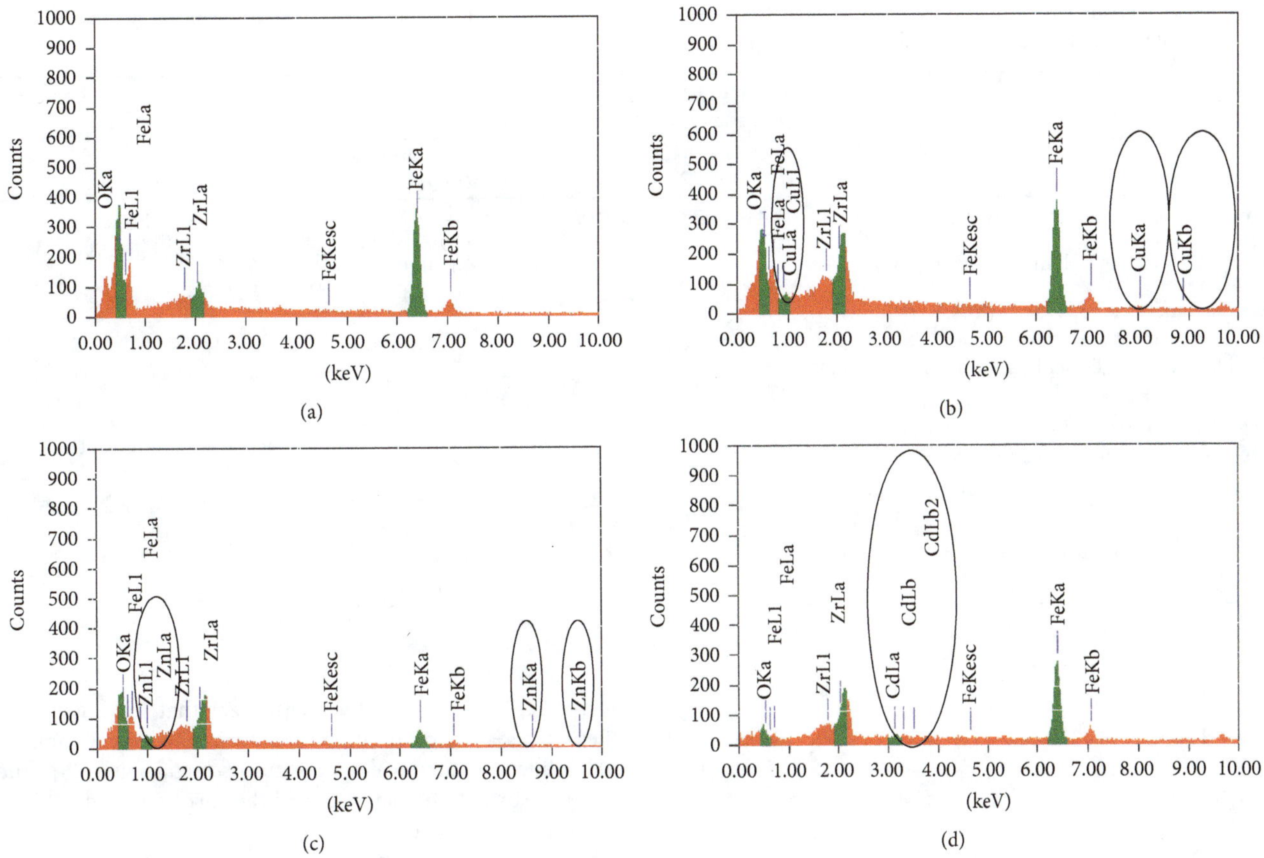

FIGURE 5: EDS spectra of the ferromagnetic sorbent (a) before and after adsorption of (b) Cu^{2+}, (c) Zn^{2+}, and (d) Cd^{2+} [experimental conditions: sorbent dosage = 1 g/L, initial concentration = 200 mg/L, and contact time = 24 h].

existence of Cu^{2+}, Zn^{2+}, and Cd^{2+} on the sorbent surface after the adsorption process could be observed accordingly comparing the EDS peaks with that of unloaded one, as presented in Figures 5(b)–5(d). It was suggested that the heavy metal ions have been adsorbed on the surface of the magnetic particles successfully.

Potentiometric titration was used to elucidate the functional groups on the ferromagnetic sorbent responsible for metal ions sorption. The acid-base nature of the zirconium-based magnetic sorbent and its equivalence point were identified. Based on Figure 6, the equivalence point of the titration curve was around pH 7 and the buffering zone was around pH 4 to pH 10. It was discovered that the ferromagnetic sorbent exhibited amphoteric behaviour where it could resist changes in pH upon the addition of small amount of acid or base. In order to further examine their behaviour, the effect of initial solution pH on the sorption process was performed and discussed in the section below.

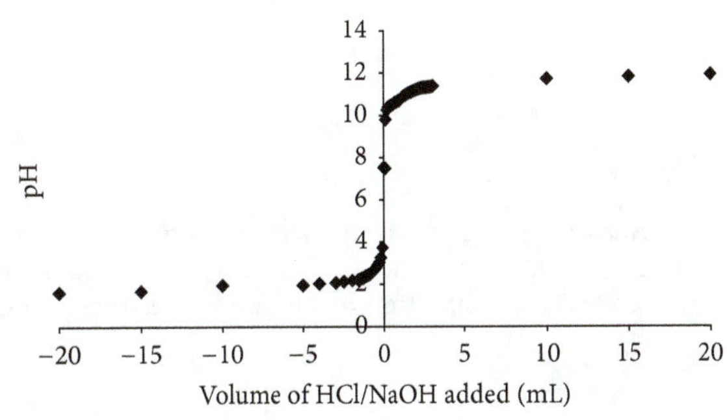

FIGURE 6: Titration curve of the zirconium-based ferromagnetic sorbent.

4.2. Effect of Initial Metal Ion Concentration. The maximum sorption capacity obtained from the equilibrium studies was in the following order: Cu^{2+} (48.4 mg/g) > Cd^{2+} (30.9 mg/g) > Zn^{2+} (29.2 mg/g) as presented in Figure 7, where the different sorption capacity may be due to disparity in cations radius and interaction enthalpy values [23]. Comparison of maximum adsorption capacity of the zirconium-based ferromagnetic sorbent with various magnetic sorbents reported in literatures for Cu^{2+}, Zn^{2+}, and Cd^{2+} adsorption is presented in Table 1. The adsorption capacity of the ferromagnetic sorbent was found to increase with the equilibrium concentrations of the metal ions in solution, progressively reaching saturation of the sorbent. The results indicated that when the metal ion concentration in the solution increased, the difference in concentration between bulk solution and surface also

TABLE 1: Comparison of maximum adsorption capacity of the zirconium-based ferromagnetic sorbent with various magnetic sorbents reported in literatures for Cu^{2+}, Zn^{2+}, and Cd^{2+} adsorption.

Magnetic sorbents	Maximum adsorption capacity (mg/g)			Experimental conditions		References
	Cu^{2+}	Zn^{2+}	Cd^{2+}	Initial concentration (mg/L)	Adsorbent dosage (g/L)	
Zirconium-based magnetic sorbent	48.4	29.2	30.9	5–350	1	Present study
Fe_3O_4/cyclodextrin polymer nanocomposites	—	—	27.7	50–400	12	Badruddoza et al. [12]
Fe_3O_4 magnetic nanoparticles (MNPs) modified with 3-aminopropyltriethoxysilane (APS) and copolymers of acrylic acid (AA) and crotonic acid (CA)	126.9	43.4	29.6	20–450	1	Ge et al. [13]
Magnetic ferric oxide doped vinyl benzene chloride divinylbenzene-based polymer	25.6	—	—	30	0.05–2	Simsek et al. [14]
Magnetic polymer beads with abundant amino groups	51.7	—	—	100–400	—	Lin et al. [15]
Magnetic Fe_3O_4 baker's yeast biomass	—	—	41.0	40–300	1	Xu et al. [16]
Iron oxide nanoparticles	—	—	18.5	10–600	10	Nassar [17]

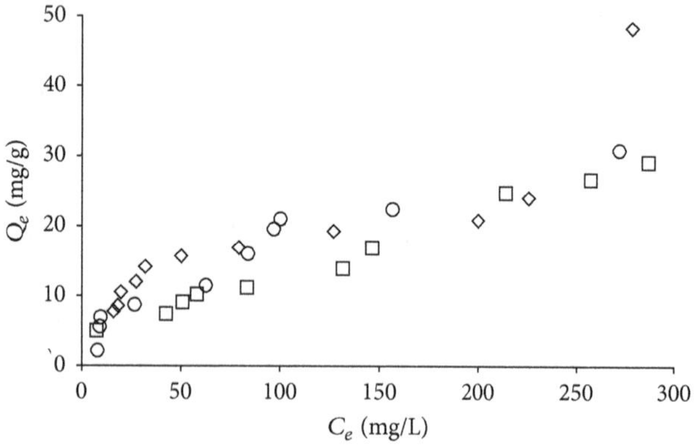

FIGURE 7: Adsorption capacity of the zirconium-based ferromagnetic sorbent for Cu^{2+}, Zn^{2+}, and Cd^{2+} [experimental conditions: sorbent dosage = 1 g/L, contact time = 24 h, and temperature = 25°C].

increased hence intensifying the mass transfer processes [24]. This sorption characteristic suggested that surface saturation is dependent on the initial metal ion concentrations. At low concentrations, adsorption sites took up the available metal more rapidly, whereas at higher concentrations, metal ions need to diffuse to the sorbent surface by intraparticle diffusion and greatly hydrolyzed ions were diffused at a slower rate [25]. In the adsorption process, the metal ions were initially diffused from the boundary layer film onto the sorbent surface and finally diffuse into the porous structure of the sorbent.

4.3. Effect of Initial Solution pH. The solution pH of the sorbate solutions has been known as the most important parameter governing sorption of metal ions on different sorbents. This may be due to the fact that hydrogen ions themselves are strong competing sorbate and the solution pH influences chemical speciation of metal ions [26]. The effect of surface characteristics of the sorption of metal ions from aqueous solution was correlated with the pH of the solution, as this affects the surface charge of the sorbents, solubility of metal ions, concentration of counterions on sorbent functional groups, and the degree of ionization of the sorbate [27].

The effect of initial solution pH on the heavy metal ions sorption by the zirconium-based ferromagnetic sorbent demonstrated in Figure 8(a) was strongly pH dependent. It was observed that the uptake of metal ions increased as the solution pH increased until the maximum uptakes were noted at pH around 5. As the pH of the solution increased to approximately 7, precipitation begins in the form of insoluble hydroxide, $M(OH)_2$. Hence, it was believed that the sorption process became increasingly masked by precipitation and led to the increase of sorption capacity of Zn^{2+} and Cd^{2+} detected at pH around 7. Consequently, experiments were not conducted over pH 7 to avoid precipitation of metal complexes and hinder the distinction between adsorption and precipitation as heavy metals removal process [28]. The idea of working in a condition close to that naturally established by the medium without requiring significant modifications was also considered; the optimum pH for metal ions sorption was therefore determined to be 5 and applied in all experiments. Optimal sorption of inorganic divalent metal ions at this pH has been reported in numerous studies for rice bran, black gram husk, and Agbani clay [29–31].

The effect of solution pH on metal cation sorption is principally the result of changes in the net proton charge on particles. At low solution pH, the minimal sorption may be due to the higher concentration and high mobility of H^+, which are preferentially adsorbed rather than metal ions. The increase in positive charge density on the sorbent surface

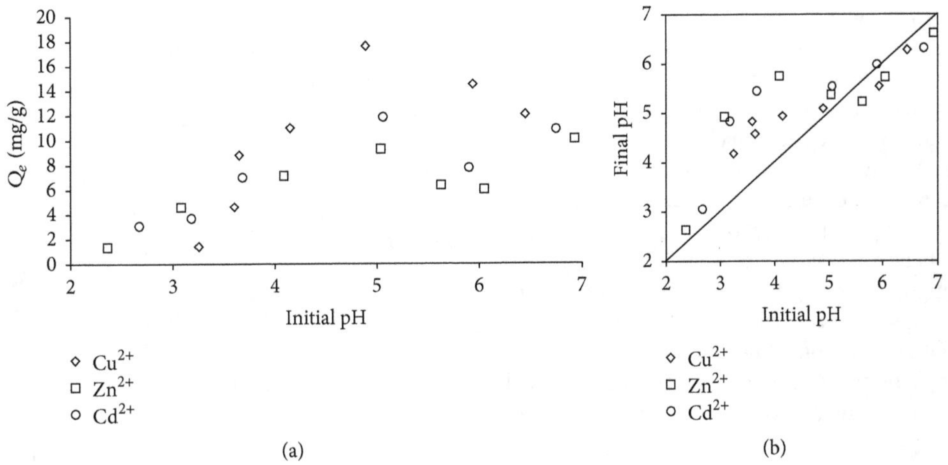

FIGURE 8: (a) Effect of initial solution pH and (b) initial solution pH versus final solution pH for the removal of Cu^{2+}, Zn^{2+}, and Cd^{2+} by the zirconium-based ferromagnetic sorbent [experimental condition: initial concentration 50 mg/L].

sites resulted in electrostatic repulsion between the metal ion and positive charge on the sorption edges. In contrast, electrostatic repulsion decreases with increasing pH due to the increase in negative charge on the surface of sorbent thus enhancing metal cations sorption [32].

Similarly, the sorption would be retarded if the acidic environment was too strong to disturb the complex interaction of metal ions and functional group of the sorbent. This phenomenon has suggested that treatment with acid would enable regeneration of the ferromagnetic sorbent. Correspondingly, increasing of the solution pH value after saturation does not increase the metal ions sorption, but a partial metal ions desorption may occur due to the reaction of metal ions with OH^- ions and precipitate as a metal hydroxide [33].

The metal ions in the aqueous solution may undergo hydrolysis. Below pH 6, the dominant species of all metals are M^{2+} and $M(OH)^+$ [12]. Ghaffar [34] mentioned that simple hydrolysis of most divalent metal ions is written as

$$M^{2+}(aq) + H_2O(l) \longleftrightarrow M(OH)^+(aq) + H^+(aq) \quad (3)$$

where M^{2+} represents Cu^{2+}, Zn^{2+}, and Cd^{2+}. This reaction generates monovalent cations $M(OH)^+$ and protons which contribute to the increased acidity of M^{2+} solutions. If M^{2+} is being taken up by the sorbent, the reaction above shifts to the left, leading to the depletion of protons and accordingly a rise in pH value. The addition of NaOH to raise the pH before equilibration enhanced sorption of metal ions by the sorbent, suggesting that metal ion binding was occurring as carboxyl or hydroxyl ligands [35].

Additionally, Horsfall Jr. and Spiff [36] claimed that chemical bonding may be involved in metal adsorption; for instance, the H^+ ions compete effectively with M^{2+} ions for sorption sites at low pH and the ion exchange mechanism may be represented in two ways as shown in

$$2B^- + M^{2+} = M(B)_2 \quad (4)$$

$$2HB + M^{2+} = M(B)_2 + 2H^+, \quad (5)$$

where B^- and B are polar sites on the sorbent surface.

Furthermore, it was notable from Figure 8(b) that the reactions occurring during the sorption process caused slight fluctuation in the equilibrium pH. The equilibrium pH slightly increased as initial pH was below 5, whereas it decreased to lower values for initial pH above 5. These observations might be due to some buffering phenomena that occur during the sorption reaction. Good buffering capacity was shown at pH around 5 as mentioned in previous section which might be contributed by metal hydrous oxides. It can be explained that zirconium and iron hydrous oxides exhibit amphoteric surface hydroxyl sites in water [19]. Hydroxyl groups on metal oxide in water which can be protonated or deprotonated by the addition or deletion of hydrogen are the sites for ion exchange [37].

At a lower solution pH, the hydrous oxides could be protonated as shown in

$$\equiv \text{Zr-OH} + H^+ \longleftrightarrow \equiv \text{Zr-OH}_2{}^+ \quad (6)$$

$$\equiv \text{Fe-OH} + H^+ \longleftrightarrow \equiv \text{Fe-OH}_2{}^+ \quad (7)$$

It is noteworthy that the sorbent surface binding sites are protonated; thereby H^+ ions compete for active sorption sites on the surface of sorbent; this condition does not favour the uptake of cations leading to electrostatic repulsion. Due to the phenomena of competitive adsorption, the sorption of cations onto the surface of sorbent is decreased.

At a higher solution pH, the hydrous oxides could be deprotonated as shown in

$$\equiv \text{Zr-OH} + OH^- \longleftrightarrow \equiv \text{Zr-O}^- + H_2O \quad (8)$$

$$\equiv \text{Fe-OH} + OH^- \longleftrightarrow \equiv \text{Fe-O}^- + H_2O \quad (9)$$

Nevertheless, an increase in pH produces a decrease in the positive surface charge as OH^- ions predominate. This condition leads to electrostatic force of attraction between the negatively charged surface of sorbent and cations as the sorbent becomes less repulsive to cationic species. This phenomenon favours the adsorption of cations, thus resulting in increased binding of positively charged metal ions.

In short, the adsorption mechanism is mainly contributed by the chemical interactions between the metal ions and the surface functional groups of the ferromagnetic sorbent. The increase in the adsorption of divalent ions with increasing pH indicates that an ion exchange mechanism between the H^+ ions and metal ions occurs at the oxygen-containing functional groups on the sorbent surface. Based on the FT-IR analyses, the oxygen atoms of the hydroxyl groups donate their single pair of electrons to the metal ions, thus increasing the cation exchange capacity of the sorbent. Protons in the functional groups of the sorbent are exchanged with the metal ions. Chemical bonds between the metal ions and the surface functional groups of the zirconium-based ferromagnetic sorbent are mainly responsible for adsorption.

5. Adsorption Isotherms

The adsorption isotherms were mathematical models that described the distribution of the adsorbate species among the liquid and solid phases at the equilibrium state of an adsorption system. A state-of-the-art review of adsorption two-parameter isotherms models including Langmuir, Freundlich, Jovanovic, Dubinin-Radushkevich, Temkin, Flory-Huggins, and Hill isotherm models, their fundamental characteristics, and mathematical derivations was adopted to evaluate the decontamination of heavy metal ions by the zirconium-based ferromagnetic sorbent. The applicability of Langmuir, Freundlich, Dubinin-Radushkevich, and Temkin isotherm models is desirable to be presented for the isotherm analysis.

5.1. Langmuir Isotherm Model. The Langmuir isotherm model was derived based on the assumption of monolayer adsorption on a structurally homogeneous surface where there is no interaction between molecules adsorbed on neighbouring sites [38]. The linear form of the Langmuir isotherm is given by

$$\frac{C_e}{q_e} = \frac{1}{K_L q_m} + \frac{C_e}{q_m}, \tag{10}$$

where q_e (mg/g) is the sorbate amount sorbed on sorbent at equilibrium, K_L (L/mg) is Langmuir equilibrium constant, and q_m (mg/g) is the maximum monolayer coverage capacity. The Langmuir constants (K_L and q_m) can be calculated from the plot of C_e/q_e versus C_e. K_L is related to the apparent energy of sorption and q_m reflects a complete monolayer (mg/g).

Moreover, the essential characteristics of Langmuir isotherm can be explained in terms of a dimensionless constant separation factor or equilibrium parameter R_L, which is defined by

$$R_L = \frac{1}{1 + K_L C_0}. \tag{11}$$

The value of R_L indicates the shape of the isotherm which is unfavourable ($R_L > 1$), linear ($R_L = 1$), favourable ($0 < R_L < 1$), or irreversible ($R_L = 0$).

TABLE 2: Isotherm parameters for sorption of heavy metal ions by the zirconium-based ferromagnetic sorbent.

Isotherm models	Cu^{2+}	Zn^{2+}	Cd^{2+}
Langmuir			
q_m	38.1679	43.6681	39.0625
K_L	0.0133	0.0054	0.0104
Freundlich			
$1/n_F$	0.4441	0.5078	0.5845
n_F	2.2517	1.9693	1.7109
K_F	2.5704	1.3871	1.2306
Dubinin-Radushkevich			
K_{DR}	5×10^{-5}	1×10^{-5}	2×10^{-5}
E_{fe}	0.10	0.22	0.16
q_m	22.4031	15.1500	18.4654
Temkin			
b_T	290.8	362.6	370.9
A_T	0.1398	0.1112	0.1922

Figure 9(a) depicts the plot of C_e/q_e versus C_e for the removal of heavy metal ions by using the zirconium-based ferromagnetic sorbent. From the plot, it was noticed that the sorption data were adequately fitted to the Langmuir equation with average regression coefficients ($R^2 = 0.7130$–0.8550). The Langmuir constants obtained were computed in Table 2. The deviation of experimental data from the Langmuir isotherm may be due to the moderately homogeneous distribution of active sites on the surface of the ferromagnetic sorbent as this model introduced a clear concept of the monomolecular adsorption on energetically homogenous surfaces [39].

Additionally, the calculated equilibrium parameter, R_L, of the Langmuir isotherm was in the range between 0 and 1 for all the three heavy metal ions which indicate that the sorption processes were favourable.

5.2. Freundlich Isotherm Model. The Freundlich isotherm model is an empirical equation which represents the multilayer adsorption on heterogeneous surfaces where there are interactions between adsorbed molecules [40]. Its linearized equation is illustrated as

$$\log(q_e) = \log(K_F) + \frac{1}{n_F}\log(C_e), \tag{12}$$

where K_F ($\mathrm{mg}^{1-(1/n)}\mathrm{L}^{1/n}\mathrm{g}^{-1}$) is the Freundlich constant and n_F is the constant of Freundlich equation exponent that represents the parameter characterizing quasi-Gaussian energetic heterogeneity of the adsorption surface. The K_F value is related to the adsorption capacity while the slope, $1/n_F$, ranging between 0 and 1 is a measure of adsorption intensity or surface heterogeneity, becoming more heterogeneous as its value gets closer to zero. However, a value below unity implies chemisorption process where $1/n_F$ above one is indicative of cooperative adsorption [41, 42].

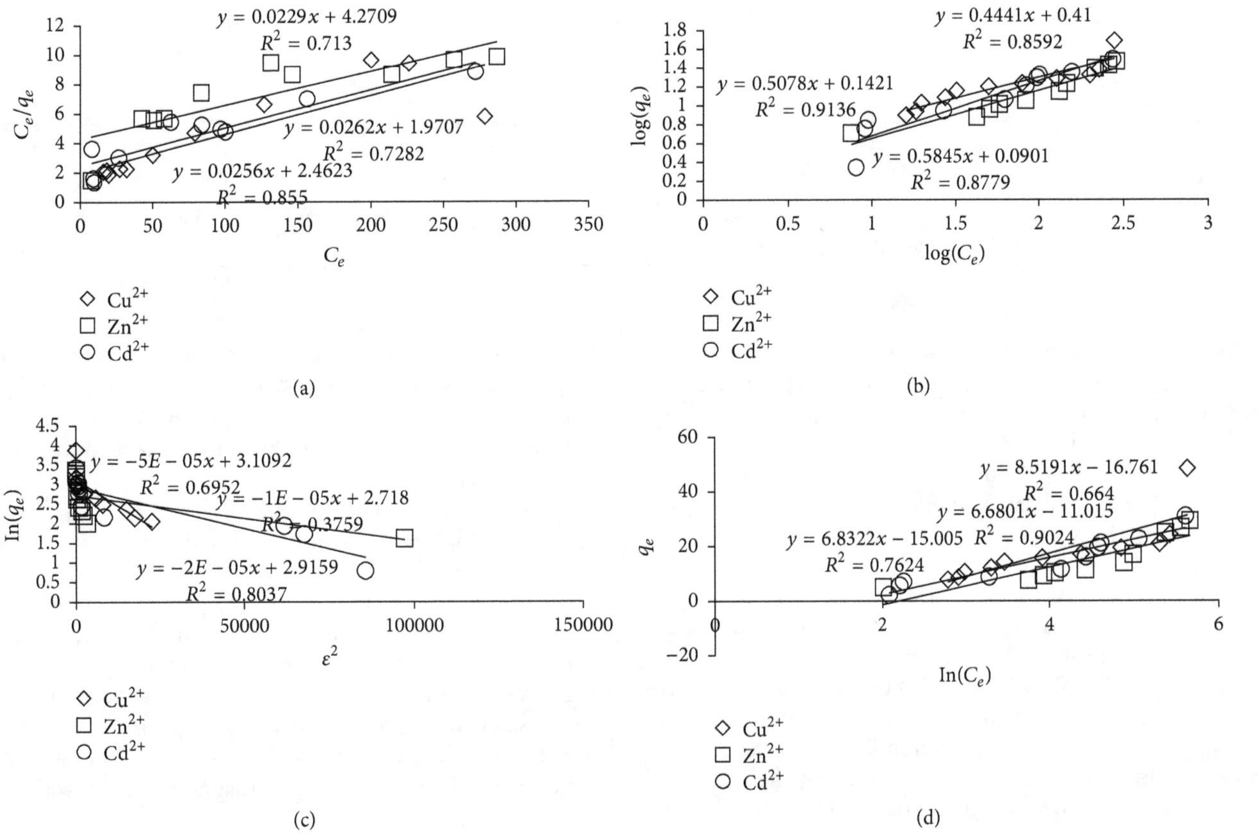

FIGURE 9: Plot for (a) Langmuir, (b) Freundlich, (c) Dubinin-Radushkevich, and (d) Temkin isotherm models for the removal of heavy metal ions by the zirconium-based ferromagnetic sorbent.

As presented in Figure 9(b), the Freundlich isotherm plot of log (C_e) against log (q_e) was fitted well with the experimental data. The n_F value indicates the degree of nonlinearity between solution concentration and adsorption where the values in the present study were found to be 1.7109–2.2517 (Table 2). This may be due to a distribution of surface sites or any factor that had caused a decrease in sorbent-sorbate interaction with increasing surface density. Besides, the n_F values lied within the range of 1–10 representing good adsorption [43].

Based on Table 2, the values of K_F obtained were 2.5704, 1.3871, and 1.2306 for Cu^{2+}, Zn^{2+}, and Cd^{2+}, respectively. Increasingly large K_F value indicates greater adsorption capacity. It can be concluded here that the studied metal ions can be reduced using the ferromagnetic sorbent for high K_F value which has a large adsorption of capacity. This isotherm does not predict any saturation of the solid adsorbent surface; therefore infinite surface coverage is mathematically feasible. Furthermore, the linearity of the plot in Figure 9(b) has revealed the applicability of the Freundlich isotherm and shown that adsorption-complexation reactions may possibly take place in the adsorption process [44].

5.3. Dubinin-Radushkevich Isotherm Model. The Dubinin-Radushkevich isotherm model predicts the nature of the sorbate sorption onto the sorbent, and it is used to calculate the mean free energy of adsorption. This model is more general than the Langmuir isotherm as its deviation is not based on ideal assumptions such as equipotential of sorption sites, absence of steric hindrances between sorbed and incoming particles, and surface homogeneity on microscopic level [45]. The isotherm has been written as

$$\ln(q_e) = \ln(q_m) - K_{DR}\varepsilon^2, \tag{13}$$

where K_{DR} (mol^2/J^2) is D-R isotherm constant related to the adsorption energy and ε is the Polanyi potential which is related to the equilibrium concentration and can be correlated as

$$\varepsilon = RT\ln\left(1 + \frac{1}{C_e}\right), \tag{14}$$

where R is universal gas constant, 8.314 J/mol·K, and T is the absolute temperature at 298 K. The mean free energy of adsorption (E_{fe}) per molecule of sorbate is calculated from the K_{DR} values using

$$E_{fe} = \frac{1}{\sqrt{2K_{DR}}}. \tag{15}$$

From the linear D-R isotherm plots for the sorption of the heavy metal ions onto the zirconium-based ferromagnetic sorbent, q_m was determined to be 22.4031 mg/g, 15.1500 mg/g, and 18.4654 mg/g for Cu^{2+}, Zn^{2+}, and Cd^{2+}, respectively. The

TABLE 3: The comparison between correlation coefficient, R^2, and average relative error (ARE) among seven two-parameter isotherm models.

Isotherm models	Cu^{2+}		Zn^{2+}		Cd^{2+}	
	R^2	ARE (%)	R^2	ARE (%)	R^2	ARE (%)
Langmuir	0.7282	1.92	0.7130	4.96	0.8550	2.65
Freundlich	0.8592	3.20	0.9136	1.59	0.8779	0.56
Dubinin-Radushkevich	0.6952	4.90	0.3759	4.81	0.8037	4.03
Temkin	0.6640	3.23	0.7624	1.89	0.9024	1.44

value of q_m reveals the adsorption capacity performance; the higher the value of q_m, the better the adsorption capacity performance. Hence, the sorption process by the ferromagnetic sorbent shows the trend: Cu > Cd > Zn. The linear plots presented in Figure 9(c) and the analysis of the data showed that the experimental data do not correlate well with the D-R equation and this was confirmed by the regression coefficient shown in Table 3.

On the contrary, the mean free energy, E_{fe} was used to ascertain the type of adsorption process under consideration. The typical range of bonding energy for ion exchange mechanisms is 8–16 kJ/mol, indicating that chemisorption may play a significant role in the sorption process. In contrast, bonding energy values lower than 8 kJ/mol indicate that the sorption process is of a physical nature. In this study, the calculated E_{fe} values were limited within the range of 0.10–0.22 kJ/mol, indicating that the removal of heavy metal ions by using the ferromagnetic sorbent occurs via physisorption. According to A. U. Itodo and H. U. Itodo [45], physisorption is also called nonspecific adsorption which occurs as a result of long range weak Van der Waals forces between adsorbates and sorbents whereas the energy released when a particle is physisorbed is of the same magnitude as the enthalpy of condensation. In physical adsorption, equilibrium is established between the adsorbate and the fluid phase resulting in multilayer adsorption. Then, again, both physisorption and chemisorption may occur on the surface at the same time in some cases where a layer of molecules may be physically adsorbed on top of an underlying chemisorbed layer [46].

5.4. Temkin Isotherm Model. Temkin isotherm encloses a factor explicitly taking into account sorbent-sorbate interactions [47]. The model assumes that the adsorption heat of all molecules decreases linearly with the increase in coverage of the adsorbent surface. As implied in the equation, its derivation is characterized by a uniform distribution of binding energies (up to some maximum binding energy). A linear form of the Temkin equation can be expressed as

$$q_e = \frac{RT}{b_T} \ln (A_T) + \frac{RT}{b_T} \ln (C_e), \quad (16)$$

where A_T (L/g) is equilibrium binding constant corresponding to the maximum binding energy and b_T (J/mol) is the variation of adsorption energy. By plotting the quantity sorbed, q_e versus ln C_e, the constants can hence be determined from the slope and intercept.

By referring to Figure 9(d), the equilibrium binding constant, A_T, was found to be 0.1398 L/g, 0.1112 L/g, and 0.1922 L/g while the variation of adsorption energy, b_T, is 290.8 J/mol, 362.6 J/mol, and 370.9 J/mol for Cu^{2+}, Zn^{2+}, and Cd^{2+}, respectively. The corresponding parameters and the regression coefficients of Temkin model were presented in Tables 2 and 3.

Apart from that, the low values of typical range of bonding energy, RT/b_T, in this study indicated a weak interaction between sorbent and sorbate supporting a mechanism of ion exchange and the adsorption energy of all the molecules in the surface layer decreased at the cover surface. Higher values of the coefficient of correlation show a good linearity regardless of the maximum capacity of adsorption used to calculate the coverage area. Consequently, the Temkin isotherm can describe reasonably well the sorption isotherms of heavy metal ions onto the ferromagnetic sorbent with R^2 = 0.6640–0.9024.

5.5. Comparison between Isotherm Models. The typical assessment of the quality of the isotherm fit to the experimental data was based on the magnitude of the correlation coefficient for the regression. The values of regression coefficient, R^2, from the linearization of the four two-parameter isotherm models and the corresponding average relative error (ARE) obtained were listed in Table 3. The ARE based on (2) gives small values when the data obtained from the model are comparable to the experimental data. It was noticed that the aforementioned isotherm models were appropriate in their own merits in describing the potential of the zirconium-based ferromagnetic sorbent for the removal of Cu^{2+}, Zn^{2+}, and Cd^{2+} from aqueous solution.

Consideration of the comparative magnitudes of the R^2 and ARE had suggested that the Freundlich isotherm model provided a better model for the sorption systems. A plausible reason of this observation is that this isotherm requires the taking of logarithms, therefore introducing similar effects to the error structure. As the Freundlich equation describes adsorption where the adsorbent has a heterogeneous surface with adsorption sites that have different energies of adsorption, the presence of zirconium dioxide in the ferromagnetic sorbent may lead to a more heterogeneous surface [48].

The results were comparable to the previous works related to the removal of metal heavy ions onto various magnetic sorbents done by other researchers. For instance, Ehrampoush et al. [49] have observed that the adsorption of cadmium ions from aqueous solution by synthesized iron oxide nanoparticles with tangerine peel extract followed Freundlich model better than Langmuir model according to the regression coefficient. Charpentier et al. [50] also claimed that Freundlich model was found to be more accurate when describing the adsorption of Cu^{2+} and Zn^{2+} on magnetic carboxymethyl chitosan nanocomposites as the mechanism is in agreement with the good fitting of the experimental data with the Freundlich adsorption isotherm. Moreover, the isotherm data of removal of Cd^{2+} from aqueous solution by shellac-coated iron oxide nanoparticles as an effective magnetic nanoparticle adsorbent were validated to be well fit with Freundlich isotherm with the maximum adsorption capacity, 18.8 mg/g at pH = 8 [51].

6. Conclusions

The present investigation demonstrated that the zirconium-based ferromagnetic sorbent fabricated by chemical coprecipitation method exhibited good performance to remove heavy metal ions (Cu^{2+}, Zn^{2+}, and Cd^{2+}) from aqueous solution. Instrumental analyses were performed using Particle Size Analyzer (PSA), Fourier Transform Infrared (FT-IR) Spectroscopy, and scanning electron microscopic-energy dispersive X-ray spectroscopy (SEM-EDS). The batch studies revealed that the sorption process was highly solution pH dependent. The sorption capacity was reduced when the solution was in strong acidic condition and reached the maximum at pH around 5. Equilibrium adsorption isotherms were measured for single component system and the experimental data were analyzed with four two-parameter isotherm models, namely, the Langmuir, Freundlich, Dubinin-Radushkevich, and Temkin isotherm equations. Two mathematically rigorous error functions, regression coefficient (R^2) and average relative error (ARE), have been addressed to evaluate the fitting ability of these models. Examination of the linear isotherm plots suggested that the Freundlich model yielded a better fit than all other models, indicating a heterogeneous surface of the zirconium-based ferromagnetic sorbent and multilayer adsorption nature of the decontamination process. Apart from that, the Freundlich constant (n_F), which was estimated to be greater than 1 proved that the sorption of heavy metal ions from the aqueous solution was favourable. Due to the undeniable advantages of the zirconium-based ferromagnetic sorbent such as inexpensive preparation method and easy separation on application of magnetic field, it was noteworthy that the sorbent can be a potential candidate for the decontamination of heavy metal ions from industrial effluents.

Conflicts of Interest

The authors declare that there are no conflicts of interest regarding the publication of this paper.

Acknowledgments

This research was funded by Special Grant Scheme (F02/SpGS/1404/16/5) from Universiti Malaysia Sarawak.

References

[1] R. Singh, N. Gautam, A. Mishra, and R. Gupta, "Heavy metals and living systems: an overview," *Indian Journal of Pharmacology*, vol. 43, no. 3, pp. 246–253, 2011.

[2] I. Uzun and F. Güzel, "Adsorption of some heavy metal ions from aqueous solution by activated carbon and comparison of percent adsorption results of activated carbon with those of some other adsorbents," *Turkish Journal of Chemistry*, vol. 24, no. 3, pp. 291–297, 2000.

[3] T. G. Merlain, N. J. Nsami, and K. J. Mbadcam, "Adsorption of copper (II) ions from aqueous solution onto synthetic goethite and two naturally available red soils from Yaoundé-Cameroon," *British Biotechnology Journal*, vol. 3, no. 3, pp. 221–235, 2013.

[4] N. Barka, S. Qourzal, A. Assabbane, A. Nounah, and Y. Ait-Ichou, "Removal of Reactive Yellow 84 from aqueous solutions by adsorption onto hydroxyapatite," *Journal of Saudi Chemical Society*, vol. 15, no. 3, pp. 263–267, 2011.

[5] K. T. Dauda, V. N. Atasie, and O. B. Adeoye, "The kinetics and equilibrium studies of the adsorption of (Cu^{2+}, Pb^{2+}) from industrial wastewater using rice husk as adsorbent," *International Journal of Advanced Research in Chemical Science*, vol. 2, no. 4, pp. 13–21, 2015.

[6] S.-F. Lim, Y.-M. Zheng, S.-W. Zou, and J. P. Chen, "Characterization of copper adsorption onto an alginate encapsulated magnetic sorbent by a combined FT-IR, XPS, and mathematical modeling study," *Environmental Science and Technology*, vol. 42, no. 7, pp. 2551–2556, 2008.

[7] G. Giakisikli and A. N. Anthemidis, "Magnetic materials as sorbents for metal/metalloid preconcentration and/or separation. A review," *Analytica Chimica Acta*, vol. 789, pp. 1–16, 2013.

[8] M. Yamaura and D. A. Fungaro, "Synthesis and characterization of magnetic adsorbent prepared by magnetite nanoparticles and zeolite from coal fly ash," *Journal of Materials Science*, vol. 48, no. 14, pp. 5093–5101, 2013.

[9] S.-F. Lim, Y.-M. Zheng, S.-W. Zou, and J. P. Chen, "Removal of copper by calcium alginate encapsulated magnetic sorbent," *Chemical Engineering Journal*, vol. 152, no. 2-3, pp. 509–513, 2009.

[10] N. Rahbar, Z. Ramezani, and Z. Mashhadizadeh, "One step in-situ formed magnetic chitosan nanoparticles as an efficient sorbent for removal of mercury ions from petrochemical waste water: batch and column study," *Jundishapur Journal of Health Sciences*, vol. 7, no. 4, pp. 13–19, 2015.

[11] R. F. De Farias, A. A. S. Do Nascimento, and C. W. B. Bezerra, "Adsorption of Co(II), Ni(II), Cu(II), and Zn(II) on hexagonal templated zirconia obtained thorough a sol-gel process: the effects of nanostructure on adsorption features," *Journal of Colloid and Interface Science*, vol. 277, no. 1, pp. 19–22, 2004.

[12] A. Z. M. Badruddoza, Z. B. Z. Shawon, W. J. D. Tay, K. Hidajat, and M. S. Uddin, "Fe_3O_4/cyclodextrin polymer nanocomposites for selective heavy metals removal from industrial wastewater," *Carbohydrate Polymers*, vol. 91, no. 1, pp. 322–332, 2013.

[13] F. Ge, M. M. Li, H. Ye, and B. X. Zhao, "Effective removal of heavy metal ions Cd^{2+}, Zn^{2+}, Pb^{2+}, Cu^{2+} from aqueous solution by polymer-modified magnetic nanoparticles," *Journal of Hazardous Materials*, vol. 211, pp. 366–372, 2012.

[14] E. B. Simsek, D. Duranoglu, and U. Beker, "Heavy metal adsorption by magnetic hybrid-sorbent: an experimental and theoretical approach," *Separation Science and Technology*, vol. 47, no. 9, pp. 1334–1340, 2012.

[15] Z. Lin, Y. Zhang, Y. Chen, and H. Qian, "Extraction and recycling utilization of metal ions (Cu^{2+}, Co^{2+} and Ni^{2+}) with magnetic polymer beads," *Chemical Engineering Journal*, vol. 200, pp. 104–112, 2012.

[16] M. Xu, Y. Zhang, Z. Zhang, Y. Shen, M. Zhao, and G. Pan, "Study on the adsorption of Ca^{2+}, Cd^{2+} and Pb^{2+} by magnetic Fe_3O_4 yeast treated with EDTA dianhydride," *Chemical Engineering Journal*, vol. 168, no. 2, pp. 737–745, 2011.

[17] N. N. Nassar, "Kinetics, equilibrium and thermodynamic studies on the adsorptive removal of nickel, cadmium and cobalt from wastewater by superparamagnetic iron oxide nanoadsorbents," *Canadian Journal of Chemical Engineering*, vol. 90, no. 5, pp. 1231–1238, 2012.

[18] S. F. Lim, *Adsorption of contaminations by magnetic sorbents [Ph.D. thesis]*, National University of Singapore, 2008, Retrieved from National University of Singapore Digital Theses.

[19] Y. M. Zheng, S. F. Lim, and J. P. Chen, "Preparation and characterization of zirconium-based magnetic sorbent for arsenate removal," *Journal of Colloid and Interface Science*, vol. 338, no. 1, pp. 22–29, 2009.

[20] J. S. Piccin, G. L. Dotto, and L. A. A. Pinto, "Adsorption isotherms and thermochemical data of FDandC RED N∘ 40 Binding by chitosan," *Brazilian Journal of Chemical Engineering*, vol. 28, no. 2, pp. 295–304, 2011.

[21] S.-F. Lim and A. Y. W. Lee, "Kinetic study on removal of heavy metal ions from aqueous solution by using soil," *Environmental Science and Pollution Research*, vol. 22, no. 13, pp. 10144–10158, 2015.

[22] F. J. Pereira, M. T. Díez, and A. J. Aller, "Effect of temperature on the crystallinity, size and fluorescent properties of zirconia-based nanoparticles," *Materials Chemistry and Physics*, vol. 152, pp. 135–146, 2015.

[23] T. Madrakian, A. Afkhami, B. Zadpour, and M. Ahmadi, "New synthetic mercaptoethylamino homopolymer-modified maghemite nanoparticles for effective removal of some heavy metal ions from aqueous solution," *Journal of Industrial and Engineering Chemistry*, vol. 21, pp. 1160–1166, 2015.

[24] B. Nagy, B. Szilagyi, C. Majdik, G. Katona, C. Indolean, and A. Măicăneanu, "Cd (II) and Zn (II) biosorption on Lactarius piperatus macrofungus: equilibrium isotherm and kinetic studies," *Environmental Progress and Sustainable Energy*, vol. 33, no. 4, pp. 1158–1170, 2014.

[25] S. E. Oh, S. H. A. Hassan, and J. H. Joo, "Biosorption of heavy metals by lyophilized cells of Pseudomonas stutzeri," *World Journal of Microbiology and Biotechnology*, vol. 25, no. 10, pp. 1771–1778, 2009.

[26] A. Rajappa, K. Ramesh, and V. Nandhakumar, "Adsorption of nickel(II) ion from aqueous solution onto $ZnCl_2$ activated carbon prepared from Delonix regia pods (flame tree)," *International Journal of Chemical and Physical Sciences*, vol. 3, no. 6, pp. 42–52, 2014.

[27] N. Rajamohan, M. Rajasimman, R. Rajeshkannan, and V. Saravanan, "Equilibrium, kinetic and thermodynamic studies on the removal of Aluminum by modified *Eucalyptus camaldulensis* barks," *Alexandria Engineering Journal*, vol. 53, no. 2, pp. 409–415, 2014.

[28] M. M. A. El-Latif, A. M. Ibrahim, M. S. Showman, and R. R. A. Hamide, "Alumina/Iron oxide nano composite for cadmium ions removal from aqueous solutions," *International Journal of Nonferrous Metallurgy*, vol. 2, pp. 47–62, 2013.

[29] S. F. Montanher, E. A. Oliveira, and M. C. Rollemberg, "Removal of metal ions from aqueous solutions by sorption onto rice bran," *Journal of Hazardous Materials*, vol. 117, no. 2-3, pp. 207–211, 2005.

[30] A. Saeed, M. Iqbal, and M. W. Akhtar, "Removal and recovery of lead(II) from single and multimetal (Cd, Cu, Ni, Zn) solutions by crop milling waste (black gram husk)," *Journal of Hazardous Materials*, vol. 117, no. 1, pp. 65–73, 2005.

[31] F. A. Dawodu, G. K. Akpomie, and I. C. Ogbu, "Isotherm modelling on the equilibrium sorption of cadmium (II) from solution by Agbani clay," *International Journal of Multidisciplinary Sciences and Engineering*, vol. 3, no. 9, pp. 9–14, 2012.

[32] Th. Bohli, A. Ouederni, N. Fiol, and I. Villaescusa, "Uptake of Cd^{2+} and Ni^{2+} metal ions from aqueous solutions by activated carbons derived from waste olive stones," *International Journal of Chemical Engineering and Applications*, vol. 3, no. 4, pp. 232–236, 2012.

[33] M. E. Goher, A. M. Hassan, I. A. Abdel-Moniem, A. H. Fahmy, M. H. Abdo, and S. M. El-sayed, "Removal of aluminum, iron and manganese ions from industrial wastes using granular activated carbon and Amberlite IR-120H," *Egyptian Journal of Aquatic Research*, vol. 41, no. 2, pp. 155–164, 2015.

[34] A. Ghaffar, "Removal of lead(II) ions from aqueous solution under different physicochemical conditions using various sorbents," *Arabian Journal for Science and Engineering*, vol. 33, no. 1A, pp. 55–61, 2008.

[35] C. I. Osu and S. A. Odoemelam, "Studies on adsorbent dosage, particle sizes and pH constraints on biosorption of Pb(II) and Cd(II) ions from aqueous solution using modified and unmodified Crasstrotrea Gasar (bivalve) biomass," *Report and Opinion*, vol. 6, no. 12, pp. 67–72, 2014.

[36] M. Horsfall Jr. and A. I. Spiff, "Studies on the effect of pH on the sorption of Pb^{2+} Cd^{2+} from aqueous solutions by bicolor Cocoyam) biomass," *Electronic Journal of Biotechnology*, vol. 7, no. 3, pp. 313–323, 2004.

[37] H. Tamura, A. Tanaka, K.-Y. Mita, and R. Furuichi, "Surface hydroxyl site densities on metal oxides as a measure for the ion-exchange capacity," *Journal of Colloid and Interface Science*, vol. 209, no. 1, pp. 225–231, 1999.

[38] K. Y. Foo and B. H. Hameed, "Insights into the modeling of adsorption isotherm systems," *Chemical Engineering Journal*, vol. 156, pp. 2–10, 2010.

[39] P. J. P. Chen, *Decontamination of Heavy Metals: Processes, Mechanisms, and Applications*, Taylor & Francis, Abingdon, UK, 2012.

[40] R. Elmoubarki, F. Z. Mahjoubi, H. Tounsadi et al., "Adsorption of textile dyes on raw and decanted Moroccan clays: Kinetics, equilibrium and thermodynamics," *Water Resources and Industry*, vol. 9, pp. 16–29, 2015.

[41] O. S. Bello, M. A. Oladipo, and A. M. Olatunde, "Sorption studies of lead ions onto activated carbon produced from Oil-Palm fruit fibre," *Stem Cell*, vol. 1, no. 1, pp. 14–29, 2010.

[42] H. Shahbeig, N. Bagheri, S. A. Ghorbanian, A. Hallajisani, and S. Poorkarimi, "A new adsorption isotherm model of aqueous solutions on granular activated carbon," *World Journal of Modelling and Simulation*, vol. 9, no. 4, pp. 243–254, 2013.

[43] M. B. Desta, "Batch sorption experiments: Langmuir and Freundlich isotherm studies for the adsorption of textile metal ions onto Teff straw (*Eragrostis tef*) agricultural waste," *Journal of Thermodynamics*, vol. 2013, Article ID 375830, 6 pages, 2013.

[44] Y. Bulut and Z. Tez, "Removal of heavy metals from aqueous solution by sawdust adsorption," *Journal of Environmental Sciences*, vol. 19, no. 2, pp. 160–166, 2007.

[45] A. U. Itodo and H. U. Itodo, "Sorption energies estimation using Dubinin-Radushkevich and Temkin adsorption isotherms," *Life Science Journal*, vol. 7, no. 4, pp. 31–39, 2010.

[46] M. A. Al-Anber, "Thermodynamics approach in the adsorption of heavy metals," in *Thermodynamics—Interaction studies—Solids, Liquids and Gases*, J. C. Moreno-Pirajan, Ed., pp. 737-764, InTech, Rijeka, Croatia, 2011.

[47] A. O. Dada, A. P. Olalekan, A. M. Olatunya, and O. Dada, "Langmuir, Freundlich, Temkin and Dubinin-Radushkevich isotherms studies of equilibrium sorption of Zn^{2+} unto phosphoric acid modified rice husk," *IOSR Journal of Applied Chemistry*, vol. 3, no. 1, pp. 38–45, 2012.

[48] Z. Ren, G. Zhang, and J. Paul Chen, "Adsorptive removal of arsenic from water by an iron-zirconium binary oxide adsorbent," *Journal of Colloid and Interface Science*, vol. 358, no. 1, pp. 230–237, 2011.

[49] M. H. Ehrampoush, M. Miria, M. H. Salmani, and A. H. Mahvi, "Cadmium removal from aqueous solution by green synthesis iron oxide nanoparticles with tangerine peel extract," *Journal of Environmental Health Science and Engineering*, vol. 13, no. 1, article 84, 2015.

[50] T. V. J. Charpentier, A. Neville, J. L. Lanigan, R. Barker, M. J. Smith, and T. Richardson, "Preparation of magnetic carboxymethylchitosan nanoparticles for adsorption of heavy metal ions," *ACS Omega*, vol. 1, pp. 77–83, 2016.

[51] J. Gong, L. Chen, G. Zeng et al., "Shellac-coated iron oxide nanoparticles for removal of cadmium(II) ions from aqueous solution," *Journal of Environmental Sciences (China)*, vol. 24, no. 7, pp. 1165–1173, 2012.

2

Experimental Research on the Impact of Ion Exchange and Infiltration on the Microstructure of Rare Earth Orebody

Xiaojun Wang,[1,2] Yulong Zhuo,[1,2] Shuqiang Deng,[1,2] Yongxin Li,[1,2] Wen Zhong,[1,2] and Kui Zhao[1,2]

[1]*Jiangxi Key Laboratory of Mining Engineering, Jiangxi University of Science and Technology, Ganzhou, Jiangxi 341000, China*
[2]*School of Resources and Environment Engineering, Jiangxi University of Science and Technology, Ganzhou, Jiangxi 341000, China*

Correspondence should be addressed to Yulong Zhuo; 879408217@qq.com

Academic Editor: Santiago Garcia-Granda

To detect the evolutional characteristics of pore structure in ore leaching process of ion-type rare earth, this paper analyzes the influence mechanism of ion exchange seepage action on the microstructure of orebody, and an experiment for remodeling rare earth saturated samples and ore leaching was designed. Using nuclear magnetic resonance technology obtains the pore structure T_2 map of H_2O and $(NH_4)_2SO_4$ solution in the ore leaching process and inverts and reconstitutes the pore structure distribution image. The results of contrastive analysis experiments indicate that impact factors of the ore leaching process on the microstructure of rare earth orebodies include two aspects: solution seepage and ion exchange. The main factor of pore structure distribution is the ion exchange action, determined by a dual effect. The sole action of solution seepage leads to an increase in pore size, which means that pore size structure is changing from small and medium to macro. Ion exchange gives rise to the movement and restructuring of particles, which results in a decrease in pore sizes. The pore structure changes from loose to compact; in the entire ore leaching process, the ion exchange action advances in a layered shape along the direction of seepage, and the chemical replacement and physical seepage alternately impact the microstructure of the orebodies.

1. Introduction

Ion-adsorption rare earth ore is one of the most precious and rare types of ore in the world and is found in the strong mantle rock of shallow ground [1–3]. At present, the in situ leaching mining method is a mature mining technology, which takes advantage of the chemical replacement reaction of leaching solutions in the process of orebody seepage to achieve the recovery of rare earth cations and thus realizes the extraction of rare earth elements [4–6]. The permeability characteristics of the leaching solution in the orebody rely heavily on the distribution of the internal microscopic pore structure. That is, the form and distribution parameters of the microstructure directly influence the diffusion and penetration of the leaching solution in rare earth orebodies. Therefore, research on the chemical replacement process of the leaching solution and the characteristics of orebody microstructures is beneficial for furthering the analysis of the leaching infiltration mechanisms and is of great significance to green extraction and effective recovery of ion-type rare earth ores.

The matrix of rare earth ore is strong, weathered granite, belonging to the same discrete heterogeneous body as ordinary rock and soil media. In recent years, with the wide application of microcomputed tomography technology [7–9], scholars have started to study the evolution and influencing mechanisms of rock and soil media microstructures. Kodali et al. conducted research on the distribution characteristics of microfissures inside copper ore particles, taking advantage of micro-CT [10]. Nosrati et al. applied micro-CT to scan the internal structure of nickel granulation ore and discovered the microstructure distribution law [11]. Yan et al. introduced microparameters to describe the pore characteristics of soil [12]. Cui et al. measured the pore characteristics of loess-adopting rate-controlled mercury and analyzed the influence of vacuum

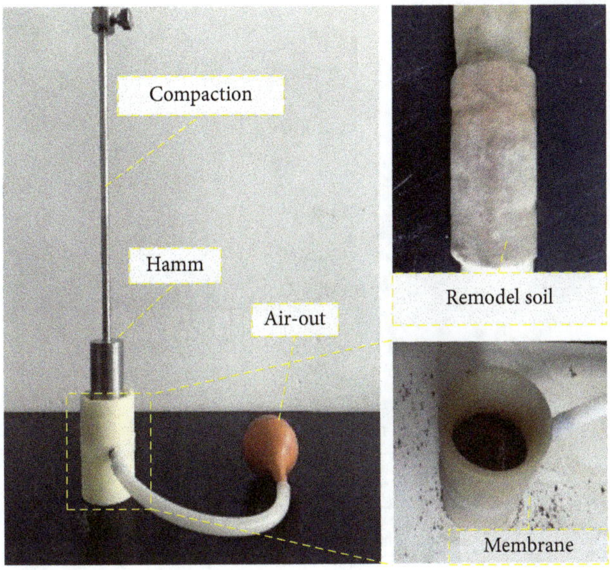

FIGURE 1: Manufacturing process of remodeling soil specimens.

freeze-drying and drying on the microstructure of loess [13]. Li et al. studied the change law of pore structures, void ratios, and moisture content in the process of soil discharge and established a corresponding relationship among them [14].

The in situ leaching of ion-type rare earth involves two important processes: ion exchange and solution infiltration. Ion exchange is a chemical replacement process, while solution infiltration is a physical migration process. At present, little research has been carried out to detect whether the coupling effect of these two processes will result in a change in the micropore structure of rare earth orebodies, and useful conclusions have not yet been reached. This paper describes the uses of nuclear magnetic resonance technology to study the microstructure evolutionary mechanisms of rare earth orebodies in the processes of ion exchange and infiltration and provides important research references for the leaching and mining of ion-type rare earth ores.

2. Experimental Section

2.1. Principles of the Experiment

2.1.1. Chemical Replacement. Rare earth elements are present in strong, weathered granite in the form of ions. $(NH_4)_2SO_4$ solution is usually adopted in the leaching process [15, 16]. The rare earth cations absorbed by the orebody will be replaced in the leaching solution after contact with NH_4^+, which has more active chemical properties, as shown in (1). This strong chemical reaction occurs throughout the entire leaching process. Ion exchange involves the adjustment and reconstruction of the mineral microstructures, leading to a change in the crystal structure. The seepage effect of leaching solution also exerts an impact on the microstructure of orebodies:

$$\begin{aligned}&2\,(\text{Kaolin})^{3-}\cdot \text{RE}^{3+}+3(\text{NH}_4^+)_2\cdot \text{SO}_4^{2-}\\&=2\,(\text{Kaolin})^{3-}\cdot(\text{NH}_4^+)_3+\text{RE}_2^{3+}(\text{SO}_4^{2-})_3.\end{aligned}\tag{1}$$

TABLE 1: Physical parameters of rare earth ore sample.

Parameter type	Geometry dimension (mm) Diameter	Geometry dimension (mm) Height	Density (g/cm^3)	Moisture content (%)
Value	40	60	1.75	15

2.1.2. Test of the Microstructure. Nuclear magnetic resonance (NMR), as a new type of detection technology, has been widely applied in the field of geotechnical engineering [17]. The most important characteristic of this technology is that it can quickly, accurately, and quantitatively measure the microstructure of pores and the pore size distributions while not destroying the structure of rock or soil. Therefore, this is a common detection technology for the analysis of the microstructure evolution of rare earth orebodies in the process of leaching.

The association of the transverse relaxation time, T_2, and the solid–liquid separation characteristics in the leaching process, as well as the specific surface area of pores, is more evident, which reflects the change in the characteristics of the pore structures in rare earth samples in the leaching process to a certain extent. According to NMR principles, free relaxation and diffusion relaxation can be neglected [18–20], and the transverse relaxation rate can be simplified as

$$\frac{1}{T_2}=\frac{\rho_2}{SV}.\tag{2}$$

In (2), ρ_2 represents the relaxation strength (μm/ms), S represents the pore surface area (cm^2), and V represents the pore volume (cm^3). In (2), the ratio between the surface area and volume determines the size of the pores, and the transverse relaxation rate reflects the frequency of proton collisions in the pores. Smaller pore sizes lead to more frequent collisions. Therefore, the transverse relaxation time, T_2, is directly proportional to the size of the pores. The spectrum analysis curve of T_2 in the leaching process of a rare earth orebody can be obtained through the detection results, which can then identify the change rule of the pore structures.

2.2. Experimental Method and Process. The sample was acquired through the remodeling of rare earth sample ore from a mine, and the sample size was consistent with the effective detection area of NMR. The diameter and height of the sample were 40 mm and 60 mm, respectively. In the remodeling process, the consistency of the density of the original soil sample and its moisture content were maintained to the greatest possible extent, as shown in Figure 1. After all phase detection, the rare earth oxide (REO) of the original sample was $0.065\pm0.003\%$, which conforms to the experimental requirements. The physical parameters of remodeling the rare earth sample are shown in Table 1.

To analyze the evolution law of the pore structure of the rare earth sample in the ion exchange process, contrastive analysis experiments were designed. The leaching experiment of remodeling the rare earth sample was conducted on

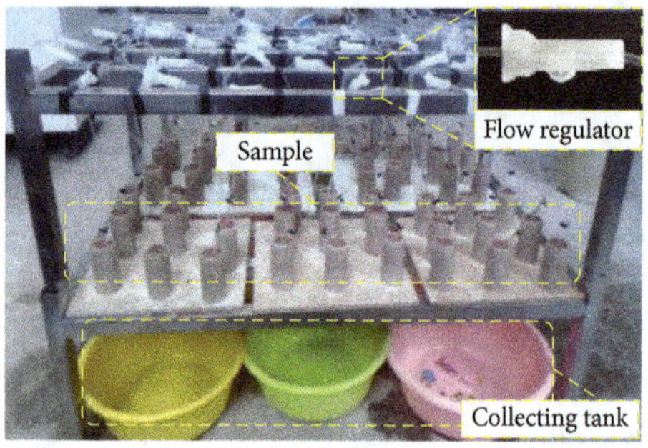

FIGURE 2: Simulation leaching.

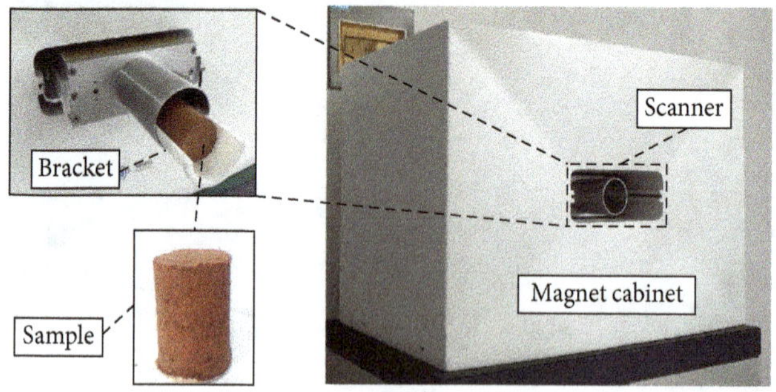

FIGURE 4: PQ-OO1-type mini-NMR spectrometer.

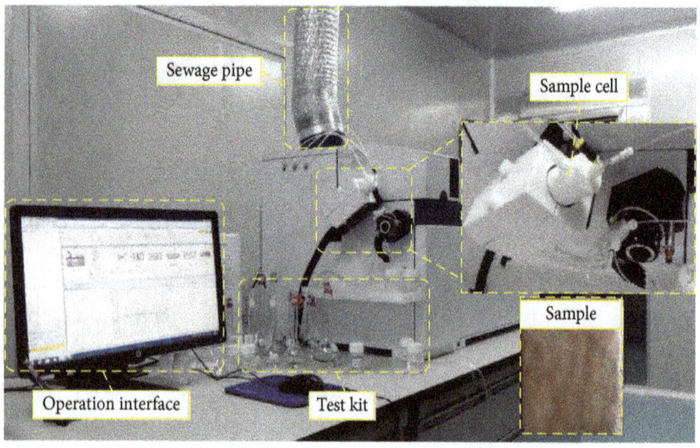

FIGURE 3: Inductively coupled plasma mass spectrometer (ICP-MS).

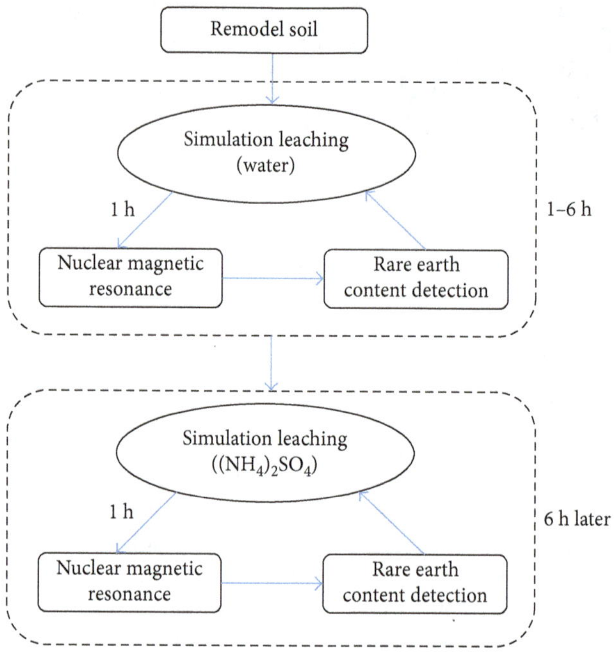

FIGURE 5: Test process.

the simulative leaching experiment setup (the flow rate is 1 mL/min), as shown in Figure 2. The entire leaching experiment was divided into two parts. In the first part, the duration of leaching was six hours, and pure water was selected as the leaching fluid. In this process, a chemical replacement reaction was not involved. In the second part, the duration of leaching was six hours, but a 2.5% $(NH_4)_2SO_4$ solution, which is commonly used in industrial experiments, was selected as the leaching agent. According to the replacement principle, the second part (7–12 h) is the ion exchange process. The two parts of the experimental results were contrastively analyzed.

To effectively identify the ion exchange rate of the chemical replacement process in the different leaching phases, all phase detections of REO were conducted every hour during leaching. When conducting the detection, two complete samples were taken and dried, and then the samples were crushed into powder. The REO in a sample was determined using an inductively coupled plasma mass spectrometer (ICP-MS), as shown in Figure 3, which represents the ion exchange process during mineral leaching.

The micropore structure of the sample was measured using a PQ-OO1-type mini-NMR spectrometer, as shown in Figure 4. The permanent magnets possess a magnetic field intensity of 0.52 T (Tesla). The valid experimental area of the sample was ϕ 60 mm × 60 mm, and the temperature of the permanent magnets was 32 ± 0.01°C, ensuring the stability and uniformity of the experimental magnetic field. Two typical samples were obtained before leaching, and the micropore structures of samples were detected every hour, based on the interval of leaching. Additionally, each sample was leached for 1 h, and its porosity was determined before continuing the leaching for another hour, using the same sample; the porosity was measured again and the process continued. The entire test process is shown in Figure 5.

3. Analyses of Experimental Results

3.1. Change in REO in the Leaching Process. During leaching, the solid sample is chemically analyzed every hour, acquiring the REO in different periods, to analyze each phase of the ion exchange reaction in the leaching process. The experimental results of each period are shown in Table 2.

Combined with the REO changing curve in the leaching experiment process (Figure 6), we know that H_2O is used as the leaching solution in the first phase. No change occurs in the REO concentrations of the solid samples, with measurements being stable within the margin of error. Thus, it can be deduced that the ion exchange reaction, as shown in (1), does not occur inside the sample and that the rare earth cations are not extracted by the leaching solution when applying H_2O as the leaching solution. After the first six

TABLE 2: REO in the leaching process.

Leaching solution	Leaching time/h	REO/10^{-6} g/t
H_2O	0	649
	1	652
	2	647
	3	657
	4	641
	5	639
	6	650
$(NH_4)_2SO_4$ (2.5%)	7	630
	8	600
	9	470
	10	290
	11	200
	12	180
	13	178
	14	181

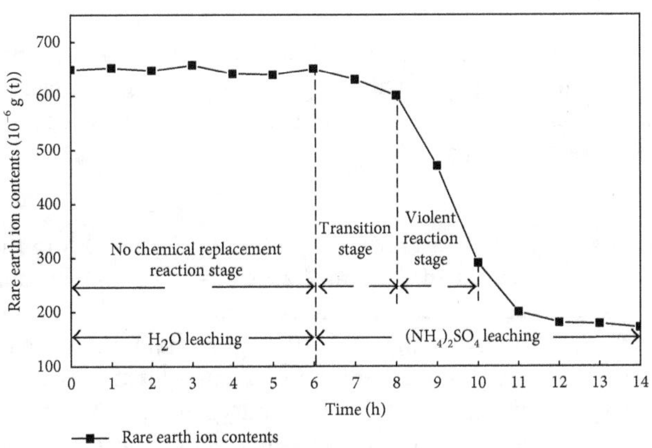

FIGURE 6: REO changing curve in the leaching experiment process.

hours, the leaching solution is replaced by 2.5% $(NH_4)_2SO_4$ solution. In the following analyses, the REO level gradually declines in the sample. The REO declines slowly between the 6th and the 8th hour, which is mainly due to the fact that leaching solutions containing rare earth cations do not extract in large amounts but remain in the sample matrix. After a prolonged time, the chemical replacement reaction occurs intensively in the sample matrix, and the leaching solution containing rare earth cations continuously extracts. We can see from Figure 6 that the REO in the sample matrix falls sharply between the 8th and 10th hour. Between the 10th and 14th hour, the REO drops to below 200 and the change begins to flatten, which proves that the ion exchange reaction is almost completed in the sample after the 10th hour. It can be deduced that, based on the above analyses, the early phase, between the 0 and 6th hour, belongs to the phase without a chemical replacement reaction; the phase between the 6th and 11th hour belongs to the intensive ion exchange phase; and the ion exchange reaction is completed after the 12th hour.

3.2. Distribution of NMR T_2. The T_2 spectrum curve is closely related to the change in the inner pores. The

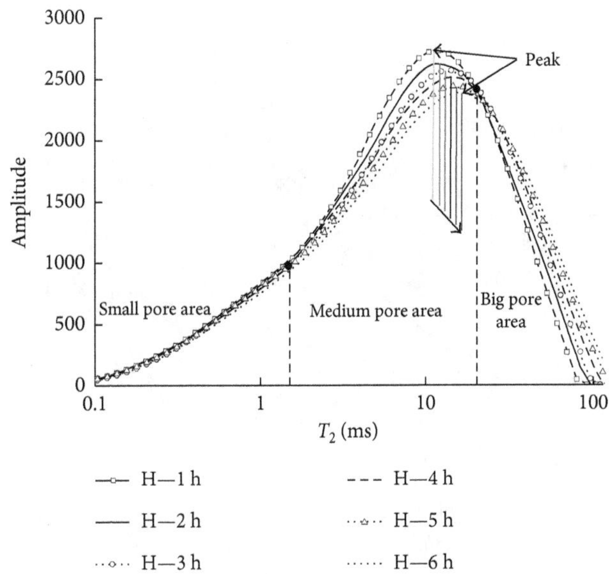

FIGURE 7: H_2O leaching (the first phase).

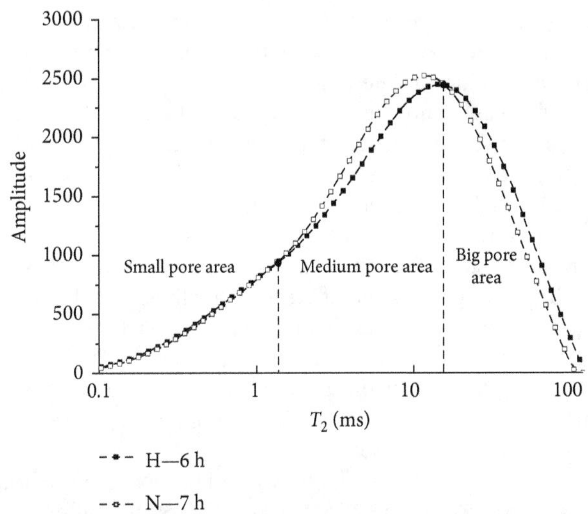

FIGURE 8: Transition phase of H_2O and $(NH_4)_2SO_4$ (the second phase).

T_2 spectrum curve can reflect the sizes of pores in the rare earth specimen and the number of pores of different sizes. The size of the T_2 value is positively related to the pore size. The amplitude represents the number of pores of different sizes.

The entire process in the mineral leaching experiment is divided into three phases. The first phase is H_2O leaching, the second phase is the transition phase of H_2O and $(NH_4)_2SO_4$, and the third phase is $(NH_4)_2SO_4$ leaching. The T_2 spectrum distribution curves of the three phases are shown in Figures 7–9, respectively.

Figure 7 presents the T_2 spectrum curve of the entire process of H_2O leaching. We can see from the area formed by the spectrum curve that the inner pore distribution of the remodeled rare earth sample is dominated by pores with medium apertures. Compared with the curve distribution between the 1st and 6th hour, we can see that an obvious

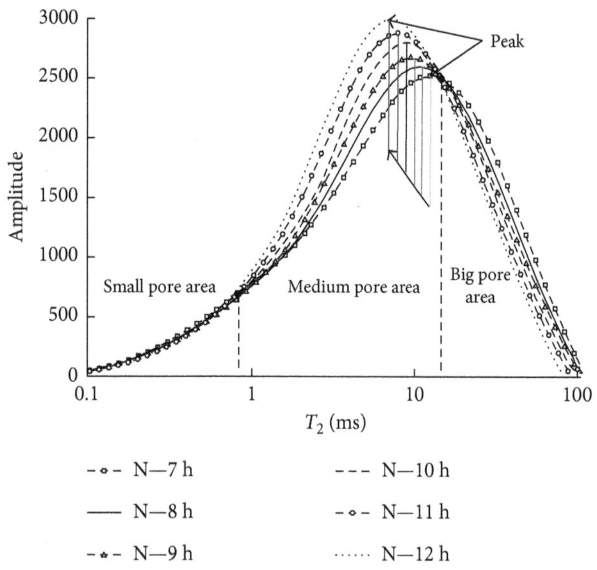

FIGURE 9: $(NH_4)_2SO_4$ leaching (the third phase).

change in the curve spectrum of the pore area with medium apertures has taken place after this prolonged amount of time. A specific manifestation is that the peak value gradually decreases and the crest moves to the right. With the analysis of the pore area with large apertures, we can discover that the large aperture segment of the T_2 spectrum curve is gradually moving to the right, which explains that the size of the pores with large apertures are also gradually increasing inside the sample in the leaching process. Under the simple seepage action of the leaching fluid, the change in internal microstructure of an ion-type rare earth orebody is that the pores with medium apertures gradually become pores with large apertures; the number of pores with medium apertures decreases, while the number of pores with large apertures increases.

According to the results shown in Figure 8, the leaching fluid is replaced by the 2.5% $(NH_4)_2SO_4$ solution after pure water leaching for six hours. During the 7th hour, we can see from the analysis of the leaching results that the change law of the T_2 spectrum curve is not consistent with that of pure water leaching. The peak value of the T_2 spectrum curve increases and the crest moves to the left, which is opposite of the previous change law.

We can see from Figure 9 that the peak value of the entire T_2 spectrum curve increases and the crest moves to the left after leaching by the $(NH_4)_2SO_4$ solution from the 7th to 12th hour, which explains that the number of pores with medium apertures is increasing and the pores with medium apertures are gradually becoming pores with small apertures after leaching by the $(NH_4)_2SO_4$ solution. We can also see from the pore area of the large apertures in the T_2 spectrum curve that after a prolonged time, the T_2 spectrum curve of the large apertures moves to the left, gradually moving toward the pore area of medium apertures. According to the analysis results presented in Figure 6, the period between the 6th and 12th hour is the ion exchange reaction phase. Not only does the infiltration process exist in this phase but also the chemical replacement process. Comparing Figure 7 with Figure 9, we can see that the main factor influencing the micropore structure of rare earth orebodies is the ion exchange reaction. Therefore, the ion exchange reaction inside the rare earth orebody will induce a decrease in the pore volume inside the orebody. Moreover, the aperture size shows the tendency of changing from large to small overall, and the entire micropore structure of the orebody tends to be more compact.

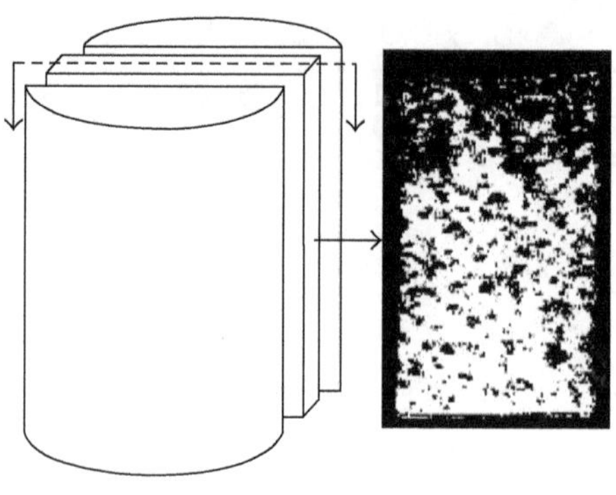

FIGURE 10: Typical cutting plane.

3.3. Imaging Analysis of Microstructure Evolution. From the analysis presented in Section 2.2, we can see that the ion exchange reaction between the rare earth orebody and the leaching solution changes the structural chain among particles, leading to the recombination of particles among inner pores and giving rise to new changes in the micropore structure. The reconstruction imaging technology of NMR is applied to conduct image reconstruction of the scanning data at regular intervals, obtaining the inversion image of the microstructure. To conduct an intuitive analysis of the micropore structure change law, the inversion image is projected. The leaching solution infiltrates along the axis of the sample, and the typical cutting plane is thus selected in the center of the sample along the generatrix, as shown in Figure 10.

Figures 11 and 12 are two-dimensional images of the cutting plane in the leaching process. Figure 11 shows the phase of pure water leaching, while Figure 12 shows the phase of 2.5% $(NH_4)_2SO_4$ solution leaching. The bright color shows the area of liquid molecules, which is the area of the pores. The activity of the water molecules in the black area is extremely weak, so this area is considered a solid region.

It can be clearly seen that after a prolonged time, the bright area gradually expands when pure water is used as the leaching solution, which explains why the number of inner pores increases after a prolonged time and why the apertures also increase under the sole action of the seepage effect, which is completely consistent with the analysis results in Section 2.2.

The most important change appears in Figure 12. This image shows the change when the leaching solution is replaced by the $(NH_4)_2SO_4$ solution. We can see from the image of the 7th hour that the bright area decreases slightly in the upper part, and the black area starts to increase. This

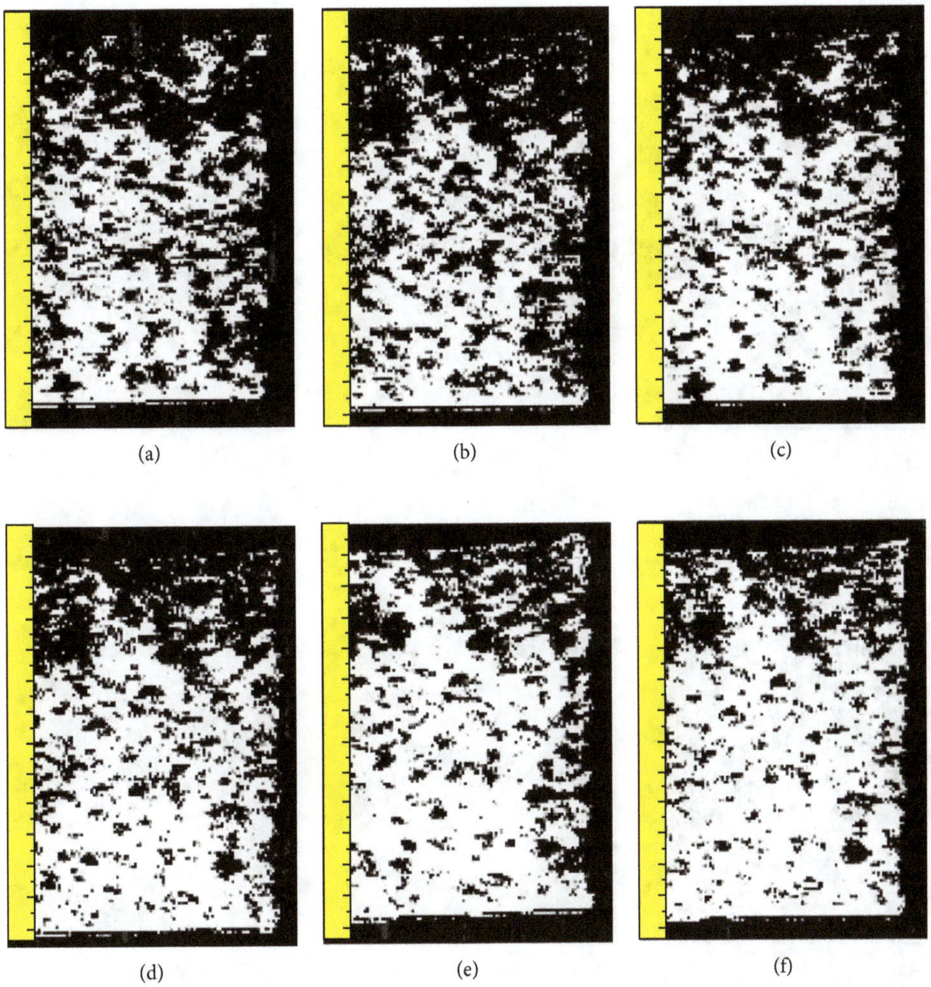

Figure 11: Phase of pure water leaching. (a) 1 h, (b) 2 h, (c) 3 h, (d) 4 h, (e) 5 h, and (f) 6 h.

area has a banding distribution in the upper part of the sample, but no obvious changes can be observed at the bottom of the sample, compared with the image at the 6th hour, which shows that the reaction occurs inside the sample when pure water is replaced by the $(NH_4)_2SO_4$ solution between the 6th and 7th hour. The ion exchange mainly occurs at the top of the sample, and the leaching chemical replacement reaction leads to the change in particle structure, giving rise to the movement and reconstruction of the pore particles and the disappearance of some pores. Pores with large apertures are transformed into pores with small and medium apertures, and the area of ion exchange tends to be more compact. The change becomes more obvious in Figure 9; the black area increases more rapidly in the middle-upper part, which shows that the pore structure is undergoing a rapid change under the effects of ion exchange. After the 9th hour, the black area starts to shift to the lower-middle part. The ion exchange reaction has been completed in the upper part, which again enters into the model of simple liquid seepage. Meanwhile, another obvious change can be observed: the bright area increases again in the middle-upper part, indicating that pores appear in large numbers in the middle-upper part. As the apertures increase, it proves that the pore structure under the simple seepage effect tends to be looser. From the 9th hour to 11th hour, the area with the black stripe moves downward, and the bright area above the black area recovers layer by layer. After 12 hours, the stripped black area completely disappears, and the white area is distributed to a great extent. According to Figure 6, the ion exchange reaction has been completed at this point in time, and only the seepage effect influences the micropore structures of orebodies.

The influence of pure water leaching on the pore structure is a seepage effect, while the influences of $(NH_4)_2SO_4$ solution leaching on the micropore structure of an orebody are both ion exchange and seepage effects. The black area moving downward in Figure 12 is completely different from that of Figure 11. Therefore, under the coupling effect of ion exchange and seepage, it is the chemical action of ion exchange, and not the physical action of seepage, that influences the microstructure. Meanwhile, through the contrastive analysis of Figures 11 and 12, the overall manifestation of the micropore structures of rare earth orebodies after $(NH_4)_2SO_4$ solution leaching is that the pore sizes shrink, the total number of pores is reduced, and the pore structure tends to be more compact.

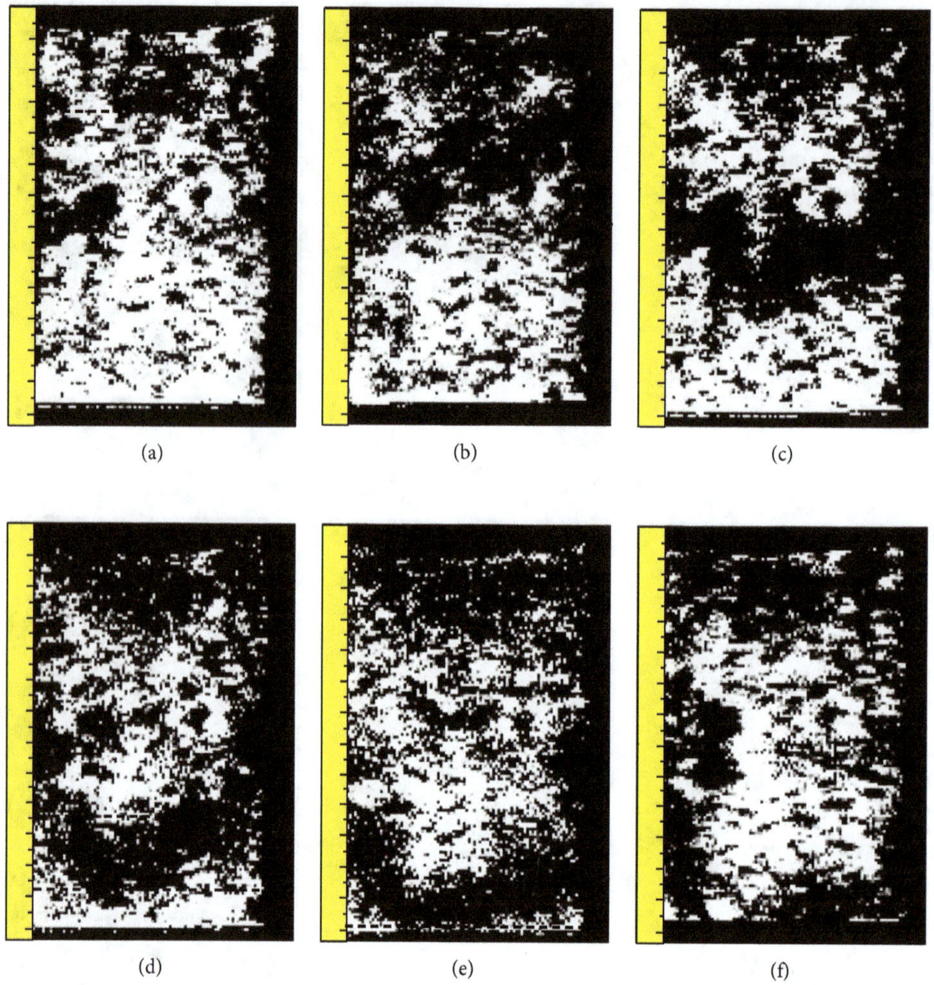

FIGURE 12: Phase of $(NH_4)_2SO_4$ solution (2.5%). (a) 7 h, (b) 8 h, (c) 9 h, (d) 10 h, (e) 11 h, and (f) 12 h.

4. Conclusions

This paper investigated a leaching experiment on ion-type rare earth orebodies and conducted a contrastive analysis of the leaching results of two different solutions. The following conclusions were obtained:

(1) The pure water leaching process does not involve ion exchange, belonging simply to a liquid seepage process, while the $(NH_4)_2SO_4$ solution leaching process is the coupling of ion exchange and liquid seepage.

(2) In the leaching process, the micropore structures of rare earth orebodies are subjected to the coupling effect of solution seepage and ion exchange. Water leaching leads to an increase of inner pores and apertures. The microstructure also tends to be looser. $(NH_4)_2SO_4$ leaching leads to the movement and reconstruction of particles, decreasing the apertures and the number of pores. The microstructure also tends to be more compact.

(3) In the leaching process, the ion exchange reaction occurs downward, layer by layer, with the solution seepage conforming to the laminar flow law of seepage theory. Under $(NH_4)_2SO_4$ solution leaching, it is the chemical action of ion exchange but not the physical action produced by the seepage effect that influences the microstructure of an orebody. Overall, the chemical replacement reaction of ion-type rare earth leaching possesses certain recovery effects on the microstructure of orebodies.

Conflicts of Interest

The authors declare that they have no conflicts of interest.

Acknowledgments

The research was financially supported by National Natural Science Foundation of China (nos. 51304083, 51364011, 51564012, and 51504102), Postdoctoral Science Foundation of China (no. 2014M552562XB), Science and Technology Support Plan of Jiangxi (no. 20141BBE50005), Science and Technology Project Founded by the Education Department of Jiangxi Province (GJJ150653), Qingjiang Excellent Young Talents of Jiangxi University of Science, and Technology and Jiangxi University of Science and Technology Excellent Doctoral Thesis Training Project (YB2017001).

References

[1] T. S. Qiu, D. M. Zhu, X. H. Fang, Q. H. Zeng, G. K. Gao, and H. L. Zhu, "Leaching kinetics of ionic rare-earth in ammonia-nitrogen wastewater system added with impurity inhibitors," *Journal of Rare Earths*, vol. 32, no. 12, pp. 1175–1183, 2014.

[2] K. Sanematsu, Y. Kon, A. Imai, K. Watanabe, and Y. Watanabe, "Geochemical and mineralogical characteristic of ion-adsorption type REE mineralization in Phuket, Thailand," *Mineralium Deposita*, vol. 48, no. 4, pp. 437–451, 2013.

[3] L. Zhang, K. X. Wu, L. K. Chen, P. Zhu, and Y. H. Ou, "Overview of metallogenic features of ion-adsorption type REE deposits in Southern Jiangxi Province," *Journal of the Chinese Society of Rare Earths*, vol. 33, no. 1, p. 10, 2015.

[4] Q. B. Zhang, Y. X. Hua, C. Y. Xu, Y. Li, J. Li, and P. Dong, "Non-haloaluminate ionic liquids for low- temperature electrodeposition of rare-earth metals—a review," *Journal of Rare Earths*, vol. 33, no. 10, pp. 1017–1025, 2015.

[5] L. Zhao, Z. Dong, G. L. Ma, and W. J. Yuan, "Solution extraction of several lanthanides from nitric acid with isohexyl-BTP in $[C_nmim][NTf_2]$ ionic liquid," *Journal of Rare Earths*, vol. 33, no. 11, pp. 1182–1188, 2015.

[6] X. Q. Sun, Y. Ji, J. Chen, and J. T. Ma, "Solvent impregnated resin prepared using task-specific ionic liquids for rare earth separation," *Journal of Rare Earths*, vol. 27, no. 6, pp. 932–936, 2009.

[7] J. Dewanckele, T. De Kock, M. A. Boone et al., "4D imaging and quantification of pore structure modifications inside natural building stones by means of high resolution X-ray CT," *Science of the Total Environment*, vol. 416, no. 2, pp. 436–448, 2012.

[8] J. R. Kyle and R. A. Ketcham, "Application of high resolution X-ray computed tomography to mineral deposit origin, evaluation, and processing," *Ore Geology Reviews*, vol. 65, no. 4, pp. 821–839, 2015.

[9] L. J. Munkholm, R. J. Heck, and B. Deen, "Soil pore characteristics assessed from X-ray micro-CT derived images and correlations to soil friability," *Geoderma*, vol. 181-182, no. 7, pp. 22–29, 2012.

[10] P. Kodali, N. Dhawan, T. Depci et al., "Particle damage and exposure analysis in HPGR crushing of selected copper ores for column leaching," *Minerals Engineering*, vol. 24, no. 13, pp. 1478–1487, 2011.

[11] A. Nosrati, K. Quast, D. F. Xu et al., "Agglomeration and column leaching behaviour of nickel laterite ores: effect of ore mineralogy and particle size distribution," *Hydrometallurgy*, vol. 146, no. 5, pp. 29–39, 2014.

[12] X Q. Yan, Y. G. Fang, and P. Zhang, "Experiment study on the effects of bentonite on the micropore structure characteristics of soil," *Chinese Journal of Geotechnical Engineering*, vol. 33, no. 8, pp. 1302–1307, 2011.

[13] D. S. Cui, W. Xiang, Q. Chen, Q. B. Liu, L. Z. Lang, and Y. P. Wu, "Pore characteristics of sliding zone soils of Huangtupo landslide by vacuum freeze-dried methods," *Earth Science-Journal of China University of Geosciences*, vol. 39, no. 10, p. 1531, 2014.

[14] X. Li, L. M. Zhang, and G. D. Ao, "Variations of pore structure, void ratio, and water content in soil drying process," *Rock and Soil Mechanics*, vol. 32, no. S1, pp. 100–105, 2011.

[15] G. Q. Deng and Y. M. Yang, "A review of the mining technologies of ion-absorbed rare earth mineral," *Chinese Rare Earths*, vol. 37, no. 3, p. 129, 2016.

[16] R. A. Chi, J. Tian, X. P. Luo, Z. G. Xu, and Z. Y. He, "The basic research on the weathered crust elution-deposited rare earth ores," *Nonferrous Metals Science and Engineering*, vol. 3, no. 4, p. 1, 2012.

[17] K. P. Zhou, J. L. Li, Y. J. Xu et al., "Experimental study of NMR characteristics in rock under freezing and thawing cycles," *Chinese Journal of Rock Mechanics and Engineering*, vol. 31, no. 4, p. 731, 2012.

[18] X. J. Li, S. D. Nie, Y. J. Wang, and P. Q. Yang, "An improved two-dimensional inversion algorithm for low-field NMR diffusion-transverse relaxation data," *Chinese Journal of Magnetic Resonance*, vol. 30, no. 3, p. 322, 2013.

[19] K. W. Wang and N. Li, "Numerical simulation on effects of reservoir characteristics and saturation on T2 spectra of nuclear magnetic resonance," *Acta Petrolei Sinica*, vol. 30, no. 3, p. 422, 2009.

[20] S. T. Bai, D. J. Cheng, J. B. Wan et al., "Quantitative characterization of sandstone NMR T2 spectrum," *Acta Petrolei Sinica*, vol. 27, no. 3, pp. 382–391, 2016.

3

Development of Electrohydraulic Forming Process for Aluminum Sheet with Sharp Edge

Ji-Yeon Shim and Bong-Yong Kang

Carbon & Light Materials Application R&D Group, KITECH, Jeonju-si, Republic of Korea

Correspondence should be addressed to Bong-Yong Kang; kanbo@kitech.re.kr

Academic Editor: Donato Sorgente

Electrohydraulic forming (EHF), high-velocity forming technology, can improve the formability of a workpiece. Accordingly, this process can help engineers create products with sharper edges, allowing a product's radius of curvature to be less than 2 mm radius of curvature. As a forming process with a high-strain rate, the EHF process produces a shockwave and pressure during the discharge of an electrical spark between electrodes, leading to high-velocity impact between the workpiece and die. Therefore, the objective of this research is to develop an EHF process for forming a lightweight materials case with sharp edges. In order to do so, we employed A5052-H32, which has been widely used in the electric appliance industry. After drawing an A5052-H32 Forming Limit Diagram (FLD) via a standard limiting dome height (LDH) test, improvements to the formability via the EHF process were evaluated by comparing the strain between the LDH test and the EHF process. From results of the combined formability, it is confirmed that the formability was improved nearly twofold, and a sharp edge with less than 2 mm radius of curvature was created using the EHF process.

1. Introduction

In the electrical appliance industry, a product layout can be designed with a relatively simple forming process. However, product design has become one of the most important factors for selecting products, so development of reproducible forming processes is needed to manufacture optimal products with sharp edges. In particular, lightweight materials such as aluminum and magnesium have been utilized for electrical appliance cases, although these materials have low formability. When attempting to form a sharp edge using conventional forming processes, these materials are often torn and wrinkled. Thus, researchers have attempted to develop new forming processes for lightweight materials with sharp edges. High-velocity forming methods have been reported to potentially increase sheet metal formability beyond the limits of conventional forming processes [1–4]. Various high-velocity forming processes have been developed, including explosive forming (EXF), electromagnetic forming (EMF), and electrohydraulic forming (EHF). Such high-strain rate forming processes result in lower tooling costs, due to the elimination of punches, which are replaced by an impulse force generated by an explosive, electromagnetic repulsion via electrical plasma channels [5]. Although EXF is one of the earliest high-velocity forming processes, handling explosives for routine research is not feasible because it is very dangerous and noisy. EMF is also limited to highly conductive materials. Therefore, EHF is the best way to form all kinds of lightweight materials.

During the EHF process, stored electrical energy is discharged to electrodes in liquid-filled chambers. A small amount of liquid between the tips of the electrodes is vaporized, forming a plasma channel. Electrical current continues to go through the channel, converting the electrical energy into the internal energy of the plasma bubble; this occurs in a very short period of time (usually less than 100 microseconds). This leads to explosive expansion of the channel by forming high-velocity shockwaves in the liquid, which forms a workpiece into a die [6]. Initial development of the EHF process began in Russia in the early 1950s by Yutkin [7]. Bruno [8] and Davies and Austin [9] reviewed early applications and research results on EHF. Specifically,

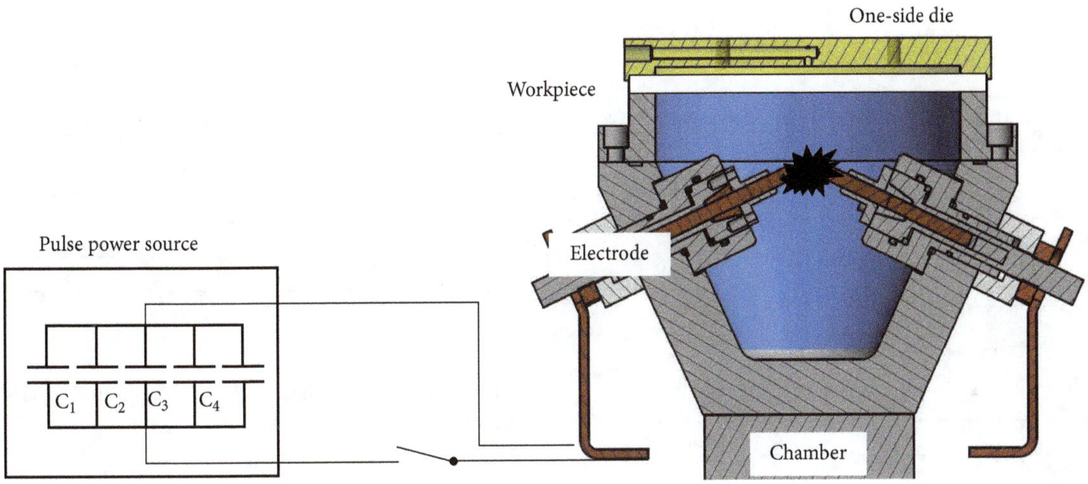

FIGURE 1: Cross-sectional diagram of an EHF chamber.

Bruno described a number of industrial examples where EHF machines storing 36 kJ, 60 kJ, 150 kJ, and 172 kJ were employed. However, due to the necessity of filling the chamber with liquid at the beginning of each forming cycle and evacuating the liquid at the end of it, the cycle time for the EHF process was approximately several minutes. Sanford [10] described a hybrid technology where static hydroforming was used to bulge sheet metal into the die cavity, followed by EHF, which provided a higher pressure level and filled in the details of the die cavity. According to this study, a typical cycle time for such a hybrid process is 10 minutes. More recently, Balanethiram and Daehn [3] reported an increase in the forming limit for A6061-T4 and Golovashchenko and Mamutov [11] reported forming limit improvements for DP590, DP780, and DP980. Rohatgi et al. [5] investigated the deformation behavior of A5182-O and DP590 during EHF using high-speed cameras and a digital image correlation (DIC) technique to understand the roles of the strain rate. Gillard et al. [12] analyzed the formability improvements of formability of DP780 and DP980 and developed an FE-model for EHF processes. Further, Vohnout et al. [13] reported results of pulsed loading in a closed volume, assuming energy equivalence in explosive forming. Additional simulation work fully modeling the discharge, pressure transmission, and part formation into the die has been conducted by Ford researchers [14–16]. Recently, Ahmed et al. [17] reported near 45–50% improvement to the formability of A5052-O via the EHF process [18]. According to this research, improvement to the formability of some materials, such as A6061-T4, DP590, and A5052-O, was achieved; however, EHF research on A5052-H32 sheets with low formability, which has been widely used for electrical appliances and automobiles, was not carried out. Therefore, the objective of this paper is to assess formability improvements and to attempt forming A5052-H32 sheets with sharp edges. In order to do so, an A5052-H32 FLD was drawn via an LDH test and combined with the experimental results of EHF processes. Importantly, sharp edge forming was attempted using the EHF process. Sharp edge forming is the filling of the die with a 2 mm radius of curvature. It is anticipated that this research will contribute to the manufacturing of lightweight material cases with sharp edges for small electrical appliances, such as cell phones and cameras.

2. EHF Forming System

The major advantages of high-strain rate forming processes include increases in metal formability, reduced wrinkling, reduced spring-back, and reduced manufacturing costs. EHF is a high-strain rate forming process that involves the conversion of electrical energy into mechanical energy in a liquid-filled chamber. The forming process is very fast, lasting no longer than 200 μs, and has a forming speed of approximately 300 m/s [17]. A high-strain rate during the EHF process is achieved by taking advantage of the electrohydraulic effect, which can be described as the rapid discharge of electric current between electrodes tips while being submerged in liquid-filled chambers and the propagation through the liquid of the resulting shockwave and reflective wave. The high-pressure propagation from the resulting shockwave and reflective wave causes a high-velocity impact between the workpiece and die, allowing the workpiece to be formed into the desired shape [19].

An EHF system consists of a pulse power source, chamber, electrodes, and one-sided die. Figure 1 shows a cross-sectional diagram of an EHF chamber. The necessary high voltage is obtained by charging a capacitor storage bank in a pulse power source. When the electronic trigger circuit is activated, energy is rapidly dumped into concentric or opposing spark-gap electrodes submerged in the liquid-filled chamber. The thickness of the chamber is 35 mm considering the 300 MPa fluid pressure and corrosion. A chamber with carbon steel, containing beryllium copper electrodes was used to avoid abrasion by shockwave.

3. Experimental Work

3.1. Conventional Formability Testing. An FLD is used during the design stage of any new sheet metal component for

Table 1: Material characteristics for A5052-H32.

(a) Tensile properties

Yield strength (Mpa)	Tensile strength (Mpa)	Elongation (%)	n
193	228	12	0.177

(b) Chemical composition

Component	Al	Cr	Cu	Fe	Mg	Mn	Si	Zn
Wt (%)	97	0.10	0.20	0.70	1.10	0.20	0.30	0.25

Table 2: FLD test conditions for this study.

Material	A5052-H32
Thickness (mm)	0.5
Specimen direction	Rolling direction
Temperature (°C)	25
Punch speed (mm/min)	80
Lubrication	Grease

tooling the shape and optimizing variables. In the sheet metal industry, this process is widely used and considered to be one of the most important factors for determining the formability of sheet metals. Every sheet metal has its own FLD that determines its formability, strain limit, and forming regions [20]. An FLD is a representation of the critical combination of two principal surface strains, major and minor, above which localized necking instability is observed [21]. The forming limit curve is plotted for varying strain ratios, from pure shear to equibiaxial tension. One can conclude that drawing has occurred in the case of a negative minor strain or negative strain ratio. Stretching is observed when positive minor strain is obtained, or the strain ratio is positive [22]. In order to construct the FLD, Hecker's simplified technique was employed in this study. This procedure comprises three main steps: grid marking on the workpiece, punch stretching of the grid marked workpiece to failure (onset of localized necking), and strain measurement on the deformed specimens [23]. In order to evaluate the formability, a 0.5 mm thick A5052-H32 (NOVELIS) sheet was considered. The tensile properties and chemical composition of the workpiece are shown in Table 1. Eight workpieces, ranging from 25 mm × 200 mm to 200 mm × 200 mm, were cut from one sheet. Square (2 mm × 2 mm) grid patterns were imprinted on the workpieces for strain measurement after forming. The formability testing equipment consists of a hemispherical punch of 101.6 mm in diameter, as well as a lower and upper die on a 20-ton hydraulic press. Table 2 shows the basic forming conditions of the FLD test. For better accuracy, experiments were conducted three times and major and minor strain values were taken into account. After forming, grids on the workpiece in and around the necked regions were imaged, and major and minor engineering strain components were calculated from the grid pattern analyzer. The grid pattern analyzer is comprised of a video camera system for data acquisition and software for data process and analysis. The basic procedure for measuring strain data with the grid pattern analyzer is the following:

(i) A grid pattern is applied to an undeformed blank.

(ii) The blank is formed.

(iii) The grid pattern analyzer is used to image a single grid element.

(iv) From the image, the computer determines major and minor strain values.

3.2. EHF. The EHF formability experiments in this study were conducted from a flat blank, and the die cavity was filled using a single EHF pulse. In this study, a conical EHF chamber was used in conjunction with the conical die to form the workpiece in a state of biaxial strain. Figure 2 shows the conical die for the EHF process. Using an electrochemical etching process, square grid patterns (2.0 mm × 2.0 mm) were marked on the workpiece to measure strain after forming.

The electrodes can be seen above, passing through the chamber walls. They were insulated from the walls by a nonconductive polymeric material with good mechanical resiliency in order to withstand the intense dynamic loading occurring within the chamber. A hemispherical chamber with a 200 mm diameter was filled with 2 L of tap water for the EHF. The electrode gap in the chamber was 20 mm, and a copper bridge wire, 0.3 mm in diameter, was placed between the electrode tips before each pulse in order to promote efficient and repeatable discharge performance.

During the EHF process, a workpiece was installed at the top of the chamber and then clamped by an o-ring and press. The air remaining in the die was removed by a vacuum pump. The chamber was filled with water up to its upper rim. The pulse power source had a maximum energy capacity of 60 kJ, and the electrical energy used in the EHF ranged from 0.3 kJ to 8 kJ. Electrical energy was stored in high voltage capacitors and charged through a transformer and a set of diodes. Each of them was connected to the discharge circuit through their respective vacuum switches. A set of vacuum switches closed the circuit and delivered the voltage stored in each module of the capacitors to the electrodes. Moreover, a Rogowski coil for measuring the currents on electrodes was set at the ends of the electrodes. The Rogowski coil measured the current linearly, from low to high current, using mutual inductance. It can also measure the current in a wide frequency range. Figure 3 shows the set-up for the EHF process.

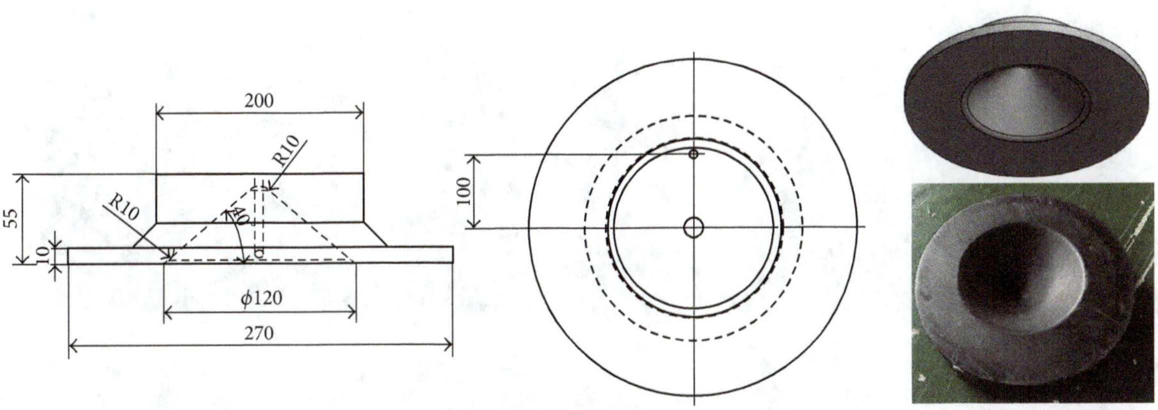

FIGURE 2: Die for EHF.

FIGURE 3: Set-up for EHF.

Generally, the parameters that define the efficiency of the EHF process are the mutual positions of the electrodes, electrical properties of the liquid, charging voltage, capacitance of the circuit, inductance and resistance of the equipment and connecting cables, volume of the chamber, and the distance between the discharge channel and workpiece [17]. In this study, all parameters except charging voltage were set under the same conditions. Table 3 shows the EHF conditions.

3.3. Results and Discussion. To draw the FLD after a dome stretching test, workpieces were measured in terms of major and minor engineering strain. Figure 4 shows the deformed workpiece after the LDH test. From these deformed grids, the FLD, consisting of major and minor strains, was plotted using a grid pattern analyzer. All of the LDH specimens were formed until failure, and the grid with the largest uniform major strain was always located near the cracks. If

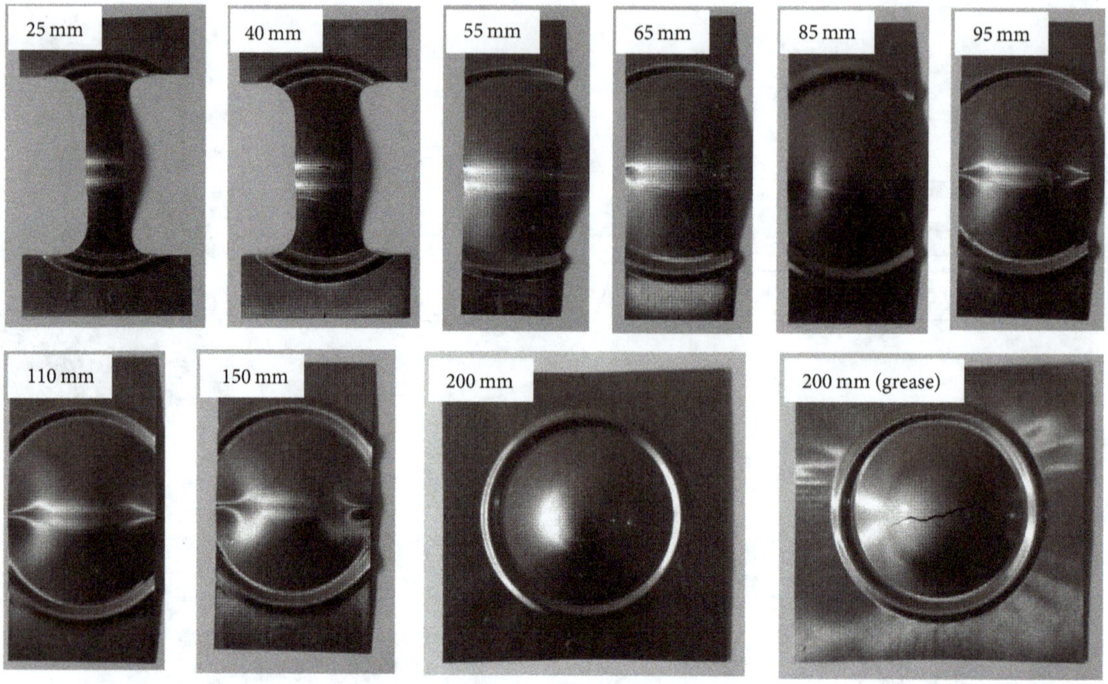

FIGURE 4: Workpieces after dome stretching test.

TABLE 3: EHF conditions.

Capacitance (μF)	Charging voltage (kV)	Charging energy (kJ)	Electrode gap (mm)
	1	0.3	
	2	1.2	
64	3	2.8	20
	4	5.1	
	5	8	

the measured square grid was located near the crack on a specimen, then that grid was marked as *Failed*. Conversely, if a square grid was located a safe distance away from the crack, then that grid was marked as *Safe*.

The typical areas with the highest strain concentrations were located on or near a crack. Figure 5 shows the static FLD for A5052-H32 from the results of the LDH test. Table 4 shows the resulting measurements of the dome height after the LDH test (repeated 3 times). The maximum dome height was 19.18 mm in workpieces with 25 mm × 200 mm dimensions and a minimum of 13.22 mm at 110 mm × 200 mm in dry conditions. When the dome stretching test was performed with lubrication, the dome height was 23.27 mm for workpieces with 200 mm × 200 mm dimensions.

Figure 6 shows the setting for measuring the waveform, wherein measured voltages can calibrate the input current of the electrodes. The discharge times were similar (25 μs) for all EHF conditions. The measured waveforms increased in direct relation to the charging voltage. Each measured voltage was 1.2 mV, 1.7 mV, 2.2 mV, 2.8 mV, and 3.2 mV at 1 kV, 2 kV, 3 kV, 4 kV, and 5 kV, respectively. Therefore, input currents for the EHF process were calculated as 30 kA, 42.5 kA, 55 kA, 70 kA, and 80 kA, by a calibration coefficient of 25 kA/mV. Based on this result, it could be determined that increasing the currents would lead to an increase in the shockwave within the chamber during the EHF process.

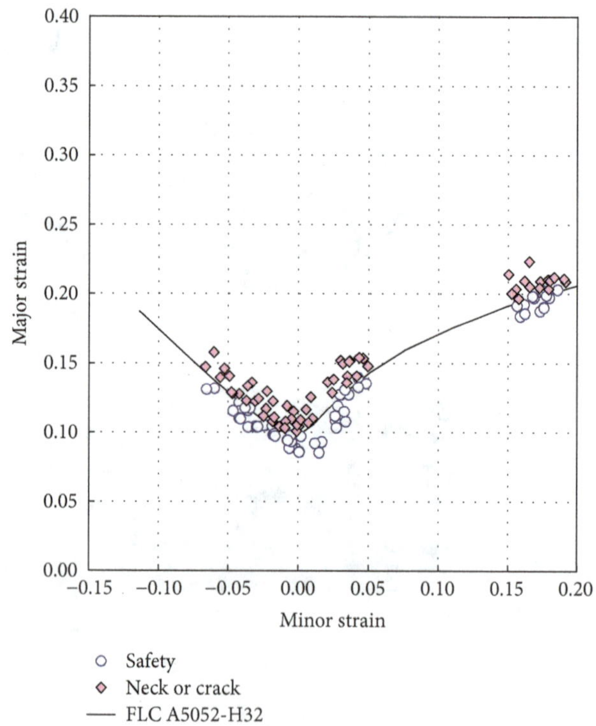

FIGURE 5: FLD of A5052-H32.

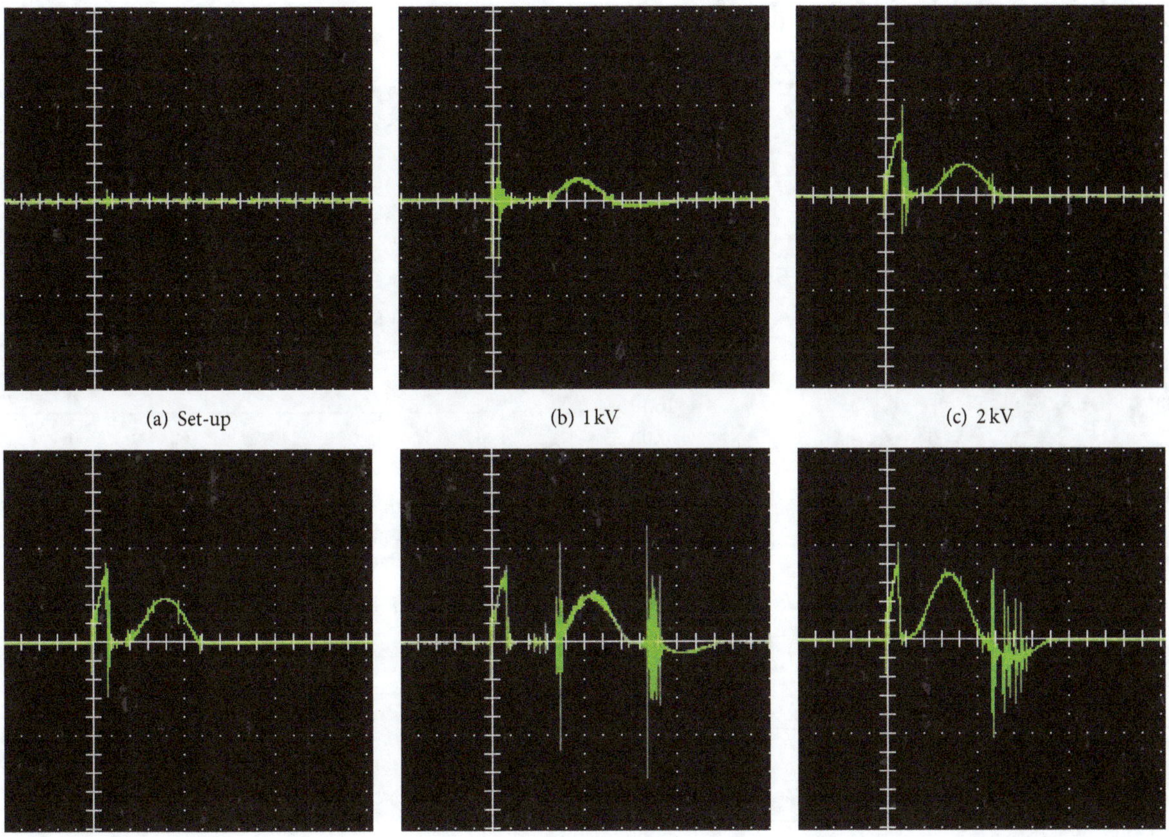

(a) Set-up (b) 1 kV (c) 2 kV

(d) 3 kV (e) 4 kV (f) 5 kV

FIGURE 6: Discharge waveform (X: 200 μs/div, Y: 500 mV/div) with a calibration coefficient of 25 kA/mV.

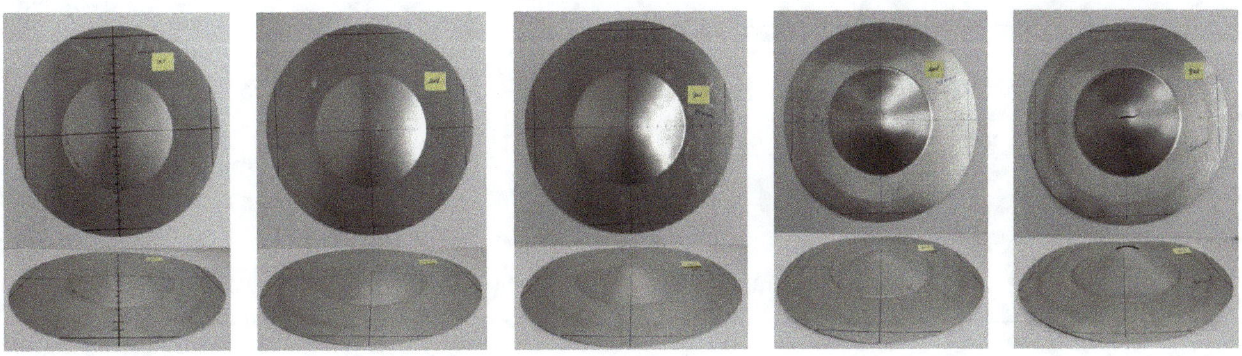

(a) 1 kV (b) 2 kV (c) 3 kV (d) 4 kV (e) 5 kV

FIGURE 7: Workpieces after EHF.

Figure 7 shows the deformed workpieces after EHF. A crack was observed under a 5 kV charging voltage at the apex of the workpiece. Table 5 shows the height of the deformed workpiece, from flat to apex, under different charging voltage conditions, after being measured precisely using a height gauge.

Dome height was measured as 5.03 mm, 8.39 mm, 12.03 mm, 19.31 mm, and 28.36 mm at 1 kV, 2 kV 3 kV, 4 kV, and 5 kV charging voltage, respectively, as shown in Table 5. The dome height was increased in direct relation to the charging voltage. As shown in Figure 8, deformed workpiece profiles at 2 kV and above increased sharply. This confirmed that the dome height of the EHF workpiece was increased in direct relation to the charging voltage, because the charging voltage provides the formation energy. Therefore, increasing the charging voltage generated an increased shockwave and reflected wave in the chamber, leading to an increase in the dome height of the workpiece.

To validate the formability improvement using EHF, strain in the workpiece after a 5 kV charging voltage was

TABLE 4: Dome height from the LDH test (unit: mm).

Size	Height
25 × 200	19.14
	19.01
	19.18
40 × 200	18.80
	19.00
	18.75
55 × 200	19.12
	19.03
	19.10
65 × 200	17.11
	16.19
	17.05
85 × 200	13.45
	14.00
	13.47
95 × 200	13.54
	13.81
	13.66
110 × 150	13.22
	13.46
	13.48
150 × 200	16.47
	16.62
	16.39
200 × 200	17.89
	17.51
	17.76
200 × 200 (Grease)	23.02
	23.17
	23.27

TABLE 5: Dome height from EHF.

Charging voltage (kV)	Charging energy (kJ)	Height (mm)
1	0.3	5.03
2	1.2	8.39
3	2.8	12.03
4	5.1	19.31
5	8	28.36

measured and plotted together, along with the static FLC via an LDH test, as shown in Figure 9. The 5 kV EHF workpiece had higher major strain than workpieces using the LDH test. Combinations of FLD for both parts showed that the workpieces using EHF had higher formability than those from the LDH test. The maximum dome height in the LDH test was measured at 23.27 mm in lubricated workpieces with dimensions of 200 mm × 200 mm, whereas the maximum dome height using an EHF test was measured at 28.36 mm with a 5 kV charging voltage. These results were mainly due to a high-velocity plasma channel leading to improved formability in the workpiece. The approximate improvement in formability was measured as 28% more than that of conventional forming processes.

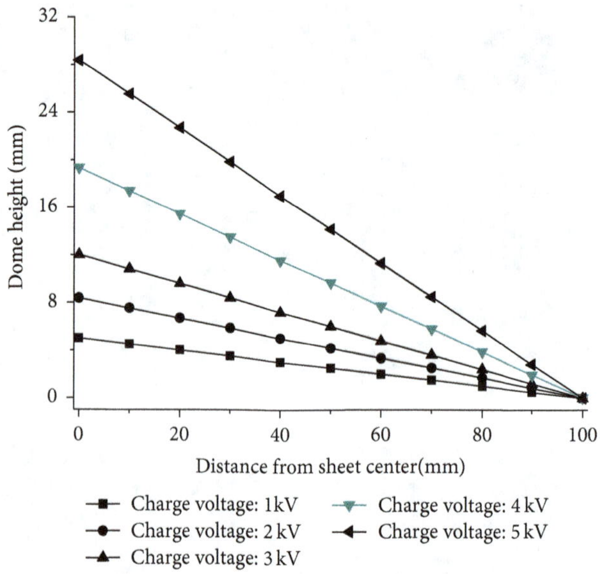

FIGURE 8: Height profile of the deformed workpieces after EHF.

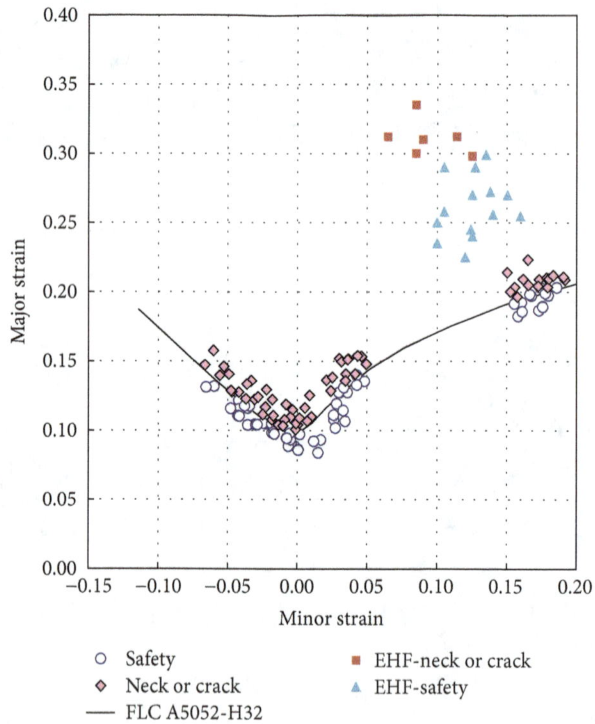

FIGURE 9: Combined formability results for A5052-H32, 0.5 mm.

4. EHF with Sharp Edges

A case with a 175 mm diameter, 5 mm height, and a less than 2 mm radius of curvature was selected as a lightweight material case with sharp edges for small electric appliances. To assess the possibility of sharp edge forming via traditional press forming, a simple press forming simulation consisting of a punch, die, and workpiece was performed using a measured FLD of A5052-H32. The press forming die's radius

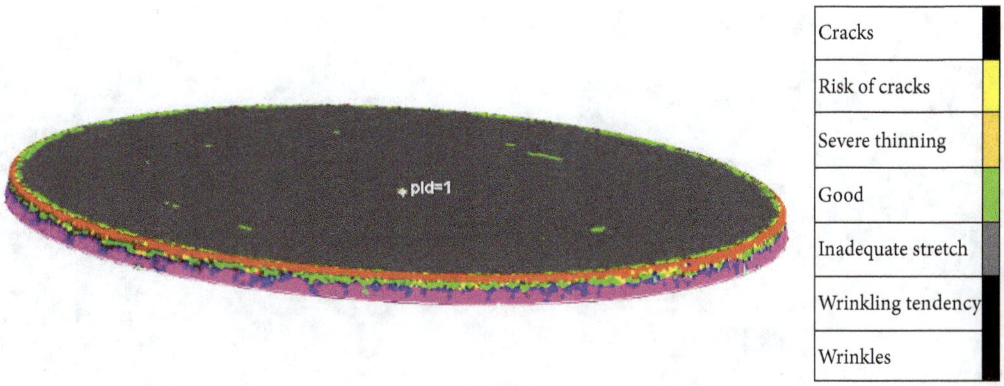

FIGURE 10: Cracks and wrinkles by press forming.

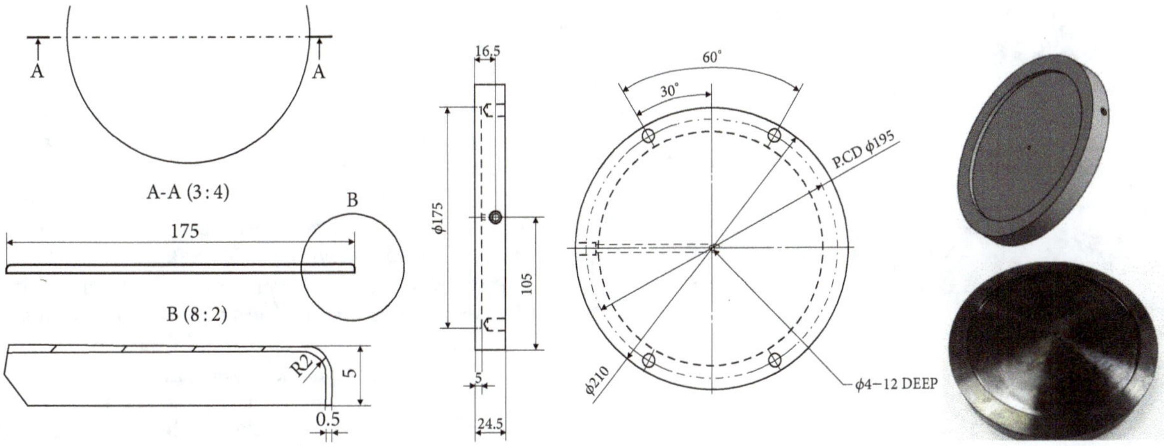

(a) Desired sharp edge

(b) Die

FIGURE 11: Target and die for sharp edge forming.

of curvature was 2.0 mm. As a result, it was predicted to crack and wrinkle along the circumference of the edge, as shown in Figure 10.

Figure 11 shows the target case and designed die for EHF. As a die material, SKD11 was employed with a 50HRC via heat treatment. In particular, the radius of curvature of the die was machined at 0.5 mm to achieve a minimum radius of curvature on the forming part. The workpiece was located at the top of the chamber, and water leakage was prevented by a rubber-ring and hydraulic press. After setting up the experiment, a 10 kV charge was run through a pulse power source; then, charged electrical energy was discharged into the electrodes. The damped sinusoidal waveform was measured as shown in Figure 6. The discharge time and peak current were measured at 25 μs and 153 kA, respectively. To evaluate the radius of curvature on the edge of the circumference, inner radius measurements were precisely performed using a 3D scanner, as shown in Figure 12. The radius of curvature was measured in 50 mm intervals. Figure 13 shows the radius of curvature on the edge of the circumference. The radius of curvature was measured to be less than 1.5 mm for all points; however, the radius of some points located adjacent to the electrodes was less than 1.0 mm. Although the radius of the edge was nonuniform, due to the nonuniform distribution of pressure applied on the workpiece by a shockwave and reflective wave, the EHF process could achieve a sharp edge of less than 2.0 mm for a lightweight materials case.

5. Conclusions

Improvements to formability using the EHF process were verified by comparing conventional forming and EHF, leading to the following conclusions:

(1) An FLD of 0.5 mm thickness A5052-H32 was drawn by an LDH test. A 28% improvement of formability was verified by comparing the measured strain between the LDH test and EHF in combinations of FLD. According to the formability improvement, a maximum dome height using EHF was approximately 5 mm higher than when using the LDH test.

(2) EHF was performed under different charging voltages via waveform measurement. Although the charging voltage was discharged for 25 μs under all EHF conditions, a peak current was increased to approximately 11 kA in direct relation to the charging voltage. In particular, dome heights were increased as charge voltage increased, meaning that

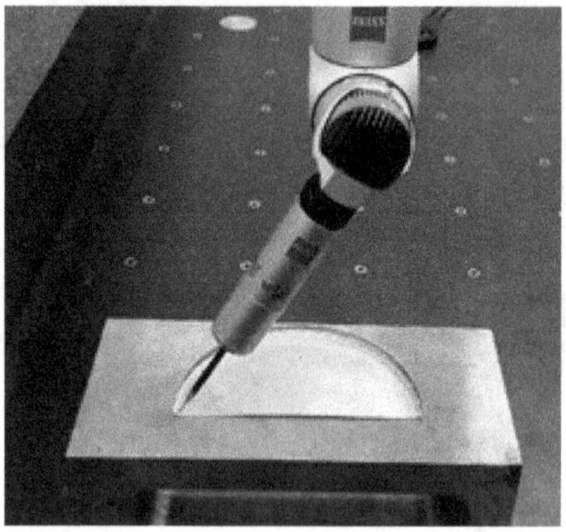

(a) 3D scanning

(b) Sharp edge

FIGURE 12: 3D scanning and forming part with sharp edge.

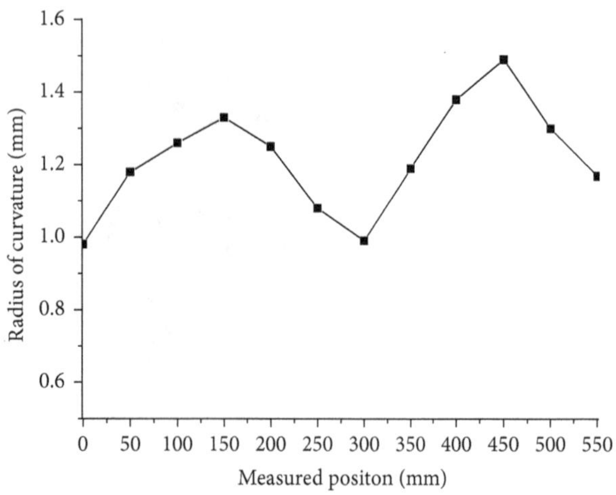

FIGURE 13: Radius of curvature on sharp edge.

increasing the charge voltage caused a greater shockwave and reflected wave in the chamber, which led to an increase in the formability of the workpiece

(3) For applying the EHF process to small electric appliances, a circular formed part with sharp edges of less than 2 mm in radius of curvature was manufactured. The radius of curvature was verified using a 3D scanner. Although all radii of curvature did not have the same radius, the radius of curvature was less than 1.5 mm for all curvatures on the manufactured components.

Conflicts of Interest

The authors declare that they have no conflicts of interest.

References

[1] V. Psyk, D. Risch, B. L. Kinsey, A. E. Tekkaya, and M. Kleiner, "Electromagnetic forming—a review," *Journal of Materials Processing Technology*, vol. 211, no. 5, pp. 787–829, 2011.

[2] D. J. Mynors and B. Zhang, "Applications and capabilities of explosive forming," *Journal of Materials Processing Technology*, vol. 125-126, pp. 1–25, 2002.

[3] V. S. Balanethiram and G. S. Daehn, "Hyperplasticity: increased forming limits at high workpiece velocity," *Scripta Metallurgica et Materiala*, vol. 30, no. 4, pp. 515–520, 1994.

[4] S. F. Golovashchenko, "Material formability and coil design in electromagnetic forming," *Journal of Materials Engineering and Performance*, vol. 16, no. 3, pp. 314–320, 2007.

[5] A. Rohatgi, E. V. Stephens, A. Soulami, R. W. Davies, and M. T. Smith, "Experimental characterization of sheet metal deformation during electro-hydraulic forming," *Journal of Materials Processing Technology*, vol. 211, no. 11, pp. 1824–1833, 2011.

[6] A. V. Mamutov, S. F. Golovashchenko, J. J. Bonnen, A. J. Gillard, S. A. Dawson, and L. Maison, "Electrohydraulic forming of light weight automotive panels," in *Proceedings of the 7th International Conference on High Speed Forming*, pp. 115–124, 2016.

[7] L. A. Yutkin, Electrohydraulic effect, Mashgiz, Translation No. AD-267-722. Armed Services Technical information Agency, Arlington Hall Station, Arlington, 1955.

[8] E. J. Bruno, "High-velocity forming of metals," *American Society of Tool and Manufacturing Engineers*, 1968.

[9] R. Davies and E. R. Austin, *Development in high speed metal forming*, Industrial Press Inc, 1970.

[10] J. E. Sanford, "High velocity takes off again," *Iron Age*, vol. N3, pp. 91–95, 1969.

[11] S. F. Golovashchenko and V. S. Mamutov, "Electrohydraulic forming of automotive panels," in *Proceedings of the 6th Global Innovations Symposium: Trends in Materials and Manufacturing Technologies for Transportation Industries*, T. R. Bieler, J. E. Carsley, H. L. Fraser, J. W. Sears, and J. E. Smugeresky, Eds., pp. 65–70, San Francisco, Calif, USA, 2005.

[12] A. J. Gillard, S. F. Golovashchenko, and A. V. Mamutov, "Effect of quasi-static prestrain on the formability of dual phase steels in electrohydraulic forming," *Journal of Manufacturing Processes*, vol. 15, no. 2, pp. 201–218, 2013.

[13] V. J. Vohnout, G. Fenton, and G. S. Daehn, "Pressure heterogeneity in small displacement electrohydraulic forming process," in *Proceedings of 4th International Conference on High Speed Forming*, G. Daehn, Ed., pp. 65–74, Columbus, Ohio, USA, 2010.

[14] S. F. Golovashchenko, A. V. Mamutov, J. J. F. Bonnen, and A. J. Gillard, "Electrohydraulic forming of sheet metal parts," in *Proceedings of the International Conference on Technology of Plasticity*, G. Hirt and A. E. Tekkaya, Eds., pp. 1170–1175, Aachen, Germany, 2011.

[15] J. J. F. Bonnen, S. F. Golovashchenko, S. A. Dawson, A. V. Mamutov, and A. J. Gillard, "Electrohydraulic sheet metal forming of aluminum panels," in *Proceedings of the TMS Light Metals 2012 Conference*, pp. 449–454, Orlando, Fla, USA, March 2012.

[16] A. Hassannejadasl, D. E. Green, S. F. Golovashchenko, J. Samei, and C. Maris, "Numerical modelling of electrohydraulic free-forming and die-forming of DP590 steel," *Journal of Manufacturing Processes*, vol. 16, no. 3, pp. 391–404, 2014.

[17] M. Ahmed, D. R. Kumar, and M. Nabi, "Enhancement of formability of AA5052 alloy sheets by electrohydraulic forming process," *Journal of Materials Engineering and Performance*, vol. 26, no. 1, pp. 439–452, 2016.

[18] T. Mane, V. Goel, and S. D. Kore, "Finite element modelling of electro-hydraulic forming of sheets," *Procedia Materials Science*, vol. 6, pp. 105–114, 2014.

[19] S. F. Golovashchenko, A. J. Gillard, D. A. Cedar, and A. M. Ilinich, "Electrohydraulic forming tool," an applicant for a patent: Ford Global Technologies, Llc, US7516634B1.

[20] S. D. Kumar, T. Amjith, and C. Anjaneyulu, "Forming limit diagram generation of aluminum Alloy AA2014 using Nakazima test simulation tool," *Procedia Technology*, vol. 24, pp. 386–393, 2016.

[21] B. Şanay, Prediction of plastic instability and forming limits in sheet metal forming, The Master degree thesis in the graduate school of Middle East Technical University (2010) 24–27.

[22] S. D. Kumar, T. R. Amjith, and C. Anjaneyulu, "Forming limit diagram generation of aluminum alloy AA2014 using Nakazima test simulation tool," *Procedia Technology*, vol. 24, pp. 386–393, 2016.

[23] S. S. Hecker, "Simple technique for determining forming limit curves," *Sheet Metal Industries*, vol. 52, no. 11, pp. 671–676, 1975.

4

Study of Tribological Behavior of Silicon Carbide Based Aluminum Metal Matrix Composites under Dry and Lubricated Environment

Kalyan Kumar Singh,[1] Saurabh Singh,[1] and Anil Kumar Shrivastava[2]

[1] *Department of Mechanical Engineering, Indian School of Mines, Dhanbad 826004, India*
[2] *Department of Mechanical Engineering, Galgotias College of Engineering & Technology, Greater Noida 201306, India*

Correspondence should be addressed to Saurabh Singh; saurabh.091638.bme@gmail.com

Academic Editor: Guru P. Dinda

Friction and wear behavior of silicon carbide based aluminum metal matrix composite and aluminum matrix alloy have been studied for sliding speeds of 3.14 m/s and 3.77 m/s and load range from 10 N to 30 N under dry and lubricated environment, respectively. The experiments were performed on pin on disk tribometer (Make: DUCOM). The composite was fabricated by stir casting process and has several challenges like inferior bonds and interfacial reaction products which will deteriorate the mechanical and tribological properties. Therefore, addition of reactive metal like magnesium (Mg) should be done which will lead to reduced solidification shrinkage, lower tendency towards hot tearing, and faster process cycles. Results have revealed that the developed composites have lower coefficient of friction and wear rates when compared with aluminum matrix alloy under dry and lubricated environment. Experimental results show that under dry condition coefficient of friction of both the matrix alloy and the composite decreases with increase in load, whereas it increases with increase in sliding speeds; on the other hand wear rates of both aluminum matrix alloy and the composites increase with increase in load as well as with sliding speeds. FESEM of worn surfaces are also used to understand the wear mechanisms.

1. Introduction

Aluminum alloy possesses excellent properties like high strength-to-weight ratio, corrosion resistance, good ductility, low density, high thermal conductivity, and so forth, thus making it suitable for wide range of applications in several industries like automobile, aerospace, structural applications, and so forth [1]. In recent times, the uses of high strength aluminum alloys are focused on structural applications in aerospace as well as on general engineering sectors. However aluminum alloys have not been well known for high performance tribological application because of their low hardness, poor wear resistance, and poor weldability. Thus, there is a greater possibility of oscillation since these materials are joined by riveting, bolting, and so forth and due to this it may often lead to sliding wear in dry environmental conditions. Therefore, to overcome this problem reinforcement of particulates, fiber, or whiskers can be used; these reinforcements are very well known for their high-specific strength and uniform distribution [2] which leads to use of metal matrix composites to complete the demand for light weight, high performance, environment friendly, wear and corrosion resistance material. Generally, Ag, Al, Be, Co, Cu, Fe, Mg, Ni, and Ti are produced and used as matrices in metal matrix composites. Due to excellent physical, mechanical, and tribological properties and good dimensional stability of aluminum matrix composite it has high commercial application compared to other available metals. The demand for aluminum composite in automobile sector has been increased drastically; this composite is used for the manufacturing of pistons, brake drums, cylinder liners, and pushrods [3, 4]. A commercial Al 7075 aluminum alloy is one of the highest strength aluminum alloys available and is having excellent strength-to-weight ratio and is formed in annealed conditions and subsequently heat treated. But due to its low wear resistance and poor tribological behavior it cannot be widely used; therefore

TABLE 1: Chemical composition (weight %) of aluminum matrix alloy (Al 7075).

Material	Si	Fe	Cu	Cr	Zn	Ti	Mn	Mg	Al	Others
Content	Max 0.4	0.5	1.2–2	0.18–0.28	5.1–6.1	0.2	0.3	2.1–2.9	87.1–91.4	0.15

reinforcement of particles has been done to improve its tribological properties. There are various particles available which are used for reinforcement such as B_4C, SiC, TiC, B, C, and Al_2O_3. With the reinforcement of high modulus ceramics particulates such as silicon carbide (SiC) having properties like more stability and chemical compatibility with aluminum matrix and low cost, wide range of available grade made aluminum alloys are suitable candidate for wider spectrum of applications [5]. The hard reinforcement of silicon carbide particle improved the properties like hardness and wear resistance of composite by acting as load bearing components [6, 7]. Many investigators have studied both parameters, that is, reinforcement size and volume fraction; Aiguo and Rack [8] observed that abrasive wear resistance in unreinforced 7091 alloys is less in comparison with both silicon carbide particle and silicon carbide whiskers. Wear resistance increases and plastic deformation decreases with the increase of reinforcement [9, 10]. Kwok and Lim [11] observed the size of silicon carbide reinforcement at high speed and suggested that when the particulate size becomes small the wear resistance rapidly deteriorates. Therefore, silicon carbide based aluminum metal matrix composites are mainly preferred for low speed applications; silicon carbide is commonly used among the reinforcement particles like carbides, nitrides, borides, and oxides, because it has several advantages like high melting point, high elastic modules, hardness, good thermal conductivity, high strength, excellent wear resistance, good availability, and low cost. Silicon carbide based aluminum metal matrix composites are widely used in automotive, marine, and other sectors because of its excellent properties such as good fluidity and low coefficient of thermal expansion and its quality depends on chemical composition, melting process, casting process, and solidification rate. As the volume fraction addition of silicon carbide in aluminum alloys increases, the value of ultimate tensile strength (UTS) increases and low degree of porosity is obtained.

There is not much improvement in strength and porosity of material, if the volume fraction of silicon carbide exceeds 10%; therefore in the present analysis 10% silicon carbide has been added in aluminum alloys as reinforcement. Due to low degree of porosity and uniform distribution of silicon carbide particles there is enhancement in tensile strength of composites and as a result it leads to effective transfer of tensile loads to the uniformly distributed silicon carbide particulates. In silicon carbide based aluminum metal matrix composites, casting properties increased by adding silicon whereas magnesium (Mg) made its heat treatable which greatly influenced the mechanical properties of composites.

There are several manufacturing techniques that are available for particle reinforcement in metal matrix composites such as liquid infiltration, spray decomposition, squeeze casting, compocasting, powder metallurgy, and mechanical alloying. Among all these available methods, stir casting method is attractive method for manufacturing of metal matrix composites (MMCs) because of low cost, near-net shape formation, and wide range of processing applications [1, 5, 12]. The mechanical strength of aluminum metal matrix composites is greatly influenced by particles size reinforcement; if size of particles ranges from micrometer to nanometric scale there is drastic increase in mechanical strength of composites but it will increase chance of agglomeration and particle clustering which will reduce the strength of composites. Therefore, it is important that reinforcement of particles should be homogeneous to achieve the improved properties of composites [13]. In stir casting method, ceramics reinforcement (i.e., silicon carbide) and metal matrix which is melted during casting are mixed homogeneously by mechanical stirrer. Fabrication of aluminum metal matrix composites has several challenges like inferior bonds and interfacial reaction products and it will directly deteriorate the mechanical properties and tribological properties of composites, which make them unsuitable for industrial application. Therefore, to overcome this problem, addition of reactive metal like magnesium has been done and the addition of this metal results in reduced solidification shrinkage, lower tendency towards hot tearing, suppression of segregation, settling or agglomeration, and faster process cycles [12].

Thus by analyzing various parameters and their effects and utilization this experiment was carried out under various environmental conditions to produce a material which can sustain a longer time. Composite fabricated has much better properties as compared with its base alloys (details and validation of which are shown in the experimental analysis). The severity of worn-out surfaces (i.e., wear scars, change of microstructure) was also analyzed using field emission scanning electron microscopes (FESEM).

2. Experimental Procedure

2.1. Fabrication of Silicon Carbide Based Aluminum Metal Matrix Composite. The composite was prepared by commercially available aluminum (Al 7075) matrix alloy and silicon carbide (10% by weight) which is used as reinforcement. The average size of aluminum matrix alloy was 150 microns to 160 microns and that of silicon carbide was 40 microns to 50 microns. In the present investigation, aluminum matrix alloy was chosen as a base matrix since it is having very high tensile strength of about 275 MPa. The addition of silicon carbide will enhance the wear resisting property of silicon carbide based aluminum metal matrix composite. The chemical composition of aluminum matrix alloys is presented in Table 1.

2.2. Composite Preparation. In order to achieve high level of mechanical properties of the composite, a good interfacial

TABLE 2: Chemical composition of silicon carbide based aluminum metal matrix composite.

Material	C	O	Mg	Al	Si	Fe	Ag	Zn	Cu
Content	24.27	20.61	1.13	36.43	10.62	1.96	0.84	3.34	0.80

bonding (wetting) between the dispatched phase and the liquid matrix has to be obtained. Fabrication of composite with discontinuous fiber can be economically done using liquid metallurgy route/technique [10]. Stir casting is a liquid state method which is the simplest and is cost effective to fabricate metal matrix composite. Care was taken to maintain optimum casting parameters of pouring temperature (670°C) stirring speed of 500 rpm, stirring time 20 minutes. The reinforcement was preheated at 600°C to the addition in aluminum matrix alloy (Al 7075). To avoid moisture from the particles, preheating of the reinforced particle should be done; otherwise, there is a chance of agglomeration of particulate which occurs due to moisture and gasses present in the particles. Degassing agent (hexachloroethane and magnesium) was used to reduce the gas porosities. The addition of magnesium will also lead to reduced solidification shrinkage, lower tendency towards hot tearing, suppression of segregation, settling or agglomeration, and faster process cycles. The depth of immersion of stirrer was maintained at about two-thirds of the depth of molten metal. The molten metal was then poured into a permanent mould preheated at 300°C of 9 mm diameter and 55 mm length. Specimen was left for curing and die was released after 6 hours; finally, on the lathe cast specimens were parted out in 2 pieces, that is, 25 mm length and 8 mm diameter, respectively. The chemical compositions of silicon carbide based aluminum metal matrix composite are presented in Table 2.

2.3. Composite and Aluminum Matrix Alloy Testing Procedure. The experiment was conducted at room temperature on a pin on disk wear and friction maintaining machine (Make: DUCOM instruments private limited, Bangalore diameter of 165 mm, thickness of 8 mm and made of material EN31 hardened to 60 HRC ground to 1.6 Ra surface roughness. Figure 1 shows the photographic view of disk). The size of cylindrical specimen which was used for testing is 8 mm diameter as well as 25 mm length. Before performing every experiment disk surface was cleaned by rubbing 600-grade Si6 emery paper and washed with acetone. Three different loads, that is, 10 N, 20 N, and 30 N, have been selected for wear and friction behavior of specimen with track diameter of 100 mm and 120 mm used for both aluminum matrix alloy and silicon carbide based aluminum metal matrix composite. The test duration was 10 minutes at a constant speed of 600 rpm for the entire test in dry and oil lubricated (SAE 20 with kinematic viscosity 25 to 30 centistokes at 50°C and oil has been used for lubrication) conditions. For oil lubricated experiments, initially, 2 drops of lubricating oil were placed over sliding surface after that oil was poured at 0.02 mL/min. All sets of experiments were repeated three times in the same manner for dry and oil lubricated conditions, respectively. Frictional forces and coefficient of friction were measured directly from the machine. For the calculation of wear loss

TABLE 3: Variation of coefficient of friction with load for aluminum matrix alloy and silicon carbide based aluminum metal matrix composite at $V = 3.14$ m/sec.

Sliding condition & load	COF for aluminum matrix alloy	COF for silicon carbide based aluminum metal matrix composite
Dry 10 N	0.65	0.42
Dry 20 N	0.57	0.37
Dry 30 N	0.47	0.30
Lubricated 10 N	0.46	0.32
Lubricated 20 N	0.43	0.25
Lubricated 30 N	0.36	0.19

Track diameter = 100 mm, $N = 600$ rpm, sliding speeds = 3.14 m/sec, and $\Delta T = 600$ sec.

FIGURE 1: DUCOM Machine.

in terms of weight, weights of specimen before and after the experiment were taken.

The sliding speed of the test specimen is calculated by using the following formula:

$$V = \frac{\pi * D * N}{60000}. \quad (1)$$

The following calculation is used to obtain wear rate:

$$K_0 = \frac{\Delta m}{\rho * F * d}. \quad (2)$$

3. Results and Discussion

3.1. Coefficient of Friction. The variations of coefficient of friction for aluminum matrix alloy (Al 7075) and composites under dry and lubricated condition at different normal loads and two sliding speeds (i.e., 3.14 m/s and 3.77 m/s) have been shown in Tables 3 and 4, respectively. Table 3 shows that there is significant decrease in the value of coefficient of friction for composite compared with aluminum matrix alloy under dry and lubricated condition. Under dry condition, 30%–55%

TABLE 4: Variation of coefficient of friction with load for aluminum matrix alloy and silicon carbide based aluminum metal matrix composite at $V = 3.77$ m/sec.

Sliding condition & load	COF for aluminum matrix alloy	COF for silicon carbide based aluminum metal matrix composite
Dry 10 N	0.72	0.56
Dry 20 N	0.63	0.43
Dry 30 N	0.50	0.35
Lubricated 10 N	0.58	0.44
Lubricated 20 N	0.47	0.32
Lubricated 30 N	0.37	0.25

Track diameter = 120 mm, N = 600 rpm, sliding speeds = 3.77 m/sec, and ΔT = 600 sec.

reductions in value of coefficient of friction for load range 10 N–30 N have been observed; this is because of uniform mixing of silicon carbide with aluminum matrix alloy and results in excellent bond between matrix alloy and silicon carbide particle. These reinforced silicon carbide particles act as load bearing members as the loads increase and reduce the touch area between specimen and rotating disk's surfaces [1] which leads to lower coefficient of friction. Similar observations have been observed by other investees [14, 15].

However, under lubricated condition, the value of coefficient of friction reduced for both aluminum matrix alloy and composites compared with dry condition; this is because of thin lubricating film formation between specimen and rotating disk surface. This thin lubricating film contains lumps of particles and as temperature and normal loads increase these particles break into a small number of particles and reorient themselves along the sliding plane and this leads to lower coefficient of friction.

Table 4 shows variation of coefficient of friction at increased sliding distance for both aluminum matrix alloy and composites. As the sliding distance increases the value of coefficient of friction increases for both aluminum matrix alloy and composites because of temperature rise of rubbing surfaces for aluminum matrix alloy and reinforcement of silicon carbide into aluminum matrix alloys makes the matrix alloy plastically constrained and it can sustain high temperature compared with aluminum matrix alloy. Silicon carbide also protects the matrix alloy from severe contact with counter surface and leads to lower coefficient of friction. Similar observations have been made by other investees [16]. As sliding distance increases, the thin lubricating film between specimen and rotating disk surface may get thickened and be responsible for breakage of small particles from specimen and these broken particles may be clogged between the rubbing surfaces and this leads to higher coefficient of friction.

Figures 2 and 3 show the variation of coefficient of friction with increasing normal loads under dry and lubricating condition at sliding speeds 3.14 m/s and 3.77 m/s, respectively. It has been observed that temperature rise is greater in case of unreinforced aluminum matrix alloys compared to those in composites irrespective of the applied load and surface conditions. The rise in temperature leads to frictional

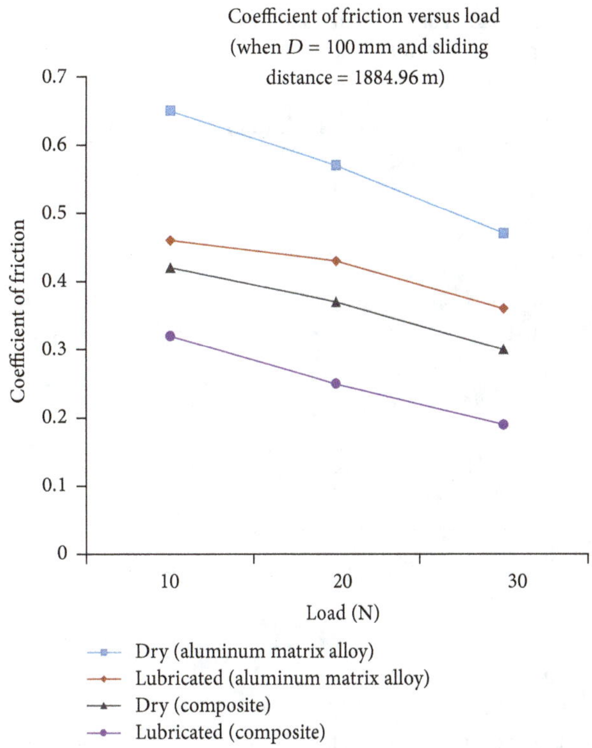

FIGURE 2: Variation of COF at V = 3.14 m/s.

heating in between the rubbing surfaces and this frictional heating is continuous because there is much less time for heat dissipation that causes higher coefficient of friction for aluminum matrix alloy compared to composites. Initially, aluminum alloys and composites particles are stronger and sharper which causes higher frictional heating and leads to higher coefficient of friction but as the time increases, there is increase in flow ability of material on specimen surface that leads to increase in slipping action and results in reduced frictional heating that leads to lower coefficient of friction at higher loads [1]. The composite shows lower coefficient of friction compared with aluminum matrix alloy because of presence of secondary phase of silicon carbide particles in aluminum matrix alloy which will restrict the flow of metal during sliding [17]. The uniform distribution of silicon carbide particle in aluminum matrix alloy would lead to sustaining higher loads with less plastic deformation and give us good load bearing capacity of composite at higher loads. The minimum values of coefficient of friction for both aluminum matrix alloy and composite have been observed at lower speeds; the same observations have been made by other investees [11].

3.2. Wear Studies

3.2.1. Weight Loss and Wear Rates. Effects of increasing loads (i.e., 10 N, 20 N, and 30 N) with sliding distance 1884.96 m (numerical values shown in Table 5) and 2262 m (numerical values in Table 6) on weight loss and wear rates for both aluminum matrix alloy and composites under dry and lubricated condition have been shown in Figures 4 and 5, respectively.

TABLE 5: Variation of weight loss and wear rate for aluminum matrix alloy and silicon carbide based aluminum metal matrix composite when sliding speeds = 3.14 m/s.

Sliding condition & load	(Weight loss)$_{alloy}$ (mg)	(Wear rate)$_{alloy}$ (mm^3/N-m)	(Weight loss)$_{composite}$ (mg)	(Wear rate)$_{composite}$ (mm^3/N-m)
Dry 10 N	3.8	4.662×10^{-5}	2.1	2.349×10^{-5}
Dry 20 N	8.6	5.275×10^{-5}	4.5	2.516×10^{-5}
Dry 30 N	15.2	6.216×10^{-5}	8.5	3.168×10^{-5}
Lubricated 10 N	3.0	3.680×10^{-5}	1.5	1.677×10^{-5}
Lubricated 20 N	6.5	3.987×10^{-5}	3.1	1.733×10^{-5}
Lubricated 30 N	13.2	5.398×10^{-5}	6.2	2.311×10^{-5}

$\rho_{alloy} = 4.324 \times 10^{-6}$ kg/mm^3, $\rho_{composite} = 4.744 \times 10^{-6}$ kg/mm^3, sliding distance (d) = 1884.96 m, and ΔT = 600 sec.

TABLE 6: Variation of weight loss and wear rate for aluminum matrix alloy and silicon carbide based aluminum metal matrix composite when sliding speeds = 3.77 m/s.

Sliding condition & load	(Weight loss)$_{alloy}$ (mg)	(Wear rate)$_{alloy}$ (mm^3/N-m)	(Weight loss)$_{composite}$ (mg)	(Wear rate)$_{composite}$ (mm^3/N-m)
Dry 10 N	6.7	6.85×10^{-5}	2.6	2.423×10^{-5}
Dry 20 N	13.7	7.003×10^{-5}	5.9	2.750×10^{-5}
Dry 30 N	22.4	7.634×10^{-5}	11.1	3.448×10^{-5}
Lubricated 10 N	4.9	5.009×10^{-5}	1.9	1.770×10^{-5}
Lubricated 20 N	11.2	5.725×10^{-5}	4.2	1.957×10^{-5}
Lubricated 30 N	17.5	5.964×10^{-5}	7.9	2.454×10^{-5}

$\rho_{alloy} = 4.324 \times 10^{-6}$ kg/mm^3, $\rho_{composite} = 4.744 \times 10^{-6}$ kg/mm^3, sliding distance (d) = 2262 m, and ΔT = 600 sec.

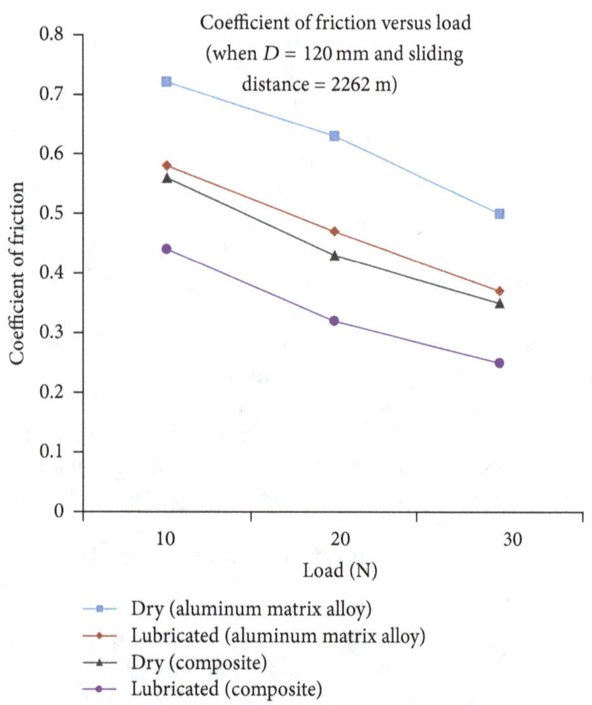

FIGURE 3: Variation of COF at V = 3.77 m/s.

Figure 4 shows the variation of weight loss and wear rates at sliding distance 1884.96 m with increasing loads. Mazahery and Shabani [18] observed that wear rates vary with normal rate and significantly it is low for composites compared to unreinforced alloys and it is indicative of Archard's law. In dry condition, rate of weight loss increases with increase in loads for both aluminum matrix alloy and composites because, with increasing loads, the rate of abrasion also increases and it will lead to detachment of metals from surfaces and leads to more loss of materials. But the rates of weight loss in composites are lower compared with aluminum matrix alloy because the presence of silicon carbide in aluminum matrix alloy controls the deformation and leads to less weight loss. The worn surfaces of composites are smoother and also shallower groove formations as the time increases which lead to higher wear resistance for composites compared with unreinforced aluminum matrix alloys [5]. As earlier we have seen that wear rates are directly proportional to weight loss, therefore wear rates will also be lower for composites compared with aluminum matrix alloy.

In lubricated condition, wear rates are lower compared with dry condition for both aluminum matrix alloy and composites because of thin lubricating film formation over rotating disk surfaces, and this thin lubricating film decreases the adhesion of small broken particles of aluminum matrix alloy and composites and leads to lower wear rates although it increases with increasing loads and if an applied load exceeds critical values, reinforced particle will break and loses its load carrying capacity properties and leads to more wear rates [5]. Critical loads basically mean that wear rates increase marginally with linear trends but suddenly these trends will

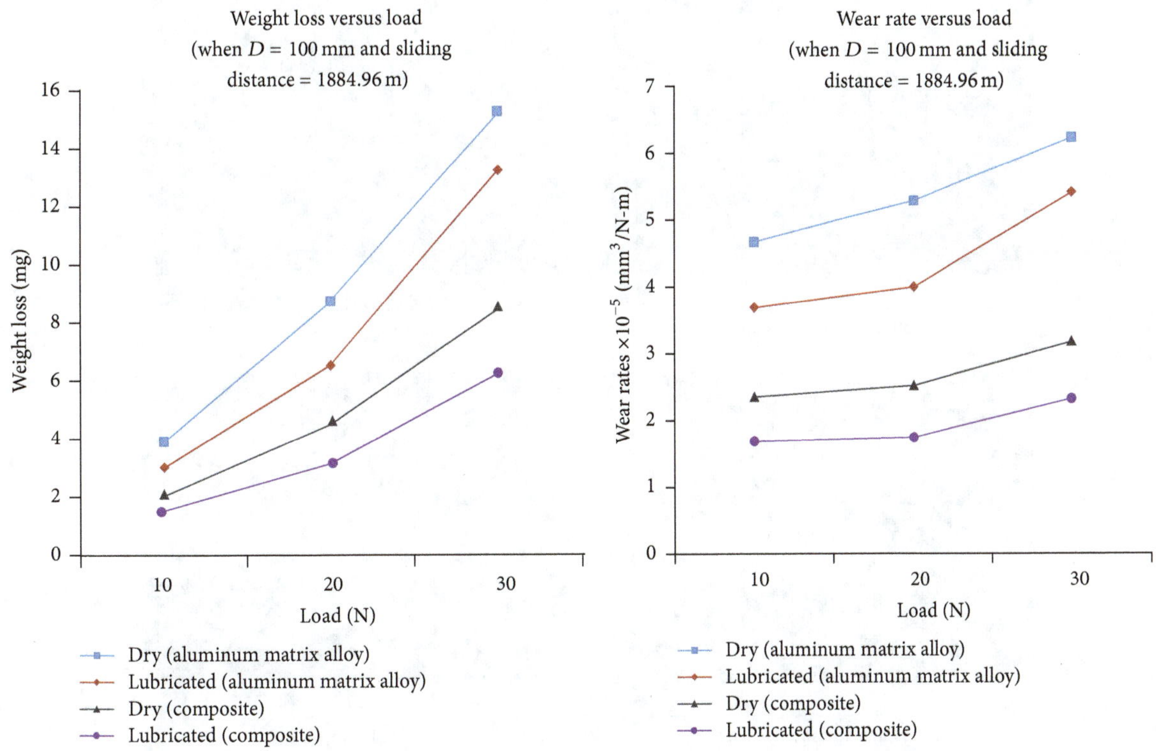

FIGURE 4: Variation of weight loss and wear rates at sliding distance 1884.96 m with increasing loads.

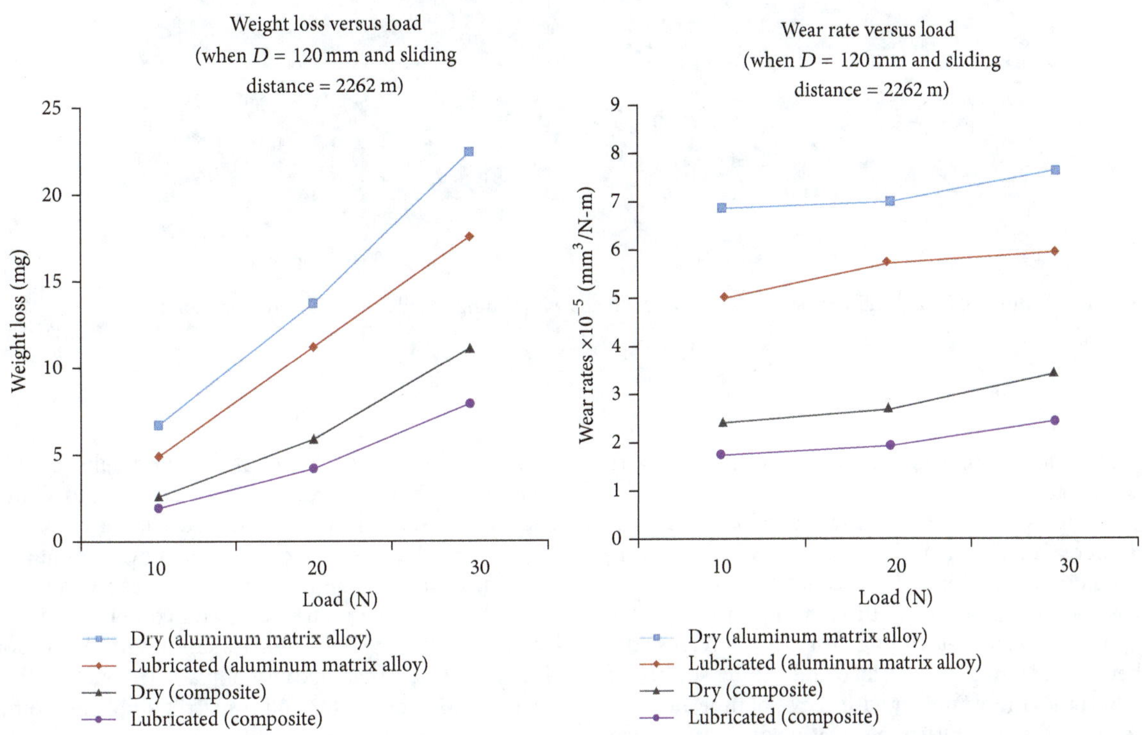

FIGURE 5: Variation of weight loss and wear rate at sliding distance 2262 m with increasing loads.

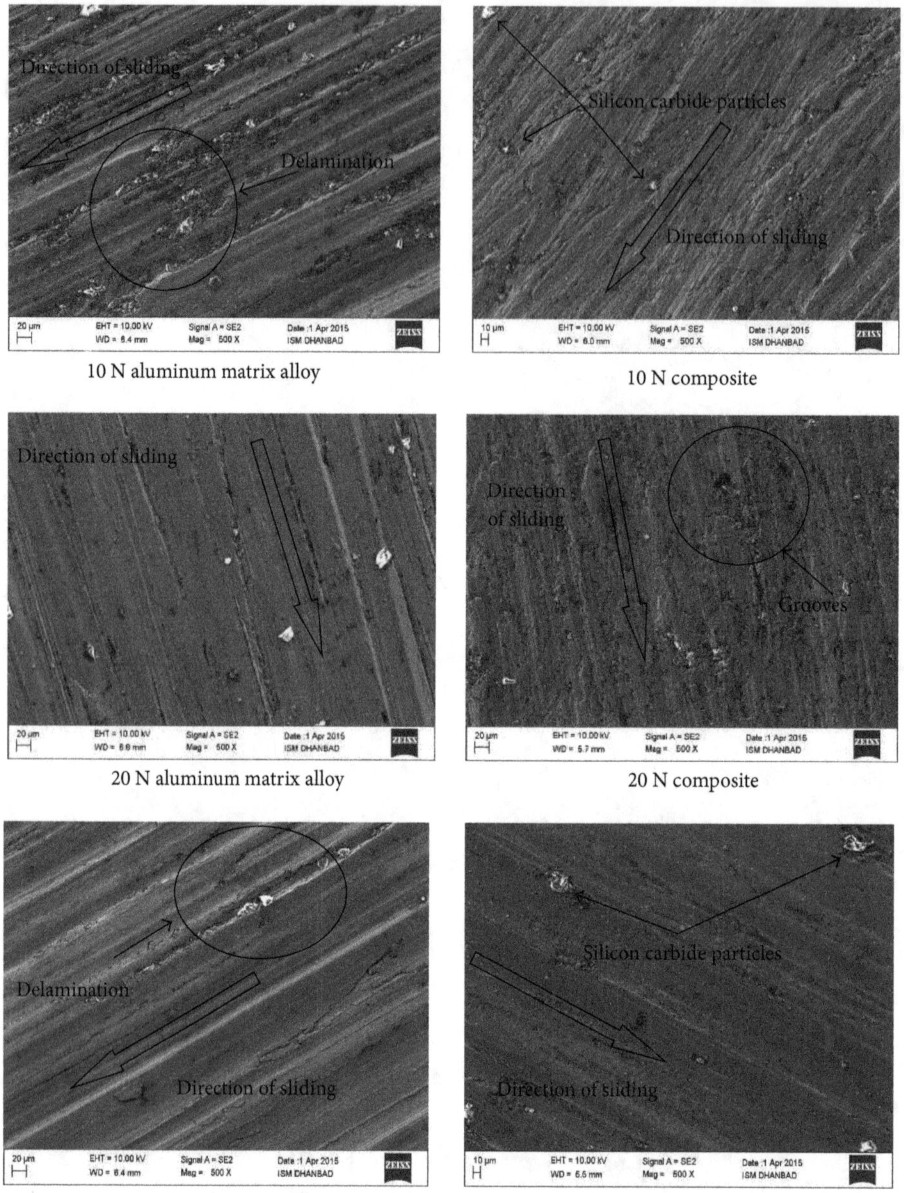

FIGURE 6: FESEM of worn surfaces at different loads with sliding speeds 3.14 m/s under dry condition.

not follow the loads up to which these trends that follow are known as critical loads.

Figure 5 shows the variation of weight loss and wear rates with increased sliding distance 2262 m with increasing loads. The increase of sliding distance leads to increase in weight loss and wear rates for both aluminum matrix alloy and composites because as sliding distance increases, the contact between rotating disk surface and sliding surface of specimen becomes more familiar with elapse of time; because of this the temperature between the rotating disk surface and sliding surface of specimen increases and leads to softening of materials and plastic state of materials occurs. Similar observations have been made by other investees [19, 20].

3.2.2. FESEM Analysis. FESEM micrographs of all worn surfaces for both aluminum matrix alloy and composites under dry and lubricated condition at sliding speeds 3.14 m/s and 3.77 m/s have been shown at magnification rate (×500).

Figure 6 shows the FESEM analysis of all worn surfaces for both aluminum matrix alloy and composites at different loads with sliding speeds 3.14 m/s under dry condition. It is clearly observed that, because of plastic flow of material, the wear rates for both aluminum matrix alloy and composites increase with increase of loads and severity of plastic flow increases at higher loads. At higher loads, the wear debris is mixture of fine and coarse size particles; this is because of breakdown of large size particles. As the load increases

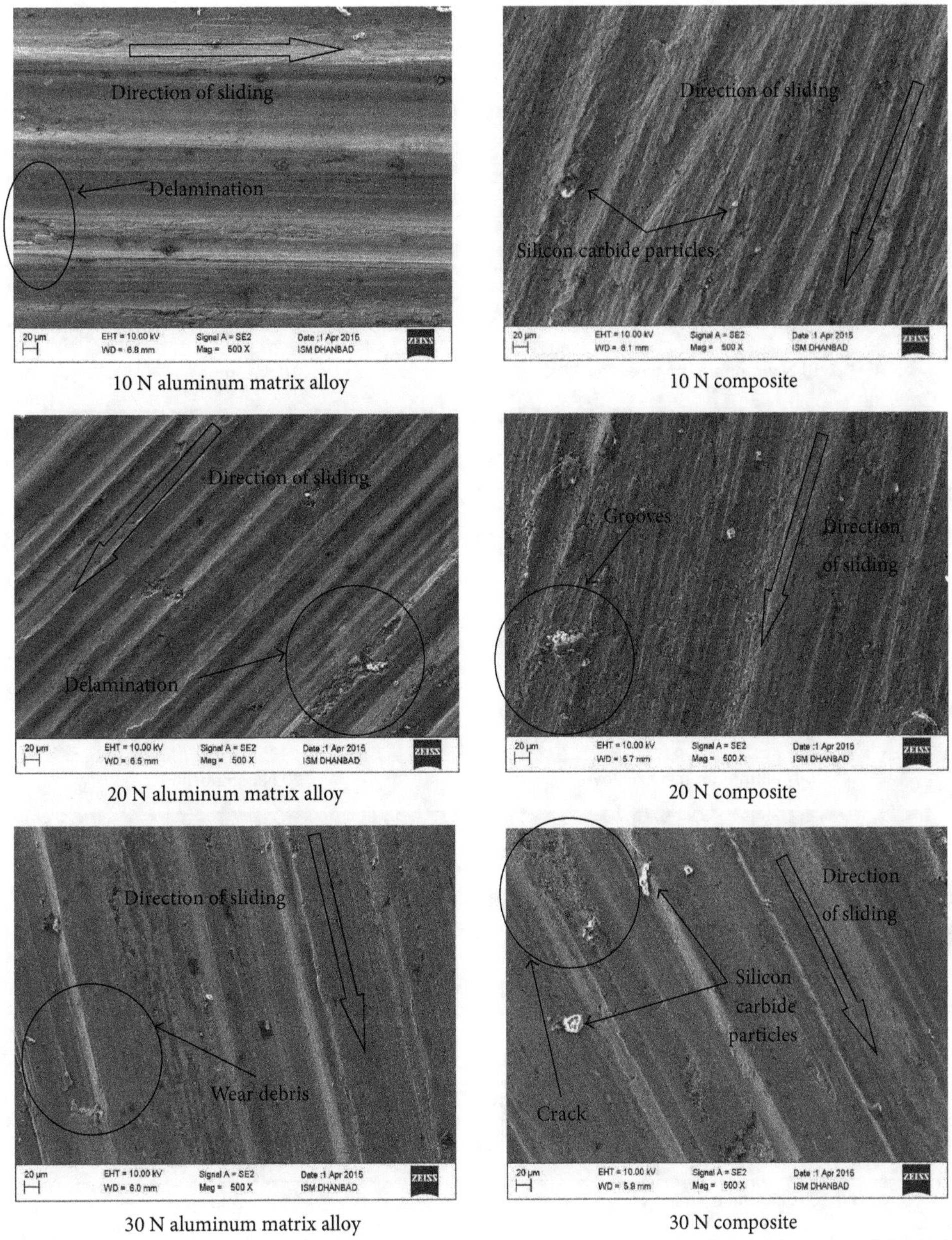

FIGURE 7: FESEM of worn surfaces at different loads with sliding speeds 3.14 m/s under lubricated condition.

the intersection between the sliding surface of specimen and rotating disk surface increases and leads to breakage of large particles into smaller number of particles. At lower loads, worn surfaces are smooth and grooves are small in size compared with higher loads; with increase in loads these grooves become severe and lead to formation of fracture spots. At these conditions reinforced particles cannot be in stable equilibrium and this leads to increase in wear rates.

Figure 7 shows the FESEM analysis of all worn surfaces for both aluminum matrix alloy and composites at different loads with sliding speeds 3.14 m/s under lubricated condition. In lubricated condition lower wear rates have been observed compared with dry condition because of thin lubricating film formation between sliding surface of specimen and rotating disk surface and it decreases the ploughing action and leads to decrease in breakage of small particles from aluminum matrix alloy and composites; therefore, lower wear rates have been observed in lubricated condition compared with dry condition.

Figure 8 shows the FESEM analysis of all worn surfaces for both aluminum matrix alloy and composites at different loads with sliding speeds 3.77 m/s under dry condition. As the sliding distance increases, there is gradual increase in wear rates for both aluminum matrix alloy and composites. The above results can be observed because of the following reason:

(i) With increase in sliding distance, the temperature at specimen interface increases and leads to softening the materials and plastic flow of materials.

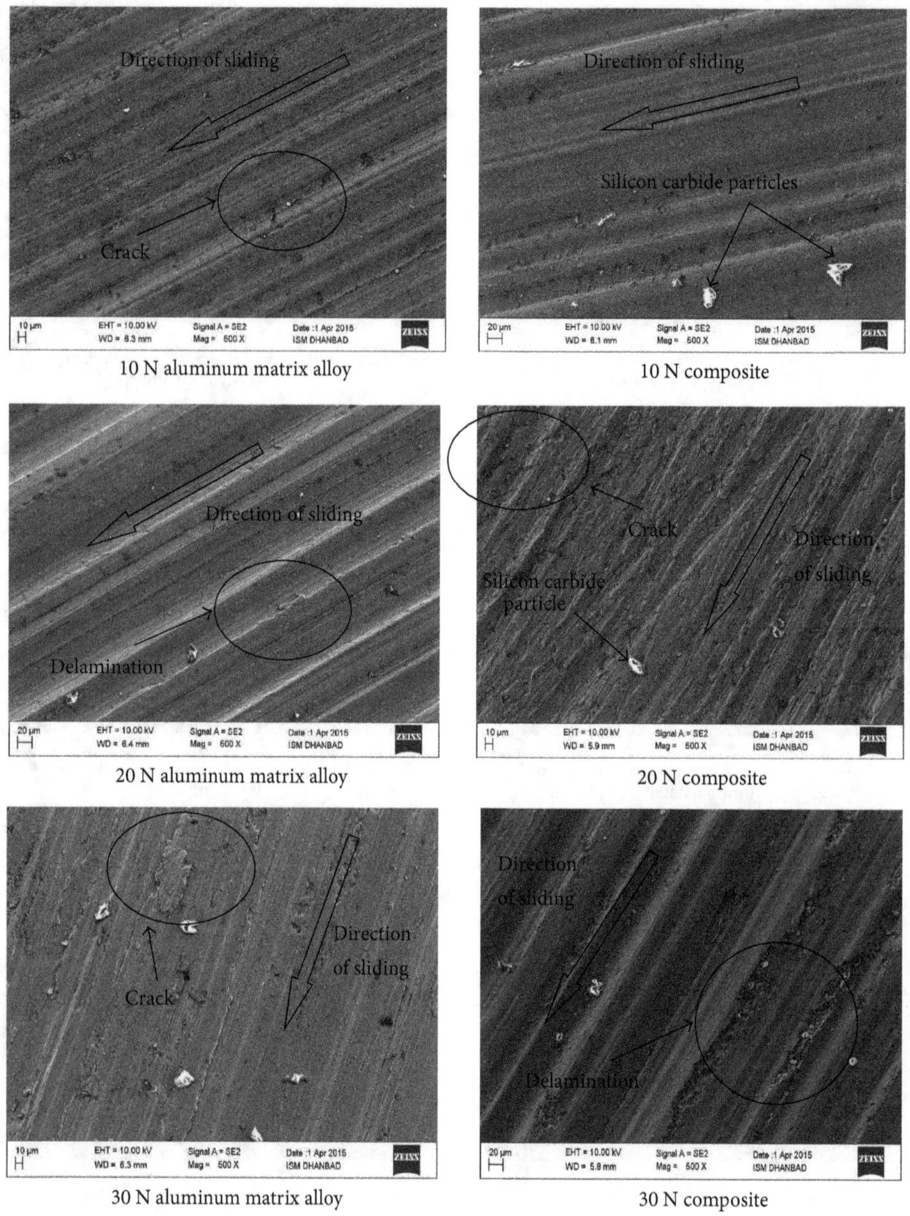

FIGURE 8: FESEM of worn surfaces at different loads with sliding speeds 3.77 m/s under dry condition.

(ii) With increase in sliding distance, the contact between sliding surface of specimen and rotating disk surface becomes intimate as time increases resulting in higher wear rates.

However, the wear rates are lower for composites compared with aluminum matrix alloy because some particles of silicon carbide break into small pieces and crushed as the disk rotates and act as abrasive medium. These powdered silicon carbide particles along with aluminum small particles made a layer which is called mechanically mixed layer (MML). These layers prevent the plastic flow and reduce the wear rates compared with aluminum matrix alloy although plastic flow of composites occurs at later stage.

Figure 9 shows the FESEM analysis of all worn surfaces for both aluminum matrix alloy and composites at different loads with sliding speeds 3.77 m/s under lubricated condition. At higher sliding speeds, wear rates are lower in lubricated condition because of thin lubricating film formation and another mechanically mixed layer (MML) formation delays the plastic flow of material and leads to lower wear rates in lubricated condition compared with dry condition.

4. Conclusions

After observing the condition, that is, dry and oil lubricated condition of aluminum matrix alloy and silicon carbide based aluminum metal matrix composite, the following conclusion can be drawn:

(1) The values of coefficient of friction are found to be maximum in case of dry condition in matrix alloy

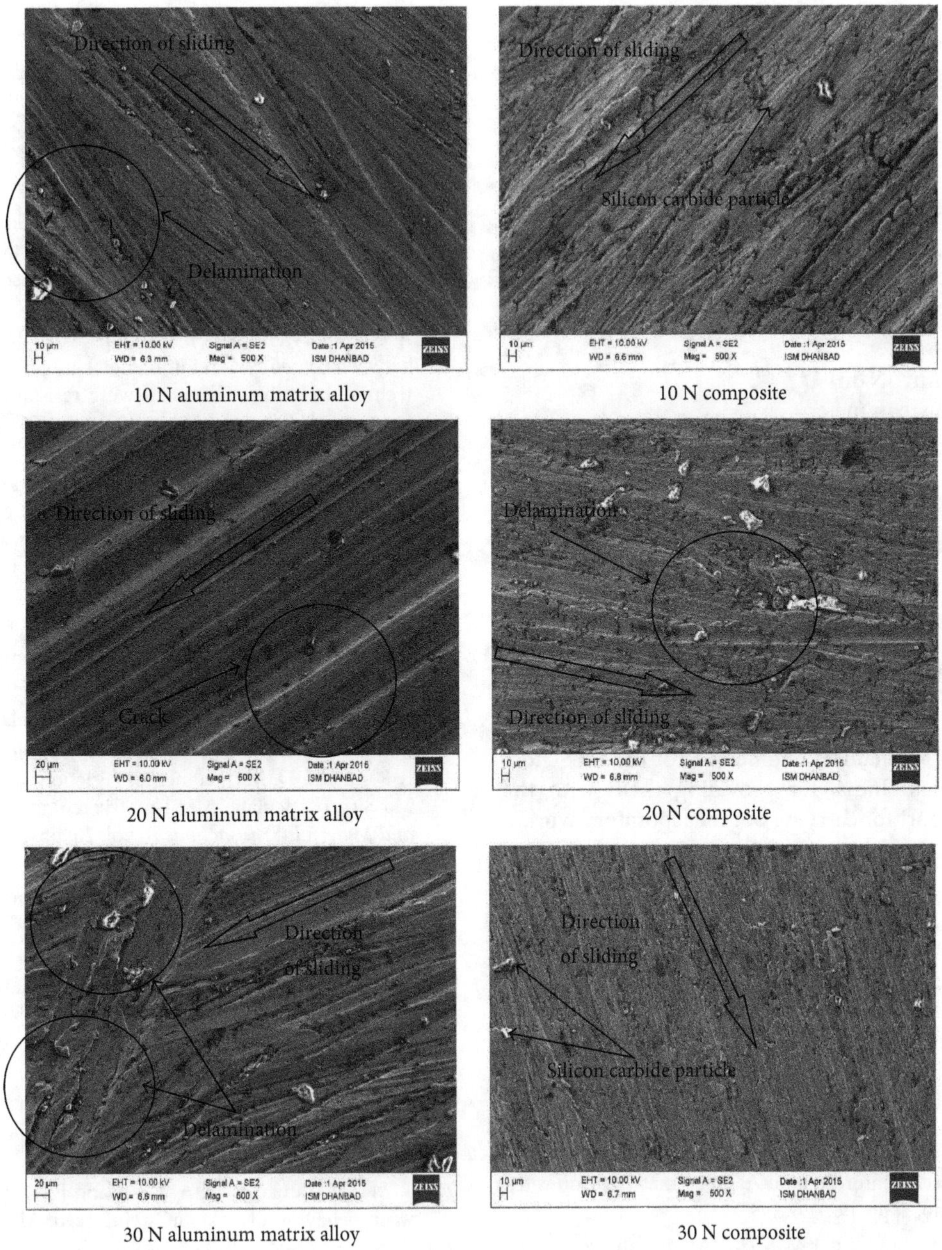

FIGURE 9: FESEM of worn surfaces at different loads with sliding speeds 3.77 m/s under lubricated condition.

at sliding speed 3.77 m/s and minimum in case of lubricated condition in composite at sliding speed 3.14 m/s. The range of coefficient of friction lies from 0.19 to 0.72.

(2) The values of coefficient of friction and wear rates are minimum for both aluminum matrix alloy and silicon carbide based aluminum metal matrix composites under lubricated condition compared with dry condition.

(3) As we increase normal load and sliding speed, the wear rate increases and it is found to be maximum in dry condition of matrix alloy at 30 N.

(4) Uniform distribution of silicon carbide in aluminum matrix alloy reduces the values of coefficient of friction and wear rates for composites compared with aluminum matrix alloy under dry and lubricated condition which prolongs the life of composites for longer duration.

(5) The reinforcements of silicon carbide in aluminum matrix alloy significantly improve the strength of material in dry and lubricated condition and act as load bearing members as the loads and sliding speeds increase and give the best results in lubricated condition at low sliding speeds and loads.

(6) From the above analysis and discussions we conclude that silicon carbide based aluminum metal matrix composite should be preferred for low speed applications as the same results are also observed by other researchers [21].

Nomenclature

V: Sliding speeds (m/sec)
D: Diameter of track (mm)
N: Rotating disk speed (rpm)
F: Normal load (N)
K_0: Wear rate (mm^3/N-m)
d: Sliding distance in meter
ρ: Density of specimen (kg/mm^3)
Δm: Weight loss of specimen (mg).

Competing Interests

The authors declare that they have no competing interests.

Acknowledgments

The authors are grateful to the Advanced Composite Lab of Department of Mechanical Engineering of Indian School of Mines, Dhanbad, India, for providing them with the experimental support for carrying out the laboratory work.

References

[1] A. Mazahery and M. O. Shabani, "Study on microstructure and abrasive wear behavior of sintered Al matrix composites," *Ceramics International*, vol. 38, no. 5, pp. 4263–4269, 2012.

[2] C. G. E. Mangin, J. A. Isaacs, and J. P. Clark, "MMCs for automotive engine applications," *JOM*, vol. 48, no. 2, pp. 49–51, 1996.

[3] G. Straffelini, F. Bonollo, A. Molinari, and A. Tiziani, "Influence of matrix hardness on the dry sliding behaviour of 20 vol.% Al_2O_3-particulate-reinforced 6061 Al metal matrix composite," *Wear*, vol. 211, no. 2, pp. 192–197, 1997.

[4] H.-S. Chu, K.-S. Liu, and J.-W. Yeh, "An in situ composite of Al (graphite, Al4C3) produced by reciprocating extrusion," *Materials Science and Engineering A*, vol. 277, no. 1-2, pp. 25–32, 2000.

[5] A. Mazahery and M. O. Shabani, "The effect of primary and secondary processing on the abrasive wear properties of compocast aluminum 6061 alloy matrix composites," *Protection of Metals and Physical Chemistry of Surfaces*, vol. 50, no. 6, pp. 817–824, 2014.

[6] S. C. Lim, M. Gupta, L. Ren, and J. K. M. Kwok, "The tribological properties of Al–Cu/SiCp metal-matrix composites fabricated using the rheocasting technique," *Journal of Materials Processing Technology*, vol. 80–90, pp. 591–596, 1999.

[7] C. García-Cordovilla, J. Narciso, and E. Louis, "Abrasive wear resistance of aluminium alloy/ceramic particulate composites," *Wear*, vol. 192, no. 1-2, pp. 170–177, 1996.

[8] W. Aiguo and H. J. Rack, "Abrasive wear of silicon carbide particulate- and whisker-reinforced 7091 aluminum matrix composites," *Wear*, vol. 146, no. 2, pp. 337–348, 1991.

[9] P. Rodrigo, M. Campo, B. Torres, M. D. Escalera, E. Otero, and J. Rams, "Microstructure and wear resistance of Al–SiC composites coatings on ZE41 magnesium alloy," *Applied Surface Science*, vol. 255, no. 22, pp. 9174–9181, 2009.

[10] Y. Sahin and V. Kilicli, "Abrasive wear behaviour of SiCp/Al alloy composite in comparison with ausferritic ductile iron," *Wear*, vol. 271, no. 11-12, pp. 2766–2774, 2011.

[11] J. K. M. Kwok and S. C. Lim, "High-speed tribological properties of some Al/SiCp composites. I. Frictional and wear-rate characteristics," *Composites Science and Technology*, vol. 59, no. 1, pp. 55–63, 1999.

[12] A. Mazahery and M. O. Shabani, "Existence of good bonding between coated B4C reinforcement and al matrix via semisolid techniques: enhancement of wear resistance and mechanical properties," *Tribology Transactions*, vol. 56, no. 3, pp. 342–348, 2013.

[13] M. O. Shabani and A. Mazahery, "Application of finite element model and artificial neural network in characterization of al matrix nanocomposites using various training algorithms," *Metallurgical and Materials Transactions A*, vol. 43, no. 6, pp. 2158–2165, 2012.

[14] Z. F. Zhang, L. C. Zhang, and Y.-W. Mai, "Particle effects on friction and wear of aluminium matrix composites," *Journal of Materials Science*, vol. 30, no. 23, pp. 5999–6004, 1995.

[15] S. Wilson and A. T. Alpas, "Wear mechanism maps for metal matrix composites," *Wear*, vol. 212, no. 1, pp. 41–49, 1997.

[16] F. Rana and D. M. Stefanescu, "Friction properties of Al-1.5% Mg/SiC particulate metal-matrix composites," *Metallurgical and Materials Transactions A*, vol. 20, pp. 1564–1566, 1989.

[17] S. Kumar and V. Balasubramanian, "Effect of reinforcement size and volume fraction on the abrasive wear behaviour of AA7075 Al/SiCp P/M composites-a statistical analysis," *Tribology International*, vol. 43, no. 1-2, pp. 414–422, 2010.

[18] A. Mazahery and M. O. Shabani, "Mechanical properties of squeeze-cast A356 composites reinforced with B_4C particulates," *Journal of Materials Engineering and Performance*, vol. 21, no. 2, pp. 247–252, 2012.

[19] I. M. Hutchings, "Tribological properties of metal matrix composites," *Materials Science and Technology*, vol. 10, no. 6, pp. 513–517, 1994.

[20] T. Ma, H. Yamaura, D. A. Koss, and R. C. Voigt, "Dry sliding wear behavior of cast SiC-reinforced Al MMCs," *Materials Science and Engineering A*, vol. 360, no. 1-2, pp. 116–125, 2003.

[21] K. R. Brown, M. S. Venie, and R. A. Woods, "The increasing use of aluminum in automotive applications," *JOM*, vol. 47, no. 7, pp. 20–23, 1995.

5

Influence of Pb Dosage on Immobilization Characteristics of Different Types of Alkali-Activated Mixtures and Mortars

Jan Koplík, Jaromír Pořízka, Lukáš Kalina, Jiří Másilko, and Matěj Březina

Faculty of Chemistry, Materials Research Centre, Brno University of Technology, Purkyňova 118, 612 00 Brno, Czech Republic

Correspondence should be addressed to Jan Koplík; koplik@fch.vutbr.cz

Academic Editor: Kedsarin Pimraksa

Alkali-activated matrices are suitable materials for the immobilization of hazardous materials such as heavy metals. This paper is focused on the comparison of immobilization characteristics of various inorganic composite materials based on blast furnace slag and on the influence of various dosages of the heavy metal Pb on the mechanical properties and fixation ability of prepared matrices. Blast furnace slag (BFS), fly ash, and standard sand were used as raw materials, and sodium water glass was used as an alkaline activator. $Pb(NO_3)_2$ served as a source of heavy metal and was added in various dosages in solid state or as aqueous solution. The immobilization characteristics were determined by leaching tests, and the content of Pb in the eluate was measured by inductively coupled plasma optical emission spectroscopy (ICP-OES). The microstructure of matrices and distribution of Pb within the matrix were determined by scanning electron microscopy (SEM) equipped with energy dispersive X-ray spectroscopy (EDS). Increasing the dosage of the heavy metal had negative impacts on the mechanical properties of prepared matrices. The leaching tests confirmed the ability of alkali-activated materials to immobilize heavy metals. With increasing addition of Pb, its content in eluates increased.

1. Introduction

Heavy metals such as Pb belong to hazardous materials, which are harmful to the human beings. Lead is highly toxic element, which attacks mainly the nervous system. Lead exposure mostly occurs through the ingestion of contaminated water or food. Therefore, it is important to prevent its leakage into the environment by stabilization or solidification of waste materials before their deposition into the landfill [1]. A possible way on how to immobilize heavy metals is using alkali-activated materials (AAMs). AAMs present a broad range of materials (geopolymers, activated blast furnace slag, etc.), but all of them are activated by high pH during their preparation. The AAMs can be divided into two big groups: high-calcium alkali-activated materials (HCAAMs) and low-calcium alkali-activated materials (LCAAMs). HCAAMs are represented mainly by activated BFS, and LCAAMs are represented mainly by activated fly ash or metakaolin. The structure of HCAAMs consists of the C-A-S-H gel as a major hydration product. This gel has a similar structure as the C-S-H gel in hydrated ordinary Portland cement and is made up of tetrahedrally coordinated silicate chains, where aluminium is located in the bridging position instead of silicon. The structure and composition of the C-A-S-H gel depend on the nature and concentration of the used activator. AFm-type phases (NaOH-activated binders), hydrotalcite (BFS with high content of MgO), and zeolites (binders with high content of Al_2O_3) are usually formed as secondary hydration products [2–4].

Heavy metals can be immobilized by two types of fixation–physical and chemical. Mostly, both types of immobilization occur together. Physical fixation is linked with mechanical properties and porosity of the matrix. Chemical fixation means that there is a chemical bond between the heavy metal and the matrix. Heavy metals can be inhibited by transformation into a less soluble form as well. After alkali activation, the heavy metals usually occur in the form of silicate or hydroxide [5–7]. The

Table 1: Particle size distribution of raw materials.

	x (10%) μm	x (50%) μm	x (90%) μm	x (99%) μm
BFS	0.88	10.73	33.97	59.57
Fly ash	4.18	44.56	236.89	486.20

Table 2: The nomenclature of samples.

Mixture	Mixture name
BFS; 1% Pb added as solid	S1
BFS; 2.5% Pb added as solid	S2
BFS; 5% Pb added as solid	S5
BFS + sand; 1% Pb added as solid	M1
BFS + sand; 2.5% Pb added as solid	M2
BFS + sand; 5% Pb added as solid	M5
BFS + fly ash; 1% Pb added as solid	P1
BFS + fly ash; 2.5% Pb added as solid	P2
BFS + fly ash; 5% Pb added as solid	P5
BFS; 1% Pb added as liquid	SR1
BFS; 2.5% Pb added as liquid	SR2
BFS + sand; 1% Pb added as liquid	MR1
BFS + sand; 2.5% Pb added as liquid	MR2
BFS + fly ash; 1% Pb added as liquid	PR1
BFS + fly ash; 2.5% Pb added as liquid	PR2

Table 3: The mixture design of prepared matrices.

	BFS matrices	Matrices with sand	Matrices with fly ash
BFS	100 wt.%	25 wt.%	50 wt.%
Fly ash	—	—	50 wt.%
Sand	—	75 wt.%	—
Activator/binder	8%	8%	8%
Water/binder	0.33	0.42	0.33

immobilization of heavy metals within AAMs can be influenced by various factors, such as the composition of the matrix, the type of alkaline activator, the immobilized heavy metal, and the leaching medium [8, 9]. Previous researches proved that Pb reached high efficiency of immobilization in AAMs, but there is only little information about the influence of Pb dosage on immobilization characteristics of AAMs [10–13].

The aim of this work was to compare the ability of three different alkali-activated matrices based on BFS to immobilize Pb within their structure. The influence of different Pb dosages on the immobilization efficiency was also investigated.

2. Materials and Methods

2.1. Materials, Sample Preparation, and Leaching Tests. BFS and high-temperature fly ash were used as raw materials. The particle size distribution of both raw materials is listed in Table 1. Liquid sodium silicate with the SiO_2/Na_2O ratio of

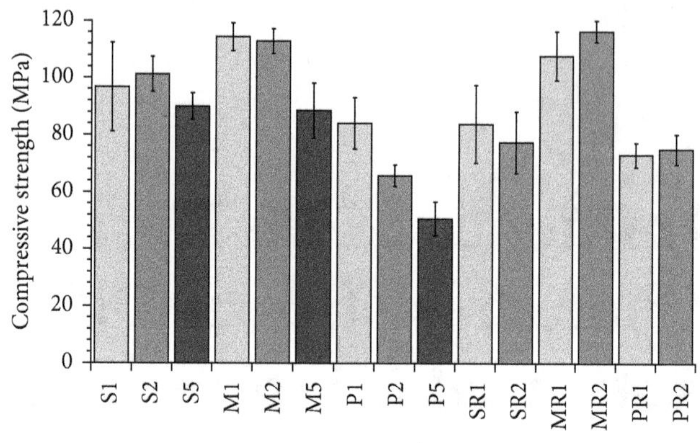

Figure 1: Compressive strength of prepared matrices (28 days).

1.85 served as an alkaline activator. As fine aggregates, three different fractions of siliceous Czech standard sand complying with ČSN EN 196-1 were used (the mass ratio of the fractions was 1 : 1 : 1). Pb was used in the form of $Pb(NO_3)_2$.

Three types of matrices were prepared. The first one was based only on BFS. In the second one, 50 wt.% of BFS was replaced by fly ash. And the third one consisted of BFS and standard sand. The sample nomenclature and the mixture designs are listed in Tables 2 and 3, respectively. Pb was added in the dosages of 1, 2.5, and 5 wt.% to binder mass and was added in two ways: as a solution and as a solid. In both cases, the water-to-binder ratios were the same.

The whole mixing procedure took four minutes. In the beginning, all materials without aggregates were put together and sand, if needed, was added after 30 s of mixing. The samples were cast in steel molds measuring 20 × 20 × 100 mm. All analyses were performed after 28 days of moist curing (98% relative humidity) at laboratory temperature (23 ± 2°C).

The leaching tests were based on ČSN EN 12457-4. The demineralized water served as leaching agent. The solid/liquid ratio was 1 : 10, and the mixture was agitated for 24 hours. After the leaching time ended, the mixture was filtered by a membrane filter with the pore size of 0.45 μm, and the concentration of Pb in solution was determined by ICP-OES.

2.2. Methods. The compressive strength was measured according to the ČSN EN 196-1 using Betonsystem Desttest 3310. To determine the porosity, prepared matrices were investigated by mercury intrusion porosimetry using Quantachrome instrument PoreMaster 33. The surface tension of mercury was 480 mN/m, and the contact angle was 140° based on the recommendation. The pore sizes were determined in the range from over 170 μm to 0.0064 μm. The ICP-OES data were obtained using Horiba Jobin Yvon Ultima II Spectrometer. The parameters of measurement were as follows: RF power, 1350 W; gas, argon; plasma gas, 13 L·min^{-1}; auxiliary gas, 0.1 L·min^{-1}; nebuliser gas, 0.85 L·min^{-1}; plasma view, radial; nebuliser, Meinhard; and nebuliser pressure, 3 bar. The scanning electron microscopy (SEM) and the energy dispersive X-ray spectroscopy (EDS)

Table 4: Compressive strength of prepared matrices (7 days).

Mixture	S1	S2	S5	M1	M2	M5	P1	P2
MPa	62.3	61.8	58.3	75.0	74.8	59.7	56.7	49.7
Mixture	P5	SR1	SR2	MR1	MR2	PR1	PR2	—
MPa	40.1	55.4	53.2	70.5	76.3	50.3	52.7	—

Table 5: The porosity of matrices.

	P1	P5	PR1	M1	M5	MR1	S1	S5	SR1
Total intruded volume (cm^3/g)	0.130	1.312	0.117	0.034	0.050	0.036	0.022	0.035	0.023

analyses were performed using Zeiss EVO LS 10 equipped with an Oxford X-Max 80 mm^2 detector in the backscattering mode. The working distance for all samples was 12 mm and the accelerate voltage was 30 kV. All samples were sputtered by carbon to obtain good surface conductivity.

3. Results and Discussion

3.1. Compressive Strength. The compressive strength of prepared matrices is shown in Figure 1 and Table 4. Following the results, the type of matrix had the highest influence on the compressive strength. The highest values had the matrix with aggregates, and the lowest values reached the matrix with fly ash. The results correspond with the findings in literature that, by replacing BFS with fly ash, the compressive strength decreases [2]. No clear behavior correlation was observed between the Pb dosage amount and mechanical properties. Increasing the dosage of Pb from 1 up to 2.5 wt.% led, in some cases, to the compressive strength increase and in some cases to the strength decrease. The addition of 5 wt.% of Pb caused the decrease of compressive strength in all types of matrices.

3.2. Porosity. The porosity of the prepared materials depended mainly on the type of the matrix. The total porosity of chosen samples is listed in Table 5. The matrix with fly ash showed the highest porosity. High-temperature fly ash consists of spherical particles, which can be filled in with similar smaller particles or can be hollow. Hence, partially reacted fly ash particles increased the porosity. The lowest porosity was observed in the matrix based on BFS. It is due to a very compact microstructure of this matrix, which was confirmed by SEM analysis (Figure 2). The dosage of Pb influenced the porosity of matrices too. The total porosity of matrices increased with increasing Pb content. As can be seen in Figure 3, around the areas, where Pb was cummulated, the matrix was not compact and the cracks and pores were formed. This caused the increase in porosity. The state of Pb addition—liquid/solid—did not affect the porosity.

3.3. Leaching Tests. To determine the efficiency of Pb immobilization, the leaching test based on ČSN EN 12457-4

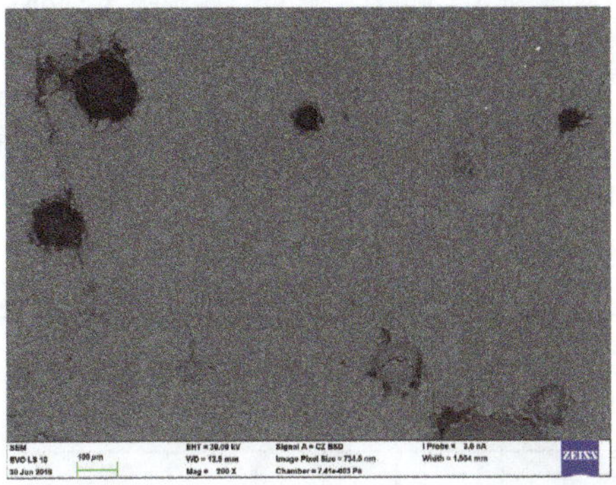

Figure 2: The microstructure of SR2 matrix.

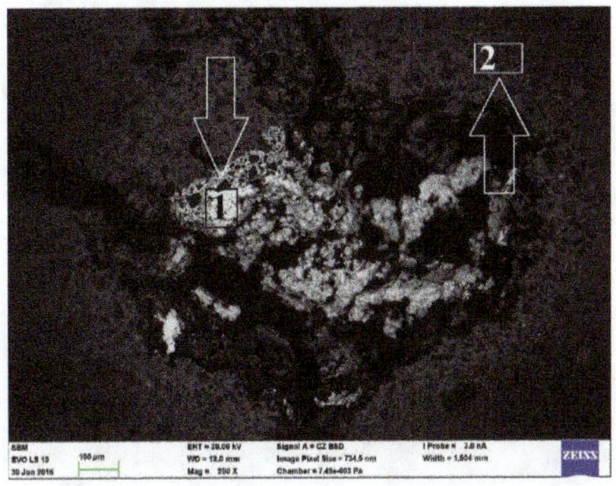

Figure 3: The microstructure of S2 matrix.

was performed. This test serves to determine whether the material is suitable for landfilling and whether it does not represent any hazard for the environment. Figure 4 shows the concentration of Pb in eluates from the prepared matrices. The immobilization of Pb in all matrices was very high and reached up to 99%. The results correspond with previous research and correlate with the total porosity of the matrices [11]. The highest Pb release occurred from matrices with fly ash, which also had the highest porosity and the lowest mechanical properties. The dosage of Pb had the influence on immobilization too. With increasing Pb addition, the content of Pb released into the leaching medium increased as well. It should be taken into consideration that when the Pb addition rose five times, the release increased only maximally three times. The efficiency of Pb immobilization was better with higher dosages of Pb.

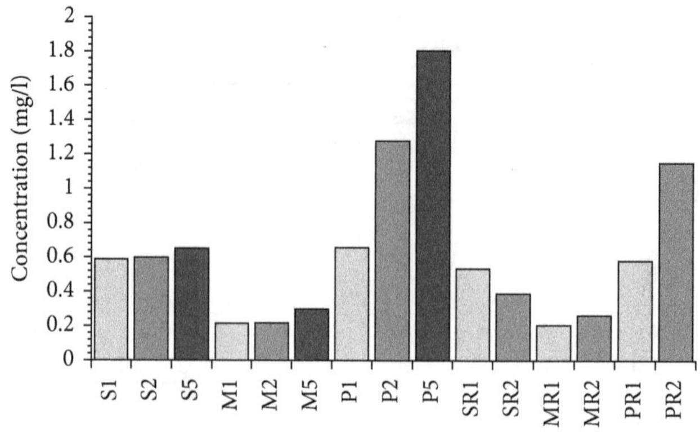

FIGURE 4: Concentration of Pb in eluates from matrices.

TABLE 6: The pH of eluates from matrices.

Mixture	S1	S2	S5	M1	M2	M5	P1	P2
pH	11.92	11.89	11.77	11.32	11.29	11.15	11.63	11.60
Mixture	P5	SR1	SR2	MR1	MR2	PR1	PR2	—
pH	11.51	11.89	11.88	11.31	11.31	11.63	11.58	—

TABLE 7: EDS analysis of S2 matrix.

					Area 1						
Element	O	Na	Mg	Al	Si	Ca	Pb	—	—	—	—
Atomic (%)	57.85	3.29	0.76	0.95	12.32	1.64	23.20	—	—	—	—
					Area 2						
Element	O	Na	Mg	Al	Si	Ca	Pb	S	K	Ti	Mn
Atomic (%)	58.13	4.62	3.93	3.00	15.23	12.80	0.97	0.88	0.19	0.07	0.18

TABLE 8: EDS analysis of P2 matrix.

					Area 3					
Element	O	Si	Ca	Al	Pb	—	—	—	—	—
Atomic (%)	81.61	3.95	3.61	1.03	9.79	—	—	—	—	—
					Area 4					
Element	O	Si	Ca	Al	Pb	Na	Mg	K	Ti	Fe
Atomic (%)	59.10	18.91	5.96	7.57	0.12	4.89	1.86	0.57	0.30	0.72

The pH of eluates is listed in Table 6. The matrices with sand reached the lowest pH, because they contained the smallest amount of the binder. The presence of fly ash led to the decrease of pH. The highest dosage of Pb (5 wt.%) caused the decrease of pH as well.

3.4. SEM. The SEM analysis showed three different microstructures depending on the type of matrix. The samples S2, P2, M2, and SR2 were investigated. The matrix based on BFS had a compact structure with unreacted particles of BFS and hydration products between them (Figure 3, Area 2; Table 7). A similar structure was observed in the matrix containing fly ash, but moreover, unreacted particles of fly ash were identified. The addition of fly ash led to the increase in the content of Si and Al and the decrease in the content of Ca (Table 8). A quite different microstructure occurred in the matrix with aggregates. The unreacted aggregates filled the major area of the sample and in between them there was the matrix with the same composition as observed in the S2 sample (Figure 5, Table 9, and Area 6). The aggregates were composed of quartz (Table 9 and Area 7).

When Pb was added as a solid, it behaved in the same manner in all types of matrices. Pb was cumulated in pores and formed specific structures (Figures 3, 5, and 6). These structures consisted mainly of Pb and O. It can be assumed that Pb transformed into its insoluble salt $Pb(OH)_2$ after the alkali activation. These findings correspond to previous research [14]. Minor amount of Pb was dispersed throughout the matrix. A quite different situation came up with the addition of Pb as a solution. Pb was dispersed

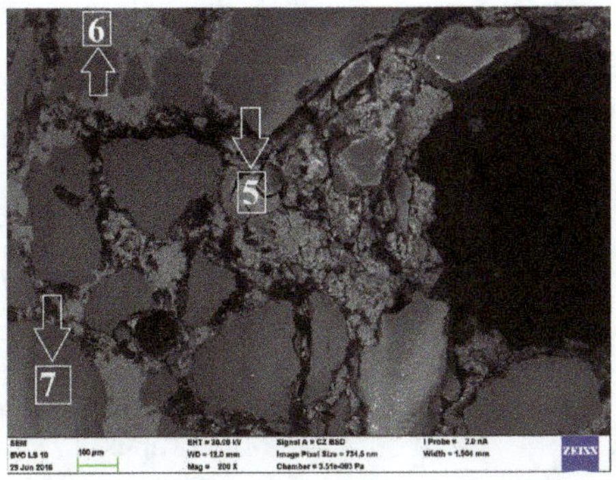

Figure 5: The microstructure of M2 matrix.

Table 9: EDS analysis of M2 matrix.

					Area 5						
Element	O	Si	Ca	Pb	—	—	—	—	—	—	—
Atomic (%)	73.26	7.61	5.21	13.92	—	—	—	—	—	—	—
					Area 6						
Element	O	Si	Ca	Pb	Na	Mg	Al	S	K	Ti	Mn
Atomic (%)	59.02	15.38	12.87	0.14	3.93	4.17	3.52	0.44	0.21	0.13	0.19
					Area 7						
Element	O	Si	—	—	—	—	—	—	—	—	—
Atomic (%)	63.65	36.35	—	—	—	—	—	—	—	—	—

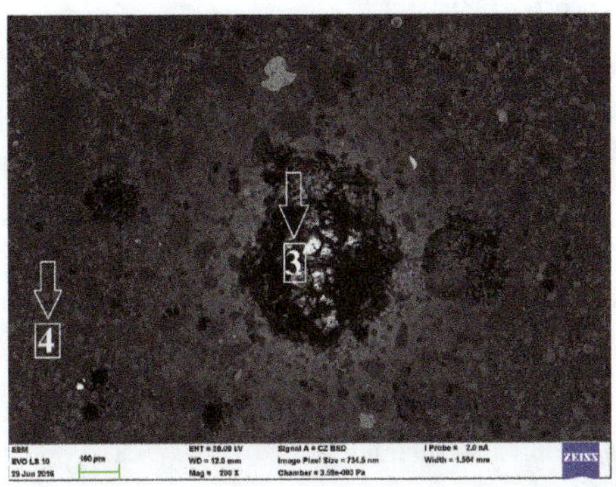

Figure 6: The microstructure of P2 matrix.

Table 10: EDS mapping of SR2 matrix.

Element	O	Si	Ca	Al	Pb	Na	Mg	K	Ti	Mn
Atomic (%)	58.96	15.39	13.06	3.08	0.27	4.65	3.86	0.18	0.07	0.15

equally throughout the matrix and did not create any specific structures (Figure 2, Table 10).

4. Conclusions

The immobilization of Pb in three different matrices was investigated. All matrices showed good ability to immobilize Pb. The increase in the Pb dosage led to the increase of Pb concentration in eluates, but the immobilization efficiency remained up to 99% in all cases. After the alkali activation, Pb formed its insoluble salt Pb $(OH)_2$. Both the compressive strength and the porosity were influenced by the Pb dosage. The decrease of mechanical properties after the addition of higher Pb dosage (5 wt.%) was observed. The concentration of Pb in eluates correlated with the porosity: when the porosity was higher, more Pb was released in eluates. It can be assumed that Pb was fixed by physical fixation, which was linked with the mechanical properties of matrices. Pb was also well immobilized, thanks to the formation of its insoluble salt. There was a difference in Pb behavior when added either as a solid or as a solution. In the case of solution, Pb was dispersed throughout the matrix equally, but when added as a solid, it formed specific structures, which cumulated in pores.

Conflicts of Interest

The authors declare that there are no conflicts of interest regarding the publication of this paper.

Acknowledgments

This outcome has been achieved with the financial support by the project Materials Research Centre at FCH BUT-Sustainability and Development (REG LO1211), National Programme for Sustainability I (Ministry of Education, Youth and Sports), and "Development of shrinkage reducing agents designed for alkali activated systems" (GA17-03670S, Czech Science Foundation).

References

[1] J. M. Stellman, *Encyclopaedia of Occupational Health and Safety*, International Labour Office, Geneva, Switzerland, 4th edition, 1998.

[2] J. L. Provis and J. S. J. van Deventer, *Alkali Activated Materials: State-of-the-Art Report*, Springer, Dorderecht, Netherlands, 2014.

[3] L. Chao, S. Henghu, and L. Longtu, "A review: the comparison between alkali-activated slag (Si + Ca) and metakaolin (Si + Al) cements," *Cement and Concrete Research*, vol. 40, no. 9, pp. 1341–1349, 2010.

[4] F. Puertas, M. Palacios, H. Manzano, J. S. Dolado, A. Rico, and J. Rodríguez, "A model for the C-A-S-H gel formed in alkali-activated slag cements," *Journal of the European Ceramic Society*, vol. 31, no. 12, pp. 2043–2056, 2011.

[5] J. G. S. van Jaarsveld, J. S. J. van Deventer, and L. Lorenzen, "Factor affecting the immobilization of metals in geopolymerized fly ash," *Metallurgical and Materials Transactions B*, vol. 29, no. 1, pp. 283–291, 1998.

[6] J. W. Phair and J. S. J. van Deventer, "Effect of silicate activator pH on the leaching and material characteristic of waste-based inorganic polymer," *Minerals Engineering*, vol. 14, no. 3, pp. 289–304, 2001.

[7] A. Palomo and M. Palacios, "Alkali-activated cementious materials: alternative matrices for the immobilisation of hazardous wastes part II. Stabilisation of chromium and lead," *Cement and Concrete Research*, vol. 33, no. 2, pp. 289–295, 2003.

[8] J. W. Phair, J. S. J. van Deventer, and J. D. Smith, "Effect of Al source and alkali activation on Pb and Cu immobilisation in fly-ash based "geopolymers"," *Applied Geochemistry*, vol. 19, no. 3, pp. 423–434, 2004.

[9] L. Zheng, W. Wang, and Y. Shi, "The effects of alkaline dosage and Si/Al ratio on the immobilization of heavy metals in municipal solid waste incineration fly ash-based geopolymer," *Chemosphere*, vol. 79, no. 6, pp. 665–671, 2010.

[10] J. Zhang, J. L. Provis, D. Feng, and J. S. J. van Deventer, "Geopolymers for immobilization of Cr^{6+}, Cd^{2+} and Pb^{2+}," *Journal of Hazardous Materials*, vol. 157, no. 2-3, pp. 587–598, 2008.

[11] J. Deja, "Immobilization of Cr^{6+}, Cd^{2+}, Zn^{2+} and Pb^{2+} in alkali-activated slag binders," *Cement and Concrete Research*, vol. 32, no. 12, pp. 1971–1979, 2002.

[12] J. G. S. van Jaarsveld and J. S. J. van Deventer, "The effect of metal contaminants on the formation and properties of waste-based geopolymers," *Cement and Concrete Research*, vol. 29, no. 8, pp. 1189–1200, 1999.

[13] L. Kalina, J. Koplík, F. Šoukal, J. Másilko, and L. Jaskowiecová, "Potential uses of geopolymers to immobilize toxic metals from by-products material," *Environmental Engineering and Management Journal*, vol. 11, no. 3, pp. 579–584, 2012.

[14] J. Koplík, L. Kalina, J. Másilko, and F. Šoukal, "The characterization of fixation of Ba, Pb, and Cu in alkali-activated fly ash/blast furnace slag matrix," *Materials*, vol. 9, no. 7, p. 533, 2016.

6

Durability Improvement of Synthetic Dry Adhesives by Metal Coatings

Gyuhe Kim, Taechang Ahn, and Hui Yun Hwang

Department of Mechanical Design Engineering, Andong National University, 1375 Gyeongdong-ro, Andong-si, Gyeongsangbuk-do 760-749, Republic of Korea

Correspondence should be addressed to Hui Yun Hwang; hyhwang@andong.ac.kr

Academic Editor: Mikhael Bechelany

Gecko-like synthetic dry adhesives (SDAs) have adhesion comparable to that of a real Gecko's foot, but with very low durability. To address this problem, self-cleaning or stiff core embedding methods have been proposed. However, the proposed methods require special locomotion or complicated manufacturing. In this study, we suggested a metal coating on synthetic dry adhesives to improve durability. SDAs were fabricated via PDMS. Then, metals such as indium, zinc, and gold were coated on the SDAs. The adhesion tests show that the indium- and zinc-coated microstructures have a higher shear adhesion strength than the noncoated ones. Also, the shear adhesion strength of noncoated SDAs was only 14.5% of the initial strength while that of the zinc-coated ones was 35.6% after 200 times of attachment and detachment. We could find PDMS debris and fractures on noncoated SDAs, which results in weakening of the adhesion strength. On the other hand, a relatively high hardness, strength, and stiffness of the zinc coating layers reduced the wear and fractures of the micropatterns, which led to an improved durability in the SDAs. From these tests, we can conclude that the metal coating method could improve the durability of the SDAs.

1. Introduction

Many researchers have engineered systems with a higher efficiency by adopting novel features that have been observed in nature. As such, synthetic dry adhesives (SDAs) inspired by Gecko's feet are a typical example of systems inspired by nature [1, 2]. Gecko's feet are widely known to have millions of microsized setae with over 1,000 nanosized spatulae in hierarchical structures, and the Van de Waals force between the spatulae and foreign surfaces gives the superb adhesion to Gecko's feet. In addition, Gecko's feet have other superior features, including directionally different and reversible adhesion, self-cleaning, and nonpollution characteristics. Therefore, there have been many attempts to mimic Gecko's feet for industrial applications in, for example, medical patches and sticky robot hands [3–16].

In the initial phase (earlier stage), micro- or nanopatterned structures with a high aspect ratio were designed for SDAs, and their adhesion characteristics were evaluated using a pull-off test, friction test, and peeling test [17–19]. Tilted or hierarchical structures with 3D shaped-tips, such as with a mushroom shape, and self-cleaning properties were tested to improve the adhesion [20–27]. These researches were focused on how to fabricate SDAs with very small and complicated structure, and various microelectronic and micromechanical processes were applied. Dip transfer, multistep photolithography, UV-assisted CFL, and multistep template are typical examples of numerous trials. Micro- and nanopillars with a higher aspect ratio are essential to achieve reasonable adhesion, but the aforementioned fabrication methods cannot be used commercially since extremely precise equipment and a long time are required for fabrication [24, 28–41]. In contrast, polymers or elastomers such as polydimethylsiloxane (PDMS) have been widely used for SDAs due to their flexibility and ease of fabrication. However, these materials have a low stiffness and strength, which reduces the adhesion strength of the micro- and nanopillars due to fiber collapse, abrasion, and fracture [1, 22, 42–44]. This limits SDAs to only a single application, such as medical patches, whereas demand for applications for robot hands and transportation equipment for clean rooms requires a more durable adhesion. Therefore, maintaining the adhesion strength of the SDAs can

Table 1: Specifications of polydimethylsiloxane (SYLGAD 184, Dow Corning Co. Ltd., USA).

Property	Unit	Value
	cP	5175
Viscosity (part A)	mPa-sec	5175
	Pa-sec	5.2
	cP	3500
Viscosity (mixed)	mPa-sec	3500
	Pa-sec	3.5
Specific gravity (cured)	—	1.04
Working time at 25°C	hr	1.4
Cure time at 25°C	hrs	48
Heat cure time @ 100°C	min	35

Table 2: Specifications of cured PDMS and coating metals.

	Purity (%)	Strength (MPa)	Stiffness (GPa)	Hardness (HV)	Company
Indium	99.99	4.5	12.7	<10	iNecus.lnc.
Zinc	99.99	37	96.5	30	iNecus.lnc.
Gold	99.99	102	77	25	iNecus.lnc.

increase their applicability to a wider range. One method that was proposed to maintain the adhesion strength was to coat it with an adhesive structure. Suh's group reported on the adhesion strength of nanohairy dry adhesives coated with a 4 nm thick platinum layer [22], and Kim's group also presented dry adhesives with micropatterns coated with a 6 nm thick gold layer [45]. However, these studies have only given the changes in the adhesion strength with and without metal coatings.

The purpose of this study is to investigate the changes not only of the adhesion strength, but also of the structural integrity by considering the mechanical and physical properties of the coating materials and their thickness. We fabricated SDAs with micropatterns using PDMS and coated three different metals on the micropatterns with different thicknesses via plasma sputtering. We then measured the adhesion strength of the fabricated dry adhesives during repeated shear tests on glass surfaces and observed the structural changes with an optical microscope as well as a scanning electron microscope (SEM). For further analysis, we compared the changes in the adhesion strength and structural integrity based on the material properties and thickness of the coating metals.

2. Materials and Methods

2.1. Materials and Specimen. Many studies have used polydimethylsiloxane (PDMS), polyurethane (PU), and polyurethane acrylate (PUA) to develop SDAs. We selected PDMS (SYLGARD 184, Dow Corning Co. Ltd., USA) due to its easy micro- and nanofabrication, relatively high durability, and flexibility. The specifications of the PDMS are shown in Table 1.

Gold, indium, and zinc were selected as coating materials because they have different stiffness, strength, and hardness, as shown in Table 2. These properties were expected to have an effect changes in the adhesion strength, durability, and structural integrity.

2.2. Sample Preparation. SDAs with micropatterns were fabricated using the process shown in Figure 1. The base material (SYLGARD 184 Base, Dow Corning Co. Ltd., USA) and curing agent (SYLGARD 184 Curing, Dow Corning Co. Ltd., USA) were mixed at a ratio of 10 : 0.8 using a planetary centrifugal mixer (ARM-310, Thinky Co., Japan) under a rotating speed of 2,000 rpm for 1 min. Then, the mixture was poured in a micropatterning mold with microholes with a diameter of 30 μm, height of 60 μm, and interval of 30 μm. The mixtures were poured in the mold and cured at 80°C for 3 hours under vacuum in a vacuum oven (VO-64, Hanyang Scientific Equipment Co. Ltd., Korea). Finally, we could obtain SDAs with micropatterns once they were separated from the mold. Subsequently, micropatterns were coated with metal materials (gold, zinc, and indium) using a compact plasma sputtering coater (GSL-1100X-SPC12-LD, MTI Co., USA). DC argon plasma sputtering was conducted under gas purity of 99.995%, gas pressure of 8 Pa, discharge power of 2.4 W, and the sputtering time ranging from 4 s to 20 s to control the coating thickness from 2 nm to 10 nm.

2.3. Test Method. The macroscopic shear adhesion strength of the SDAs was measured using a frictional adhesion test, as shown in Figure 2 since the fabricated SDAs only had simple micropatterns [29, 46]. Specimens with sizes of 10 mm × 10 mm were attached to a slide glass with a preload of 2 N. A moving stage with a load cell at 30 N (DBCM, Bongshin Load Cell Co., Ltd., Korea) pulled off the specimens at the speed of 0.03 mm/s, and the frictional forces were recorded during the tests. Upon completing these tests, the thickness of each coating material with a maximum shear adhesion strength was selected. Frictional adhesion tests were then repeated 200 times with metal-coated SDAs with selected thickness, and the changes were analyzed regarding the shear adhesion strength to evaluate their durability.

In addition, we observed changes in the surfaces and structural integrity of micropatterns for every tenth frictional adhesion test using a digital microscope (KH-7700, Hirox Co., Ltd., Japan) and a field emission scanning electron microscope (JSM-6700F, Jeol Ltd., Japan).

3. Results and Discussion

3.1. Thickness Selection of the Coated Metal Layers. Figure 3 shows the measured shear adhesion strength of the metal-coated SDAs for each material and the thickness of the metal films on the adhesives. As reference, the SDAs without metal coatings had a shear adhesion strength of 9.1 kPa. The shear adhesion strength of the SDAs coated with indium increased from 3.9 kPa to 10.9 kPa when the indium thickness increased from 2 nm to 4 nm, and the shear adhesion strength with an indium thickness ranging from 4 nm to 10 nm decreased from 10.9 kPa to 4.51 kPa. The maximum adhesion strength was 10.9 kPa for the 4 nm thick indium layer, and there was

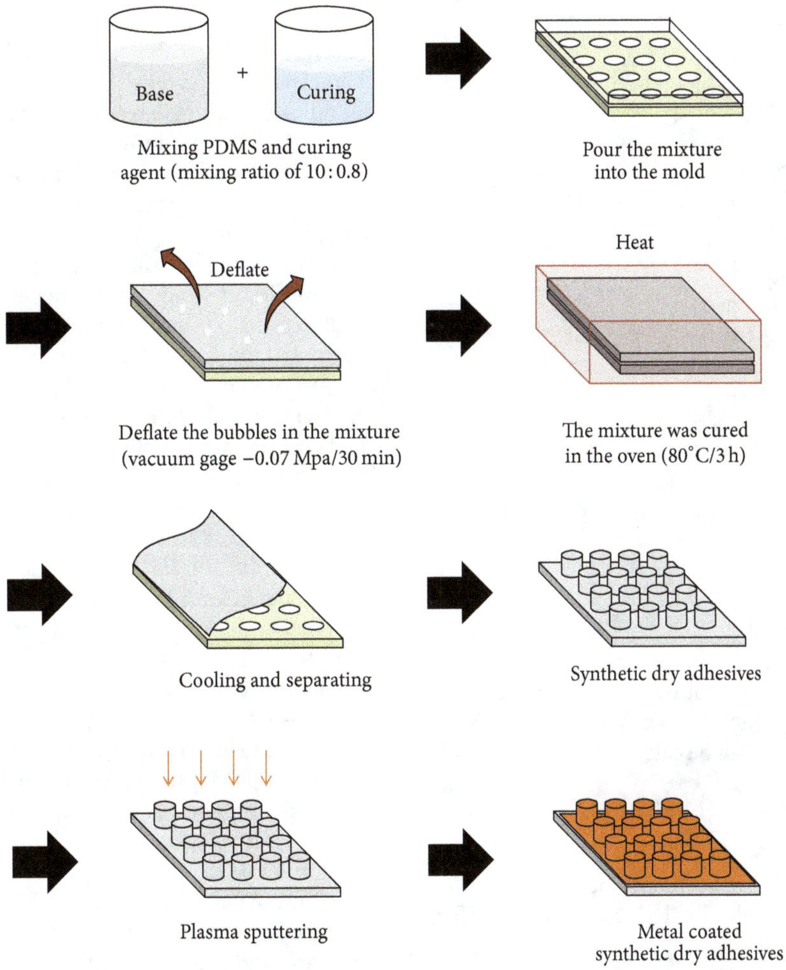

Figure 1: Fabrication process of synthetic dry adhesives coated with metal.

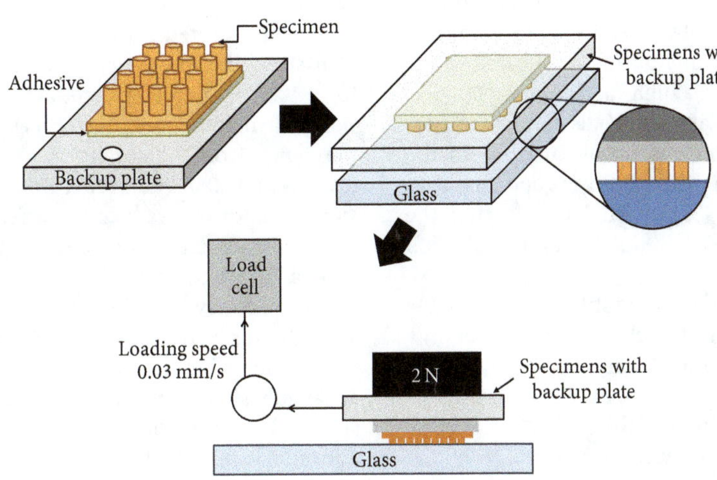

Figure 2: Schematic configuration of the frictional adhesion tests.

less than a 15% error when measuring the data. On the other hand, the zinc-coated synthetic dry adhesive displayed a different tendency. The shear adhesion strength increased monotonically from 5.5 kPa to 9.6 kPa and then decreased to 8.5 kPa. It reached maximum at 9.6 kPa when the zinc layer thickness was 8 nm, and the maximum error was 9.6%. SDAs coated with gold had a relatively low shear adhesion strength compared to the other two metals. The shear adhesion strength varied from 3.6 kPa to 6.3 kPa with a maximum error of 12.3% in the thickness range from 2 nm to 10 nm. The

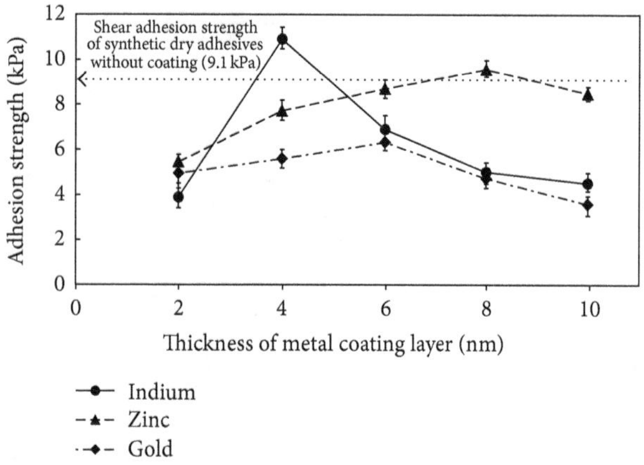

Figure 3: Shear adhesion strength of synthetic dry adhesives with respect to coating materials and thickness (the gray dotted line indicates the shear adhesion strength of synthetic dry adhesives without a metal coating).

maximum strength was 6.3 kPa when the layer of gold was 6 nm.

The adhesion characteristics of the SDAs depend on the Van der Waals forces [5]. The Van der Waals force (Fvdw) between the two flat surfaces can be calculated as follows [47]:

$$F = -\frac{A}{(6\pi D^2)}, \tag{1}$$

where A and D are the Hamaker constant (J) and cut-off distance (≈0.165 nm). A can also be derived from the surface energy as follows [46]:

$$A = 24\pi D^2 \gamma = 24\pi D^2 \sqrt{\gamma_1 \gamma_2}, \tag{2}$$

where γ_1 and γ_2 represent the surface energy of the contact materials (dyne/cm), respectively. Therefore, the adhesion characteristics of the SDAs can improve if the SDAs are fabricated using a material with a high surface energy.

On the other hand, the Van der Waals forces are proportional to bending the SDAs due to an increase in the contact area [15], and bending is inversely proportional to the effective stiffness of the SDAs. This means that the increase in stiffness might lead to a reduction in the adhesion characteristics of the SDAs.

Table 3 shows the calculated bending stiffness of the SDAs with and without metal coatings as well as the surface energy of the coating metals. The surface energies of the coating metals are much higher than those of PDMS (about 20 to 70 times), but the bending stiffness of the SDAs increased by coating metals on the PDMS (about 10~100 times).

Two different results were reported regarding the effect of the stiffness of the SDAs [22]. Regarding the material properties, SDAs made with a hard polyurethane acrylate (PUA) with 320 MPa of modulus had a higher adhesion strength compared to those with a soft PUA of 19.8 MPa. However, from the point of view of the structural stiffness, SDAs made with a soft PUA with a high effective stiffness had a lower adhesion strength compared to those with a low effective stiffness.

The results calculated in Table 3 and the reported results [22] indicate that different materials have different adhesion characteristics, and only the modulus or structural stiffness is not unique factor in determining the adhesion characteristics of the SDAs. Whether the effects of the surface energy or stiffness of SDAs are more important to the adhesion characteristics has not been reported, so we could not estimate a change in the adhesion strength by the metal coatings. However, the adhesion strength of the SDAs with metal coatings might be higher or lower compared to those without metal coatings.

Based on the maximum shear adhesion strength, SDAs coated with indium and zinc had a higher strength compared to those without coatings, but the gold-coated ones had a relatively low adhesion strength. From the results of the shear adhesion test, we selected the optimal thickness of the layer for each metal: 4 nm, 8 nm, and 6 nm for indium, zinc, and gold, respectively.

3.2. Durability of the SDAs. The durability of the SDAs was evaluated with and without metal coating by repeating frictional adhesion tests. The changes in the shear adhesion strength for each metal are shown in Figure 4.

The reference durability data of the SDAs without a coating is shown in Figure 4(a). The shear adhesion strength decreased drastically from 9.1 kPa to 4.1 kPa during the first 50 repetitions. This displays reductions in the strength of 54.3% and a reduction rate of 0.099 kPa/repetition. Until 90 repetitions, a relatively gradual reduction in the shear adhesion strength was observed, and the reduction rate was 0.023 kPa/repetition. After 90 repetitions, the shear adhesion strength decreased monotonically with a reduction rate of 0.034 kPa/repetition and then gradually reduced to 1.3 kPa at 200 repetitions with a reduction rate of 0.011 kPa/repetition. After 200 repetitions, only 14% of initial shear adhesion strength was retained.

Indium-coated SDAs had the highest initial shear adhesion strength, and their strength reduction was only 51.7% from 10.9 kPa to 5.3 kPa (reduction rate of 0.061 kPa/repetition) during the first 90 repeated tests (Figure 4(b)). In the range from 100 to 180 repetitions, the reduction rate decreased as 0.021 kPa/repetition, and then the shear adhesion strength was reduced at a reduction rate of 0.064 kPa/repetition. Finally, the shear adhesion strength became 1.8 kPa (16.3% of initial strength).

A tendency for the reduction in the shear adhesion strength for SDAs coated with zinc was similar to those with indium (Figure 4(c)). First, 90 repetitions decreased the shear adhesion strength from 9.6 kPa to 6.3 kPa, which corresponded to a reduction rate of 0.036 kPa/repetition. The shear adhesion strength decreased slowly to 5.15 kPa (reduction rate of 0.016 kPa/repetition) until 160 repetitions, and then the reduction rate increased at 0.032 kPa/repetition. In total, 36.5% of the initial shear adhesion strength (3.5 kPa) was maintained after 200 repetitions.

As shown in Figure 4(d), SDAs with a gold coating layer had a different trend in the shear adhesion strength compared to the other 3 cases. During the first 100 repetitions, the shear adhesion strength decreased from 6.4 kPa to 2.1 kPa at a reduction rate of 0.042 kPa/repetition. Even though the

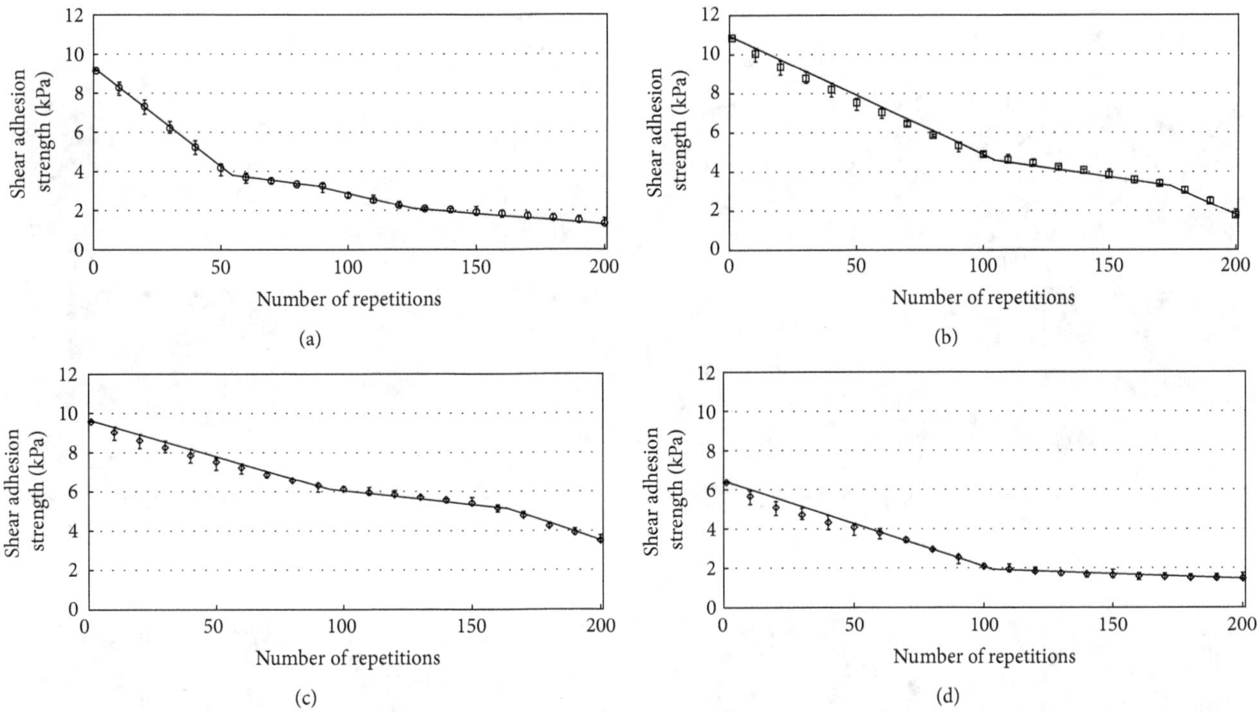

FIGURE 4: Changes in shear adhesion strength of synthetic dry adhesives (a) without coating and coated with (b) indium, (c) zinc, and (d) gold with respect to repetitions of frictional adhesion test.

TABLE 3: Comparison of the bending stiffness of SDAs with and without metal coatings and the surface energy of the coating metals.

	Stiffness (GPa)	Coating thickness (nm)	Bending stiffness (Nm^2)	Surface energy (mN/m)
Indium	12.7	4	3.22×10^{-13}	560
Zinc	96.5	8	4.14×10^{-12}	989
Gold	77	6	2.50×10^{-12}	1700
PDMS	0.00132	—	5.25×10^{-14}	23

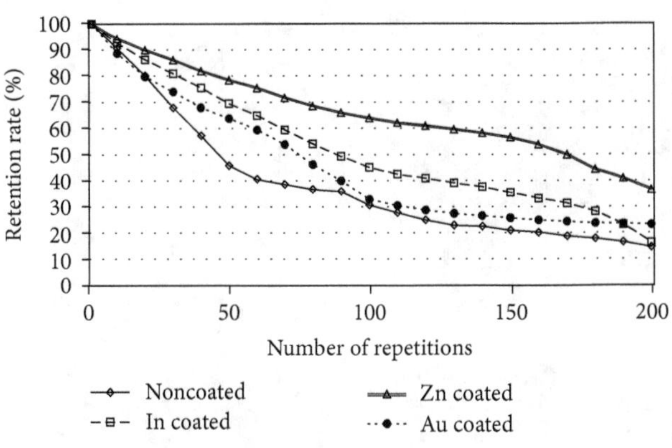

FIGURE 5: Retention rates of shear adhesion strength of synthetic dry adhesives with and without metal coatings.

reduction rate seemed to be lower than or similar to that of the other cases, the cause may be due to a relatively low initial shear adhesion strength rather than good durability. At 100 repetitions, the shear adhesion strength was only 2.1 kPa, which is close to those of reference data after 200 repetitions. Until 200 repetitions, the reduction rate was quite low at 0.005 kPa/repetition, but the shear adhesion strength eventually became 1.5 kPa (23% of initial shear strength).

We calculated the retention rate, percentage ratio of the shear adhesion strength after repeated tests to the initial strength to compare the durability of the SDAs with and without metal coating layers based on Figure 4. As shown in Figure 5, the retention rate of the reference synthetic dry adhesive degraded the fastest, and those with metal coating layers had a relatively higher retention rate. In particular, the zinc-coated SDAs had the highest retention rate (63.9% at 100 repetitions and 36.6% at 200 repetitions) as well as the highest shear adhesion strength (3.5 kPa) at 200 repetitions although initial shear adhesion strength (9.6 kPa) was lower than those of indium-coated ones (10.8 kPa).

3.3. Surface and Structural Integrity of SDAs. The changes in the surface and structural integrity of the SDAs with and without metal coatings were investigated, as shown in Figure 6.

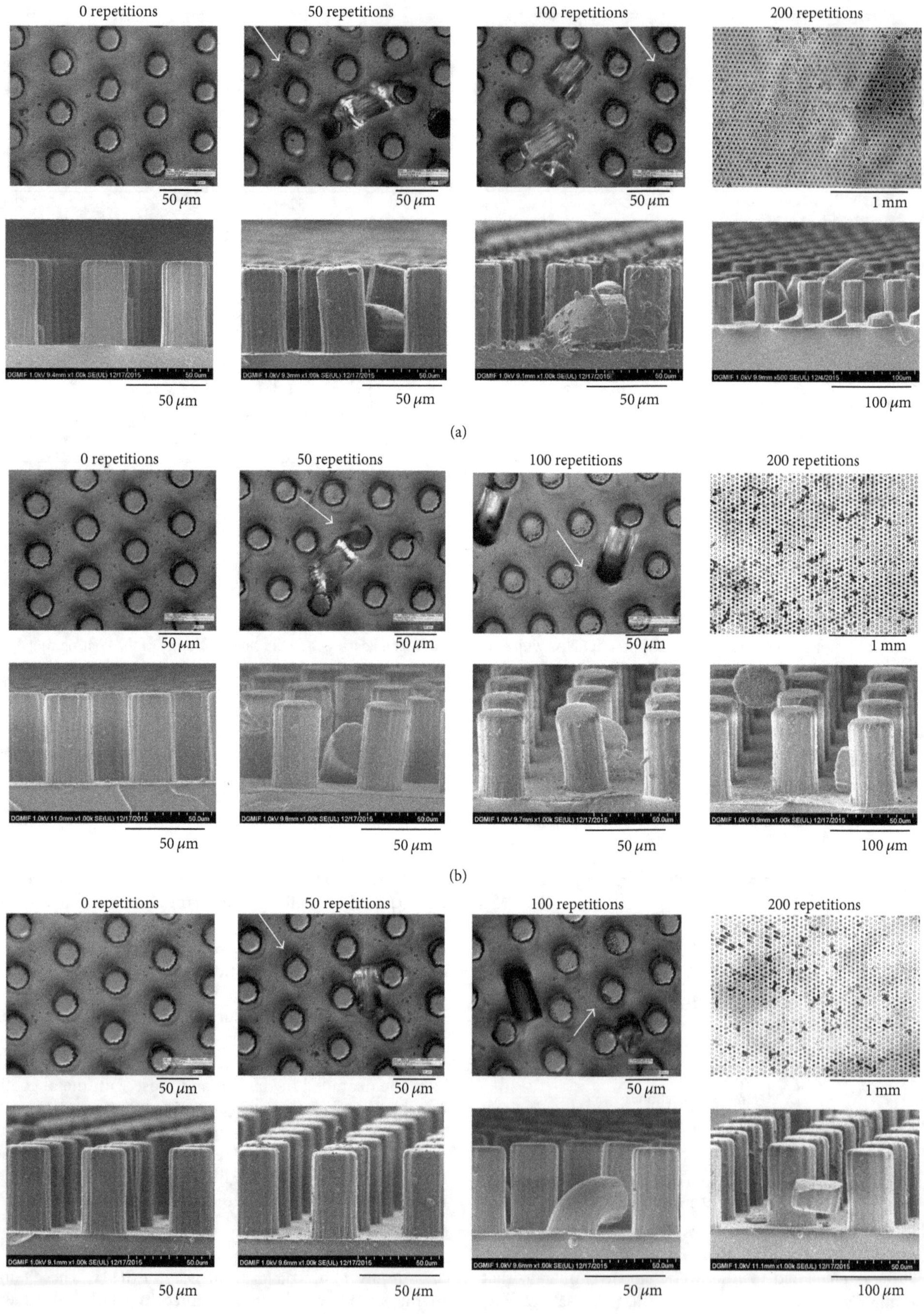

FIGURE 6: Continued.

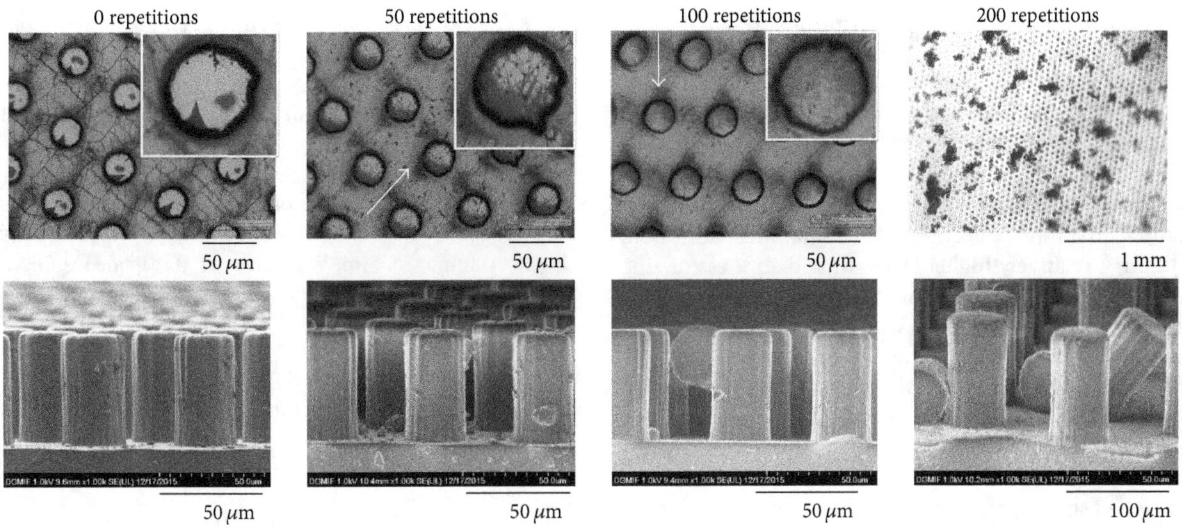

(d)

FIGURE 6: Changes of surface and structural integrity of SDAs (a) without metal coatings and with (b) indium coating, (c) zinc coating, and (d) gold coating (the inset shows the enlarged view of wear sites).

We could observe a large amount of wear on the top surfaces of the reference SDAs along the direction of the slide, as shown in Figure 6(a) (see the dark areas pointed by the arrows) after only the first 50 repetitions. The fracture on the root of the micropatterns appeared in some places at 50 repetitions because the top surface wear made a larger deformation, even though the wear was not severe until 100 repetitions. However, the fracture began to spread to the entire region after 100 repetitions and continued until 200 repetitions. At the end, 98% of the micropatterns were fractured.

Indium-coated SDAs kept the top surfaces of micropatterns relatively less worn than those of the reference until 50 repetitions due to the metal coating layer, as shown in Figure 6(b). From 50 to 100 repetitions, the wear on the top surfaces increased. At 100 repetitions, we could observe self-adhesion due to large bending deformation and some fractures at the middle of micropillars rather than the root in the reference. After 200 repetitions, fractures occurred on both the middle and root of the micropillars, and debris formed between the unfractured micropillars. At 200 repetitions, 47% of the micropatterns were in self-adhesion or large bending, and 15% were fractured.

Figure 6(c) presents the surfaces of SDAs coated with zinc. Unlike the above two cases, small abrasive particles of the zinc layer were attached on the micropillars at 50 repetitions. This means that the wear tendency of the zinc-coated SDAs is abrasion, but the above two cases represent plastic deformation, which led to the contact area becoming wider. However, the degree of wear was not severe compared to the other cases until 100 repetitions. After 100 repetitions, a large bending deformation was observed, and only 12% of the micropillars were fractured on the root at 200 repetitions. A tendency in the bending deformation and fractures was similar to that of the indium-coated ones.

Cracks were observed on a gold-coated back-up plate of the SDAs (see scratch-shaped lines in Figure 6(d)), and the top surfaces of the micropatterns were relatively bright due to the gold coating. After 50 repetitions, cracks appeared on the top surfaces, and the coated gold partially wore due to the repetitive contact. The gold-coated layer kept peeling until 100 repetitions, and it totally wore out on the top surfaces at 100 repetitions. From 100 to 200 repetitions, fractures in the micropatterns continued, similarly to the reference. Finally 89% of the SDAs were removed.

3.4. Discussion. An investigation of the surfaces and structural integrity of the SDAs could explain the tendency to reduce the shear adhesion strength through repeated use.

For the reference SDAs, wear on the top surfaces accelerated the degradation of the shear adhesion strength during the first 50 repetitions. A partial fracture was observed from 50 to 100 repetitions, and a total fracture observed by 200 repetitions resulted in a further reduction in the shear adhesion strength as a different trend. Fractured micropatterns occupied 98% of the entire area, which made the shear adhesion strength only 14% of the initial strength. The tendency for a reduction in strength implies that the most effective factors are the wear of the contact surfaces and the fracture of the micropatterns. Therefore, we can improve the durability of the SDAs if the contact surface is protected, wear is reduced, and strength of the micropillars increased. In this aspect, fabricating thin metal layers on the contact surfaces is an effective counterplan because the thin metal layers enhance the wear resistant contact surfaces and strengthen the micropillars by covering weak polymer materials.

The indium-coated layer could decrease the wear of the top surfaces until 100 repetitions compared to those of the reference, as shown in Figure 6(b). This leads to an increase in the retention rate of the share adhesion strength, as shown in Figure 5. However, a relatively low strength and stiffness for indium were not sufficient to decrease the bending deformation and reinforce the micropillars. After

180 repetitions, an abrupt reduction in the shear adhesion strength was observed, as shown in Figure 4(b).

For comparison, the zinc-coated SDAs had a higher durability though a reduction in the shear adhesion strength similar to those with indium, as shown in Figure 6(c). The wear of the top surfaces was not noticeable until 100 repetitions as hardness of zinc-coated layer is much higher than that of indium, and it retained 63.9% of the initial shear adhesion. In addition, a relatively higher strength and stiffness of zinc than indium resulted in less bending deformation and more strengthening of the micropillars. A retention rate of 36.5% compared to the initial shear adhesion strength was achieved by good mechanical properties of the zinc-coated layer.

Gold-coated SDAs showed a different tendency even though the strength and stiffness of gold are the highest among three coating metals where the hardness is comparable to zinc. As shown in Figure 6(d), the main reason for the reduction of shear adhesion strength was exfoliation of gold-coated layer on SDAs. In general, the adhesion between the gold and polymers is relatively weaker than that of the other metals [48]. Therefore, peeling the gold-coated layer resulted in a relatively large drop in the shear adhesion strength. After 100 repetitions, the gold-coated SDAs had a trend close to reference since no gold layer remained.

4. Conclusions

We fabricated four different SDAs with and without metal-coated layers to investigate the effects of the metal coating on the durability of the shear adhesion strength. Among the coating materials that were considered, zinc-coated SDAs had not only the highest shear strength (10.8 kPa), but also durability in repeated tests (retention rate of 63.9% at 100 repetitions and 36.5% at 200 repetitions). These results are more than double the strength of the SDAs without any coating. Surface observations with respect to the number of repetitions revealed important factors of their durability, including the mechanical properties (hardness, strength, and stiffness) of the coating metal and adhesion characteristics of the coating metals to the micropillars.

Conflicts of Interest

The authors declare that they have no conflicts of interest.

Acknowledgments

This work was supported by the National Research Foundation of Korea Grant funded by the Korean Government (NRF-2012S1A2A1A01030569) and in part by the National Research Foundation of Korea Grant funded by the Korean Government (NRF-2016R1A2B4014603).

References

[1] L. F. Boesel, C. Cremer, E. Arzt, and A. D. Campo, "Gecko-inspired surfaces: A path to strong and reversible dry adhesives," *Advanced Materials*, vol. 22, no. 19, pp. 2125–2137, 2010.

[2] M. K. Kwak, C. Pang, H.-E. Jeong et al., "Towards the next level of bioinspired dry adhesives: New designs and applications," *Advanced Functional Materials*, vol. 21, no. 19, pp. 3606–3616, 2011.

[3] K. Autumn, Y. A. Liang, S. T. Hsieh et al., "Adhesive force of a single gecko foot-hair," *Nature*, vol. 405, no. 6787, pp. 681–685, 2000.

[4] K. Autumn, S. T. Hsieh, D. M. Dudek, J. Chen, C. Chitaphan, and R. J. Full, "Dynamics of geckos running vertically," *Journal of Experimental Biology*, vol. 209, no. 2, pp. 260–272, 2006.

[5] K. Autumn, M. Sitti, Y. A. Liang et al., "Evidence for van der Waals adhesion in gecko setae," *Proceedings of the National Academy of Sciences of the United States of America*, vol. 99, no. 19, pp. 12252–12256, 2002.

[6] D. Campolo, S. Jones, and R. S. Fearing, "Fabrication of gecko foot-hair like nano structures and adhesion to random rough surfaces," in *Proceedings of the 2003 3rd IEEE Conference on Nanotechnology, IEEE-NANO 2003*, pp. 856–859, usa, August 2003.

[7] W. R. Hansen and K. Autumn, "Evidence for self-cleaning in gecko setae," *Proceedings of the National Academy of Sciences of the United States of America*, vol. 102, no. 2, pp. 385–389, 2005.

[8] RB. Huey and PE. Hertz, "Effects of body size and slope on sprint speed of a lizard (Stellio(Agama) Stellio). J Exp Biol," in *Hertz PE. Effects of body size and slope on sprint speed of a lizard (Stellio(Agama) Stellio). J Exp Biol*, pp. 97–401, 97, 401-9, 1982.

[9] C. Greiner, A. Del Campo, and E. Arzt, "Adhesion of bioinspired micropatterned surfaces: Effects of pillar radius, aspect ratio, and preload," *Langmuir*, vol. 23, no. 7, pp. 3495–3502, 2007.

[10] R. Spolenak, S. Gorb, and E. Arzt, "Adhesion design maps for bio-inspired attachment systems," *Acta Biomaterialia*, vol. 1, no. 1, pp. 5–13, 2005.

[11] A. Del Campo, C. Greiner, I. Álvarez, and E. Arzt, "Patterned surfaces with pillars with controlled 3D tip geometry mimicking bioattachment devices," *Advanced Materials*, vol. 19, no. 15, pp. 1973–1977, 2007.

[12] A. Mahdavi, L. Ferreira, C. Sundback et al., "A biodegradable and biocompatible gecko-inspired tissue adhesive," *Proceedings of the National Academy of Sciences of the United States of America*, vol. 105, no. 7, pp. 2307–2312, 2008.

[13] H. Gao and H. Yao, "Shape insensitive optimal adhesion of nanoscale fibrillar structures," *Proceedings of the National Academy of Sciences of the United States of America*, vol. 101, no. 21, pp. 7851–7856, 2004.

[14] E. Arzt, S. Gorb, and R. Spolenak, "From micro to nano contacts in biological attachment devices," *Proceedings of the National Academy of Sciences of the United States of America*, vol. 100, no. 19, pp. 10603–10606, 2003.

[15] Y. Tian, N. Pesika, H. Zeng et al., "Adhesion and friction in gecko toe attachment and detachment," *Proceedings of the National Academy of Sciences of the United States of America*, vol. 103, no. 51, pp. 19320–19325, 2006.

[16] N. J. Glassmaker, A. Jagota, C.-Y. Hui, W. L. Noderer, and M. K. Chaudhury, "Biologically inspired crack trapping for enhanced adhesion," *Proceedings of the National Academy of Sciences of the United States of America*, vol. 104, no. 26, pp. 10786–10791, 2007.

[17] A. K. Geim, S. V. Dubonos, I. V. Grigorieva, K. S. Novoselov, A. A. Zhukov, and S. Y. Shapoval, "Microfabricated adhesive mimicking gecko foot-hair," *Nature Materials*, vol. 2, no. 7, pp. 461–463, 2003.

[18] S. Reddy, E. Arzt, and A. Del Campo, "Bioinspired surfaces with switchable adhesion," *Advanced Materials*, vol. 19, no. 22, pp. 3833–3837, 2007.

[19] H. E. Jeong, S. H. Lee, P. Kim, and K. Y. Suh, "Stretched polymer nanohairs by nanodrawing," *Nano Letters*, vol. 6, no. 7, pp. 1508–1513, 2006.

[20] B. Aksak, M. P. Murphy, and M. Sitti, "Adhesion of biologically inspired vertical and angled polymer microfiber arrays," *Langmuir*, vol. 23, no. 6, pp. 3322–3332, 2007.

[21] J. Lee, R. S. Fearing, and K. Komvopoulos, "Directional adhesion of gecko-inspired angled microfiber arrays," *Applied Physics Letters*, vol. 93, no. 19, Article ID 191910, 2008.

[22] T.-I. Kim, H. E. Jeong, K. Y. Suh, and H. H. Lee, "Stooped nanohairs: Geometry-controllable, unidirectional, reversible, and robust Gecko-like dry adhesive," *Advanced Materials*, vol. 21, no. 22, pp. 2276–2281, 2009.

[23] A. del Campo and C. Greiner, "SU-8: a photoresist for high-aspect-ratio and 3D submicron lithography," *Journal of Micromechanics and Microengineering*, vol. 17, no. 6, pp. R81–R95, 2007.

[24] M. P. Murphy, S. Kim, and M. Sitti, "Enhanced adhesion by gecko-inspired hierarchical fibrillar adhesives," *ACS Applied Materials and Interfaces*, vol. 1, no. 4, pp. 849–855, 2009.

[25] H. E. Jeong, J.-K. Lee, H. N. Kim, S. H. Moon, and K. Y. Suh, "A nontransferring dry adhesive with hierarchical polymer nanohairs," *Proceedings of the National Academy of Sciences of the United States of America*, vol. 106, no. 14, pp. 5639–5644, 2009.

[26] MP. Murphy, *Biologically-inspired synthetic dry adhesives for wall-climbing robots*, Doctor of Philosophy, Carnegie Mellon University, 2008.

[27] B. Yurdumakan, N. R. Raravikar, P. M. Ajayan, and A. Dhinojwala, "Synthetic gecko foot-hairs from multiwalled carbon nanotubes," *Chemical Communications*, no. 30, pp. 3799–3801, 2005.

[28] L. Ge, S. Sethi, L. Ci, P. M. Ajayan, and A. Dhinojwala, "Carbon nanotube-based synthetic gecko tapes," *Proceedings of the National Academy of Sciences of the United States of America*, vol. 104, no. 26, pp. 10792–10795, 2007.

[29] L. Qu, L. Dai, M. Stone, Z. Xia, and Z. L. Wang, "Carbon nanotube arrays with strong shear binding-on and easy normal lifting-off," *Science*, vol. 322, no. 5899, pp. 238–242, 2008.

[30] M. T. Northen and K. L. Turner, "A batch fabricated biomimetic dry adhesive," *Nanotechnology*, vol. 16, no. 8, pp. 1159–1166, 2005.

[31] T. S. Kustandi, V. D. Samper, W. S. Ng, A. S. Chong, and H. Gao, "Fabrication of a gecko-like hierarchical fibril array using a bonded porous alumina template," *Journal of Micromechanics and Microengineering*, vol. 17, no. 10, article no. N02, pp. N75–N81, 2007.

[32] L. Qu and L. Dai, "Gecko-foot-mimetic aligned single-walled carbon nanotube dry adhesives with unique electrical and thermal properties," *Advanced Materials*, vol. 19, no. 22, pp. 3844–3849, 2007.

[33] B. Aksak, *A design methodology for biologically inspired dry fibrillar adhesives*, Doctor of Philosophy, Carnegie Mellon University, 2008.

[34] B. N. J. Persson and S. Gorb, "The effect of surface roughness on the adhesion of elastic plates with application to biological systems," *Journal of Chemical Physics*, vol. 119, no. 21, pp. 11437–11444, 2003.

[35] K. A. Daltorio, A. D. Horchler, S. Gorb, R. E. Ritzmann, and R. D. Quinn, "A small wall-walking robot with compliant, adhesive feet," in *Proceedings of the IEEE IRS/RSJ International Conference on Intelligent Robots and Systems, IROS 2005*, pp. 4018–4023, can, August 2005.

[36] S. Kim and M. Sitti, "Biologically inspired polymer microfibers with spatulate tips as repeatable fibrillar adhesives," *Applied Physics Letters*, vol. 89, no. 26, Article ID 261911, 2006.

[37] S. Gorb, M. Varenberg, A. Peressadko, and J. Tuma, "Biomimetic mushroom-shaped fibrillar adhesive microstructure," *Journal of the Royal Society Interface*, vol. 4, no. 13, pp. 271–275, 2007.

[38] J. Davies, S. Haq, T. Hawke, and J. P. Sargent, "A practical approach to the development of a synthetic Gecko tape," *International Journal of Adhesion and Adhesives*, vol. 29, no. 4, pp. 380–390, 2009.

[39] D. Sameoto and C. Menon, "A low-cost, high-yield fabrication method for producing optimized biomimetic dry adhesives," *Journal of Micromechanics and Microengineering*, vol. 19, no. 11, Article ID 115002, 2009.

[40] D. Sameoto and C. Menon, "Deep UV patterning of acrylic masters for molding biomimetic dry adhesives," *Journal of Micromechanics and Microengineering*, vol. 20, no. 11, Article ID 115037, 2010.

[41] H. E. Jeong, M. K. Kwak, and K. Y. Suh, "Stretchable, adhesion-tunable dry adhesive by surface wrinkling," *Langmuir*, vol. 26, no. 4, pp. 2223–2226, 2010.

[42] A. Parness, D. Soto, N. Esparza et al., "A microfabricated wedge-shaped adhesive array displaying gecko-like dynamic adhesion, directionality and long lifetime," *Journal of the Royal Society Interface*, vol. 6, no. 41, pp. 1223–1232, 2009.

[43] S. S. Chary, *Microfabrication of Bio-Inspired Adhesive Systems*, Doctor of Philosophy, University of California, 2013.

[44] E. Cheung, *Biologically inspired fibrillar adhesion and friction on biological tissues: Modeling, Fabrication and applications [PhD thesis]*, Carnegie Mellon University, 2008.

[45] G. H. Kim, T. C. An, and H. Y. Hwang, "Effects of metal coatings on adhesive characteristics of gecko-like micro structures," *Transactions of the Korean Society of Mechanical Engineers, A*, vol. 39, no. 11, pp. 1099–1103, 2015.

[46] M. Sitti and R. S. Fearing, "Synthetic gecko foot-hair micro/nano-structures as dry adhesives," *Journal of Adhesion Science and Technology*, vol. 17, no. 8, pp. 1055–1073, 2003.

[47] J. N. Israelachivili, *Intermolecular and Surface Forces*, Elsevier, 3rd edition, 2011.

[48] Y. Yamada, DR. Wheeler, and DH. Bcukley, "Adhesion between polymers and evaporated gold and nickel films. National Aeronautics and Space Administration," Tech. Rep., 1984.

7

A Review on Melt-Pool Characteristics in Laser Welding of Metals

Behzad Fotovvati, Steven F. Wayne, Gladius Lewis, and Ebrahim Asadi

Department of Mechanical Engineering, The University of Memphis, Memphis, TN 38152, USA

Correspondence should be addressed to Ebrahim Asadi; easadi8@yahoo.com

Academic Editor: Paolo Ferro

Laser welding of metals involves with formation of a melt-pool and subsequent rapid solidification, resulting in alteration of properties and the microstructure of the welded metal. Understanding and predicting relationships between laser welding process parameters, such as laser speed and welding power, and melt-pool characteristics have been the subjects of many studies in literature because this knowledge is critical to controlling and improving laser welding. Recent advances in metal additive manufacturing processes have renewed interest in the melt-pool studies because in many of these processes, part fabrication involves small moving melt-pools. The present work is a critical review of the literature on experimental and modeling studies on laser welding, with the focus being on the influence of process parameters on geometry, thermodynamics, fluid dynamics, microstructure, and porosity characteristics of the melt-pool. These data may inform future experimental laser welding studies and may be used for verification and validation of results obtained in future melt-pool modeling studies.

1. Introduction

Laser is a coherent single-phase beam of lights from a single wavelength (monochromatic) with low beam divergence and high energy content, which creates heat when it strikes a metal surface. The advent of high-power (multi-kW) lasers in the 1970s [1] opened the door to many metal working applications, which, previously, had been done using conventional high-flux heat sources, such as reacting gas jets, electric discharges, and plasma arcs. One metal working application of lasers is laser welding, which requires power density > 10^3 kW·cm^{-2} [2]. In laser welding, two adjacent or stacked metal pieces are fused together by melting the parts at the weld line; usually, the process is conducted under an inert gas flow with or without addition of material to the weld line. The moving melted volume is called the melt-pool (Figure 1). The size of this pool, which is on the order of 1 mm, is influenced by many variables, such as the material, laser power, and welding speed.

The deep volume directly under the laser focus area is called the keyhole, within which the high energy of the laser creates heating rates >10^9 K·s^{-1} [3]. Thus, the material in the keyhole is rapidly melted and even boiled, thereby creating a metallic plasma around it. Boiling of the material maximizes the absorption of the laser energy by the material because it turns the keyhole to a black body [4]. The amount of absorbed energy in the material decreases exponentially through the thickness, as predicted by the Beer–Lambert law. A smaller portion of the absorbed energy is conducted away through reradiation and convection from the surface, while the rest is conducted into the substrate. An intense recoil pressure created by evaporation of the material in the keyhole generates a vapor jet and a fluid flow in the keyhole and the melt-pool (Figure 1) [5]. In addition, the surrounding area of the melt-pool that is still in the solid state will reach temperatures high enough to change the microstructure of the material or to cause solid-state phase transformation, depending on the material thermodynamics. This area is called the heat-affected zone (HAZ). Hereafter, we use the term "melt-pool" to refer to the combination of the keyhole, the molten-metal area (MMA), and the HAZ. Laser welding mechanisms can be divided into two categories based on the existence of the keyhole: keyhole mode and conduction mode. Keyhole-mode welding is more common because it produces narrow HAZs. However, keyhole oscillations and closures result in instabilities of the melt-pool, leading to creation of pores in the welded zones. On the other hand, there is more stability

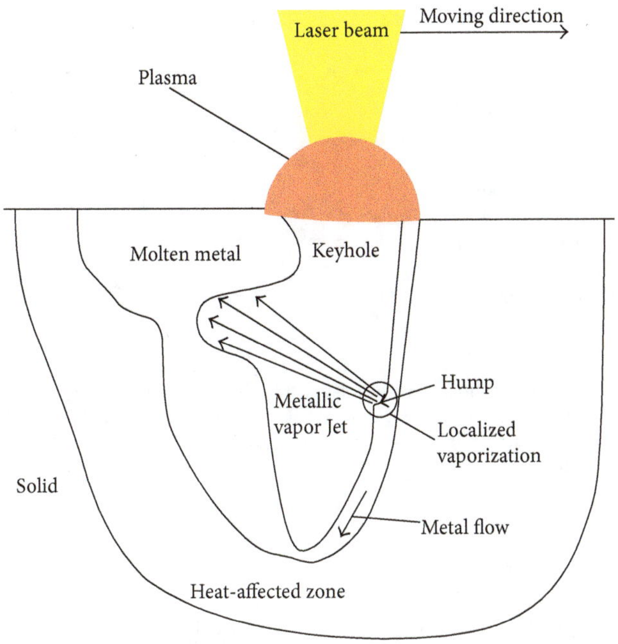

Figure 1: A cross-sectional schematic of a side view (the interface between two solids) of the melt-pool formed during a laser welding process.

in the conduction mode since vaporization is minimal. Conduction-mode welds are produced using low-power laser beams; as such, these welds are shallower than keyhole-mode welds [6]. The focus of this review is keyhole-mode laser welding.

Melt-pool characteristics directly control the quality of the weld, for example, porosity in the weld through keyhole thermodynamics and residual stresses through HAZ thermomechanics. As a result, one of the main goals of many research studies is to understand the relationships between weld quality and laser welding process parameters (such as laser type and laser power), substrate temperature, and melt-pool characteristics [7–12]. There are different types of lasers, three widely used ones being neodymium-doped yttrium aluminum garnet (Nd:YAG), CO_2, and argon lasers. Lasers differ in characteristics, such as maximum output power and pulse repetition rate that they can provide, and, as such, the choice of the laser should be based on the application being considered. For instance, Morgan et al. [13] conducted density analysis experiments on 316L stainless steel and chose Nd-YAG laser over CO_2 laser due to increased absorption of a 1.064 μm wavelength by metallic powder compared to a longer wavelength (10.64 μm) [14].

Locke et al. [15] carried out one of the early experiments on laser welding of metals. They used laser power levels of 8 kW and 20 kW, leading to penetration depth and speed that had not been possible previously. The penetration depth achieved was 12.7 mm at a ratio of 2.54 m·min^{-1} in a 5-to-1 depth-to-average-width fusion zone in 304 stainless steel at a 20 kW laser power level. The state of the art of laser welding of metals and associated melt-pool characteristics in those early days of research was reviewed by Mazumder in 1982 [2] and in 1987 [16]. Since then, laser welding of metals has advanced significantly in many aspects, such as welding materials, process monitoring, computational modeling, and quality. There are a few review papers in the literature that deal with recent advancements in laser welding of metals. In 1989, David and Vitek [17] focused on the solidification behavior of the melt-pool and investigated the correlation between the weld metal microstructure and solidification parameters such as crystal growth rates and the consequent interface cooling rates. They presented a diagram showing the variation of weld microstructure as a function of cooling rate, growth rate, and combination(s) of these variables. In 2003, Cao et al. [18, 19] reviewed research and progress in laser welding of wrought Al alloys. They reviewed findings regarding the influence of an assortment of parameters, which they divided into three categories (laser-, process-, and material-related parameters), on weld quality. They quantified the weld quality by metallurgical microstructures and defects, such as porosity, cracking, oxide inclusions, and loss of alloying elements, as well as mechanical properties of the weld, such as hardness, tensile strength, fatigue strength, and formability. In 2005, Shao and Yan [20] reviewed on-the-fly monitoring techniques for inspecting the laser welding process, highlighting the advantages and limitations of acoustic, optical, visual, thermal, and ultrasonic techniques. In 2006, Cao et al. [21] conducted a similar review but focused on Mg alloys. In 2014, Liu et al. [22] reviewed laser welding studies of dissimilar Mg and Al alloys. Their review also included discussion of progress on research on other welding techniques applied to these alloys, including solid-state processes and fusion welding. The authors stated that a challenge in welding dissimilar Mg and Al alloys is the formation of brittle intermetallic compounds, which can be addressed by eliminating or reducing the Mg-Al intermetallic reaction layer through careful selection of process parameters.

Modeling the laser welding process has been another major research focus. This is challenging due to the multiphysics nature of the problem (Figure 1); that is, it involves laser-material interaction, fluid flow, large temperature variation, plasma formation, vapor-liquid-solid coexistence, and possible solid-state phase transformation [4, 23–25]. As analytical solution of the laser welding process is not possible (except in the case of a simplified physics and geometry model), numerical/computational approaches have been taken. In 2005, Mackwood and Crafer [26] reviewed the literature on thermal modeling up to 2002. They divided the work conducted into categories based on the welding method, such as arc, resistance, and friction, as well as welding processes, such as alloying, cladding, and surface hardening. The review covered basic analytical solutions, such as a moving/fixed-point heat source combined with a line source of heat, along with numerical solutions, including standard heat transfer solutions, welding dissimilar metals, multipass welding, melt-pool models, and keyhole models. Also, in 2005, Yaghi and Becker [27] reviewed thermal and mechanical welding simulations in which finite element analysis (FEA) was used. The simulations included heat flow processes and solid-state phase transformations occurring in the welding process. They discussed several relevant modeling considerations (such as parametric studies of residual stresses), influence of material properties on residual stresses, and combination of welding simulation with other heat transfer engineering processes. In 2012, He [28] updated the review of FEA studies on laser welding, with special attention to the simulation of defect formation. He discussed numerical problems in FEA of laser welding, including materials modeling, meshing procedure, and failure criteria. He concluded that establishing an accurate and reliable finite element model of laser welding is very difficult because the process is a complex phenomenon that comprises many interrelated mechanisms and metallurgical processes. In 2015, Svenungsson et al. [29] conducted a review of modeling investigations of the keyhole and categorized these models based on the considerations and assumptions used in constructing the models.

Most reviews on laser welding are limited to either a specific material or method of study, namely, experimental, on-the-fly monitoring, or computational. In other words, there is no review on the state of the art on the properties of the melt-pool and their relationship to the welding process parameters and weld quality. In the present review, we focus on these aspects. Thus, the present review focuses on the following melt-pool characteristics: (1) geometrical features, such as the penetration depth, width, HAZ geometry, keyhole geometry, and MMA geometry; (2) thermodynamic characteristics, such as laser energy absorption, surface temperature, cooling rates, and temperature map in the melt-pool; (3) fluid dynamic characteristics, such as fluid flow in the melt-pool and vaporization in the keyhole; (4) resulting microstructures; and (5) porosity characteristics, including factors that influence porosity formation and methods to avoid it. Due to the multiphysics and integrated nature of the melt-pool, in some literature reports, there may be some overlap coverage of the abovementioned aspects. In each section of the present review, we critically discuss the state of the art in determination of the considered melt-pool characteristic, its variation with process parameters, and its influence on commonly used weld quality quantifiers, such as the microstructure and mechanical properties. In the final section, we summarize the key points made and identify some gaps in the reviewed models of laser welding of metals that hinder full characterization of the process.

2. Melt-Pool Geometry

The magnitude and distribution of cooling rates, temperature, and the maximum thickness that can be achieved by a single welding pass are determined by the melt-pool geometry [7]. Additionally, the microstructure of the fusion zone is also influenced by the melt-pool geometry. There are several studies on the influence of laser power and welding speed on melt-pool geometrical features (Table 1). A summary of optimum laser welding process parameters, for a selection of metals and alloy materials, is given in Table 2. Summaries of studies on the influence of laser welding process variables on melt-pool and keyhole features are presented in Appendix.

Table 1: Influence of two laser welding parameters on various melt-pool geometry parameters.

Geometry parameter	Process parameter	
	Laser power	Welding speed
Melt-pool depth	$+^{a}$	$-^{b}$
Melt-pool width	$+^{c}$	$-^{d}$
Melt-pool depth/width ratio	NS	$+^{e}$
Melt-pool length	$+^{f}$	$-^{g}$
Keyhole radius	$+^{g}$	$-^{g}$
Cooling rate	$-^{h}$	$+^{i}$
Melt-pool surface area	$+^{j}$	$-^{k}$
Vaporization rate	$+^{l}$	NS

$^{+}$Direct relationship; $^{-}$inverse relationship; NS: not stated in the report; [a][11, 30–37]; [b][11, 31, 33, 38–41]; [c][11, 33–37, 42–44]; [d][11, 12, 31, 33, 38, 41, 42, 45]; [e][30]; [f][32, 42]; [g][42, 45]; [h][7, 8]; [i][8]; [j][43, 46]; [k][38, 41, 47]; [l][44, 46].

In laser welding applications, the maximum achievable welding speed is limited by the maximum available laser power. For economic reasons, it is desirable to apply the highest possible speed during laser welding, while a full penetration is achieved at the same time. A feedback controller, in which an optical sensor measures the intensity of the melt-pool radiation and exports it to a feedback control system, has been described by Postma et al. [48]. They also proposed a dynamic model, which describes the sensor and laser source dynamics, using system identification techniques. This model, which uses laser power as the input and the modeled sensor signal as the output, is capable of maintaining full penetration in the presence of artificial power variations and speed changes. This procedure optimizes the welding speed without risking lack of penetration.

Surface structure and hardness of the substrate after laser-related processes are also affected by process parameters. Ashby and Easterling [52] conducted experiments and combined the equations of heat flow and kinetics to evaluate the near-surface structure and hardness after laser treatment. Using a Gaussian heat source for their model, they presented diagrams, which show the structure and the hardness of the surface, as a function of process parameters. Using these diagrams, the maximum achievable surface hardness without surface melting that results from using an optimum combination of process parameters is identified.

In the process of deep penetration welding, a high energy density is transferred to the workpiece through the keyhole; therefore, the flows of metal vapor inside the keyhole and the molten material around it play an important role in the welding process, and the shape of the keyhole would highly affect the weld quality. One approach that is employed to evaluate the shape of the keyhole in this process is to estimate its cross-sectional area in each depth. The model presented by Dowden et al. [53] utilized this approach. They assigned a single temperature to vapor materials. Therefore, by obtaining the temperature distribution in each depth, a border between the vapored materials and the materials that are not vapored could be distinguished as the keyhole. The keyhole shape that they obtained was one that has a circular cross section with a curved axis. Steen et al. [54]

TABLE 2: Optimum laser welding parameters for a selection of metals and alloys.

Metal/alloy	Parameter			
	Laser power (kW)	Welding speed ($m \cdot min^{-1}$)	Focal position (relative to the surface) (mm)	Shielding gas
Mg alloy, WE43 [11]	2	2	0 or −1	Helium
Mg alloy, AZ91 [11]	2	3	0	Helium
Several Mg alloys [49]	1.5[a] 2–2.5[b]	2.5–3[a] 1-2[b]	−2	Helium
Ti-6Al-4V alloy [50]	NS	0.8	NS	Helium
Stainless steel 304L [30]	4	3	0.2	Helium
Stainless steel 347 [30]	5	2	0.4	Helium
Stainless steel 304 [33]	1.25	0.75	NS	NS
Galvanized steel [34]	1.3	1	NS	Argon
Inconel 625 [39]	1.5	2	NS	NS
Zn and Sn [51]	1.6	1.5	5	NS
Stainless steel 440 and 416 [35]	Any combination of parameters that produces energy density in the range of 20.8 to 27.7 $J \cdot mm^{-2}$			

NS: not stated in the report; [a]for thinner plates (2.5 mm and 3.0 mm); [b]for thicker plates (5.0 mm and 8.0 mm).

employed another approach to estimate the keyhole shape by the combination of a point heat source and a line heat source. They found a simple analytical form for the temperature distribution, and possible weld profiles were found numerically with specific choices for the strengths of the line and point sources and specific locations for the point source. Comparison of the obtained profiles with the actual measured profile led to the profile that gave the best fit, leading to magnitudes of the line and point sources and the point source location.

Beck et al. [55] used the equations of continuity, motion, and energy to obtain the velocity and temperature fields and, hence, the keyhole shape and the maximum velocity of the melt flow at the sides of the keyhole. Kaplan [56] employed another technique to obtain the keyhole shape. He used the energy balance point-by-point through the substrate thickness to find the position of each point of the keyhole profile. This work showed that most of the laser beam heat was absorbed at the front wall of the keyhole rather than at the rear wall. Lampa et al. [57] simplified Kaplan's model and used it for calculating the penetration depth, by applying a correction factor for the material conductivity at the top of the melt-pool. The correction factor was calculated as 2.5, which makes the material more conductive.

Amara and Bendib [58] solved the Navier–Stokes equations for an incompressible fluid flow concentrated on the vapor pressure in the keyhole and confirmed that the vapor pressure works against forces, such as surface tension, that tend to close the keyhole opening. They used the ray-tracing code, which allows calculation of the energy deposited on the keyhole wall after each reflection of the laser beam. However, Fabbro et al. [59] showed that due to the high absorptivity of the front keyhole surface (60% to 80%), only one reflection of the laser beam is necessary for the modeling. Tenner et al. [60] used another method to estimate the keyhole shape, that is, by relating it to the plasma plume. Through experiments, they showed that the keyhole dynamic behavior is well correlated with the plume when a threshold laser power is reached, with this power being ~80% of the power required for full penetration (in this case, 2.5 kW). They concluded that, at low laser powers, the stability of the keyhole is determined by the evaporation process in the keyhole. Fabbro [61] defined some regimes in laser welding and investigated the keyhole shape, particularly, the keyhole front wall tilting, in the different regimes. They observed that, at low welding speeds, the keyhole is more unstable, and the intensity, which is absorbed by the keyhole front wall, depends on the welding speed, not the intensity itself. Postacioglu et al. [62] estimated the shape of the surface of the weld as elevated or depressed. Kar and Mazumder [63] used mass conservation, momentum, and energy equations along with the heat transfer in the solid and vapor phases of the materials to predict the shape of the melt-pool by calculating the surface velocity and temperature distribution. They observed that the axial velocity at the beginning of the laser melting process is negligible, compared to velocities in other directions; however, after keyhole development, the dominant velocity is in the axial direction. They asserted that the moving speed of the solid-liquid interface is much higher than that of the liquid-vapor interface; in other words, the melt-pool depth increases more rapidly than does the keyhole depth.

3. Melt-Pool Thermodynamics

Heat transfer in the melt-pool during laser welding significantly affects the melt-pool shape, melt flow inside the melt-pool, and cooling rates of the melt-pool and, thus, the quality and microstructure of the weld [64]. Since, in laser welding, dimensions of the melt-pool are on the order of mm and timescales are on the order of fractions of a second, measuring the temperature profiles and cooling rates of each point is costly and time-consuming. Thus, many of the studies reported on the thermodynamic characteristics of the melt-pool during laser welding are modeling works.

Mazumder and Steen [4] developed a three-dimensional quasi-steady heat transfer model of laser melting and compared their results using this model to those obtained from experiments. For simplicity, they assumed that there is no reflectivity at the material surface, where the temperature exceeds the boiling temperature of the material and the thermal properties of the material are constant and independent of temperature changes. Using this model, some process parameters were predicted, including the temperature profile, maximum welding speed, HAZ width, thermal cycle at any location or speed, and the effect of supplementary heating or cooling, thickness, reflectivity, and thermal conductivity on the melt-pool shape.

Goldak et al. [65] presented a more sophisticated model of weld heat sources that consisted of two combined ellipsoid shapes and is flexible enough to change the size and the shape of the heat source so that it could be used for shallow or deep penetration welding with various types of heat sources, such as arc, laser, and electron beam. They observed some differences between FEA and experimental results and suggested the reason to be due to neglecting the heat flow in the longitudinal direction. They found that the energy losses due to radiation and convection near the heat source are negligible.

Wang et al. [66] considered continuity, energy, and momentum equations and employed the volume of fluid method [67], which can be used to calculate the free surface shape of the keyhole, to solve a three-dimensional model of the temperature distribution. Using assumptions that the material properties are constant and the fluid flow is laminar and incompressible during the process, they found large temperature gradients in the front region of the keyhole. They claimed that the recoil pressure is the main driving force for the keyhole formation. Akbari et al. [68] used the same sets of equations and assumed the melt-pool surface to be flat and the fluid flow to be transient, laminar, and incompressible. They found that regardless of the welding speed, the temperature distribution decreased sharply at the laser beam center and then decreased slightly far away from the center of the laser beam. Frewin and Scott [69] considered temperature dependence of material properties and stated that the temperature profile is a function of absorptivity and laser beam energy distribution. De et al. [70] assumed a double-ellipsoidal model for the heat source to present a two-dimensional conduction heat transfer model. They investigated the effect of varying the penetration depth and absorptivity on the exactness of the model results.

Vaporization is important in laser welding of alloys that contain one or more volatile constituents because vaporization determines the thermodynamic characteristics of the melt-pool. Khan and Debroy [44] concluded that the relative rate of vaporization of any two elements from the melt-pool was an indicator of melt-pool temperature, irrespective of the element pair selected. Increasing the laser power will increase the vaporization rate by increasing the temperature and the surface area of the melt-pool. Collur et al. [71] conducted several experiments to examine the role of gas-phase mass transfer in the vaporization of alloying elements and modeled the role of plasma in the vaporization of alloying elements. The melt-pool is surrounded by a plasma during laser welding, allowing molten metal drops to vaporize, both in the presence and absence of plasma, isothermally. They found that, under various shielding gas environments, the rate of vaporization of alloying elements is independent of the flow rate and also of the nature of the shielding gas but is controlled by plasma-influenced intrinsic vaporization at the melt-pool surface.

To simplify modeling convection in the melt-pool, many researchers assumed that the thermal conductivity of the material is isotropic in that region [24, 72, 73]. In contrast, Safdar et al. [74] took into account the anisotropic thermal conductivity and stated that the anisotropy-enhanced thermal conductivity approach leads to a more accurate result in the prediction of melt-pool temperature distribution.

There are many parameters that affect the shape of the melt-pool. One of them is the Marangoni convection, also called surface tension-driven convection, which is convection along an interface between two fluids due to a surface tension gradient [75]. Tsotridis et al. [76] presented a simplified model of the melt-pool considering the Marangoni convection. For simplicity, they assumed that all the physical properties of the solid and the liquid are the same, and they claimed that the Marangoni flow dominated over the buoyancy flow. Tsai and Kou [73] presented a two-dimensional heat transfer fluid flow model to describe the Marangoni convection in the melt-pool that is dependent on the surface tension temperature coefficient. They asserted that when this parameter has a negative value, the Marangoni convection direction is radially outward, and the pool center is depressed but the outer part is elevated. However, a positive coefficient results in a convex melt-pool.

Limmaneevichitr and Kou [43] conducted experiments to investigate the effect of the Marangoni convection on the melt-pool shape. In this work, they used $NaNO_3$ and Ga for welding as these two materials have extremely high and low Prandtl number (Pr), respectively; $Pr = C_p\mu/k$, where C_p is the specific heat, μ is the dynamic viscosity, and k is the thermal conductivity. The Peclet number (Pe) is also important in determining the effect of the Marangoni convection. Pe expresses the ratio of heat transport by convection to heat transport by conduction; that is, $Pe = LV/\alpha$, where L is the pool surface radius, V is the maximum outward surface velocity, and α is the thermal diffusivity. For the melt-pools of $NaNO_3$, a material with a high Pr, α is very low and V is high (strong Marangoni convection). Therefore, Pe is very high, and heat transport in the melt-pool is dominated by the Marangoni convection. Increasing the beam power increases the Marangoni convection, the melt-pool size, V, L, and, hence, Pe. Experimental results showed that if a strong outward surface flow carries the heat outward to the melt-pool edge, it makes a concave pool bottom wide and flat. Reducing the beam diameter also increases the Marangoni convection and Pe. The return flow penetrates the pool bottom close to the pool edge and turns the flat pool bottom to a convex one. Both the convex and flat pool bottoms indicate that the Marangoni convection dominates over buoyancy convection, which is induced by gravity in the pools. On the other hand, for melt-pools of Ga, Pe is very low, and conduction dominates heat transport in the melt-pool. Heat is conducted downward and outward and, thus, makes the pool bottom concave. Reducing the beam diameter makes the melt-pool more hemispherical, which confirms the domination of conduction over heat transport in the pool.

Yang et al. [9] presented a model by combining continuity, momentum, and energy equations for liquid and solid phases and reported that the thermal properties of the material and the Marangoni flow in the melt-pool could significantly influence the melt-pool shape such that more Marangoni flow results in a wider and shallower melt-pool. Abderrazak et al. [10] utilized their experimental and finite volume simulation results obtained from Mg alloy specimens to assert that a negative Marangoni effect, due to the absence of the surface active agent in the alloy composition, makes the melt-pool wider and shallower.

The physical origin of the enhanced energy transfer from a laser to a material may be explained on the basis of two alternative mechanisms, namely, Fresnel absorption and inverse bremsstrahlung (IB) absorption [77]. The Fresnel equation describes the behavior of light when moving between media of differing refractive indices. The absorption of light that the equations predict is known as Fresnel absorption. IB absorption is one of the important mechanisms for transferring energy from laser light to matter. In the very intense field used in a laser fusion program, processes involving multiphoton absorption and emission are very important [78]. There have been a number of different formalisms suggested for treating IB in intense fields [79] together with a few numerical calculations [79]. For example, Zhang et al. [80] presented a sandwich model to observe the keyhole in deep penetration laser welding, thus providing an effective way to analyze both Fresnel and IB absorptions. By increasing the thickness of Al films between two glass pieces, higher densities of the keyhole plasma are achieved, leading to deep keyholes. By continuing to thicken the Al films, the aperture of the keyhole continues to widen. However, above a critical thickness, the depth of the keyhole reduces (in Al films, this critical thickness is 0.3 mm). This is due to the excess density of the keyhole plasma, which prevents the transmission of the laser beam to the keyhole. The density of the keyhole plasma creates similar effects on changes in welding depth compared to keyhole depth. Cheng et al. [77] computed the laser intensity absorbed on the keyhole wall using Fresnel and IB absorptions of the keyhole plasma. They concluded that IB absorption of the keyhole plasma plays a more important role than does Fresnel absorption. They asserted that the temperature of the keyhole plasma decreases from the top to the bottom of the keyhole and decreases from the center to the edge of the keyhole. Tan et al. [81] found that almost invariably, the maximum temperature in the keyhole wall is located at the bottom of the keyhole.

4. Melt-Pool Fluid Dynamics

In order to obtain a high-quality laser-welded metal product, it is necessary to prevent defects before they occur. When a metal is in the liquid state, the probability of collapsing of the melt-pool or its partial penetration is high. Therefore, the dynamics of the melt-pool and the fluid flow patterns are important.

Zacharia et al. [82] proposed that surface active elements may alter the flow field in the melt-pool and, hence, affect weld penetration. They showed that a combination of the concentration of the surface active elements and the temperature distribution has an important role in determining weld penetration. It was also shown that the melt-pool flow can be simulated more realistically considering not only the coefficient of surface tension as a function of temperature but also the concentration of the surface active elements. In laser welding (unlike gas tungsten arc welding [83, 84]), the latter factor makes the temperature coefficient of surface tension largely negative, causing the flow to be radially outward at the melt-pool surface. This flow transfers the heat out from the center of the melt-pool and makes the pool shallow.

Semak et al. [85] investigated the dynamics of the melt-pool by conducting experiments using three different types of pulses: a single 20 or 30 ms pulse, continuous wave pulse, and repetitive 20 ms pulses. They observed that the vaporization pressure exceeds surface tension and hydraulic pressure in the melt-pool, creating a high-velocity melt flow and, thus, a melt crown around the keyhole at the melt-pool edge. Also, significant variations in the shape of the keyhole opening were observed, which were attributed to the instability of the vapor pressure. Semak et al. [86] presented a model to simulate the fluid flow in the melt-pool during pulsed laser welding. In a later contribution, they modified their simulation to include the effect of surface tension [87]. They asserted that the effect of this force could be an ejection or a retention, depending on the distribution of the beam intensity.

Cho et al. [88] simulated the fluid flow in the melt-pool during the transition from conduction laser spot welding to keyhole laser spot welding and showed an upward and downward oscillation in the fluid flow in the center of the melt-pool in the direction of normal to the surface. They attributed this oscillation to interaction of competing pressures, including recoil pressure and surface tension pressure. Using a sandwich model, Zhang et al. [89] observed the dynamics of the keyhole and showed that the hydrodynamics at the keyhole wall have a dominant effect on defects in the weld. Geiger et al. [90] used continuity, heat conduction, and the Navier–Stokes equations to show how pores form at high welding speeds (such as $12\,m{\cdot}min^{-1}$). A higher welding speed results in a higher pressure at the keyhole front and, thus, higher velocities of melt flow around the keyhole, which lead to a depression outside the keyhole. A combination of this phenomenon with the surface tension leads to formation of pores.

The coupling between the melt-pool and the keyhole is complicated. It has been shown that the sideways liquid displacement around the front keyhole wall is the main process for generating high velocities of the fluid that enters the melt-pool [25, 91]. Basu and DebRoy [92] found a threshold for the melt-pool surface temperature above which the vaporization-induced recoil pressure overcomes the surface tension pressure, causing an outward flow to the sides [93]. The recoil pressure is one of the three main mechanisms responsible for expelling melt from the keyhole. The other two, which are particularly important at higher melt surface temperatures, are melt evaporation and the shielding gas interaction with molten metal. The melt flow generated by the recoil pressure has a direction in which the recoil pressure gradient is the highest. Therefore, in laser welding, the melt flow is ejected by the recoil pressure to the sides of the melt-pool [25]. Fabbro et al. [40] discussed the effects of the interaction between the vapor, which is generated by the ablation process occurring on the front keyhole wall, and the surrounding melt-pool. They showed that an efficient control of the dynamics of the melt-pool can be achieved using a side gas jet, which can be localized in the front or the rear position. This gas jet decouples the interaction zone inside the keyhole and the melt-pool. Therefore, the melt-pool flow can be well stabilized, resulting in a high-quality weld and improved penetration at low welding speeds. Amara and Fabbro [94] modeled the fluid flow in the melt-pool, considering the interaction between the vapor and the liquid and between the liquid and the air. Fabbro et al. [95] showed that the escaping vapor, which is generated in the keyhole, creates friction forces, which, in turn, play an important role in fluid flow in the melt-pool. Experiments [96] showed that these forces generate humping instabilities on the melt-pool above a critical welding speed. Amara et al. [97] considered the friction effects of the vapor flow with the liquid walls as an important factor to numerically solve the hydrodynamic equations, obtaining the shear stress distribution on the keyhole walls. Further investigations [98] lead to three-dimensional calculations of the molten metal flow velocity induced by the friction phenomenon and the thickness of the boundary layer. The friction force, which is induced on the melt-pool wall, results in a drag force expelling the flow towards the surface. The other main driving forces for the molten material in the melt-pool result from surface tension, recoil pressure, and buoyancy forces [5, 85, 94]. By solving a combination of the Navier–Stokes, energy conservation, and ideal gas equations, using the finite volume method, it was confirmed that using a gas jet during deep penetration laser welding results in better weld joints because the melt flow in the melt-pool is enhanced.

Insufficient metal flow in the melt-pool may be due to excessive welding speed or incorrect laser power, which leads to hump formation, a phenomenon that produces variation in weld penetration [99]. Once the hump starts to be solidified, further melt flows upwards and resolidifies, causing the hump to grow [100]. The travel angle between the laser beam and the welding direction has been found to affect the onset of humping. Forehand welding has been shown to suppress hump formation to higher welding speeds [101, 102]. Gratzke et al. [103] defined a critical ratio of the width to the length of the melt-pool, which determined the likelihood of hump formation, such that maximizing this ratio during welding decreases the possibility of hump formation. Another way of reducing humping defects in laser welding is by using a tandem dual beam [104]. When the beams are far apart, the second beam suppresses the humps formed by the first one, and when the beams are close, the following beam stabilizes the keyhole, thereby preventing hump formation after the leading beam. According to Beck et al. [55], any reduction in flow velocities in the rearward direction avoids hump formation. Kern et al. [105] used this concept in their experiments of CO_2 laser welding of steel by applying a magnetic field transverse to the welding direction, thus altering the melt flow profile within the melt-pool and suppressing hump formation. Matsunawa and Semak [106] simulated the keyhole during high-speed laser welding and found that hump formation frequency was increased with increasing welding speed. However, Kawahito et al. [107] defined a process window of welding speed and laser beam diameter, in which humping occurred over a particular range of laser power densities. In a later work [108], these authors found hump formation to be caused by several dynamic and static factors, including flow velocity, surface tension, solidification, and melt volume. They asserted that hump formation could be avoided in fully penetrated welds by decreasing melt volume so that the formation of the convex surface at the rear end of the melt-pool was suppressed. According to the model of Matsunawa and Semak [106], when the component of the keyhole velocity that is parallel to the surface was higher than the beam translation speed, the instability of the keyhole resulted in hump formation on the weld surface. A hump may also form on the keyhole wall surface when the upper part of the keyhole wall moves away from the laser beam axis and the lower part continues to move towards the axis. Ilar et al. [109] introduced root humping, which was different from top surface humping, being formed due to a gravity effect. Root humping was initiated by the increase in the amount of the material flowing in the melt-pool that originated from the bottom of the melt-pool. Amara and Fabbro [110] presented a 3D model based on the numerical resolution of the fluid flow and the heat transfer equations showing hump formation at high welding speeds in deep penetration laser welding. Pang et al. [111] found significant differences between melt-pool dynamics of an unstable keyhole and a stable one and that, by controlling the welding speed and surface tension, they could prevent the formation of humps on the keyhole wall, thus reducing keyhole instability. They stated that, under certain low-heat input welding conditions, collapse of the keyhole wall could be avoided.

Ki et al. [45, 112] presented a three-dimensional laser keyhole welding model and used the Navier–Stokes and energy equations to simulate the movements of the liquid-vapor interface and the solid-liquid interface as well as the heat transfer. In addition, they simulated the transition from conduction-mode welding to keyhole-mode welding. For the sake of simplicity, they extrapolated material properties at high temperatures from values obtained at lower temperatures. They did not take plasma into account, assumed the gas was incompressible, and neglected recondensation of the vapor after interacting with the hole surface. They confirmed that one of the main differences between the two types of laser welding (keyhole mode and conduction mode) is the recoil pressure, which is generated by evaporation during the laser keyhole welding. There is a fluctuation in the amount of laser energy absorbed in the keyhole, which, in turn, leads to fluctuation of the shape of the keyhole, and this fluctuation affects the recoil pressure and the flow field in the melt-pool. Their model also allows prediction of the microstructure and property evolution in laser-welded joints. Chakraborty and Chakraborty [72] developed a three-dimensional model of laser welding using conservation of mass, momentum, and energy equations to evaluate the influence of turbulence in the melt-pool on the process parameters and found that the velocity and temperature gradients are smaller in the turbulent melt-pool, a finding that agrees with the experimental results.

5. Weld Microstructure

A high cooling rate typically is experienced by the melt-pool immediately after laser welding. Solidification takes place usually in a few tens of milliseconds, and metastable microstructures are produced that influence the final mechanical properties of the weld. Therefore, microstructure characterization is vital in the determination of weld quality [113, 114]. Solidification of molten weld metal depends on the kinetics of the liquid-solid interface. Kou [115] described this by using values of the thermal gradient (G) (usually, they are in the range of 100–1000 $K{\cdot}m^{-1}$) and the travel speed of the liquid-solid interface (R) (usually, they are in the range of 10–103 $m{\cdot}s^{-1}$). Kou identified four possible modes of solidification: (1) planar (high G and low R), (2) cellular, (3) columnar dendritic, and (4) equiaxed dendritic (low G and high R) (Figure 2). The ratio of G to R determines the mode of solidification. Kou showed that the product of G and R indicated the cooling rate, so these two parameters determined the fineness of the solidified microstructure (Figure 2). Kou also noted that solidification of the melt-pool could take place in one of two ways, namely, (a) epitaxial and (b) nonepitaxial, depending on the composition of the weld metal.

The microstructure of rapidly solidified laser-molten Al-4.5 wt.% Cu alloyed surfaces was studied, and melted regions were found to resolidify epitaxially onto unmolten crystalline substrates [116]. Solidification proceeded as follows: a plane front mode, then cellular, and, finally, continuing in a columnar competitive manner. The major impact of the rapid solidification was a refinement of the surface microstructure. Kou [115] found that melt convection was not sufficiently vigorous to produce a homogeneous melt. Evidence of epitaxial resolidification was also found in a nickel-based superalloy (Udimet 700) when laser melted [117]. This face-centered cubic (fcc) material showed a strong preference for dendritic growth along (100) directions. The consequence of the rapid cooling rate was evident by fine dendritic regrowth, with a spacing of ~2.5 mm. The dendrites grew nearly parallel to the local direction of maximum heat flow [117].

Many researchers have investigated the microstructures of welds in laser welding of stainless steels and other ferrous alloys. Zambon and Bonollo [118] characterized the microstructure of weld beads and the HAZ of austenite and duplex stainless steels. They stated that high cooling rates might result in formation of nonequilibrium microstructures, which contain larger amounts of δ-ferrite in duplex steels than predicted both by the Fe-Ni-Cr pseudobinary phase diagram and by the Schaeffler diagram.

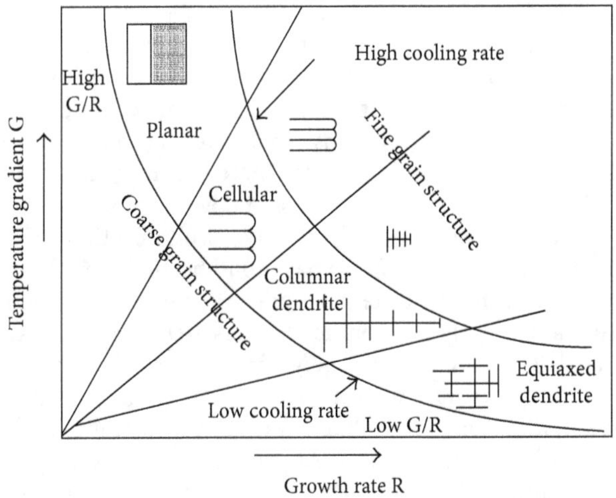

FIGURE 2: Influence of temperature gradient (G) and growth rate (R) on the mode of solidification and grain structure [115].

They concluded that nonequilibrium microstructures decreased the corrosion resistance of the welded joints. The rate at which a ferrous metal/alloy weldment cools significantly influences the ferrite morphology and distribution [119]. Zacharia et al. [8] presented a model to obtain the complex temperature distribution and the cooling rates and showed that, in pulsed laser welding, at low speeds, the weld metal remains molten, even during the time when the laser beam is not being applied. They confirmed that the microstructure is dependent on the cooling rates and ranged from duplex austenite + ferrite to fully austenitic or fully ferritic. These authors conducted another study in two parts (analytical and experimental) [82, 120] and by employing the equations of momentum, energy, and mass continuity concluded that the dominant force for the fluid flow is the surface tension gradient. They found the cooling rates at the solidification temperature to be the highest at the edge of the melt-pool rather than at the bottom or the top center of it. In another study by Zacharia et al. [82], their observed microstructural evaluation of laser-welded 304 stainless steel fusion zones revealed a fine dispersion of chromium oxide inclusions and a continuous oxide layer. The observed microstructures were sensitive to the cooling rates, with the decrease in the cooling rate resulting in a coarser solidification substructure with a widely spaced ferrite network. The rapid solidification of the laser beam-welded metal resulted in a fully austenitic microstructure with a fine solidification substructure. Lippold [121] determined the susceptibility of weld solidification cracking in austenitic stainless steels during pulsed laser welding. The author found that a shift in weld solidification behavior occurred under rapid solidification conditions. Solidification as primary austenite was found to be the most detrimental, and cracking depended mainly on composition, whereas pulsed laser welding process parameters had only a small influence. A solidification model was discussed that related the transition in primary solidification from ferrite to austenite to dendrite tip undercooling at high solidification growth rates.

Lippold [121] also found out that the available predictive microstructure diagrams and solidification models (the Suutala weldability diagram [122] and the Welding Research Council constitution diagram [123]) are not accurate under rapid solidification conditions, which happens during pulsed laser welding of stainless steels. Therefore, regarding rapid solidification, they proposed a predictive diagram for weld solidification cracking susceptibility, a solidification model relating the transition in primary solidification from ferrite to austenite to dendrite tip undercooling, and a microstructural map for austenitic stainless steel welds. Brooks et al. [124] studied high-energy stainless steel welds and concluded that minimal solid-state diffusion occurs during the solidification and cooling of primary austenite solidified welds, whereas structures which solidify as ferrite may become almost completely homogenized as a result of diffusion. A nearly segregation-free, single-phase austenite structure, which appears to be unique to the rapid solidification velocities and cooling rates of high-energy welds, was also observed. They suggested that this structure was a product of a marked phase transformation in which ferrite was transformed to austenite.

Recently, marked transformations were identified in the selective electron beam melting of Ti-6Al-4V. Thus, Lu et al. [125] concluded that the β (body-centered cubic (bcc)) to α_m (hexagonal close-packed) transformation led to the formation of a variety of patch-shaped massive grains, including large grain-boundary-crossing grains with misorientations being as much as 30°. Marked transformations have been identified in laser welding of stainless steels where the influence of composition and cooling rate on the solid-state transformation to γ-austenite was studied [126]. An analysis by D'amato et al. [127] showed that grain refinement at the weld area occurred and that δ-ferrite was present in the as-welded samples. The authors also concluded that the welds solidified by primary ferrite solidification with some chromium carbide precipitates in the weld area. The microstructure of the weld metal of a duplex stainless steel made with Nd:YAG pulsed laser was studied by Mirakhorli et al. [128]. They found the weld microstructure to be composed of two distinct zones: (1) at high overlapping factors, an array of continuous axial grains at the weld centerline was formed, and (2) at low overlapping factors, in the zone of higher cooling rate, a higher percentage of ferrite was transformed to austenite. They concluded that the high cooling rates involved in pulsed laser welding led to low overlapping, thus limiting the ferrite-to-austenite transformation to the grain boundaries only.

Concerning other ferrous-based alloys, Babu et al. [129] studied the primary solidification phase of Fe-C-Al-Mn steel welds under rapid and slow cooling rates. They found nonequilibrium austenite solidification during rapid cooling in contrast to equilibrium δ-ferrite solidification that occurs under slow cooling conditions. Nakao et al. [130] studied the effects of rapid solidification by CO_2 laser surface melting of Fe-Cr-Ni ternary alloys. They found rod-like eutectic microstructures that first increased and then decreased with increasing cooling rates. The so-called "massively solidified structures" were formed when the cooling rate exceeded a critical value, which, in turn, is markedly influenced by the chemical composition of the alloy. Microstructurally, the δ-ferrite contents were influenced by the cooling rate.

El-Batahgy [30] evaluated the fusion zone shape and solidification structure as a function of laser welding parameters. He found that the type of the fusion zone microstructure does not depend on the change in heat input, and it is always austenite, with ~2-3 vol.% ferrite. However, a finer solidification structure could be obtained by lowering the heat input.

Mohanty and Mazumder [131] observed the solidification behavior of the melt-pool during laser melting and stated that the keyhole shape influences the flow pattern in the melt-pool, which may change the microstructure characteristics. Even under constant scanning speed conditions, they observed an unsteady motion of the solid-liquid interface, resulting in fluctuation in growth rates and in thermal fields, which makes a solidified zone remelt and resolidify. This leads to discrete structural bands in the solidified bead. Using time-resolved X-ray diffraction, David et al. [132] analyzed the instabilities at the solid-liquid interface and confirmed that, on slowly cooled spot welds, the equilibrium primary solidification phase is δ-ferrite but, in rapid solidification, primary austenite was observed. Using momentum, continuity, and energy equations for incompressible, laminar, and Newtonian flow, Roy et al. [133] developed a model to simulate the temperature and velocity fields during pulsed laser welding and verified it using experimental results [134]. The computed cooling rates and weld bead dimensions were consistent with experimental results. However, the ratio of the temperature gradient to the solidification rate indicated that conditions for plane front solidification of stainless steel were not satisfied for the pulsed laser welding parameters. Therefore, these workers suggested that numerical calculations could improve understanding of solidification during pulsed laser welding.

The role of the shape of the melt-pool on the weld microstructure has been studied by Rappaz et al. [135], who created a three-dimensional reconstruction of the electron beam weld-pool shape and measured dendrite spacing as a function of growth velocity. The dendrites were found to grow parallel to three <100> crystallographic directions, which indicated that dendrites that occurred from the single crystal portion remained solid during the welding process. The weld microstructure contained dendrites that were only slightly branched and had a cell-like structure. David and Vitek [136] were the first one to observe the effect of cooling rate on the modification of microstructure from austenite + ferrite to fully austenitic structure in austenitic stainless steels. They determined that this was due to a large undercooling encountered by the liquid under rapid cooling conditions encountered during electron and laser beam welding. Here, two phenomena occur as solidification growth velocities increase: (1) partitioning of the solute between solid and liquid and (2) nonequilibrium phase formation. Kelly et al. [137] made similar observations in their study of rapid solidification of 303 stainless steel droplets and found that solute elements were more completely trapped in the bcc structures. The crystal-to-liquid nucleation temperatures showed that bcc nucleation was favored at large liquid supercooling. More recently, Siefert and David [138] studied the weldability of austenitic stainless steels and attributed changes in the microstructure to large undercooling in the liquid and partitionless solidification.

Hu and Richardson [139] evaluated the cracking behavior in welds of high-strength Al alloys and found out that cracking happens when the fusion zone is in the semisolid state and it is related to the temperature distribution, which is elongated in the welding direction. These workers confirmed that this temperature distribution during the cooling phase causes a transverse tensile strain in the fusion zone. To avoid cracking, they suggested three solutions: decrease the scanning speed in order to decrease the longitudinal strain, alter the composition in the fusion zone to improve the strength and ductility of the weld, and add a heating or cooling source to modify the thermal history of the fusion zone in the semisolid temperature range. Rai et al. [140] stated that the values of solidification parameters at the trailing edge of the melt-pool depend on the physical properties of the material, with some very influential ones being thermal diffusivity, absorption coefficient, melting temperature, and boiling temperature. Materials with a lower thermal conductivity are expected to have a fusion zone, which is spread near the top. A number of workers have combined various existing models that consider multiple beam reflections in the keyhole to calculate temperature and velocity fields, weld geometry, and solidification parameters during laser welding of tantalum, Ti-6Al-4V, 304L stainless steel, and vanadium [4, 25, 53, 56, 62, 65, 141–150]. In addition, these researchers used a turbulence model to calculate the thermal conductivity and effective viscosity in the melt-pool. They confirmed that the main mechanism of heat transfer for all four materials was convective heat transfer that depends on the thermal diffusivity and temperature coefficient of surface tension. The smallest melt-pool was observed in tantalum, a consequence of its high boiling temperature, melting temperature, and solid-state thermal diffusivity.

Ghaini et al. [151] conducted experiments to examine the influence of process parameters on the microstructure and hardness during overlap laser bead-on-plate spot welding. They defined the effective peak power density that takes into account the effect of overlapping. They presented two approaches for full-penetration welding: high peak power densities with high travel speeds that have low overlapping and medium peak power densities with medium travel speeds. In the first approach, due to the higher cooling rates and the nature of the thermal effects of the next pulse on the previous weld spot, the weld metal has high hardness and displayed large hardness variation, while opposite results were obtained when the latter approach was used. Combining these two approaches and having the optimum power density with overlapping factor enhances prediction of the weld microstructure and hardness. In a bid to understand the hot cracking phenomena in laser overlap spot welding of Al alloys, Ghaini et al. [151] investigated the interdependency of solidification cracking in the weld metal with liquation cracking in the base metal and concluded that the liquation cracks act as initiation sites for solidification cracks. However, at low laser pulse energies, liquation grain boundary cracks occur less frequently and solidification cracks initiate independently from the fusion lines between subsequent weld spots. These workers stated that cracks could only occur when the rate of induced strains was greater than the rate of backfilling.

Kadoi et al. [152] studied the influence of welding speed on solidification cracking susceptibility in laser welding of type 310S stainless steel and found that an increase in welding speed decreases the critical strain for solidification crack initiation. They suggested the reason to be the distribution morphology of the residual liquid at the weld bead center that depends on the microstructure at the rear of the melt-pool. Tan and Shin [153] presented a multiscale model of solidification and microstructure development during laser keyhole welding of austenite stainless steel. On a macroscale, a model was utilized to predict the fluid flow, thermal history, and solidification conditions of the melt-pool, which is influenced by the welding speed. The mesoscale model was used to predict the grain growth in welds, and the macroscale model was developed to simulate the dendrite growth. These workers observed that grain growth direction varies according to the melt-pool complex shape. The maximum temperature gradient controls the dendrite orientation, while the dendrite morphology is influenced by the cooling rate. Increasing the cooling rate reduces the spacing of the primary dendrite arms and suppresses the growth of the secondary dendrite arms. A summary of microstructural development as a function of cooling rate is presented in Table 3.

6. Weld Porosity

Porosity is especially important in Mg and Al alloys, and researchers have conducted several experiments on these two metal alloys in order to determine the porosity characteristics of the melt-pool in the keyhole laser welding process. In recent years, Mg and its alloys have gained increasing interest in industry, mainly due to their low density [155]. Furthermore, liquid Mg has a much larger solubility of hydrogen than solid Mg [156]. Therefore, hydrogen porosity is an important concern for the welding of Mg alloys [157]. Galun and Mordike [49] observed a large number of pores in welds of high-pressure die-cast alloys, such as AZ 91 and AM 60, due to escaping gas entrapped in the material during the die casting process. Through experiments, Pastor et al. [41] showed that overfill on the AM60B alloy weld was caused by the displacement of liquid metal by the pores. Therefore, any parameter that reduces porosity in the melt-pool decreases overfill. They showed that expansion of the initial pores in the base metal is the most important mechanism of porosity formation. For this alloy, Zhao and DebRoy [47] came to the same conclusion. They observed that coalescence and expansion of the initial pores, due to heating and reduction of internal pressure, play a key role in increasing the porosity in the fusion zone. They asserted that a balance between surface tension pressure and vapor pressure determines the stability

Table 3: Effect of the solidification rate on the microstructure of metals and alloys.

Cooling rate ($K{\cdot}s^{-1}$)	10^5	10^4	10^3	10^2	10^1
Microstructural features	Amorphous	Fine grains	Fine dendrites	Martensite [154]	Dendrites [132]
Comments	Metastable	Nonequilibrium	None	None	Follows equilibrium-phase diagram

of the keyhole. However, pore formation during laser welding of alloy AM60B does not depend on the keyhole instability.

The 2000, 5000, and 6000 series Al alloys are used in many automotive applications, such as body panels, because of the combination of high specific strength, good crashworthiness, and excellent corrosion resistance [158]. These attributes make laser beam welding an attractive joining process for such applications [159, 160]. However, porosity, hot cracking, and weld metal composition change are major concerns in the welding of Al alloys [161]. The formation of the keyhole leads to a deep penetration weld, and a hole created in a liquid is unstable by its nature, causing the formation of porosity in the weld metal. Since porosity is one of the serious problems in very high-power laser welding, Matsunawa et al. [5] observed that, in pulsed laser spot welding of Al alloys, the keyhole opening collapses within one-tenth of the time that the melt-pool solidifies and a large cavity forms at the bottom of the keyhole. Fluctuation of the keyhole opening was less unstable in continuous-wave laser than that in pulsed laser. However, the shape and the size of the melt-pool change with time. By observing the keyhole using optical and X-ray methods, they found that a deep depression is formed on the rear wall of the keyhole, moving from the top to the bottom periodically. They also observed a large bubble in the melt-pool, resulting in the formation of pores. The bubble is composed of evaporated metal vapor and entrained shielding gas. These workers observed two types of porosity in laser-welded parts: porosity induced by hydrogen and a large cavity caused by the fluctuation of the keyhole by intense evaporation of the metal. They also found two effective methods for reducing porosity in Al alloys: use of a low-dew point shielding gas below 250°C and removal of the oxide layer from the surface. The cavity formation in pulsed laser spot welding can be suppressed by adding a proper tailing pulse to avoid collapse of the keyhole opening. In continuous-wave laser welding of Al alloys, Matsunawa et al. [5] found N_2 shielding to be effective in suppressing large pores. This is because of the formation of aluminum nitride (AlN) on the liquid surface, which suppresses the perturbation of both the melt-pool and the keyhole. Moreover, entrained N_2 in the keyhole is consumed, forming AlN; therefore, the number of shielding gas-filled pores is reduced.

Mizutani et al. [51] irradiated a laser beam to the surface of a solid metal and to an already molten metal and observed that the keyhole initiates much earlier in the molten metal than it does in the solid metal. They presented a simplified numerical calculation demonstrating that the formation of bubbles is influenced by surface tension. They showed that the deepest location of the keyhole tends to collapse more easily. Therefore, formation of the bubbles in deep and narrow keyholes is expected. Courtois et al. [162] confirmed that, in pulsed laser welding with a high laser power, when the laser beam is not being applied, the keyhole wall collapses and entraps some gas, creating bubbles, which, in turn, lead to pores. In addition to resolidification microstructures, defects, such as voids, form. Kim and Weinman [163] irradiated samples of 2024-T3-51 Al with a pulsed Nd-glass 1.06 μm laser at an incident energy density of 440 $J{\cdot}cm^{-2}$, with and without a protective helium gas flow over the surface. A cooling rate of 105 to 107 $K{\cdot}s^{-1}$ was estimated. They found that many elongated and small voids formed at the melt-matrix interface due to a combination of shrinkage and gas expulsion and that void presence reduced fracture resistance. These authors determined that gas ejection in the melt affected dendrites' growth patterns.

In a computational fluid dynamics study, Zhao et al. [164] considered the existence of the three phases of the material and employed continuity, energy, and momentum equations. They extrapolated the material properties for high temperatures and assumed the fluid to be Newtonian and the flow to be laminar and incompressible. They reported the main cause of the porosity defect to be the oscillation of the keyhole depth, while the depth of the melt-pool is steady. The keyhole oscillates due to the opposition of the dynamic forces and the melt flow. Courtois et al. [165] confirmed the findings by Zhao et al. by calculating the laser reflections in the keyhole during laser welding. They used Maxwell equations, coupled with continuity, energy, and momentum equations, to develop a model for calculating the laser reflections in the process. Moreover, they showed that the shear stress at the keyhole surface has a marked influence on the melt-pool dynamics. Cho et al. [166] simplified the laser welding process by assuming a void region for the region of gas or plasma. They modified the laser beam model they used in their previous work [167], in which an infinitesimal point was considered as the focal point on the surface. In the later model, the focal point was calculated and the reflections were taken into account. Through the use of mass, energy conservation, and the Navier–Stokes equations, they considered buoyancy and Marangoni forces as well as recoil pressure. They confirmed that using a beam with a Gaussian profile could lead to reliable results, and they observed that consideration of the shear stress on the keyhole wall, which is generated by the metal vapor, does not play a significant role in the shape of the HAZ.

7. Summary

The present work is a state-of-the-art literature review on the properties of the melt-pool in laser welding and the relationship between welding process parameters and melt-pool characteristics. The characteristics considered were geometry, thermodynamics, fluid dynamics, microstructure, and porosity. Furthermore, the optimum laser welding parameters for a selection of metals and alloys are presented in this review. Several experimental studies have been conducted on melt-pool characterization in laser welding. However, direct experimental observation of melt-pool characteristics remains a challenge because of the high temperatures in the melt-pool and the difficulty of monitoring the metal vapor in the keyhole. Thus, there is scope for developing more sophisticated experimental techniques. A number of models, having varying degrees of sophistication, have been used. Four common shortcomings of many of these models are identified. First, simplifications were used; for instance, the temperature dependence of the thermophysical properties of materials is either neglected or extrapolated for high temperatures. Second, the influence of consideration of the three heat transfer modes, namely, conduction, convection, and radiation, in both the radial and the axial directions in the melt-pool, has received little attention. Third, fluid flow in the melt-pool is considered incompressible and laminar. Fourth, the agreement between model and experimental results is not very good. These observations suggest a number of areas for future study. For example, models may be improved by taking into account the compressibility of the vapor in the keyhole and the turbulence of the fluid flow in the melt-pool. In terms of models, multiscale models, which integrate nanostructures and microstructures of materials with multiphysics macroscale models, are needed. Additionally, more experimental results are needed on a wide collection of alloys and welding parameters, yielding results that would enhance verification and validation of models.

Appendix

Several experimental and modeling studies have been performed to understand the influence of process parameters on melt-pool and keyhole features. This appendix contains summaries of a number of these studies.

Chande and Mazumder [7] evaluated the influence of process parameters on the melt-pool shape and cooling rates. They used a finite difference model for the heat source and assumed a quasi-steady state model and observed that when a surface reflectivity is very high, there is no melting; however, as surface melting occurs, surface reflectivity variation has no influence on other process parameters. Therefore, in their model, at the temperatures higher than the melting temperature, the surface reflectivity is considered zero. They concluded that the depth of the penetration is more affected than the width of the weld by the absorption of laser energy in the keyhole. Lankalapalli et al. [31] presented a two-dimensional heat conduction model (heat conduction in the axial direction is neglected) to estimate the dependence of penetration depth on process parameters. They obtained the depth of penetration by equating the conducted heat in the substrate to the absorbed laser power.

The interdependency between the melt-pool and the keyhole has been investigated. Ducharme et al. [42] presented an integrated model of laser welding, taking into account the conditions in the keyhole as well as in the melt-pool, interactively. These authors investigated the influence of process parameters on the melt-pool shape. Whether penetration was full, partial, or blind, the melt-pool shape was different. However, their model is applicable only for full penetration. They concluded that a change in process parameters has more influence on the length of the melt-pool than on its width or shape.

To simulate the laser penetration welding process, Sudnik et al. [150] considered the keyhole, the melt-pool, and the solid substrate as a single nonlinear thermodynamic continuum and divided the whole process to submodels for laser beam, plasma formation, radiation absorption, vapor channel, melt-pool, and solid substrate. This allowed them to calculate the keyhole and melt-pool geometries and temperature distribution, as well as energy losses due to, for example, reflection, vaporization, and radiation. In a later contribution, Sudnik et al. [32] enhanced this model by suggesting a correlation between the depth and the length of the melt-pool. They added the consideration of heat transport due to the moving flow in the radial direction. In the case of a constant welding speed with a varying laser power, they suggested a linear correlation between the depth and the length of the melt-pool. These authors also investigated laser welding of overlap joints [168] and suggested a low welding speed in cases of larger gap widths so that there would be time for the heat to expand through the gap more uniformly.

Butt welding is a technique used to connect parts that are nearly parallel and do not overlap. Benyounis et al. [12] investigated laser butt welding and developed linear and quadratic-fitted polynomial equations for predicting the heat input and the weld bead geometry. They asserted that achieving the maximum penetration is possible by using the maximum laser power with a focused beam while the welding speed is minimum. They confirmed that the most important factors affecting the welded zone width are the welding speed and the laser focal point position. Shanmugam et al. [33] carried out experiments in which they obtained excellent weld bead geometry by selecting an effective combination of input parameters and radiating the laser beam with different angles to the specimen surface.

To better understand the behavior of steel during the welding process, Mei et al. [34] constructed a setup to avoid most of the defects such as pores and cracks in the HAZ by optimizing the process parameters. They also determined various mechanical properties of the alloy and the welded joints. Based on the results of these tests, they confirmed that both the yield strength and the tensile strength of the welded joints are higher than those of the base metal. They stated that, by moving the focal point position down to the depth, melt-pool depth increases at first and then decreases. To understand the effects of laser power, welding speed, and fiber diameter on bead geometry and mechanical properties of the weld, Khan et al. [35] conducted an experimental investigation of laser beam welding in a constrained overlap configuration. They found that welding speed and laser power are the most significant factors that influence weld bead geometry. By increasing the energy density input, the bead profile shape changes from conical to cylindrical. In another study, Khan et al. [169] presented an experimental design approach to process parameter optimization. They developed a set of mathematical models to obtain the graphical optimization of the results and thus the optimal parameters.

The low density, excellent high-temperature mechanical properties, and good corrosion resistance of Ti and its alloys have led to successful applications of these materials in a variety of fields, such as the medical, aerospace, automotive, petrochemical, nuclear, and power generation industries [50, 170]. Fusion welding of Ti has been performed principally using inert gas-shielded arc and high-energy beam welding processes. Laser welding of Ti-6Al-4V alloy is widely used in aerospace and other applications. Casalino et al. [50] investigated laser welding of Ti-6Al-4V alloy using either lap or butt configurations and obtained the process parameters that lead to welds with the minimal number of imperfections. A pulsed and continuous-wave mode laser has been used to weld Ti alloys. In pulse-mode laser welding, the most important parameter affecting the penetration depth is the peak power of the pulsed laser [36]. If it is too high, it creates vapors on the surface of the material, preventing the laser beam from reaching the material, and the penetration depth remains constant. Therefore, for increasing the penetration depth while preventing creation of vapor craters, the peak power should be kept constant and the pulse duration should be increased. These researchers illustrated the relationship between peak power, HAZ width, and melt-pool width: the higher the peak power, the higher the transfer of heat energy to the keyhole walls and the higher the proportion between the HAZ width and the melt-pool width [36]. In order to determine the influence of the heat input on the quality of the welded joint, Quan et al. [37] carried out experiments and showed that, by increasing the heat input, the widths at the top and at the bottom of the weld become equal and more craters and pores are created. Combining various models and concepts, such as multiple reflections of the laser beam in the keyhole, Al-Kazzaz et al. [38] calculated the geometry of the keyhole and weld profiles as well as the temperature gradient in the melt-pool.

Shercliff and Ashby [171] developed a model involving both Gaussian and non-Gaussian heat sources. This enhanced model is applicable for all practical beam speeds. They also presented process diagrams, in a combined form called a "Master Diagram," for rectangular heat sources so that process variables could be selected to achieve optimum results. Steen et al. [172] presented a simple relation between the penetration depth and the process parameters using a one-dimensional conduction balance without radiative heat transfer. They assumed that the material properties are constant for all temperature ranges and claimed that the state of the convection in the melt-pool does not affect the prediction of the depth of the pool. Ahmed et al. [39] used three heat sources to investigate the effect of heat source on the melt-pool shape in laser welding of Inconel 625. The heat sources were a single circular Gaussian beam and two superimposed multiple Gaussian heat sources forming a rectangular beam and one made up of three laser beams and the other of ten beams. The melt-pool profiles modeled using rectangular beams agreed with the experimental results, considering the dependence on the scanning speed. These profiles have a top-hat shape at higher speeds and a crescent shape at lower speeds, as is seen experimentally.

Chan et al. [24] used nondimensional forms of the energy, continuity, and momentum equations and found the highest fluid velocity and the solidification start position at the edge of the beam due to the maximum temperature gradient at this point. The width-to-depth ratio of the melt-pool increases with the increase in Pr. The increase of this ratio with the increase of the surface tension number was not uniform; specifically, it increased up to surface tension number of 55,000 and then it decreased. To evaluate the shape of the melt-pool, Sonti and Amateau [173] solved a nonlinear heat conduction model using FEA and calculated the temperature distribution in the melt-pool. The results were comparable to the results of the experiments that Sonti [174] had carried out to evaluate the influence of the process parameters on laser welding of Al alloys.

Conflicts of Interest

The authors declare that there are no conflicts of interest.

Acknowledgments

The authors thank the FedEx Institute of Technology, The University of Memphis, Memphis, TN, USA, for partial funding of this work under the DRONES cluster.

References

[1] Lasers—high-power and low-power," *Optics & Laser Technology*, vol. 3, no. 3, pp. 181–185, 1971.

[2] J. Mazumder, "Laser welding: state of the art review," *JOM*, vol. 34, no. 7, pp. 16–24, 1982.

[3] M. S. Brown and C. B. Arnold, "Fundamentals of laser-material interaction and application to multiscale surface modification," *Laser Precision Microfabrication. Springer Series in Materials Science*, vol. 135, pp. 91–120, Springer, Berlin, Germany, 2010.

[4] J. Mazumder and W. M. Steen, "Heat transfer model for cw laser material processing," *Journal of Applied Physics*, vol. 51, no. 2, pp. 941–947, 1980.

[5] A. Matsunawa, J. D. Kim, N. Seto, and M. Mizutani, "Dynamics of keyhole and molten pool in laser welding," *Journal of Laser Applications*, vol. 10, no. 6, pp. 247–254, 1998.

[6] P. Okon, G. Dearden, K. Watkins, M. Sharp, and P. French, "Laser welding of aluminium alloy 5083," in *Proceedings of the 21st International Congress Applications of Lasers Electro-Optics*, vol. 53, pp. 1689–1699, Scottsdale, AZ, USA, October 2002.

[7] T. Chande and J. Mazumder, "Estimating effects of processing conditions and variable properties upon pool shape, cooling rates, and absorption coefficient in laser welding," *Journal of Applied Physics*, vol. 56, no. 7, pp. 1981–1986, 1984.

[8] T. Zacharia, S. A. David, J. M. Vitek, and T. Debroy, "Heat transfer during Nd-Yag pulsed laser welding and its effect on solidification structure of austenitic stainless steels," *Metallurgical Transactions A*, vol. 20, no. 5, pp. 957–967, 1989.

[9] L. X. Yang, X. F. Peng, and B. X. Wang, "Numerical modeling and experimental investigation on the characteristics of molten pool during laser processing," *International Journal of Heat and Mass Transfer*, vol. 44, no. 23, pp. 4465–4473, 2001.

[10] K. Abderrazak, S. Bannour, H. Mhiri, G. Lepalec, and M. Autric, "Numerical and experimental study of molten pool formation during continuous laser welding of AZ91 magnesium alloy," *Computational Materials Science*, vol. 44, no. 3, pp. 858–866, 2009.

[11] M. Dhahri, J. E. Masse, J. F. Mathieu, G. Barreau, and M. Autric, "Laser welding of AZ91 and WE43 magnesium alloys for automotive and aerospace industries," *Advanced Engineering Materials*, vol. 3, no. 7, pp. 504–507, 2001.

[12] K. Y. Benyounis, A. G. Olabi, and M. S. J. Hashmi, "Effect of laser welding parameters on the heat input and weld-bead profile," *Journal of Materials Processing Technology*, vol. 164-165, pp. 978–985, 2005.

[13] R. Morgan, C. J. Sutcliffe, and W. O'neill, "Density analysis of direct metal laser re-melted 316L stainless steel cubic primitives," *Journal of Materials Science*, vol. 39, no. 4, pp. 1195–1205, 2004.

[14] W. M. Steen, *Laser Material Processing*, Springer Verlag, London, UK, 1st edition, 1991.

[15] E. Locke, E. D. Hoag, and R. Hella, "Deep penetration welding with high-power CO_2 lasers, quantum electron," *IEEE Journal of Quantum Electronics*, vol. 8, no. 2, pp. 132–135, 1972.

[16] J. Mazumder, "An overview of melt dynamics in laser processing," in *International Society for Optics and Photonics*, E.-W. Kreutz, A. Quenzer, and D. Schuoecker, Eds., pp. 228–241, 1987.

[17] S. A. David and J. M. Vitek, "Correlation between solidification parameters and weld microstructures," *International Materials Reviews*, vol. 34, no. 1, pp. 213–245, 1989.

[18] X. Cao, W. Wallace, and C. Poon, "Research and progress in laser welding of wrought aluminum alloys. I. Laser welding processes," *Materials and Manufacturing Processes*, vol. 18, no. 1, pp. 1–22, 2003.

[19] X. Cao, W. Wallace, and J. P. Immarigeon, "Research and progress in laser welding of wrought aluminum alloys. II. Metallurgical microstructures, defects, and mechanical properties," *Materials and Manufacturing Processes*, vol. 18, no. 1, pp. 23–49, 2003.

[20] J. Shao and Y. Yan, "Review of techniques for on-line monitoring and inspection of laser welding," *Journal of Physics: Conference Series*, vol. 15, pp. 101–107, 2005.

[21] X. Cao, M. Jahazi, J. P. Immarigeon, and W. Wallace, "A review of laser welding techniques for magnesium alloys," *Journal of Materials Processing Technology*, vol. 171, no. 2, pp. 188–204, 2006.

[22] L. Liu, D. Ren, and F. Liu, "A review of dissimilar welding techniques for magnesium alloys to aluminum alloys," *Materials*, vol. 7, no. 5, pp. 3735–3757, 2014.

[23] W. M. Steen, "Laser surface treatment," in *Laser Material Processing*, pp. 172–219, Springer, London, UK, 1991.

[24] C. Chan, J. Mazumder, and M. M. Chen, "A two-dimensional transient model for convection in laser melted pool," *Metallurgical Transactions A*, vol. 15, no. 12, pp. 2175–2184, 1984.

[25] V. V. Semak and A. Matsunawa, "The role of recoil pressure in energy balance during laser materials processing," *Journal of Physics D: Applied Physics*, vol. 30, no. 18, pp. 2541–2552, 1997.

[26] A. P. Mackwood and R. C. Crafer, "Thermal modelling of laser welding and related processes: a literature review," *Optics & Laser Technology*, vol. 37, no. 2, pp. 99–115, 2005.

[27] A. Yaghi and A. A. Becker, "Weld Simulation using Finite Element Methods," in *Proceedings of the NAFEMS World Congress*, St Julians, Malta, May 2005.

[28] X. He, "Finite element analysis of laser welding: a state of art review," *Materials and Manufacturing Processes*, vol. 27, no. 12, pp. 1354–1365, 2012.

[29] J. Svenungsson, I. Choquet, and A. F. H. Kaplan, "Laser welding process–a review of keyhole welding modelling," *Physics Procedia*, vol. 78, pp. 182–191, 2015.

[30] A. M. El-Batahgy, "Effect of laser welding parameters on fusion zone shape and solidification structure of austenitic stainless steels," *Materials Letters*, vol. 32, no. 2-3, pp. 155–163, 1997.

[31] K. N. Lankalapalli, J. F. Tu, and M. Gartner, "A model for estimating penetration depth of laser welding processes," *Journal of Physics D: Applied Physics*, vol. 29, no. 7, pp. 1831–1841, 1996.

[32] W. Sudnik, D. Radaj, S. Breitschwerdt, and W. Erofeew, "Numerical simulation of weld pool geometry in laser beam welding," *Journal of Physics D: Applied Physics*, vol. 33, no. 6, p. 662, 2000.

[33] S. Shanmugam, "Numerical and experimental investigation of laser beam welding of AISI 304 stainless steel sheet," *Advances in Production Engineering & Management*, vol. 3, pp. 93–105, 2008.

[34] L. Mei, G. Chen, X. Jin, Y. Zhang, and Q. Wu, "Research on laser welding of high-strength galvanized automobile steel sheets," *Optics and Lasers in Engineering*, vol. 47, no. 11, pp. 1117–1124, 2009.

[35] M. M. A. Khan, L. Romoli, M. Fiaschi, F. Sarri, and G. Dini, "Experimental investigation on laser beam welding of martensitic stainless steels in a constrained overlap joint configuration," *Journal of Materials Processing Technology*, vol. 210, no. 10, pp. 1340–1353, 2010.

[36] E. Akman, A. Demir, T. Canel, and T. Sinmazçelik, "Laser welding of Ti6Al4V titanium alloys," *Journal of Materials Processing Technology*, vol. 209, no. 8, pp. 3705–3713, 2009.

[37] Y. J. Quan, Z. H. Chen, X. S. Gong, and Z. H. Yu, "Effects of heat input on microstructure and tensile properties of laser welded magnesium alloy AZ31," *Materials Characterization*, vol. 59, no. 10, pp. 1491–1497, 2008.

[38] H. Al Kazzaz, M. Medraj, X. Cao, and M. Jahazi, "Nd:YAG laser welding of aerospace grade ZE41A magnesium alloy: modeling and experimental investigations," *Materials Chemistry and Physics*, vol. 109, no. 1, pp. 61–76, 2008.

[39] N. Ahmed, K. T. Voisey, and D. G. McCartney, "Investigation into the effect of beam shape on melt pool characteristics using analytical modelling," *Optics and Lasers in Engineering*, vol. 48, no. 5, pp. 548–554, 2010.

[40] R. Fabbro, S. Slimani, I. Doudet, F. Coste, and F. Briand, "Experimental study of the dynamical coupling between the induced vapour plume and the melt pool for Nd–Yag CW laser welding," *Journal of Physics D: Applied Physics*, vol. 39, no. 2, pp. 394–400, 2006.

[41] M. Pastor, H. Zhao, and T. DebRoy, "Continuous wave-Nd: yttrium–aluminium–garnet laser welding of AM60B magnesium alloys," *Journal of Laser Applications*, vol. 12, no. 3, pp. 91–100, 2000.

[42] R. Ducharme, K. Williamss, P. Kapadia, J. Dowden, B. Steent, and M. Glowackit, "The laser welding of thin metal sheets: an integrated keyhole and weld pool model with supporting experiments," *Journal of Physics D: Applied Physics*, vol. 27, no. 8, pp. 1619–1627, 1994.

[43] C. Limmaneevichitr and S. Kou, "Experiments to simulate effect of Marangoni convection on weld pool shape," *Welding Journal*, vol. 79, pp. 231s–237s, 2000.

[44] P. A. A. Khan and T. Debroy, "Alloying element vaporization and weld pool temperature during laser welding of AlSl 202 stainless steel," *Metallurgical Transactions B*, vol. 15, no. 4, pp. 641–644, 1984.

[45] H. Ki, J. Mazumder, and P. S. Mohanty, "Modeling of laser keyhole welding: part II. simulation of keyhole evolution, velocity, temperature profile, and experimental verification," *Metallurgical and Materials Transactions A*, vol. 33, no. 6, pp. 1831–1842, 2002.

[46] H. Zhao and T. Debroy, "Weld metal composition change during conduction mode laser welding of aluminum alloy 5182," *Metallurgical and Materials Transactions B*, vol. 32, no. 1, pp. 163–172, 2001.

[47] H. Zhao and T. DebRoy, "Pore formation during laser beam welding of die-cast magnesium alloy AM60B-mechanism and remedy," *Welding Journal*, vol. 80, pp. 204–210, 2001.

[48] S. Postma, R. G. K. M. Aarts, J. Meijer, and J. B. Jonker, "Penetration control in laser welding of sheet metal," *Journal of Laser Applications*, vol. 14, no. 4, p. 210, 2002.

[49] A. Weisheit, R. Galun, and B. L. Mordike, "CO_2 laser beam welding of magnesium-based alloys," *Welding Journal*, vol. 77, pp. 149–154, 1998.

[50] G. Casalino, F. Curcio, and F. M. C. Minutolo, "Investigation on Ti6Al4V laser welding using statistical and Taguchi approaches," *Journal of Materials Processing Technology*, vol. 167, no. 2-3, pp. 422–428, 2005.

[51] M. Mizutani, S. Katayama, and A. Matsunawa, "Observation of molten metal behavior during laser irradiation—basic experiment to understand laser welding phenomena," in *Proceedings of the First International Symposium on High-Power Laser Macroprocessing*, pp. 208–213, Osaka, Japan, May 2002.

[52] M. F. Ashby and K. E. Easterling, "The transformation hardening of steel surfaces by laser beams-I. Hypo-eutectoid steels," *Acta Metallurgica*, vol. 32, no. 11, pp. 1935–1948, 1984.

[53] J. Dowden, N. Postacioglu, M. Davis, and P. Kapadia, "A keyhole model in penetration welding with a laser," *Journal of Physics D: Applied Physics*, vol. 20, no. 1, pp. 36–44, 1987.

[54] W. M. Steen, J. Dowden, M. Davis, and P. Kapadia, "A point and line source model of laser keyhole welding," *Journal of Physics D: Applied Physics*, vol. 21, no. 8, pp. 1255–1260, 1988.

[55] M. Beck, P. Berger, F. Dausinger, and H. Hugel, "Aspects of keyhole/melt interaction in high speed laser welding," in *Proceedings of the 8th International Symposium of Gas Flow Chemical Lasers*, pp. 769–774, Madrid, Spain, February 1991.

[56] A. Kaplan, "A model of deep penetration laser welding based on calculation of the keyhole profile," *Journal of Physics D: Applied Physics*, vol. 27, no. 9, pp. 1805–1814, 1994.

[57] C. Lampa, A. F. H. Kaplan, and J. Powell, "An analytical thermodynamic model of laser welding," *Journal of Physics D: Applied Physics*, vol. 30, no. 9, pp. 1293–1299, 1997.

[58] E. H. Amara and A. Bendib, "Modelling of vapour flow in deep penetration laser welding," *Journal of Physics D: Applied Physics*, vol. 35, no. 3, pp. 272–280, 2002.

[59] R. Fabbro, F. Coste, S. Slimani, and F. Briand, "Study of keyhole geometry for full penetration Nd-Yag CW laser welding," *Journal of Physics D: Applied Physics*, vol. 38, no. 12, pp. 1881–1887, 2005.

[60] F. Tenner, C. Brock, F. Klampfl, and M. Schmidt, "Analysis of the correlation between plasma plume and keyhole behavior in laser metal welding for the modeling of the keyhole geometry," *Optics and Lasers in Engineering*, vol. 64, pp. 32–41, 2015.

[61] R. Fabbro, "Melt pool and keyhole behaviour analysis for deep penetration laser welding," *Journal of Physics D: Applied Physics*, vol. 43, no. 44, p. 445501, 2010.

[62] N. Postacioglu, P. Kapadia, and M. Davis, "Upwelling in the liquid region surrounding the keyhole in penetration welding with a laser," *Journal of Physics D: Applied Physics*, vol. 20, no. 3, pp. 340–345, 1987.

[63] A. Kar and J. Mazumder, "Mathematical modeling of keyhole laser welding," *Journal of Applied Physics*, vol. 78, no. 11, pp. 6353–6360, 1995.

[64] X. Ye and X. Chen, "Three-dimensional modelling of heat transfer and fluid flow in laser full-penetration welding," *Journal of Physics D: Applied Physics*, vol. 35, no. 10, pp. 1049–1056, 2002.

[65] J. Goldak, A. Chakravarti, and M. Bibby, "New finite element model for welding heat sources," *Metallurgical Transactions B*, vol. 15, no. 2, pp. 299–305, 1984.

[66] R. Wang, Y. Lei, and Y. Shi, "Numerical simulation of transient temperature field during laser keyhole welding of 304 stainless steel sheet," *Optics & Laser Technology*, vol. 43, no. 4, pp. 870–873, 2011.

[67] C. W. Hirt and B. D. Nichols, "Volume of fluid (VOF) method for the dynamics of free boundaries," *Journal of Computational Physics*, vol. 39, no. 1, pp. 201–225, 1981.

[68] M. Akbari, S. Saedodin, D. Toghraie, R. Shoja Razavi, and F. Kowsari, "Experimental and numerical investigation of temperature distribution and melt pool geometry during pulsed laser welding of Ti6Al4V alloy," *Optics & Laser Technology*, vol. 59, pp. 52–59, 2014.

[69] M. R. Frewin and D. A. Scott, "Finite element model of pulsed laser welding," *Welding Journal*, vol. 78, pp. 15s–22s, 1999.

[70] A. De, S. K. Maiti, C. A. Walsh, and H. K. D. H. Bhadeshia, "Finite element simulation of laser spot welding," *Science and Technology of Welding and Joining*, vol. 8, no. 5, pp. 377–384, 2003.

[71] M. M. Collur, A. Paul, and T. Debroy, "Mechanism of alloying element vaporization during laser welding," *Metallurgical Transactions B*, vol. 18, no. 4, pp. 733–740, 1987.

[72] N. Chakraborty and S. Chakraborty, "Modelling of turbulent molten pool convection in laser welding of a copper-nickel dissimilar couple," *International Journal of Heat and Mass Transfer*, vol. 50, no. 9-10, pp. 1805–1822, 2007.

[73] M. C. Tsai and S. Kou, "Marangoni convection in weld pools with a free surface," *International Journal for Numerical Methods in Fluids*, vol. 9, no. 12, pp. 1503–1516, 1989.

[74] S. Safdar, A. J. Pinkerton, L. Li, M. A. Sheikh, and P. J. Withers, "An anisotropic enhanced thermal conductivity approach for modelling laser melt pools for Ni-base super alloys," *Applied Mathematical Modelling*, vol. 37, no. 3, pp. 1187–1195, 2013.

[75] C. R. Heiple, "Mechanism for minor element effect on GTA fusion zone geometry," *Welding Journal*, vol. 61, pp. 975–1025, 1982.

[76] G. Tsotridis, H. Rother, and E. D. Hondros, "Marangoni flow and the shapes of laser-melted pools," *The Science of Nature*, vol. 76, no. 5, pp. 216–218, 1989.

[77] Y. Cheng, X. Jin, S. Li, and L. Zeng, "Fresnel absorption and inverse bremsstrahlung absorption in an actual 3D keyhole during deep penetration CO_2 laser welding of aluminum 6016," *Optics & Laser Technology*, vol. 44, no. 5, pp. 1426–1436, 2012.

[78] L. Schlessinger and J. Wright, "Inverse-bremsstrahlung absorption rate in an intense laser field," *Physical Review A*, vol. 20, no. 5, pp. 1934–1945, 1979.

[79] G. J. Pert, "Inverse bremsstrahlung absorption in large radiation fields during binary collisions-classical theory. II(b). Summed rate coefficients for Coulomb collisions," *Journal of Physics A: Mathematical and General*, vol. 9, no. 10, pp. 1797–1800, 1976.

[80] Y. Zhang, G. Chen, H. Wei, and J. Zhang, "A novel "sandwich" method for observation of the keyhole in deep penetration laser welding," *Optics and Lasers in Engineering*, vol. 46, no. 2, pp. 133–139, 2008.

[81] W. Tan, N. S. Bailey, and Y. C. Shin, "Investigation of keyhole plume and molten pool based on a three-dimensional dynamic model with sharp interface formulation," *Journal of Physics D: Applied Physics*, vol. 46, no. 5, p. 55501, 2013.

[82] T. Zacharia, S. A. David, J. M. Vitek, and T. Debroy, "Weld pool development during GTA and laser beam welding of type 304 stainless steel, part II—experimental correlation," *Welding Journal*, vol. 68, pp. 510–519, 1989.

[83] C. R. Heiple and J. R. Roper, "Effect of selenium on GTAW fusion zone geometry. Small additions of selenium to a stainless steel dramatically increase the depth/width ratio of GTA welds," *Welding Journal*, vol. 60, p. 143, 1981.

[84] S. A. David and T. Debroy, "Current issues and problems in welding science," *Science*, vol. 257, no. 5069, pp. 497–502, 1992.

[85] V. V. Semak, J. A. Hopkins, M. H. Mccay, and T. D. McCay, "Melt pool dynamics during laser welding," *Journal of Physics D: Applied Physics*, vol. 28, no. 12, pp. 2443–2450, 1995.

[86] V. V. Semak, G. A. Knorovsky, and D. O. MacCallum, "On the possibility of microwelding with laser beams," *Journal of Physics D: Applied Physics*, vol. 36, no. 17, p. 2170, 2003.

[87] V. V. Semak, G. A. Knorovsky, D. O. MacCallum, and R. A. Roach, "Effect of surface tension on melt pool dynamics during laser pulse interaction," *Journal of Physics D: Applied Physics*, vol. 39, no. 3, pp. 590–595, 2006.

[88] J.-H. Cho, D. F. Farson, J. O. Milewski, and K. J. Hollis, "Weld pool flows during initial stages of keyhole formation in laser welding," *Journal of Physics D: Applied Physics*, vol. 42, no. 17, p. 175502, 2009.

[89] M. Zhang, G. Chen, Y. Zhou, and S. Li, "Direct observation of keyhole characteristics in deep penetration laser welding with a 10 kW fiber laser," *Optics Express*, vol. 21, no. 17, pp. 19997–20004, 2013.

[90] M. Geiger, K. H. Leitz, H. Koch, and A. Otto, "A 3D transient model of keyhole and melt pool dynamics in laser beam welding applied to the joining of zinc coated sheets," *Production Engineering*, vol. 3, no. 2, pp. 127–136, 2009.

[91] R. Fabbro and K. Chouf, "Dynamical description of the keyhole in deep penetration laser welding," *Journal of Laser Applications*, vol. 12, no. 4, pp. 142–148, 2000.

[92] S. Basu and T. DebRoy, "Liquid metal expulsion during laser irradiation," *Journal of Applied Physics*, vol. 72, no. 8, pp. 3317–3322, 1992.

[93] T. Debroy and S. A. David, "Physical processes in fusion welding," *Reviews of Modern Physics*, vol. 67, no. 1, pp. 85–112, 1995.

[94] E. H. Amara and R. Fabbro, "Modelling of gas jet effect on the melt pool movements during deep penetration laser welding," *Journal of Physics D: Applied Physics*, vol. 41, no. 5, p. 55503, 2008.

[95] R. Fabbro, M. Hamadou, and F. Coste, "Metallic vapor ejection effect on melt pool dynamics in deep penetration laser welding," *Journal of Laser Applications*, vol. 16, no. 1, pp. 16–19, 2004.

[96] R. Fabbro, F. Coste, S. Slimani, and F. Briand, "Analysis of the various melt pool hydrodynamic regimes observed during Cw Nd-YAG deep penetration laser welding," in *Proceedings of the Conference on Lasers and Electro-Optics (CLEO)*, Europe, June 2007.

[97] E. H. Amara, R. Fabbro, L. Achab, F. Hamadi, and N. Mebani, "Modeling of friction on keyhole walls," in *Proceedings of the 23rd International Congress on Applications of Lasers in Electro-Optics*, San Francisco, CA, USA, 2004.

[98] E. H. Amara, R. Fabbro, and F. Hamadi, "Modeling of the melted bath movement induced by the vapor flow in deep penetration laser welding," *Journal of Laser Applications*, vol. 18, no. 1, p. 2, 2006.

[99] L. Li, D. J. Brookfield, and W. M. Steen, "Plasma charge sensor for in-process, non-contact monitoring of the laser welding process," *Measurement Science and Technology*, vol. 7, no. 4, pp. 615–626, 1996.

[100] P. Norman, H. Engström, and A. F. H. Kaplan, "Theoretical analysis of photodiode monitoring of laser welding defects by imaging combined with modelling," *Journal of Physics D: Applied Physics*, vol. 41, no. 19, 2008.

[101] T. C. Nguyen, D. C. Weckman, D. A. Johnson, and H. W. Kerr, "High speed fusion weld bead defects," *Science and Technology of Welding and Joining*, vol. 11, no. 6, pp. 618–633, 2006.

[102] W. Shimada and S. Hoshinouchi, "A study on bead formation by low pressure TIG arc and prevention of under-cut bead," *Journal of Japan Welding Society*, vol. 51, pp. 280–286, 1982.

[103] U. Gratzke, P. D. Kapadia, J. Dowden, J. Kroos, and G. Simon, "Theoretical approach to the humping phenomenon in welding processes," *Journal of Physics D: Applied Physics*, vol. 25, no. 11, pp. 1640–1647, 1992.

[104] J. Xie, "Dual beam laser welding," *Welding Journal*, vol. 81, pp. 223s–230s, 2002.

[105] M. Kern, P. Berger, and H. Hugel, "Magneto-fluid dynamic effects," *Welding Journal*, vol. 79, p. 72s, 2000.

[106] A. Matsunawa and V. V. Semak, "The simulation of front keyhole wall dynamics during laser welding," *Journal of Physics D: Applied Physics*, vol. 30, no. 5, pp. 798–809, 1997.

[107] Y. Kawahito, M. Mizutani, S. Katayama, and S. Steel, "Investigation of high-power fiber laser welding phenomena of stainless steel," *Transactions of JWRI*, vol. 36, pp. 11–16, 2007.

[108] Y. Kawahito, M. Mizutani, and S. Katayama, "High quality welding of stainless steel with 10 kW high power fibre laser," *Science and Technology of Welding and Joining*, vol. 14, no. 4, pp. 288–294, 2009.

[109] T. Ilar, I. Eriksson, J. Powell, and A. Kaplan, "Root humping in laser welding-an investigation based on high speed imaging," *Physics Procedia*, vol. 39, pp. 27–32, 2012.

[110] E. H. Amara and R. Fabbro, "Modeling of humps formation during deep-penetration laser welding," *Applied Physics A*, vol. 101, no. 1, pp. 111–116, 2010.

[111] S. Pang, L. Chen, J. Zhou, Y. Yin, and T. Chen, "A three-dimensional sharp interface model for self-consistent keyhole and weld pool dynamics in deep penetration laser welding," *Journal of Physics D: Applied Physics*, vol. 44, no. 2, p. 25301, 2010.

[112] H. Ki, J. Mazumder, and P. S. Mohanty, "Modeling of laser keyhole welding: Part I. Mathematical modeling, numerical methodology, role of recoil pressure, multiple reflections, and free surface evolution," *Metallurgical and Materials Transactions A*, vol. 33, no. 6, pp. 1817–1830, 2002.

[113] J. Charles and B. Bonnefois, "Super duplex stainless steels: properties and weldability," in *Proceedings of the Conference in Applications of Stainless Steels*, pp. 1108–1121, Stockholm, Sweden, 1992.

[114] D. J. Kotecki, "Ferrite control in duplex stainless steel weld metal," *Welding Journal*, vol. 65, pp. 273s–278s, 1986.

[115] S. Kou, *Welding Metallurgy*, John Wiley & Sons Inc., Hoboken, NJ, USA, 2003.

[116] A. Munitz, "Microstructure of rapidly solidified laser molten AI-4.5 Wt Pct Cu surfaces," *Metallurgical Transactions B*, vol. 16, no. 1, pp. 149–161, 1985.

[117] S. Narasimhan, S. Copley, E. Van Stryland, and M. Bass, "Epitaxial resolidification in laser melted superalloys," *IEEE Journal of Quantum Electronics*, vol. 13, no. 9, p. 814, 1977.

[118] A. Zambon and F. Bonollo, "Rapid solidification in laser welding of stainless steels," *Materials Science and Engineering: A*, vol. 178, pp. 203–207, 1994.

[119] J. C. Lippold and W. F. Savage, "Solidification of austenitic stainless steel weldments: part 2—the effect of alloy composition on ferrite morphology a transition in ferrite morphology occurs as a function of composition and weld cooling rate," *Welding Journal*, vol. 59, pp. 48s–58s, 1980.

[120] T. Zacharia, S. A. David, J. M. Vitek, and T. Debroy, "Weld pool development during GTA and laser beam welding of type 304 stainless steel, part I-theoretical analysis," *Welding Journal*, vol. 68, pp. 499–509, 1989.

[121] J. C. Lippold, "Solidification behavior and cracking susceptibility of pulsed-laser welds in austenitic stainless steels, a shift in solidification behavior under rapid solidification conditions promotes an increase in cracking susceptibility," *Welding Journal*, vol. 73, pp. 129–139, 1994.

[122] V. Kujanpaa, N. Suutala, T. Takalo, and T. Moisio, "Correlation between solidification cracking and microstructure in austenitic and austenitic-ferritic stainless steel welds," *Welding Journal*, vol. 9, pp. 55–75, 1979.

[123] D. J. Kotecki and T. A. Siewert, "WRC-1992 constitution diagram for stainless steel weld metals: a modification of the WRC-1988 diagram," *Welding Journal*, vol. 71, pp. 171–178, 1992.

[124] J. A. Brooks, M. I. Baskes, and F. A. Greulich, "Solidification modeling and solid-state transformations in high-energy density stainless steel welds," *Metallurgical Transactions A*, vol. 22, no. 4, pp. 915–926, 1991.

[125] S. L. Lu, M. Qian, H. P. Tang, M. Yan, J. Wang, and D. H. StJohn, "Massive transformation in Ti-6Al-4V additively manufactured by selective electron beam melting," *Acta Materialia*, vol. 104, pp. 303–311, 2016.

[126] M. J. Perricone, J. N. Dupont, T. D. Anderson, C. V. Robino, and J.R. Michael, "An investigation of the massive transformation from ferrite to austenite in laser-welded Mo-bearing stainless steels," *Metallurgical and Materials Transactions A*, vol. 42, no. 3, pp. 700–716, 2011.

[127] C. D'amato, M. Fenech, S. Abela, J. C. Betts, and J. Buhagiar, "Autogenous laser keyhole welding of AISI 316LTi," *Materials and Manufacturing Processes*, vol. 25, no. 11, pp. 1269–1277, 2010.

[128] F. Mirakhorli, F. M. Ghaini, and M. J. Torkamany, "Development of weld metal microstructures in pulsed laser welding of duplex stainless steel," *Journal of Materials Engineering and Performance*, vol. 21, no. 10, pp. 2173–2176, 2012.

[129] S. S. Babu, J. W. Elmer, J. M. Vitek, and S. A. David, "Time-resolved X-ray diffraction investigation of primary weld solidification in Fe-C-Al-Mn steel welds," *Acta Materialia*, vol. 50, no. 19, pp. 4763–4781, 2002.

[130] Y. Nakao, K. Nishimoto, and W. P. Zhang, "Effects of rapid solidification by laser surface melting on solidification modes and microstructures of stainless steels," *Transactions in Japan Welding Society*, vol. 19, pp. 100–106, 1988.

[131] P. S. Mohanty and J. Mazumder, "Solidification behavior and microstructural evolution during laser beam—material interaction," *Metallurgical and Materials Transactions B*, vol. 29, no. 6, pp. 1269–1279, 1998.

[132] S. A. David, S. S. Babu, and J. M. Vitek, "Welding: solidification and microstructure," *JOM*, vol. 55, no. 6, pp. 14–20, 2003.

[133] G. G. Roy, J. W. Elmer, and T. DebRoy, "Mathematical modeling of heat transfer, fluid flow, and solidification during linear welding with a pulsed laser beam," *Journal of Applied Physics*, vol. 100, no. 3, p. 34903, 2006.

[134] Y. F. Tzeng, "Effects of operating parameters on surface quality for the pulsed laser welding of zinc-coated steel," *Journal of Materials Processing Technology*, vol. 100, no. 1–3, pp. 163–170, 2000.

[135] M. Rappaz, S. A. David, J. M. Vitek, and L. A. Boatner, "Development of microstructures in Fe−15Ni−15Cr single crystal electron beam welds," *Metallurgical Transactions A*, vol. 20, no. 6, pp. 1125–1138, 1989.

[136] S. A. David and J. M. Vitek, "Laser in metallurgy," in *Proceedings of Metallurgical Society of AIME*, Chicago, IL, USA, February 1981.

[137] T. F. Kelly, M. Cohen, and J. B. Sande, "Rapid solidification of a droplet-processed stainless steel," *Metallurgical Transactions A*, vol. 15, no. 5, pp. 819–833, 1984.

[138] J. A. Siefert and S. A. David, "Weldability and weld performance of candidate austenitic alloys for advanced ultra-supercritical fossil power plants," *Science and Technology of Welding and Joining*, vol. 19, no. 4, pp. 271–294, 2014.

[139] B. Hu and I. M. Richardson, "Mechanism and possible solution for transverse solidification cracking in laser welding of high strength aluminium alloys," *Materials Science and Engineering: A*, vol. 429, no. 1-2, pp. 287–294, 2006.

[140] R. Rai, J. W. Elmer, T. A. Palmer, and T. DebRoy, "Heat transfer and fluid flow during keyhole mode laser welding of tantalum, Ti–6Al–4V, 304L stainless steel and vanadium," *Journal of Physics D: Applied Physics*, vol. 40, no. 18, pp. 5753–5766, 2007.

[141] D. T. Swift Hook and A. E. F. Gick, "Penetration welding with lasers," *Welding Journal*, vol. 52, no. 11, pp. 492s–499s, 1973.

[142] J. G. Andrews and D. R. Atthey, "Hydrodynamic limit to penetration of a material by a high-power beam," *Journal of Physics D: Applied Physics*, vol. 9, no. 15, pp. 2181–2194, 1976.

[143] P. G. Klemens, "Heat balance and flow conditions for electron beam and laser welding," *Journal of Applied Physics*, vol. 47, no. 5, pp. 2165–2174, 1976.

[144] J. Dowden, M. Davis, and P. Kapadia, "Some aspects of the fluid dynamics of laser welding," *Journal of Fluid Mechanics*, vol. 126, no. 1, p. 123, 1983.

[145] P. S. Wei and C. Y. Ho, "Axisymmetric nugget growth during resistance spot welding," *Journal of Heat Transfer*, vol. 112, no. 2, p. 309, 1990.

[146] J. Kroos, U. Gratzke, and G. Simon, "Towards a self-consistent model of the keyhole in penetration laser beam

welding," *Journal of Physics D: Applied Physics*, vol. 26, no. 3, pp. 474–480, 1993.
[147] H. Zhao and T. DebRoy, "Macroporosity free aluminum alloy weldments through numerical simulation of keyhole mode laser welding," *Journal of Applied Physics*, vol. 93, no. 12, pp. 10089–10096, 2003.
[148] R. Rai and T. DebRoy, "Tailoring weld geometry during keyhole mode laser welding using a genetic algorithm and a heat transfer model," *Journal of Physics D: Applied Physics*, vol. 39, no. 6, pp. 1257–1266, 2006.
[149] R. Rai, G. G. Roy, and T. DebRoy, "A computationally efficient model of convective heat transfer and solidification characteristics during keyhole mode laser welding," *Journal of Applied Physics*, vol. 101, no. 5, p. 54909, 2007.
[150] W. Sudnik, D. Radaj, and W. Erofeew, "Computerized simulation of laser beam welding, modelling and verification," *Journal of Physics D: Applied Physics*, vol. 29, no. 11, pp. 2811–2817, 1996.
[151] F. M. Ghaini, M. J. Hamedi, M. J. Torkamany, and J. Sabbaghzadeh, "Weld metal microstructural characteristics in pulsed Nd: YAG laser welding," *Scripta Materialia*, vol. 56, no. 11, pp. 955–958, 2007.
[152] K. Kadoi, A. Fujinaga, M. Yamamoto, and K. Shinozaki, "The effect of welding conditions on solidification cracking susceptibility of type 310S stainless steel during laser welding using an in-situ observation technique," *Welding in the World*, vol. 57, pp. 383–390, 2013.
[153] W. Tan and Y. C. Shin, "Multi-scale modeling of solidification and microstructure development in laser keyhole welding process for austenitic stainless steel," *Computational Materials Science*, vol. 98, pp. 446–458, 2015.
[154] S. K. Dhua, D. Mukerjee, and D. S. Sarma, "Effect of cooling rate on the as-quenched microstructure and mechanical properties of HSLA-100 steel plates," *Metallurgical and Materials Transactions A*, vol. 34, no. 11, pp. 2493–2504, 2003.
[155] D. Thomas, "Increased magnesium usage: not just due to density," *Automobile Engineering*, vol. 99, p. 47, 1991.
[156] H. Baker, *Magnesium and Magnesium Alloys*, ASM international, Materials Park, OH, USA, 1999.
[157] J. D. Shearouse and B. A. Mikucki, *The Origin of Microporosity in Magnesium Alloy A291*, SAE International, Warrendale, PA, USA, 1994.
[158] L. Griffing, "Metals and their weldability," in *Welding Handbook*, American Welding Society, Miami, FL, USA, 1972.
[159] B. Irving, "Blank welding forces automakers to sit up and take notice," *Welding Journal*, vol. 71, pp. 39–45, 1991.
[160] P. A. Hilton, *Optical and Quantum Electronics*, vol. 27, Springer, Berlin, Germany, 1995.
[161] H. Zhao, D. White, and T. DebRoy, "Current issues and problems in laser welding of automotive aluminium alloys," *International Materials Reviews*, vol. 44, no. 6, pp. 238–266, 1999.
[162] M. Courtois, M. Carin, P. Le Masson, S. Gaied, and M. Balabane, "A complete model of keyhole and melt pool dynamics to analyze instabilities and collapse during laser welding," *Journal of Laser Applications*, vol. 26, no. 4, p. 42001, 2014.
[163] C. Kim and L. S. Weinman, "Porosity formation in laser surface melted aluminum alloys," *Scripta Metallurgica*, vol. 12, no. 1, pp. 57–60, 1978.
[164] H. Zhao, W. Niu, B. Zhang, Y. Lei, M. Kodama, and T. Ishide, "Modelling of keyhole dynamics and porosity formation considering the adaptive keyhole shape and three-phase coupling during deep-penetration laser welding," *Journal of Physics D: Applied Physics*, vol. 44, no. 48, p. 485302, 2011.
[165] M. Courtois, M. Carin, P. Le Masson, S. Gaied, and M. Balabane, "A new approach to compute multi-reflections of laser beam in a keyhole for heat transfer and fluid flow modelling in laser welding," *Journal of Physics D: Applied Physics*, vol. 46, no. 50, p. 505305, 2013.
[166] W. I. Cho, S. J. Na, C. Thomy, and F. Vollertsen, "Numerical simulation of molten pool dynamics in high power disk laser welding," *Journal of Materials Processing Technology*, vol. 212, no. 1, pp. 262–275, 2012.
[167] W. I. Cho, S. J. Na, M. H. Cho, and J. S. Lee, "Numerical study of alloying element distribution in CO_2 laser-GMA hybrid welding," *Computational Materials Science*, vol. 49, no. 4, pp. 792–800, 2010.
[168] W. Sudnik, D. Radaj, and W. Erofeew, "Computerized simulation of laser beam weld formation comprising joint gaps," *Journal of Physics D: Applied Physics*, vol. 31, no. 24, pp. 3475–3480, 1998.
[169] M. M. A. Khan, L. Romoli, M. Fiaschi, G. Dini, and F. Sarri, "Experimental design approach to the process parameter optimization for laser welding of martensitic stainless steels in a constrained overlap configuration," *Optics & Laser Technology*, vol. 43, no. 1, pp. 158–172, 2011.
[170] S. H. Wang, M. D. Wei, and L. W. Tsay, "Tensile properties of LBW welds in Ti-6Al-4V alloy at evaluated temperatures below 450°C," *Materials Letters*, vol. 57, no. 12, pp. 1815–1823, 2003.
[171] H. R. Shercliff and M. F. Ashby, "The prediction of case depth in laser transformation hardening," *Metallurgical Transactions A*, vol. 22, no. 10, pp. 2459–2466, 1991.
[172] P. H. Steen, P. Ehrhard, and A. S. Ssler, "Depth of melt-pool and heat-affected zone in laser surface treatments," *Metallurgical and Materials Transactions A*, vol. 25, no. 2, pp. 427–435, 1994.
[173] N. Sonti and M. F. Amateau, "Finite-element modeling of heat flow in deep-penetration laser welds in aluminum alloys," *Numerical Heat Transfer, Part A: Applications*, vol. 16, no. 3, pp. 351–370, 1989.
[174] N. Sonti, *Influence of Process Parameters on Laser Weld Characteristics in Aluminum Alloys*, Pennsylvania State University, State College, PA, USA, 1988.

8

Synchrotron Radiation X-Ray Absorption Spectroscopy and Spectroscopic Ellipsometry Studies of InSb Thin Films on GaAs Grown by Metalorganic Chemical Vapor Deposition

Yingda Qian,[1,2] Yuanlan Liang,[1,2] Xuguang Luo,[1,2] Kaiyan He,[1,2] Wenhong Sun,[1] Hao-Hsiung Lin,[3] Devki N. Talwar,[4,5] Ting-Shan Chan,[6] Ian Ferguson,[7] Lingyu Wan,[1,2] Qingyi Yang,[1,2] and Zhe Chuan Feng[1,2]

[1]*College of Physical Science and Technology, Guangxi University, Nanning 530004, China*
[2]*Laboratory of Optoelectronic Materials and Detection Technology, Guangxi Key Laboratory for the Relativistic Astrophysics, Guangxi University, Nanning 530004, China*
[3]*Graduate Institute of Electronics and Department of Electrical Engineering, National Taiwan University, Taipei 106-17, Taiwan*
[4]*Department of Physics, Indiana University of Pennsylvania, Indiana, PA 15705-1087, USA*
[5]*Department of Physics, University of North Florida, Jacksonville, FL 32224, USA*
[6]*National Synchrotron Radiation Research Center, Hsinchu 300-76, Taiwan*
[7]*Department of Electrical and Computer Engineering, Missouri University of Science and Technology, Rolla, MO 65409, USA*

Correspondence should be addressed to Kaiyan He; gredhky@gxu.edu.cn, Wenhong Sun; youzi7002@gxu.edu.cn, and Zhe Chuan Feng; fengzc@gxu.edu.cn

Academic Editor: Si Joon Kim

A series of ultrathin InSb films grown on GaAs by low-pressure metalorganic chemical vapor deposition with different V/III ratios were investigated thoroughly using spectroscopic ellipsometry (SE), X-ray diffraction, and synchrotron radiation X-ray absorption spectroscopy. The results predicted that InSb films on GaAs grown under too high or too low V/III ratios are with poor quality, while those grown with proper V/III ratios of 4.20–4.78 possess the high crystalline quality. The temperature-dependent SE (20–300°C) and simulation showed smooth variations of SE spectra, optical constants (n, k, e_1, and ε_2), and critical energy points (E_1, $E_1 + \Delta_1$, E_0', E_2, and E_1') for InSb film when temperature increased from 20°C to 250°C, while at 300°C, large changes appeared. Our study revealed the oxidation of about two atomic layers and the formation of an indium-oxide (InO) layer of ~5.4 nm. This indicates the high temperature limitation for the use of InSb/GaAs materials, up to 250°C.

1. Introduction

As a member of the III-V compound semiconductors family, the growth of ultrathin films of indium antimonide has attracted a great deal of attention for its use in midwavelength infrared detectors (viz., thermal imaging cameras and forward looking infrared systems), magnetic sensors, magnetoresistors, field-effect transistors, photoconductors, and high-speed electronic devices [1–33]. Among other semiconductors, the intrinsic binary InSb possesses the highest electron mobility ($\sim 7.8 \times 10^4$ cm^2/V·s), higher breakdown field ($\sim 10^3$ V/cm), high saturated electron drift velocity ($\sim 5 \times 10^7$ cm/s), small effective mass ($m_e^* \sim 0.013\ m_o$), lowest energy band gap E_g ~0.18 eV (at 300 K), and ballistic length up to ~0.7 μm (at 300 K). The InSb material can be used as a lattice-matched substrate for epitaxial growth of CdTe [34] and other relevant heterostructures and superlattices [35]. The InSb-based ternary as well as quaternary alloys are equally valuable for realizing midwavelength infrared detectors [36–39]. More recently, there has been a growing challenge to explore the possibility of developing InSb-based quantum dots and nanostructured laser diodes [40–42] for optoelectronic as well as photonic device applications in the 3–5 μm wavelength range for room temperature operations.

In optoelectronic devices, one must grow ultrathin layers of InSb on semi-insulating infrared transparent substrates to prevent current leakage. While CdTe is the only semi-insulating lattice-matched substrate available for InSb growth, it is difficult, however, to avoid the formation of the In_2Te_3 precipitates at the InSb/CdTe interface [43, 44]. Hence, many alternative materials (viz., Si, GaAs, InP, sapphire, and mica) have been chosen as substrates for preparing InSb epifilms [2, 15–17, 19–32] by molecular beam epitaxy, liquid-phase epitaxy [18, 30], metalorganic chemical vapor deposition (MOCVD) [20–22, 31], metal-organic magnetron sputtering [32, 45], and two-step growth process [46] methods. Despite a large (14.6%) lattice mismatch between InSb and GaAs, the semi-insulating GaAs is considered as an attractive substrate due to high chemical stability and resistivity. Efforts have been made to grow InSb thin films on four-inch GaAs substrates [31, 32, 47]. Also many attempts made by MOCVD have shown improvements in InSb film growth on large area semi-insulating GaAs substrates [47–49]. These efforts have certainly helped provided the increasing mass production needs of isolated InSb/GaAs (001) thin films for integration with III-V semiconductor platform to achieve better performance of both the electronic and photonic devices.

However, the incompatibility of the lattice constants between the two materials strongly limits the quality of InSb epilayers grown on GaAs [23]. In general, the epifilms suffer from high density of dislocations or defects near the film/substrate interface. These defects propagating throughout the entire material create antiphase domains and cause autodoping effects. The layers with intrinsic defects may also affect the carrier mobility and leakage current in both the electronic and photonic devices. To circumvent these difficulties and improve the crystallinity of InSb films, several strategies are pursued—especially in optimizing the growth conditions by controlling the V/III ratios, pressure, growth temperature, growth rate, and film thickness [20–22, 31, 47, 48].

Accurate estimation of InSb epifilm thickness is of paramount importance using them for device engineering. Traditionally, the thin semiconducting films are characterized by using Hall measurements, XRD, ultraviolet-visible spectrophotometry, cross-sectional scanning electron microscopy [20, 21, 23, 49, 50], and so on. Being destructive, most of these methods are not convenient for assessing material quality required in electronic industry for device productions. Due to the narrow band gap of InSb, no reflectance interference fringes appear in the ultraviolet-visible wavelength range from the InSb/GaAs heterostructure. Therefore, a variety of nondestructive and penetrative tools have been exploited, providing nanoscale resolution for evaluating film thickness with greater degree of accuracy. Some of these techniques used in characterizing large area wafers include the atomic force microscopy, energy-dispersive X-ray spectroscopy, secondary-ion-mass spectrometry, and SE. RSS is another valuable and nondestructive tool to offer useful information on the crystalline quality and other parameters necessary for optimizing the InSb epifilm growth [31].

Table 1: Thicknesses and MSE of InSb thin film samples from SE.

Sample no.	S1	S2	S3	S4	S5
V/III ratio	3.91	3.98	4.20	4.78	5.38
Layer thickness (nm)	56.4	63.5	52.0	34.3	51.4
MSE	7.20	6.14	4.58	5.90	3.82

In this study, we present the results of our comprehensive investigations using SE, XRD, and SR-XAS, respectively, to assess the InSb film thickness and report the effects of V/III source ratios on the films crystalline quality for optimizing the MOCVD growth parameters, which were not exactly achieved from previous reports [31]. The SR-XAS was used to acquire the structural properties of the material at the atomic scale. Furthermore, temperature-dependent optical constants of InSb thin films between 25°C and 300°C (n, k, and ε) were probed by exploiting SE. This work provides a helpful guide to thin film characterization procedures required to monitor the growth processes, understanding the chemical and physical properties of materials and guiding the designs of high-performance InSb thin film devices.

2. Experimental

2.1. Material Growth. The growth of ultrathin uniform InSb thin epilayers was carried out on 4-inch semi-insulating (SI) GaAs (100) substrates using a low-pressure MOCVD method in vertical configuration by exploiting a high-speed rotating disk (180 mm diameter) reactor [31, 47]. The substrate orientation of 4-inch SI GaAs (100) was 2–4° off towards <110>. The trimethyl indium and trimethyl antimony sources were used as vapor-phase constituents of In and Sb into a reaction chamber at approximately ~20°C and with the bubbler pressures set at 400 Torr for trimethyl indium and 323 Torr for trimethyl antimony, respectively. Hydrogen acted as carrier gas, and the pyrometer growth temperature was 395°C. In this paper, we have studied a series of five InSb/GaAs samples grown with different V/III source flow ratios and thickness. Table 1 summarizes the growth relevant parameters of the samples used from the MOCVD growth [31] and their thicknesses determined in this study. For InSb growth using trimethyl indium and trimethyl antimony, the surface morphology was found to be very sensitive to the V/III ratio and closely reflected the crystallinity of as-grown films. All InSb films with V/III ratio varied between 3.91 and 5.38 exhibit mirror-like surface morphology having thickness ranging from 34 to 64 nm. To examine the effects of V/III ratio on the InSb film characteristics, we named the five samples S1 to S5 in the order of V/III ratio corresponding to their growth run numbers (IA376, 380, 373, 370, and 371, resp. [31]).

2.2. Optical Measurements

2.2.1. Spectroscopic Ellipsometry. SE is a valuable nondestructive technique commonly used for determining

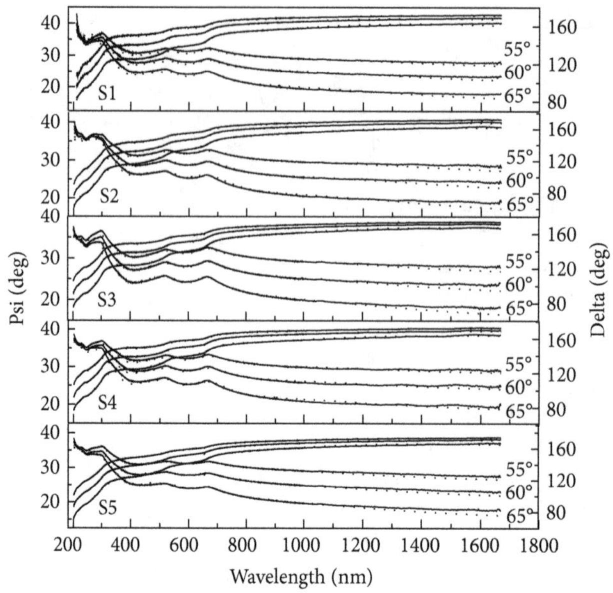

FIGURE 1: Experimental SE data and model fitting for five MOCVD grown InSb/GaAs samples.

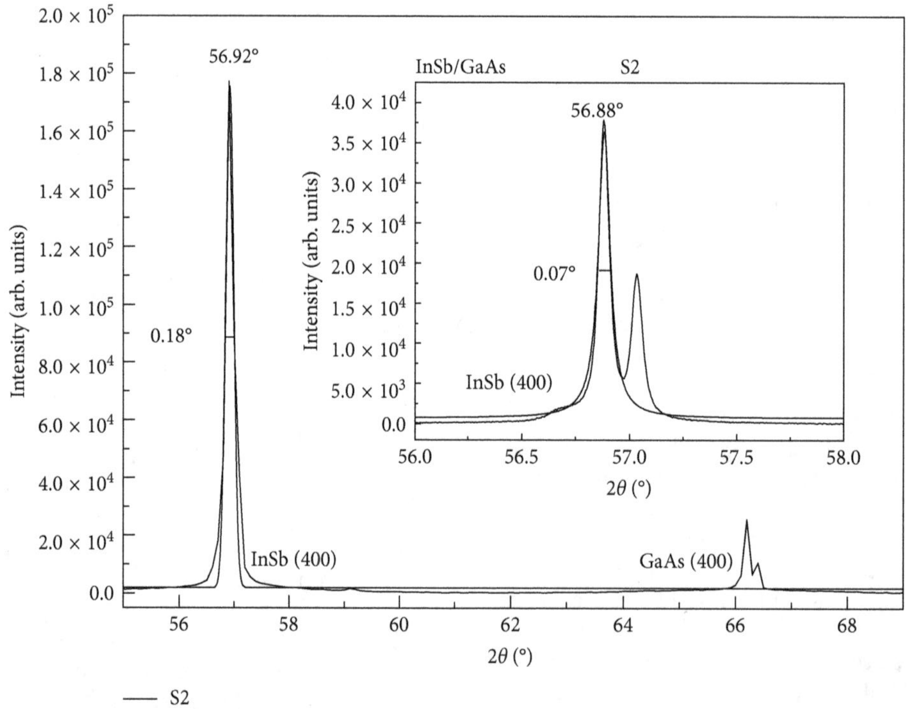

FIGURE 2: XRD spectrum and Lorentz fit of S2 in the angle range of 55°–69°.

thickness, optical constants of materials; the method has also been employed for a complete depth profiling in semiconducting epitaxially grown ultrathin or thin films. While the conventional SE approach suffers from the drawback of slow data acquisition process and covers a limited spectral range, the phase-modulated method that we have employed here offers the fast and precise data acquisition over a large wavelength range. A dual rotating-compensator Mueller matrix ellipsometer (ME-L ellipsometer, Wuhan Eoptics Technology Co. Ltd., China), equipped with Linkam Scientific heating and cooling stage device (THMSG600E), was used for the acquisition of ellipsometric spectra on ultrathin InSb epifilms. At room temperature, we acquired (Figure 1) two parameters Ψ and Δ, represented the ratio of amplitude decay between and p and s polarization of the reflected light and their phase difference, respectively, as a function of wavelength (λ) from 200 nm to 1670 nm at 1 nm step (or energy from 0.74 eV to 6.2 eV) at three angles of incidence 55°, 60°, and 65°.

TABLE 2: The fitting and calculation results from XRD of five InSb samples with different V/III ratios.

Sample no.	S1 (376)	S2 (380)	S3 (373)	S4 (370)	S5 (371)
V/III ratio	3.91	3.98	4.2	4.78	5.38
XRD FWHM ($\Delta 2\theta$, degree)	0.21	0.07	0.05	0.11	0.1
Peak position (2θ, degree)	56.95	56.92	56.92	56.96	56.96
Layer thickness (nm) (SE measured)	56.4	63.5	52.0	34.3	51.4
Calculated β ($\Delta 2\theta$, degree)	0.29	0.26	0.32	0.48	0.32

2.2.2. Raman Scattering Spectroscopy. RSS is a powerful and nondestructive technique for providing valuable information on materials' characteristics—especially for assessing epilayer thickness, strain, disorder, and site selectivity of defects. The RSS method is particularly suited for probing the local atomic and/or nanoscale structural changes in the InSb ultrathin films grown on SI GaAs while making the careful analysis of its subtle spectral variations. At the earlier time, we had quickly measured RSS on our InSb samples using a Raman spectrometer with the excitation from a HeNe laser 633 nm line. Later on, after more years, we performed further Raman measurements using 514-nm laser excitation on these samples, with similar results obtained. To avoid duplication, these data were not presented here.

2.2.3. X-Ray Diffraction Spectroscopy. As a common technique for material characterization, the XRD is widely used to evaluate the quality of crystal structure. It is sensitive for estimating stress/strains in epitaxially grown thin films. Furthermore, the peak position and the full width at half maximum (FWHM) of X-ray diffraction spectra give the information of crystal orientation and crystal quality. The diffractometer we used in experiments is Rigaku MiniFlex 600, Japan. The five InSb samples were measured by this X-ray diffractometer using a Cu Kα radiation ($\lambda = 1.5406$ Å).

2.3. Synchrotron Radiation X-Ray Absorption Spectroscopy. XAFS spectra were collected for the InSb thin film samples in X-ray fluorescence yield mode at the beam line 01C1 of the National Synchrotron Radiation Research Center in Hsinchu, Taiwan. The photon energy for the XAFS covered the range from 27,640 to 28,840 eV. The intensity of X-ray was monitored by a liquid N_2–filled ionization chamber, and fluorescence emitted from the sample was measured by an argon-filled Stern-Held-Lytle detector. A Si (111) double-crystal monochromater with a 0.5-mm entrance slit was used. A filter was inserted between the sample and the detector window to reduce the noise from scattering and to improve the spectrum quality. The incident photon direction was 45 degree to surface of the sample, and the flux I_0 of incident photon was monitored simultaneously by an ion chamber located just before the sample chamber. All measurements were made at RT, and all X-ray absorption spectra were normalized to I_0.

3. Results and Discussion

3.1. Spectroscopic Ellipsometry at Room Temperature. The measured SE spectra are fitted empirically by minimizing the

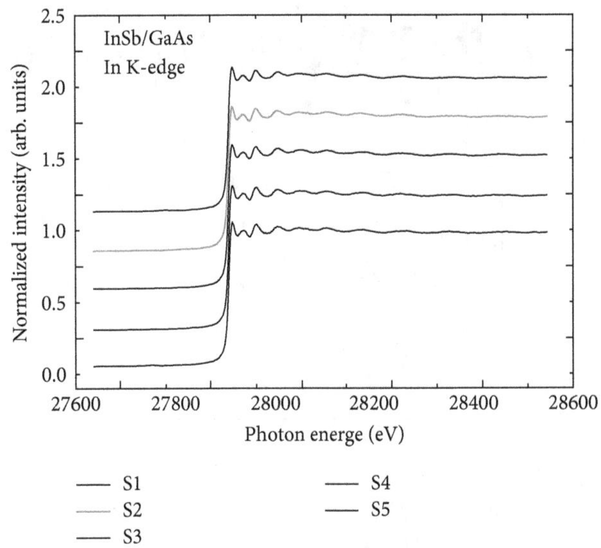

FIGURE 3: X-ray absorption spectra of In-K edge (27,940 eV) for five InSb samples.

squared differences between the observed and calculated Ψ and Δ values, which generated from the fitting model at the corresponding wavelengths. The quality of the fit can be judged by the mean square error (MSE) defined as follows:

$$\mathrm{MSE} = \frac{1}{2n-m}\sum_{i}^{n}\left[\left(\frac{\Psi_i^{\mathrm{mod}}-\Psi_i^{\mathrm{exp}}}{\sigma_{\Psi,i}^{\mathrm{exp}}}\right)^2+\left(\frac{\Delta_i^{\mathrm{mod}}-\Delta_i^{\mathrm{exp}}}{\sigma_{\Delta,i}^{\mathrm{exp}}}\right)^2\right], \tag{1}$$

where n stands for the total number of experimental measurements of Ψ and Δ for each chosen wavelength, m represents the number of fitting parameters used, the superscripts "mod" and "exp" represent the appropriate values of model and experimental data, and σ signifies the MSE between the calculated and experiment data at each wavelength.

In this study, the SE data were fitted using Tauc–Lorentz multiple oscillator modes [48]:

$$\varepsilon_2(E)\begin{cases} = \dfrac{AE_0C\left(E-E_g\right)^2}{\left(E^2-E_0^2\right)^2+C^2E^2}\cdot\dfrac{1}{E}, & E > E_g, \\ = 0, & E \le E_g, \end{cases} \tag{2}$$

where E is photon transition energy, A and C are fitting constants, and estimates are obtained by fitting Ψ and Δ. Equation (2) is useful for evaluating the dielectric function ε

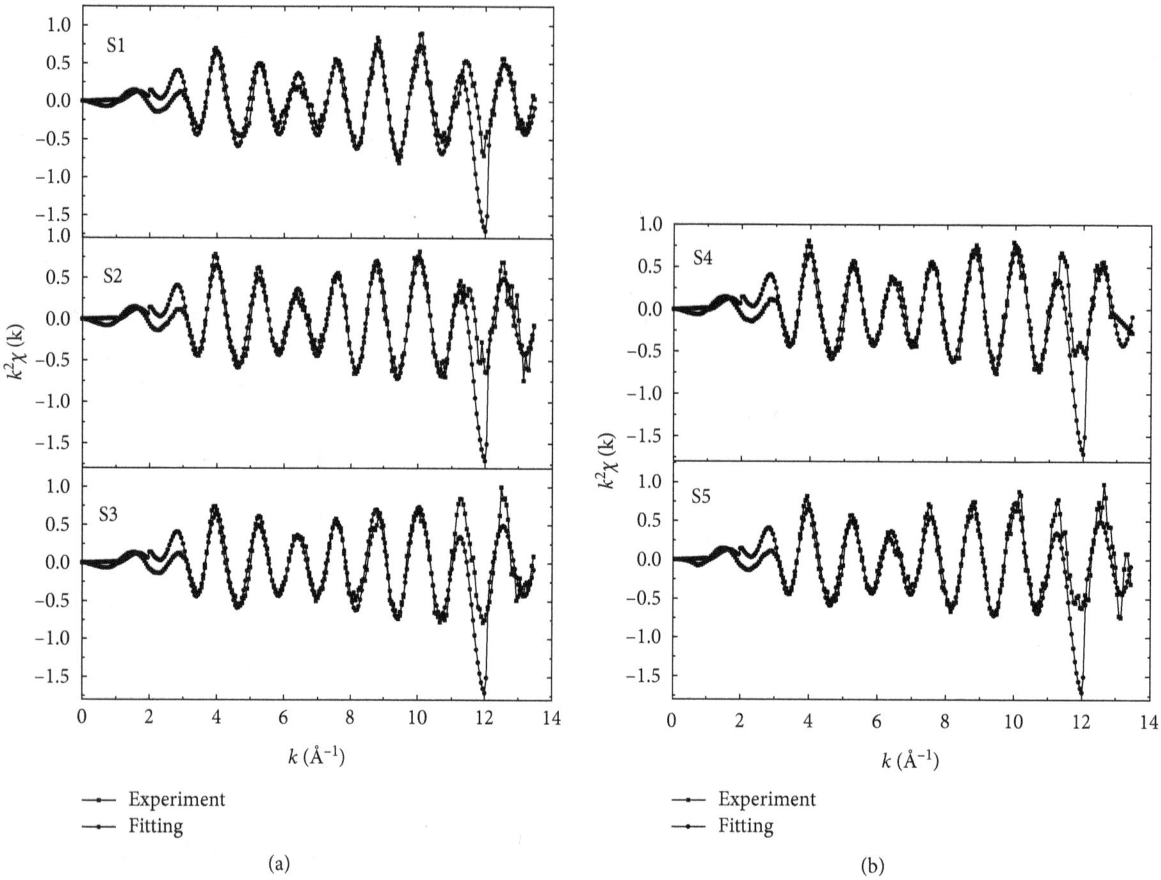

FIGURE 4: The In K-edge oscillation $k^2\chi$ (k) for all the measured InSb thin film samples.

(or n, k). The real part of the dielectric function ε_1 is obtained by exploiting the Kramers–Kronig integrations, that is,

$$\varepsilon_1(E) = \varepsilon_1(\infty) + \frac{2}{\pi} P \int_{E_g}^{\infty} \frac{\xi \varepsilon_2(\xi)}{\xi^2 - E^2} d\xi. \quad (3)$$

In (3), the term P is the main part of the Cauchy integral, where $\varepsilon_1(\infty)$ is added as a fitting parameter.

Figure 1 shows experimental SE Ψ and Δ spectra of five InSb/GaAs samples, measured with three incident angels at room temperature (RT). The spectral oscillations of the five samples are similar, indicating that they have similar optical properties. All of the Ψ and Δ data with variable angles were considered in the calculation of the optical parameters like n and k. The thicknesses of InSb films were extracted through analyzing the SE data. The three-layer model, substrate/film/oxidation, or surface roughness was first tested for modeling. The simulation results showed the surface oxidation layer with the thickness of 0.01 nm or less for all sample. This is physically meaningless, that is, the thickness of the surface oxidation layer should be zero. Therefore, the two-layer model, substrate/film, was used for modeling on five InSb/GaAs samples here. Fitting results are listed in Table 1.

3.2. The X-Ray Diffraction Data and Fits. Figure 2 shows the wide-scan XRD spectrum of a typical InSb/GaAs sample (S2) in the angle range (2θ) from 55° to 69° at 0.1° step. The intensity of the GaAs (400) peak was observed to be weaker than that of the InSb (400) peak. This phenomenon was due to that the semi-insulating GaAs wafer cut 2–4° off (100) towards <110> was used as the substrate. The surface of GaAs wafer was deviated from the real (100) orientation. The GaAs wafer was not the real GaAs (100) substrate or not with the real (100) orientation surface. This type of semi-insulating GaAs (100) 2° off towards <110> substrate was successfully used for the growth of InSb thin films [47, 49]. During the growth process, it could lead the heteromismatch-induced dislocations to spread along the InSb/GaAs interface plane but not along the film growth normal direction, that is, not to the InSb film surface, so as to greatly decrease the dislocation density at the surface area of InSb film.

From this rough scan, the InSb (400) peak appeared as a single peak with the FWHM of 0.18°. Inset of Figure 2 shows fine-scan XRD spectrum in the angle (2θ) of 56°–58° at 0.005° step. The observation of the peak doublet splitting is caused due to the X-ray source Cu Kα_1 and Cu Kα_2 radiation. The well-separation of the InSb (400) peak doublet and the narrow FWHM of 0.07° from the InSb (400) Kα_1 peak indicated good crystallinity of the film. A sharp diffraction peak was observed at $2\theta = 56.88°$, which corresponds to the (400) crystalline plane of InSb (S2). The other diffraction peak located at 66.20° can indexed to the (400) crystal plane of GaAs substrate. Table 2 lists the values of

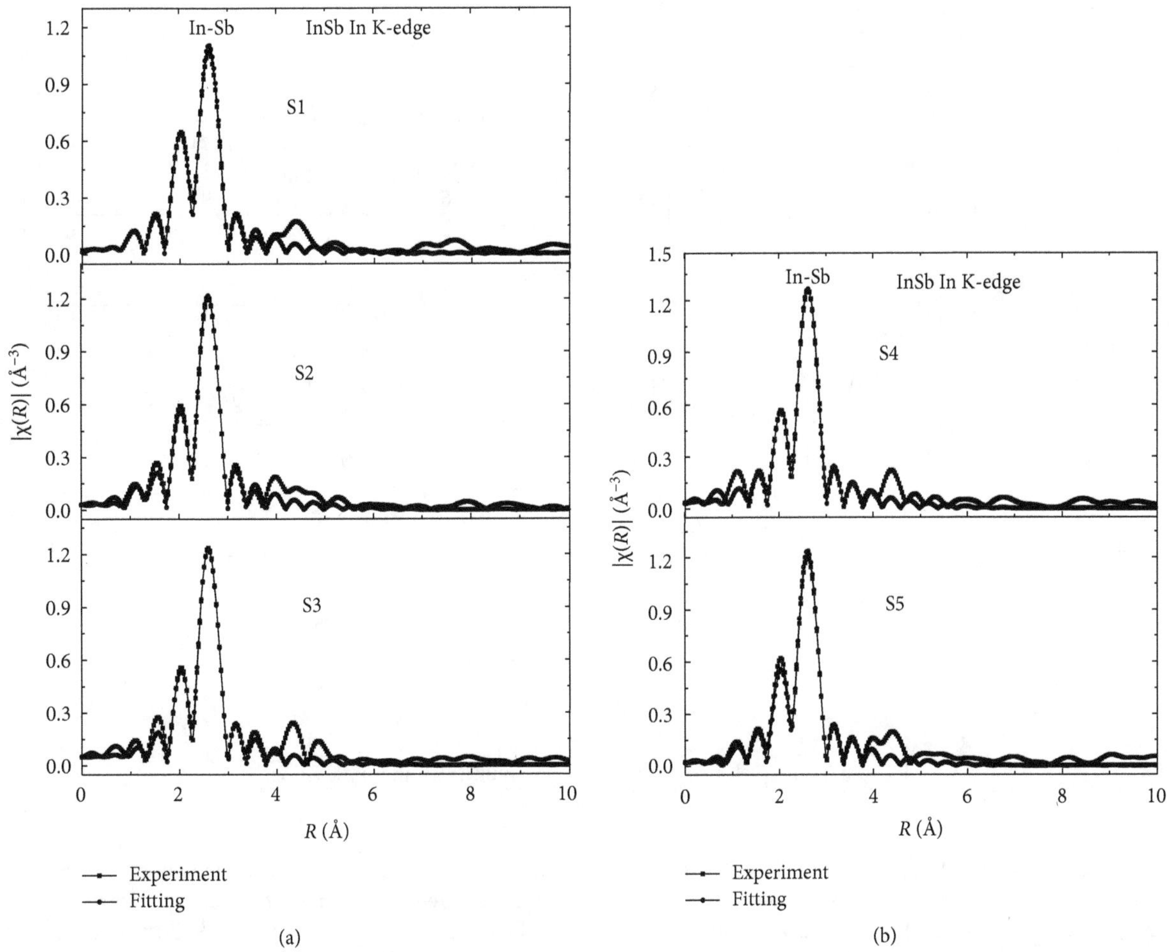

FIGURE 5: The magnitude of the Fourier transforms of measured samples (black lines) and the results of fitting (red lines).

XRD peak and FWHM obtained by Lorentz fitting, for five InSb samples with different V/III ratios.

Farag et al. [51] calculated the crystallite size (D) of the InSb film grown on GaAs from the broadening of few XRD peaks using the Debye-Scherrer equation (4):

$$D = \frac{K_S \cdot \lambda}{\beta \cos \theta}, \tag{4}$$

where λ is the X-ray wavelength of Cu Kα (0.15418 nm), β is the width of the peak at half maximum intensity for the thin film, K_S is the Scherrer constant of the order of unity (0.95 for powder and 0.89 for film), and θ is the corresponding Bragg's angle.

Usually, finite size gives broader peaks. The contribution of our InSb film thickness, D, to the XRD broad peak width, β, can be estimated by modified Debye-Scherrer formula (5),

$$\beta = \frac{0.89\lambda}{D \cos \theta}. \tag{5}$$

The so-calculated β values using the InSb film thickness obtained from SE are also listed in Table 2. It is seen that these β values are much larger than the XRD FWHMs measured in our experiments. This predicts the high crystalline quality of our MOCVD grown InSb films on GaAs.

3.3. Simulations of SR-XAS Data. Indium K-edge extended X-ray absorption fine structure (EXAFS) was employed to study the local structure of InSb thin films on GaAs. Figure 3 plots the four In K-edges EXAFS of InSb, with all of In absorption edges near 27,940 eV. Replicate EXAFS scans were coadded to improve the signal-to-noise ratio. Due to numerous diffraction peaks in the In-EXAFS spectra, the samples were spun to average out the diffraction peaks in the EXAFS spectra. All EXAFS spectra were collected to 1,200 eV beyond the In K-edges. All EXAFS data were analyzed for wave vectors (k) from 3.0 Å^{-1} to 11.5 Å^{-1}. The chi data were $k^2\chi$ (k) weighted as shown in Figure 4 and Fourier transformed with a window width of $k = 0.5\,\text{Å}^{-1}$ to yield the R-space data. As shown in Figure 5, the simulated EXAFS spectra were generated, based on the documented crystallographic properties for In and Sb using *ab initio*-based theory.

We used ATHENA program to remove the background and extract the EXAFS oscillations from the k-space signals. To extract the bond length of $R_{In\text{-}Sb}$, a structural model was built using the package IFEFFIT. Fitting results of the Fourier transforms show a good agreement with the measurement shown in Figure 5. With the first neighbor peak being well defined in the In absorption, we extracted the $R_{In\text{-}Sb}$ from the In data listed in Table 3.

TABLE 3: The fitting results of the EXAFS data for first coordination shell atom around In atom.

Sample no.	V/III ratio	Amp	CN	R_{In-Sb} (Å)	R-factor
S1	3.91	0.9500 (0)	3.80	2.8027 (9)	0.0052 (5)
S2	3.98	0.9805 (6)	3.92	2.8044 (9)	0.0050 (9)
S3	4.20	0.9807 (3)	3.92	**2.8000 (7)**	0.0089 (1)
S4	4.78	0.9950 (0)	**3.98**	2.8014 (3)	0.0078 (5)
S5	5.38	0.9898 (6)	3.96	2.7989 (9)	0.0055 (2)

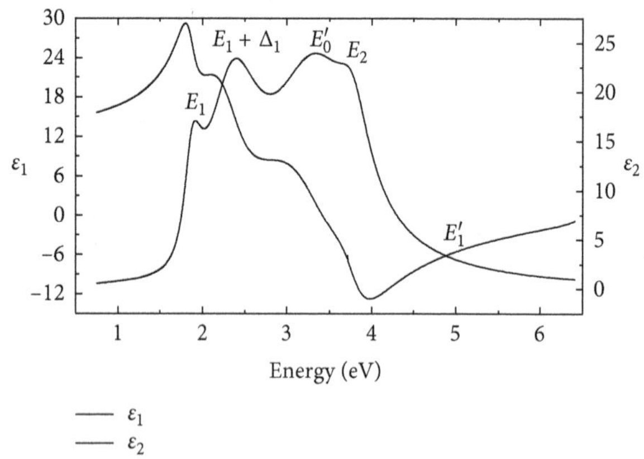

(a)

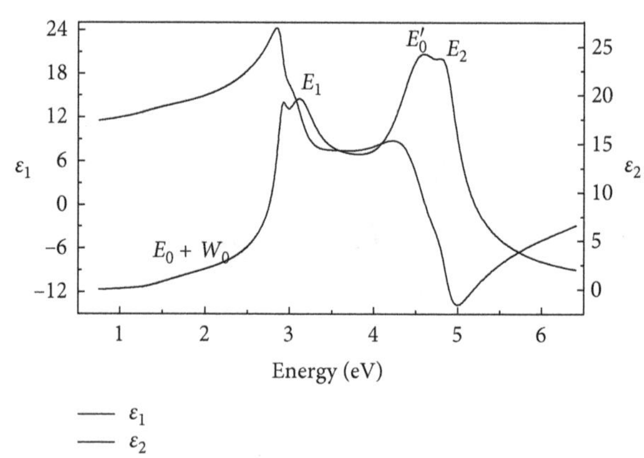

(b)

FIGURE 6: The dielectric functions of the bulk InSb and bulk GaAs.

Fitting results show that the bond length and coordination number of the first In-Sb shell are about 2.80 Å and 4, respectively. The bond length and the coordination number of unstrained InSb are 2.80 Å and 4, respectively. The EXAFS fitting for five InSb/GaAs gave bond length values around 2.80 Å, and the sample S5 with the highest V/III ratio of 5.38 has its R_{In-Sb} slightly less than 2.80 Å. The coordination number (CN) for all samples is close to but slightly below 4 (the regular value) with samples S4 and S5 nearest to 3.96–3.98, samples S2 and S3 at 3.92, and only sample S1 with the lowest V/III ratio of 3.91 significantly lower at 3.80. These variations could be caused by residual strain and too low/too high might cause higher lattice mismatch in the InSb/GaAs heterointerface, leading to higher residual strain within the epitaxial InSb film. This finding suggests that V/III ratios of 4.20 (S3) and 4.78 (S4) are the better growth parameter values, which is in consistent with the results from Raman scattering.

3.4. Further Analyses and Temperature-Dependent SE. The real part ε_1 and imaginary part ε_2 of the dielectric function of the bulk InSb and bulk GaAs as function of photon energy is shown in Figure 6. The peaks in these spectra reflect the high-energy-state transitions of bulk InSb and can be used to obtain several related critical energy points (E_1, $E_1 + \Delta_1$, E_0', E_2, and E_1') [52]. The high-energy-state transitions of bulk GaAs are $E_0 + W_0$, E_1, E_0', and E_2 [52]. Therefore, Figure 6 shows the peaks those are the high-energy-state transitions of the InSb film and none of GaAs substrate. D'Costa et al. [15] reported the SE study at room temperature for an InSb/GaAs grown by molecular beam epitaxy, showing similar critical energy points from InSb.

From the simulation of the SE Ψ and Δ data in Figure 1, the reflective index (n) and extinction coefficient (k) versus wavelength or energy can be extracted. Figures 7(a) and 7(b) exhibit the comparative n and k variation with energy (eV), for five InSb/GaAs, respectively. The critical energy points (E_1, $E_1 + \Delta_1$, E_0', E_2, and E_1') are marked as discussed in last paragraph.

Temperature-dependent (TD) SE measurements were performed in the temperature range from 20°C to 300°C. Figure 8 presents a set of TD-SE experimental data for sample S2, which were measured at three different angles, 55°, 60°, and 65°. Simulation fits were performed for all SE Ψ and Δ data. As mentioned in Section 3.1 for RT-SE, the three-layer model—substrate/film/surface oxide—was first used for modeling. The best fitting simulation results showed the surface oxide layer with the thickness of 0.01 nm or less, that is, zero, for all SE data measured from 20°C to 250°C. However, for the SE spectra (Ψ and Δ) data measured at 300°C, the three-layer model—substrate/film/surface oxide—has to be used, resulting in a surface oxide thickness of 5.4 nm for sample S2. Other samples had similar results, that is, SE data between RT and 250°C showed no surface oxide layer but SE at 300°C revealed a near 5-nm surface oxide layer, corresponding to about two atomic layer of indium oxides.

Figure 9 shows the variation of refractive index n and extinction coefficient k of InSb films in a temperature range from 20°C to 300°C. As seen, the InSb film contains several absorption peaks in the wavelength range from 200 to

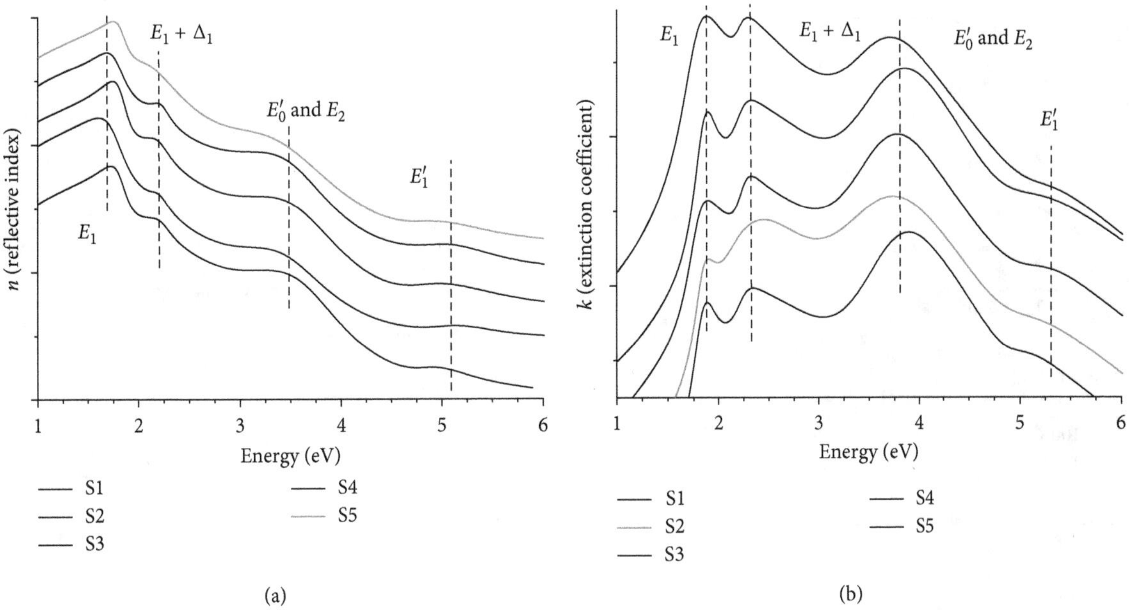

FIGURE 7: (a) Comparative reflective index (n) and (b) comparative extinction coefficient (k) for five InSb/GaAs samples, deduced from SE Ψ and Δ data in Figure 1.

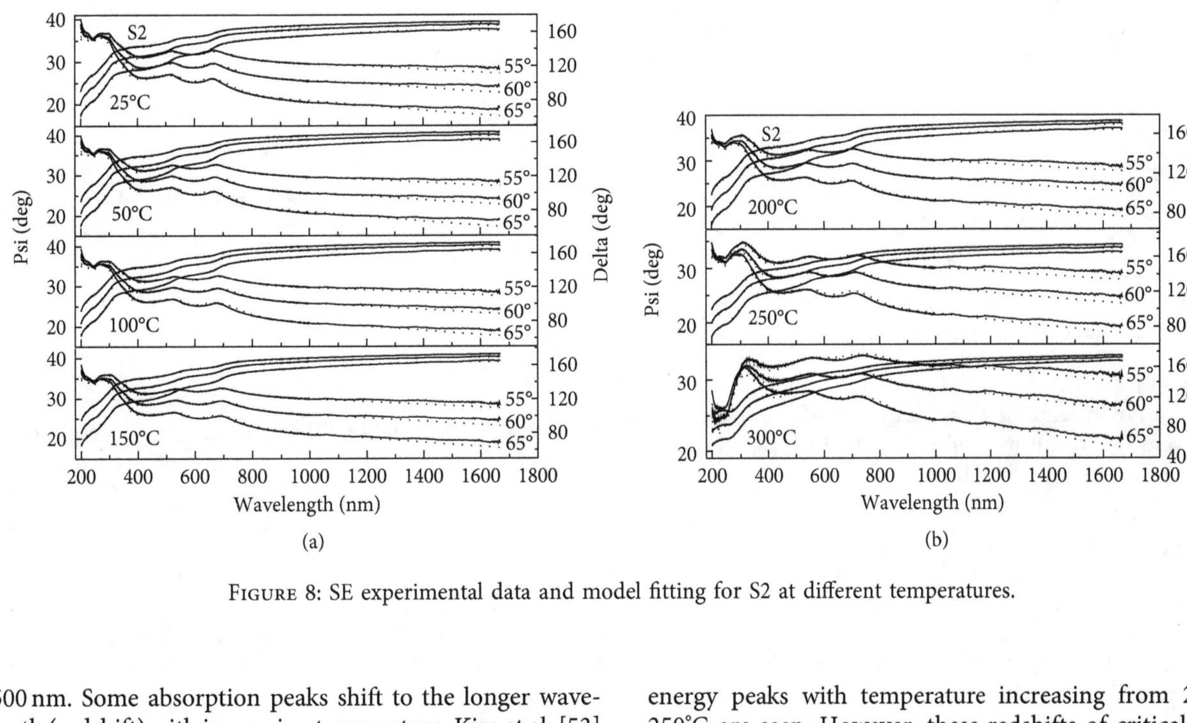

FIGURE 8: SE experimental data and model fitting for S2 at different temperatures.

1500 nm. Some absorption peaks shift to the longer wavelength (redshift) with increasing temperature. Kim et al. [53] performed TD-SE on bulk InSb over a temperature of 31 K to 675 K. A similar redshift was observed in our experimental InSb films on GaAs over the temperature range of 20°C to 300°C, that is, 293 K to 573 K, comparable with that observed in other study [53].

Figure 10 exhibits the complex dielectric function, real and imaging parts, ε_1 and ε_2, calculated by $\varepsilon_1 = n^2 - k^2$ and $\varepsilon_2 = 2nk$, of InSb film S2 with different temperatures. The high-energy critical points of E_1, $E_1 + \Delta_1$, E_0', E_2, and E_1' are indicated in the figure. The gradual redshifts of these critical energy peaks with temperature increasing from 20°C to 250°C are seen. However, these redshifts of critical energy peaks in Figure 10 and n and k spectra in Figure 9 become much more pronounced as temperature increases to 300°C. These large variations in n and k and ε_1 and ε_2 are due to the formation of an InO surface layer from the severe oxidation of the surface of InSb film at 300°C. This oxidation could be ignored for InSb film measured in 20°C–250°C because the SE fits with three-layer model leaded to an InO surface thickness of <0.01 nm, much thinner than one monoatomic layer. But the oxide layer cannot be ignored in sample heated at 300°C, that is, higher than 250°C, the severe oxidation

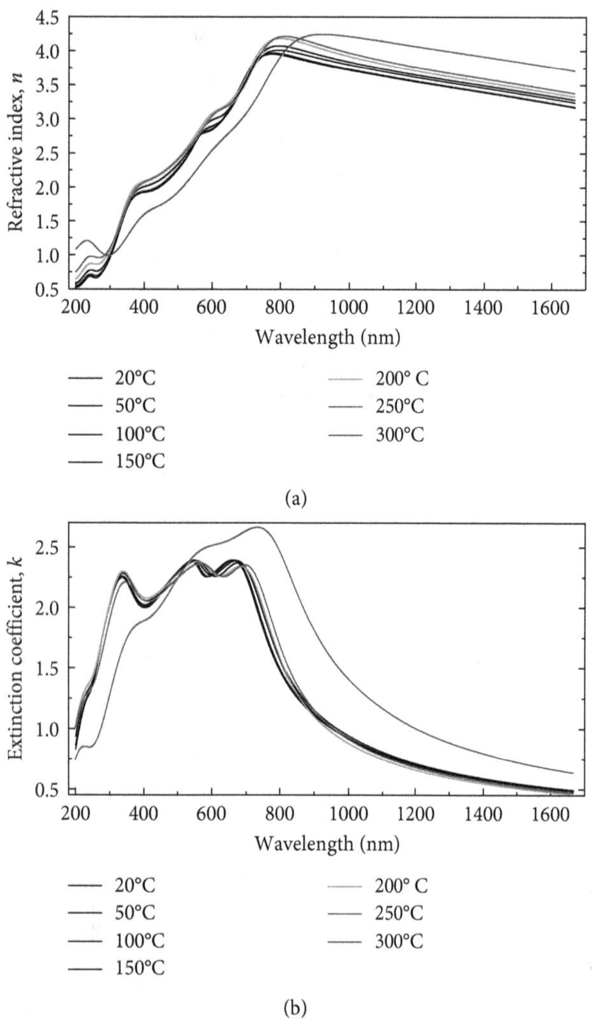

FIGURE 9: The refraction index n (a) and the extinction coefficient k (b) of InSb thin film (S2) with varied temperatures of 20°C–300°C.

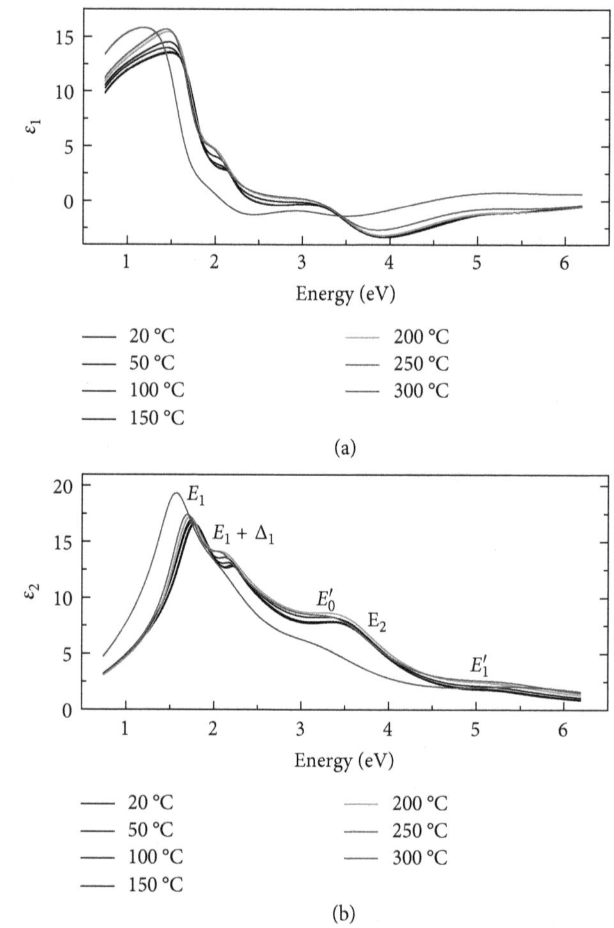

FIGURE 10: The real and imaginary part of the dielectric function ε of the InSb thin film (S2) at varied temperatures of 20°C–300°C.

appeared on InSb, leading to the formation of a thin indium-oxide layer. This result may predicate the high temperature limitation for the use of InSb/GaAs materials.

4. Conclusions

In summary, a series of InSb thin films grown on 4-inch GaAs substrates by MOCVD technique, with different V/III ratios, were investigated by SE, XRD, and SR-XAS, respectively. InSb thin film thickness was extracted by fitting the experimental SE data. The crystallinity of the films was high quality extracted from the well separation of the InSb (400) peak doublet and the narrow FWHM of the InSb (400) $K\alpha_1$ peak in the XRD spectrum. Through advanced synchrotron radiation technique of XAS and data simulation, the atomic scale bonding length and coordination number were obtained.

Through combined multiple technological analyses, the results showed that InSb films on GaAs grown under too high or too low V/III ratios are with poor quality, while those grown with proper V/III ratios of 4.20 and 4.78 possess the high crystalline quality. These results are useful to the material growers for improving the growth processing.

The temperature-dependent (TD)-SE measurements (20–300°C) and simulation revealed a significant phenomenon: the SE spectra, optical constants (n, k, ε_1, ε_2), and critical energy points (E_1, $E_1 + \Delta_1$, E_0', E_2, and E_1') of the InSb thin films varied with temperature (T) gradually and smoothly as T increased from 20°C to 250°C, and all SE spectra can be fitted well by the two-layer model (substrate/film). However, up to 300°C, the SE spectra appeared to show large changes, and simulation revealed the existence of an indium-oxide (InO) layer of ~5.4 nm, that is, about two atomic layers. The optical constants (n, k, ε_1, ε_2) and critical energy points (E_1, $E_1 + \Delta_1$, E_0', E_2, and E_1') had showed transilient changes from 250°C to 300°C, which are due to the top about two atomic layers oxidized. This indicates the high temperature limitation for the use of InSb/GaAs materials, up to 250°C, which provides a hinder to the device designers using InSb materials.

Conflicts of Interest

The authors declare that there are no conflicts of interest regarding the publication of this paper.

Acknowledgments

This work was supported by National Natural Science Foundation of China (nos. 61367004 and 61504030) and special funding (nos. T3120097921 and T3120099202) for Guangxi distinguished professors (Bagui Rencai and Bagui Xuezhe). The authors acknowledge the support and help from Prof. Enwei Liang, Prof. Gu Xu, Prof. Jingli Chen, Ms. Qingxuan Li, Ms. Fangze Wang, Prof. Shiyuan Liu, and Dr. Chuanwei Zhang.

References

[1] Y. Cheng, J. Yang, Q. Jiang et al., "New insight to InSb-based thermoelectric materials: from the divorced eutectic design to remarkable high thermoelectric performance," *Journal of Materials Chemistry A*, vol. 5, no. 10, pp. 5163–5170, 2017.

[2] B. Wen Jia, K. Hua Tan, W. Khai Loke, S. Wicaksono, and S. F. Yoon, "Effects of surface reconstruction on the epitaxial growth of III-Sb on GaAs using interfacial misfit array," *Applied Surface Science*, vol. 399, pp. 220–228, 2017.

[3] M. Hilal, B. Rashid, S. Haider Khan, and A. Khan, "Investigation of electro-optical properties of InSb under the influence of spin-orbit interaction at room temperature," *Materials Chemistry and Physics*, vol. 184, pp. 41–48, 2016.

[4] D. Seo, J. Na, S. Lee, and S. Lim, "Behavior of GaSb (100) and InSb (100) surfaces in the presence of H_2O_2, in acidic and basic cleaning solutions," *Applied Surface Science*, vol. 399, pp. 523–534, 2017.

[5] A. Arlauskas, L. Subačius, A. Krotkus, and V.L. Malevich, "Terahertz emission from InSb illuminated by femtosecond laser pulses," *Journal of Physics D: Applied Physics*, vol. 50, no. 5, article 055101, 2017.

[6] R. Giulian, D. J. Manzo, J. B. Salazar et al., "Structural and electronic characterization of antimonide films made by magnetron sputtering," *Journal of Physics D: Applied Physics*, vol. 50, no. 7, article 075106, 2017.

[7] T. D. Subash, T. Gnanasekaran, C. Divya, and J. Jagannathan, "Design and simulated characteristics of nanosized InSb based heterostructure devices," *Advances in Materials Science and Engineering*, vol. 2014, Article ID 196732, 5 pages, 2014.

[8] L. Zhang, W. Tian, Q. Meng, M. Sun, N. Li, and Z. Lei, "Analysis on structural stress of 64 × 64 InSb IRFPAs with temperature dependent elastic underfill," *Journal of Sensors*, vol. 2014, Article ID 952323, 7 pages, 2014.

[9] J. Abautret, J. P. Perez, A. Evirgen, J. Rothman, A. Cordat, and P. Christol, "Characterization of midwave infrared InSb avalanche photodiode," *Journal of Applied Physics*, vol. 117, no. 24, article 244502, 2015.

[10] A. G. Hernández, Y. Kudriavtsev, S. Gallardo, M. Avendaño, and R. Asomoza, "Formation of self-organized nano-surfaces on III–V semiconductors by low energy oxygen ion bombardment," *Materials Science in Semiconductor Processing*, vol. 37, pp. 190–198, 2015.

[11] M. Shafa, H. Ji, L. Gao et al., "Mid-infrared photodetectors based on InSb micro/nanostructures grown on low-cost mica substrates," *Materials Letters*, vol. 169, pp. 77–81, 2016.

[12] E. B. Elkenany, "Optoelectronic and mechanical properties of InSb semiconductor under the effect of temperature," *Silicon*, vol. 8, no. 3, pp. 391–396, 2016.

[13] J. L. He, W. D. Hu, Z. H. Ye, Y. Q. Lv, X. S. Chen, and W. Lu, "Joint FDTD-optical/FEM-electrical numerical simulation of reflection-type subwavelength-microstructure insb infrared focal-plane arrays," *Journal of Electronic Materials*, vol. 45, no. 9, pp. 4552–4556, 2016.

[14] J. A. del Alamo, "Nanometre-scale electronics with III-V compound semiconductors," *Nature*, vol. 479, no. 7373, pp. 317–323, 2011.

[15] V. R. D'Costa, K. H. Tan, B. W. Jia, S. F. Yoon, and Y.-C. Yeo, "Mid-infrared to ultraviolet optical properties of InSb grown on GaAs by molecular beam epitaxy," *Journal of Applied Physics*, vol. 117, no. 22, article 223106, 2015.

[16] B. Wen Jia, K. Hua Tan, W. Khai Loke, S. Wicaksono, and S. F. Yoon, "Epitaxial growth of low threading dislocation density InSb on GaAs using self-assembled periodic interfacial misfit dislocations," *Materials Letters*, vol. 158, pp. 258–261, 2015.

[17] W. Sun, H. Fan, Z. Peng et al., "Photodiode properties of molecular beam epitaxial InSb on a heavily doped substrate," *Infrared Physics & Technology*, vol. 62, pp. 143–146, 2014.

[18] Y. Sato, Y. Morita, and I. Kanno, "Performance estimation of InSb compound semiconductor detectors as a function of active area using alpha particles," *Nuclear Instruments and Methods in Physics Research Section A: Accelerators, Spectrometers, Detectors and Associated Equipment*, vol. 737, pp. 1–4, 2014.

[19] M. Mori, Y. Yasui, K. Nakayama, K. Nakatani, and K. Maezawa, "Effects of initial in coverage for preparation of InSb bilayer on electrical properties of InSb films grown by surface reconstruction controlled epitaxy," *Japanese Journal of Applied Physics*, vol. 51, no. 2, pp. 165–176, 2012.

[20] Y. J. Jin, D. H. Zhang, X. Z. Chen, and X. H. Tang, "Sb antisite defects in InSb epilayers prepared by metalorganic chemical vapor deposition," *Journal of Crystal Growth*, vol. 318, no. 1, pp. 356–359, 2011.

[21] S. Yamaguchi and M. Matsumoto, "Stress reduction and electric properties of InSb thin films grown by metalorganic vapor phase epitaxy on sapphire substrates with an InAs buffer layer," *Vacuum*, vol. 84, no. 11, pp. 1323–1326, 2010.

[22] T. Iwasugia, M. Moria, H. Igarashia, K. Murataa, M. Saitob, and K. Maezawa, "Heteroepitaxial growth of InSb films on the patterned Si (001) substrate," *Physics Procedia*, vol. 3, no. 2, pp. 1329–1333, 2010.

[23] J. A. A. Engelbrecht and M. C. Wagener, "Characterization of InSb layers on GaAs substrates using infrared reflectance and a modified oscillator formula," *Physica B: Condensed Matter*, vol. 404, no. 22, pp. 4397–4401, 2009.

[24] M. Li, Y. Qiu, G. Liu, Y. Wang, B. Zhang, and L. Zhao, "Distribution of dislocations in GaSb and InSb epilayers grown on GaAs (001) vicinal substrates," *Journal of Applied Physics*, vol. 105, no. 9, article 094903, 2009.

[25] B. T. Wysocki and M. A. Marciniak, "Discrimination between electronic and optical blooming in an InSb focal-plane array under high-intensity excitation," *Infrared Physics & Technology*, vol. 51, no. 3, pp. 137–145, 2008.

[26] K. Yaemsunthorn, T. Thongtem, S. Thongtem, and C. Randorn, "Synthesis of InSb nanocrystals in an air atmosphere and their photocatalytic activity from near-infrared to ultra-violet," *Materials Science in Semiconductor Processing*, vol. 68, pp. 53–57, 2015.

[27] T. Zhang, J. J. Harris, W. R. Branford, S. K. Clowes, L. F. Cohen, and S. A. Solin, "Exploration of the inherent magnetoresistance in InSb thin films," *Semiconductor Science and Technology*, vol. 21, no. 12, pp. 1543–1546, 2006.

[28] W. R. Branford, A. Husmann, S. A. Solin et al., "Geometric manipulation of the high-field linear magnetoresistance in InSb epilayers on GaAs (001)," *Applied Physics Letters*, vol. 86, no. 20, article 202116, 2005.

[29] I. Kimukin, N. Biyikli, and E. Ozbay, "InSb high-speed photodetectors grown on GaAs substrate," *Journal of Applied Physics*, vol. 94, no. 8, pp. 5414–5416, 2003.

[30] V. K. Dixita, B. V. Rodriguesa, H. L. Bhata, R. Venkataraghavanb, K. S. Chandrasekaranb, and B. M. Arorab, "Growth of InSb epitaxial layers on GaAs (001) substrates by LPE and their characterizations," *Journal of Crystal Growth*, vol. 235, no. 1–4, pp. 154–160, 2002.

[31] Z. C. Feng, C. Beckham, P. Schumaker et al., "Optical characterization and mapping of four inch InSb epitaxial thin films grown on GaAs by turbo disk metalorganic chemical vapor deposition," *MRS Proceedings*, vol. 450, pp. 61–66, 1996.

[32] K. Li, A. T. S. Wee, J. Lin et al., "A surface and interface study on the InSb/GaAs heterostructures," *Thin Solid Films*, vol. 302, no. 1-2, pp. 111–115, 1997.

[33] M. A. Abdulsattar, T. R. Sultan, and A. M. Saeed, "Shape and size dependence of electronic properties of InSb diamondoids and nanocrystals: a density functional theory study," *Advances in Condensed Matter Physics*, vol. 2013, Article ID 713267, 2013.

[34] Z. C. Feng, Q. X. Li, L. Y. Wan, and G. Xu, "Variation of phonon coupling factors in the photoluminescence of cadmium telluride by variable excitation power," *Optical Materials Express*, vol. 7, no. 3, pp. 808–816, 2017.

[35] S. Liu, X.-H. Zhao, C. M. Campbell, M. B. Lassise, Y. Zhao, and Y.-H. Zhang, "Carrier lifetimes and interface recombination velocities in CdTe/MgxCd1−xTe double heterostructures with different Mg compositions grown by molecular beam epitaxy," *Applied Physics Letters*, vol. 107, no. 4, article 041120, 2015.

[36] C.-J. Wu, Z.-C. Feng, W.-M. Chang, C.-C. Yang, and H.-H. Lin, "Bond lengths and lattice structure of InP0.52Sb0.48 grown on GaAs," *Applied Physics Letters*, vol. 101, no. 9, article 091902, 2012.

[37] N. Yamamoto, K. Akahane, S.-I. Gozu, A. Ueta, and M. Tsuchiya, "Low-temperature growth of nanostructured InGaSb semiconductors on silicon substrates," *Physica E: Low-Dimensional Systems and Nanostructures*, vol. 40, no. 6, pp. 2195–2197, 2008.

[38] G. Tsai, D. L. Wang, and H. H. Lin, "Photoluminescence of $InAs_{0.04}P_{0.67}Sb_{0.29}$," *Journal of Applied Physics*, vol. 104, no. 2, article 023535, 2008.

[39] S. A. Cripps, T. J. C. Hosea, A. Krier et al., "Midinfrared photoreflectance study of InAs-rich InAsSb and GaInAsPSb indicating negligible bowing for the spin orbit splitting energy," *Applied Physics Letters*, vol. 90, no. 17, article 172106, 2017.

[40] M. Shafa, S. Akbar, L. Gao, M. Fakhar-e-Alam, and Z. M. Wang, "Indium antimonide nanowires: synthesis and properties," *Nanoscale Research Letters*, vol. 11, no. 1, p. 164, 2016.

[41] D. Wang and L. M. Tang, "Electronic structure and magnetism of doped wurtzite InSb nanowire," *Journal of Physics D: Applied Physics*, vol. 49, no. 17, article 175303, 2016.

[42] C. C. Ahiaa, N. Tile, Z. N. Urgessa, J. R. Botha, and J. H. Neethling, "An investigation of near-infrared photoluminescence from AP-MOVPE grown InSb/GaSb quantum dot structures," *Journal of Crystal Growth*, vol. 458, pp. 53–59, 2017.

[43] K. Sugiyama, "Molecular beam epitaxy of InSb films on CdTe," *Journal of Crystal Growth*, vol. 60, no. 2, pp. 450–452, 1982.

[44] R. Venkataraghavan, K. S. R. K. Rao, M. S. Hegde, and H. L. Bhat, "Influence of growth parameters on the surface and interface quality of laser deposited InSb/CdTe heterostructures," *Physica Status Solidi (a)*, vol. 163, no. 1, pp. 93–100, 1997.

[45] Z. C. Feng, S. Perkowitz, T. S. Rao, and J. B. Webb, "Resonance Raman scattering from epitaxial InSb films grown by metalorganic magnetron sputtering," *Journal of Applied Physics*, vol. 68, no. 10, pp. 5363–5365, 1990.

[46] M. C. Debnath, T. Zhang, C. Roberts, L. F. Cohen, and R. A. Stradling, "High-mobility InSb thin films on GaAs (001) substrate grown by the two-step growth process," *Journal of Crystal Growth*, vol. 267, no. 1-2, pp. 17–21, 2004.

[47] M. A. McKee, B.-S. Yoo, and R. A. Stall, "Uniform growth of InSb on GaAs in a rotating disk reactor by LP-MOVPE," *Journal of Crystal Growth*, vol. 124, no. 1–4, pp. 286–291, 1992.

[48] G. E. Jellison Jr. and F. A. Modine, "Parameterization of the optical functions of amorphous materials in the interband region," *Applied Physics Letters*, vol. 69, no. 3, pp. 371–373, 1996.

[49] B.-S. Yoo, M. A. McKee, S.-G. Kim, and E.-H. Lee, "Structural and electrical properties of InSb epitaxial films grown on GaAs by low-pressure MOCVD," *Solid State Communications*, vol. 88, no. 6, pp. 447–450, 1993.

[50] E. Woelk, H. Jurgensen, R. Rolph, and T. Zielinski, "Large scale production of indium antimonide film for position sensors in automobile engines," *Journal of Electronic Materials*, vol. 24, no. 11, pp. 1715–1718, 1995.

[51] A. A. M. Farag, F. S. Terra, G. M. Mahmoud, and A. M. Mansour, "Study of Gaussian distribution of inhomogeneous barrier height for n-InSb/p-GaAs heterojunction prepared by flash evaporation," *Journal of Alloys and Compounds*, vol. 481, no. 1-2, pp. 427–433, 2009.

[52] S. Adachi, *Optical Constants of Crystalline and Amorphous Semiconductors Numerical Data and Graphical Information*, Springer, Berlin, Germany, 1999, ISBN: 9780792385677.

[53] T. J. Kim, S. Y. Hwang, J. S. Byun et al., "Temperature dependence of the dielectric functions and the critical points of InSb by spectroscopic ellipsometry from 31 to 675K," *Journal of Applied Physics*, vol. 114, no. 10, article 103501, 2013.

9

Microstructures and Mechanical Properties of Al/Fe and Cu/Fe Joints by Continuous Drive Friction Welding

Yanni Wei[1] and Fu Sun[2]

[1]Department of Materials Science and Engineering, Xi'an University of Technology, 5 South Jinhua Road, Xi'an 710048, China
[2]Shaanxi Zhituo Solid-State Additive Manufacturing Technology Company, 70 West of Chaoyang Street, Weinan 714000, China

Correspondence should be addressed to Yanni Wei; weiyanni@xaut.edu.cn

Academic Editor: Paolo Ferro

The dissimilar pure metals Al/Fe and Cu/Fe with different metallurgical compatibility were joined by continuous drive friction welding. The friction weldability was investigated. The microstructure of the joining interface was analyzed by scanning electron microscopy, and the chemical compositions were tested by energy-dispersive spectroscopy. The joining strength was evaluated by tensile test, and the fracture was detected by X-ray diffraction analysis. The results show that sound joints of Al/Fe and Cu/Fe can be obtained by continuous drive friction welding process. A discontinuous reaction layer was formed on Al/Fe interface, and no obvious reaction layer appeared on Cu/Fe interface. The tensile strength of the joints increased with increasing friction pressure, and the highest strength could reach up to 70 MPa for Al/Fe joint and 222 MPa for Cu/Fe joint. All the Al/Fe friction-welded samples failed at the friction interface, while the Cu/Fe joint under 36 and 44 MPa friction pressure failed at Cu matrix during the tensile test.

1. Introduction

The need for joints between dissimilar materials often arises in extensive industrial applications [1]. A sound joining technique can provide interesting combinations of structural and functional properties, which can offer many benefits [2]. As widely used materials, aluminum/steel structure can obviously reduce weight, and copper/steel structure can provide effectively combinations of conductivity and strength [3, 4]. Welding process is the key technology to manufacture the dissimilar materials composite structures. But it is difficult to use the traditional fusion welding method because of the tremendous difference in mechanical and metallurgical properties between dissimilar materials [5]. Continuous drive friction welding (CDFW), as the most promising method to join dissimilar materials, is importantly demanded in the manufacture of a new generation of aircraft, rocket, and shipboard equipment because of the advantages such as high material savings and low production time [6–8].

CDFW is a solid-state joining method that utilizes friction-induced thermomechanical deformation to create high-quality welds. In previous works, dissimilar metals such as Al/Fe [6, 7], Al/Ti [8], Al/Cu [9], Cu/Fe [4], Ti/Fe [10], and Cu/alumina ceramic [11] of the friction welding have been carried out by many researchers. In particularly, aluminum/steel [7] and copper/steel [12] friction welding joints present good performance with tensile strength higher than the matrix. It is generally known that the excellent mechanical property of the friction welding joint is due to its reliable metallurgical bonding quality and microstructure characteristics of ultrafine crystals and even nanocrystals [13]. During the joint formation, the materials near the interface of dissimilar metal have simultaneously encountered high temperature and severe plastic deformation. The metallurgical interactions on the dissimilar interface are critical for the joint strength and stability, which attract a lot of attention by researchers [14, 15]. Correlational studies were mainly the analysis of the relationship between the parameters and the joint microstructures and performance [16].

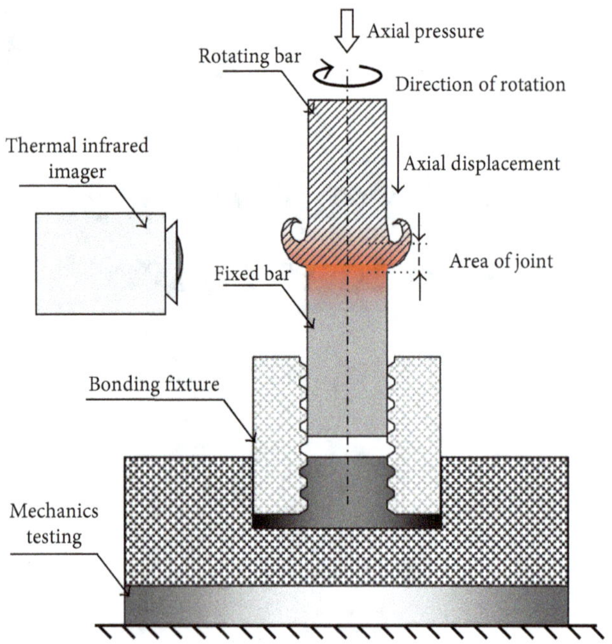

FIGURE 1: Schematic diagram of continuous drive friction welding.

However, the metallurgical compatibility between the dissimilar metals and its influence on joint microstructure evolution have not been clarified.

In this paper, the dissimilar pure metals Al/Fe and Cu/Fe with different metallurgical compatibility were designed, and a physical simulation apparatus of rotation friction welding was used to investigate the friction weldability. The microstructures of the joining interface were analyzed by scanning electron microscopy (SEM) with EDS, the joining strength was evaluated by tensile test, and the fracture was tested by X-ray diffraction analysis. The microhardness of the interface region was tested.

2. Experimentation Procedure

Commercially available bars of Al (99.6 wt.%), Cu (99.9 wt.%), and Fe (99.95 wt.%) were used. The specimens were machined with a dimension of 20 mm in the joining part and 14 mm in the clamping part. Before joining, the specimens' surfaces were milled and degreased with acetone. The experimental schematic is shown in Figure 1. Before welding, the rotating bar was rotated around its axis while the fixed bar was irrotational. When the rotation speed reached the preestablished parameter, the fixed bar moved along the axis and then contacted with the rotating bar under the action of hydraulic pressure, and then, the friction process began. The rotation speeds were 1200 rpm for Al/Fe and 1900 rpm for Cu/Fe couples. The different axial pressure was used as 20, 28, 36, 44 MPa, with friction time 5 s. In the meanwhile, the interface temperature was observed using an infrared thermographic imager (InfraTec, VarioCAM®hr head-HS) with the frame rate of 60 fps and thermocouple synchronously.

After joining, the welded sample for metallographic examination was cut perpendicular to the welding interface. The samples were subjected to metallographic characterization employing scanning electron microscopy (SEM, JEOL, JSM-6700) equipped with an energy-dispersive spectrum (EDS) and transmission electron microscope (TEM, JEM-3010). The mechanical properties of the joints were evaluated by tensile tests (Instron 5880) at a constant displacement rate of 1 mm/min at room temperature. The microhardness of the interface region was measured by the microhardness tester.

3. Results and Discussion

3.1. Joints Appearance and the Burn-Off Length. Figure 2 shows the appearance of the welded joints with different friction pressure. It is evident that the upset metal of the welded Al/Fe and Cu/Fe joints increases gradually with the friction pressure being increased. During the CDFW process, the accumulated friction heat is sufficient to soften the matrix. The upset metal subsequently forms under the squeezing action of friction pressure when the matrix reaches the plastic state. The deformation mainly occurs in Al side for Al/Fe joints and Cu side for Cu/Fe joints. The burn-off length increases with the increase of friction pressure in both joints, as shown in Table 1.

3.2. The Interface Temperature. During the CDFW process, the temperature of the interface region was detected by the thermal infrared imager and the thermocouple embedded in the fixed bar 1.5 mm from the interface simultaneously. The test results are shown in Figure 3. Figure 3(a) is a typical infrared thermal image of the welding interface, and Figure 3(b) shows the temperature changes over time during the whole welding process by the thermocouple. The area between two dashed lines in Figure 3(b) indicated the actual friction time. The temperature curves show an unsteady ascent stage and then attain a relatively stable stage which shows that the heat production and exothermic reaction reach equilibrium. The average temperatures of this equilibrium state were regarded as the welding temperature. The welding temperatures of both Al/Fe and Cu/Fe joints increase with the increase of the welding pressure.

3.3. Microstructure of Joints. The defect-free joints were obtained by the continuous drive friction welding experiments. The typical interface morphology of Al/Fe and Cu/Fe CDFW joints is shown in Figure 4. The Al/Fe interface formed a discontinuous reaction layer when the gray scale was between Al and Fe, as shown in Figure 4(a). Fe side near the interface presented relatively flat, and the reaction products were all embedded in the Al matrix. The Cu/Fe interface presented good bonding although no obvious reaction layer appeared which is shown in Figure 4(b). It showed good deformation coordination on both sides of the interface because of the similar elastic modulus between pure Cu and Fe metals. The elementary compositions on the interface region were tested by energy-dispersive X-ray spectroscopy (EDS) analysis of the interface. The EDS results for different regions in Figure 4 are shown in Table 2.

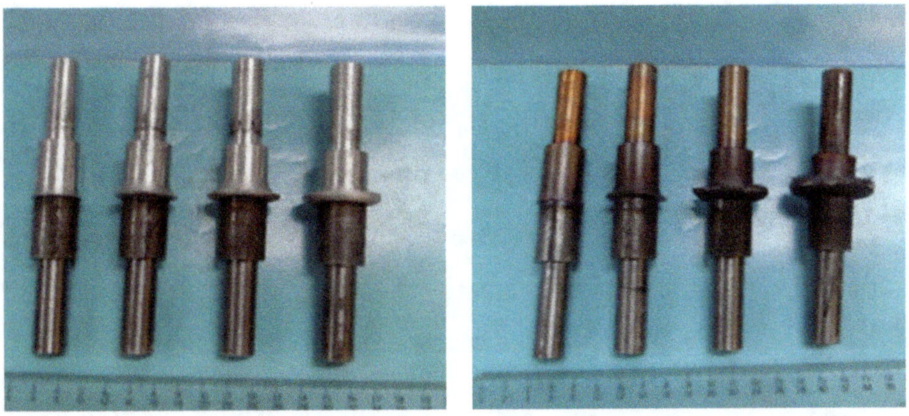

(a) (b)

FIGURE 2: Appearance of (a) Al/Fe and (b) Cu/Fe joints at different welding pressure.

TABLE 1: Friction welding parameters in the experiments and the burn-off length of joints.

Dissimilar joint	Friction pressure (MPa)	Rotation speeds (rpm)	Burn-off (mm)
Al/Fe	20	1200	2.1
	28		4.4
	36		5.4
	44		8.3
Cu/Fe	20	1900	2.9
	28		4.5
	36		6.4
	44		11.5

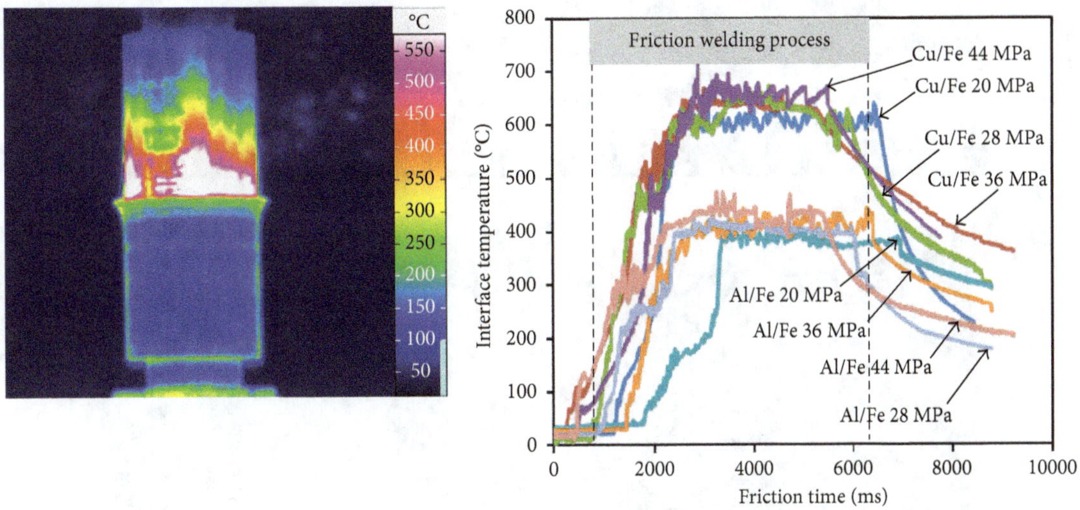

(a) (b)

FIGURE 3: The interface temperature during the CDFW process shown as (a) temperature pattern by thermal infrared imager and (b) temperature variation curve by thermocouple.

The molar ratio of Al to Fe in the reaction layer (points 01 and 02) is 3 : 1 approximately, and it could be speculated that this might be the Al_3Fe intermetallic. The region of point 03 is the Fe matrix. The compositions of the region points 04 and 05 indicated the metallurgical bonding occurred on the Cu/Fe joint. It might be nonequilibrium solid solution in region point 04 and possible phase $FeCu_4$ in region point 05.

Figure 5 shows the micromorphology of the Al/Fe joining interfaces under different welding pressure. There was no reaction product on the interface when the welding pressure was 20 MPa as shown in Figure 5(a). The amount of

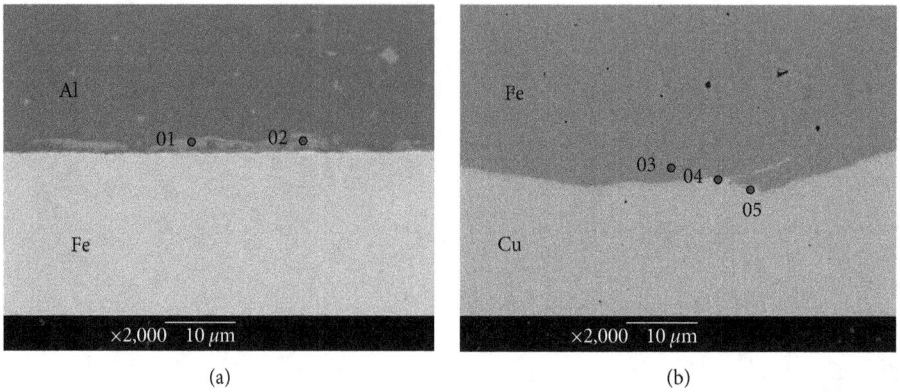

FIGURE 4: The typical interface morphology: (a) Al/Fe interface and (b) Cu/Fe interface.

TABLE 2: Composition analyses of the points indicated in Figure 4 (at.%).

Test positions	Al	Fe	Cu	O	Possible phase
01	73.78	24.02	—	2.20	Al_3Fe
02	75.67	23.58	—	0.75	Al_3Fe
03	—	99.94	—	0.06	Fe matrix
04	—	72.21	25.34	2.45	Solid solution
05	—	20.13	78.27	1.60	$FeCu_4$

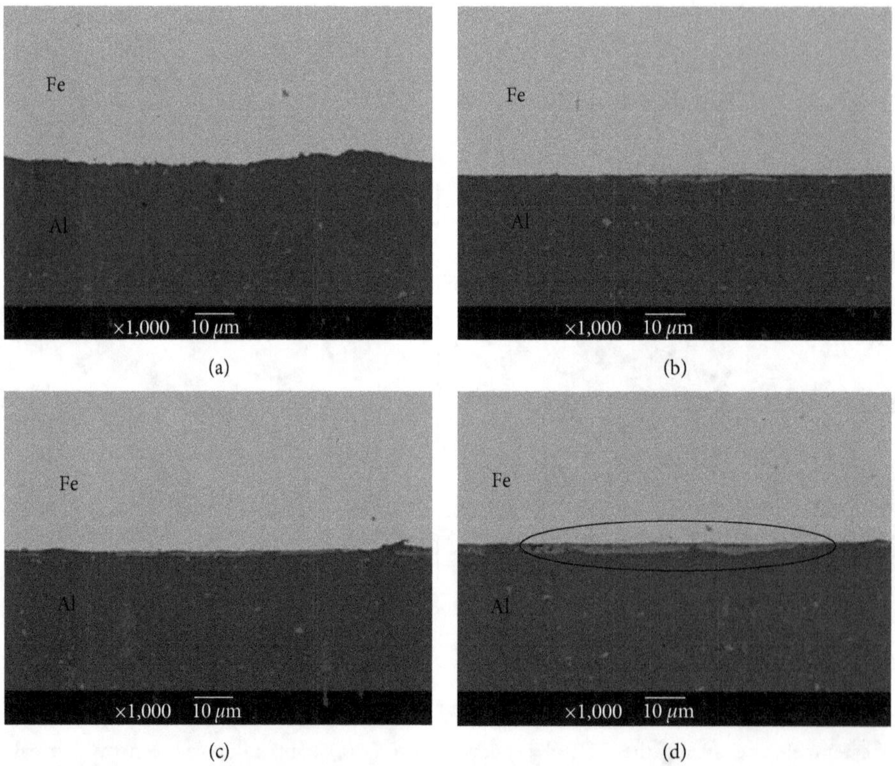

FIGURE 5: Microstructures of Al/Fe welding region metallurgical reaction layer under: (a) 20; (b) 28; (c) 36; (d) 44 MPa.

the reaction product increases with the increase of welding pressure (Figures 5(b)–5(d)). The reaction layer presented discontinuous distribution and uneven thickness characteristics. The thickest region reached 4 μm.

There was no visible difference in the morphology of the Cu/Fe joining interfaces under different welding pressure. Thus, the detailed microstructures of the interfacial region on the Cu/Fe joining interface were detected. Figure 6 shows the

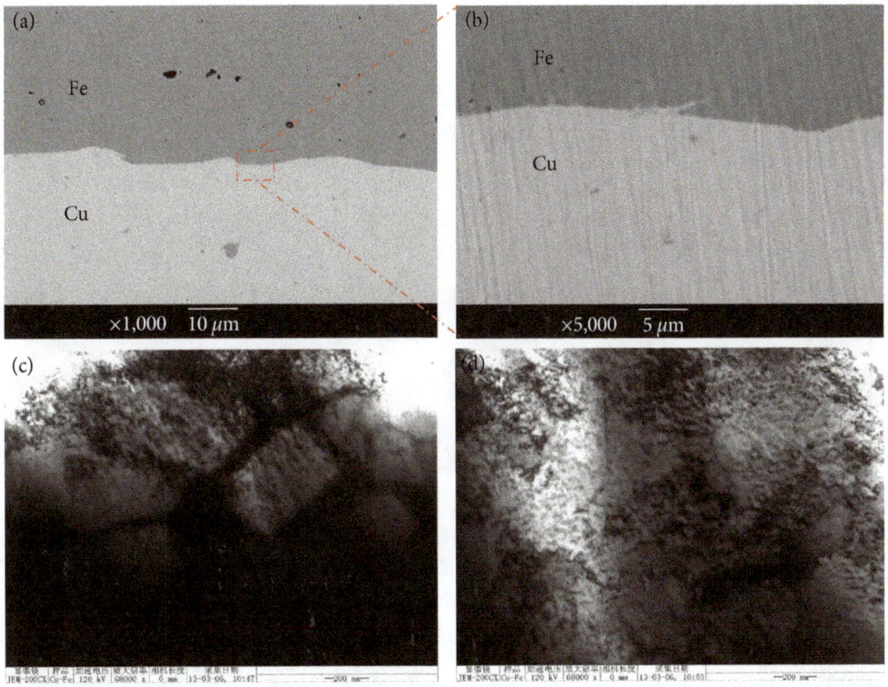

FIGURE 6: The typical interface morphology of the Cu/Fe interface shown as (a, b) SEM images, (c) TEM image of welding interface, and (d) dislocation line.

typical micromorphology of the Cu/Fe joining interface under 36 MPa welding pressure. The images of larger magnification (shown in Figure 6(b)) presented gradient transition characteristics in gray scale with different element content.

Further analysis obtained by TEM shows better solution on Cu/Fe joining interface as shown in Figure 6(c). This indicated that the mutual diffusion between Cu and Fe occurred though Cu–Fe was an immiscible system in thermodynamic equilibrium state. Figure 6(d) shows the high-density dislocation in severe deformed Cu matrix near the interface. For forced mixing, alloy components are thus assumed to flow down gradients in their concentrations, while for thermally activated diffusion atoms flow down gradients in their chemical potentials. So, the diffusion in immiscible Cu/Fe couples was most attributed to the forced mixing along the concentration gradient.

3.4. Mechanical Properties of the Joints. Figure 7 shows the tensile strength of the Al/Fe and Cu/Fe joints as a function of the welding pressure. It can be found that the tensile strength increases along with the welding pressure and reaches a maximum value as 70 MPa for Al/Fe joint and 222 MPa for Cu/Fe joint. The dashed lines in Figure 7 denote the tensile strength of Al matrix with 88 MPa and Cu matrix with 218 MPa obtained by a tensile test. Such result implies that joining strength of Cu/Fe CDFW joints falls around that of the Cu matrix at the welding pressure of 36 MPa and the ultimate fracture happens at Cu matrix. The results indicated that the sound joint with defect-free interface microstructure and excellent joint strength was obtained for Cu/Fe dissimilar metals. However, the maximum joining strength of Al/Fe CDFW joints was slightly lower than that of Al matrix.

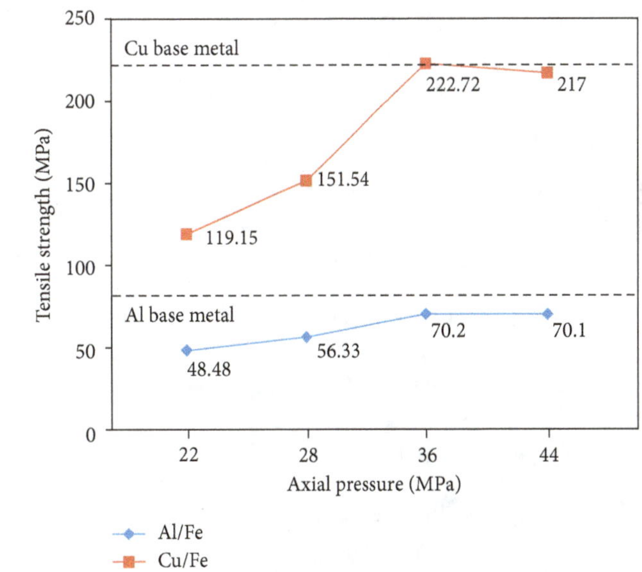

FIGURE 7: The tensile strength of the Al/Fe and Cu/Fe joints as a function of welding pressure.

This might be due to the existence of the discontinuous reaction layer and the differences of Young's modulus between Al and Fe. All the joints were fractured at the Al/Fe interface. Figure 8 shows the representative stress-strain curve of the Al/Fe and Cu/Fe joints during the tensile test.

The microhardness across the interface of the Al/Fe and Cu/Fe joints was tested. The interface microhardness distributions of different joints are shown in Figure 9. The hardness value increased near the interface. This is mainly due to grain refinement in the deformation area.

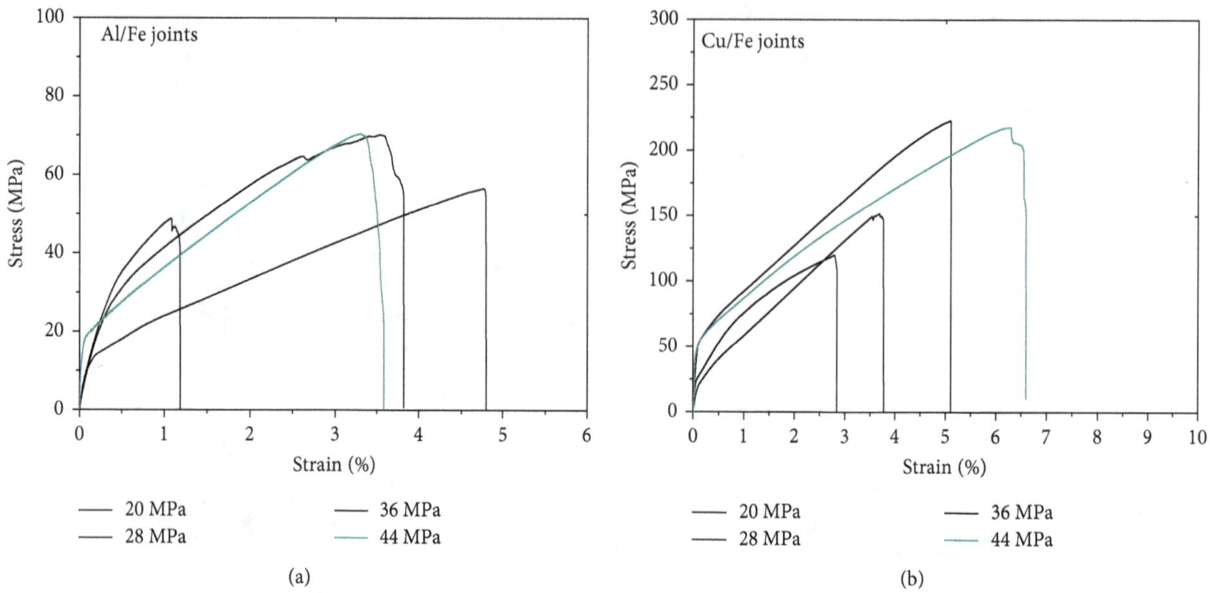

Figure 8: The representative stress-strain curve of the Al/Fe and Cu/Fe joints during the tensile test.

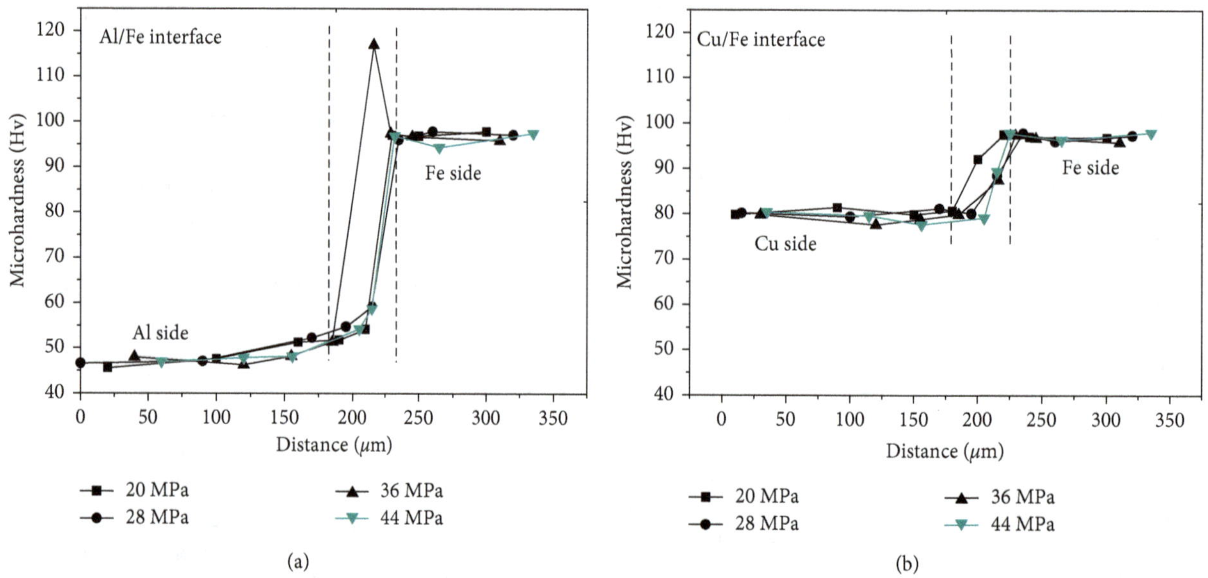

Figure 9: The interface microhardness distributions of different joints: (a) Al/Fe interface and (b) Cu/Fe interface.

3.5. Fracture Morphology and Phase Analysis. Figure 10 shows the XRD phase analysis and morphologies of the fractured surfaces of the Al/Fe joints. The XRD phase analysis result indicated that there was small amount of Fe on the Al side fractured surfaces and small amount of Al on the Fe side fractured surfaces. There are no obvious peaks of intermetallic compounds. Figures 10(b) and 10(c) show the morphologies of the Al side and Fe side fractured surfaces of the Al/Fe joint. It can be seen that the morphologies of Al side fractured surfaces showed some gap ups and downs. However, the morphologies of Fe side were mainly the flat region partially covered by a mixed layer.

Figure 11 shows the XRD phase analysis and morphologies of the fractured surfaces of the Cu/Fe joints. It can be found that there was only Cu phase on the fractured surfaces of Cu side, while Fe, Cu, and $FeCu_4$ on the fractured surfaces of Fe side, as shown in Figure 11(a). Figure 11(b) shows the morphologies of the Cu side fractured surfaces of the Cu/Fe joint. It was a typical ductile fracture of Cu matrix. Figure 11(c) shows the morphologies of the Fe side fractured surfaces of the Cu/Fe joint. Some small particles appeared in the vortices. The result indicated that the sample was fractured in Cu matrix near the interface. Then, the tensile strength was close to Cu metal. There was a small amount of $FeCu_4$ intermetallic compounds formed on the interface. The solid solubility of Cu and Fe element was achieved on the interface.

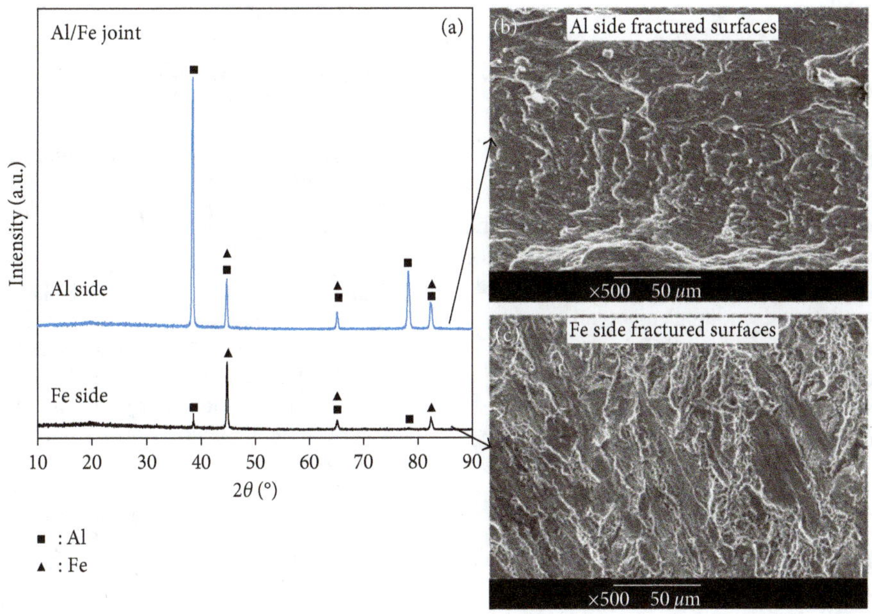

FIGURE 10: XRD phase analysis and morphologies of the fractured surfaces of the Al/Fe joints.

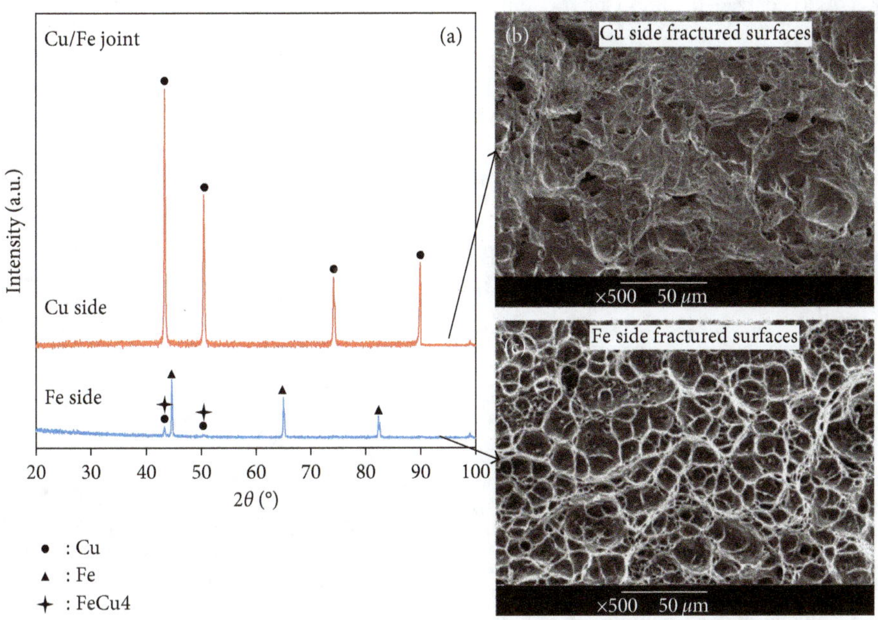

FIGURE 11: XRD phase analysis and morphologies of the fractured surfaces of the Cu/Fe joints.

4. Conclusions

The dissimilar pure metals Al/Fe and Cu/Fe with different metallurgical compatibility were joined by continuous drive friction welding. The microstructure of the joining interface was analyzed, the joining strength was evaluated by tensile shear test, and the fracture was tested by X-ray diffraction analysis. The main conclusions are as follows:

(1) The sound joints of Al/Fe and Cu/Fe can be obtained by continuous drive friction welding process. A discontinuous reaction layer was formed on Al/Fe friction interface and no obvious reaction layer appeared on Cu/Fe interface.

(2) The tensile strength of the joints increased with increasing friction pressure, and the highest strength could reach up to 70 MPa for Al/Fe joint and 222 MPa for Cu/Fe joint. All the Al/Fe friction-welded samples failed at the friction interface, while the Cu/Fe joint under 36 and 44 MPa friction pressures failed at Cu matrix during tensile test.

Conflicts of Interest

The authors declare that they have no conflicts of interest.

Acknowledgments

This work is financially supported by the National Natural Science Foundation of China under Grant no. 51701154, the Key Research and Development Project of Shaanxi Province under Grant no. 2017ZDXM-GY-033, the China Post-doctoral Science Foundation-Funded Project under Grant no. 2016M592823, the Scientific Research Project of Shaanxi Provincial Department of Education under Grant no. 17JK0563, and the Scientific Research Foundation of Xian University of Technology under Grant no. 101-451115015.

References

[1] W. Guo, G. Q. You, G. Y. Yuan, and X. L. Zhang, "Microstructure and mechanical properties of dissimilar inertia friction welding of 7A04 aluminum alloy to AZ31 magnesium alloy," *Journal of Alloys and Compounds*, vol. 695, pp. 3267–3277, 2017.

[2] A. Yazdipourn and A. Heidarzadeh, "Effect of friction stir welding on microstructure and mechanical properties of dissimilar Al 5083-H321 and 316L stainless steel alloy joints," *Journal of Alloys and Compounds*, vol. 680, pp. 595–603, 2016.

[3] Y. Kusuda, "Honda develops robotized FSW technology to weld steel and aluminum and applied it to a mass production vehicle," *Industrial Robot: An International Journal*, vol. 40, pp. 208–212, 2013.

[4] M. Sahin, E. Cil, and C. Misirli, "Characterization of properties in friction welded stainless steel and copper materials," *Journal of Materials Engineering and Performance*, vol. 22, no. 3, pp. 840–847, 2013.

[5] Y. N. Wei, J. L. Li, J. T. Xiong, F. Huang, F. S. Zhang, and S. H. Raza, "Joining aluminum to titanium alloy by friction stir lap welding with cutting pin," *Materials Characterization*, vol. 71, pp. 1–5, 2012.

[6] A. Ambroziak, M. Korzeniowski, P. KustroN, M. Winnicki, P. SokoBowski, and E. HarapiNska, "Friction welding of aluminium and aluminium alloys with steel," *Advances in Materials Science and Engineering*, vol. 2014, Article ID 981653, 15 pages, 2014.

[7] M. Kimura, M. Kusaka, K. Kaizu, K. Nakata, and K. Nagatsuk, "Friction welding technique and joint properties of thin-walled pipe friction-welded joint between type 6063 aluminum alloy and AISI 304 austenitic stainless steel," *International Journal of Advanced Manufacturing Technology*, vol. 82, no. 1–4, pp. 489–499, 2016.

[8] M. Meisnar, S. Baker, J. M. Bennett et al., "Microstructural characterisation of rotary friction welded AA6082 and Ti-6Al-4V dissimilar joints," *Materials and Design*, vol. 132, pp. 188–197, 2017.

[9] M. N. Avettand-Fènoël, G. Racineux, L. Debeugny, and R. Taillard, "Microstructural characterization and mechanical performance of an AA2024 aluminium alloy—pure copper joint obtained by linear friction welding," *Materials and Design*, vol. 98, pp. 305–318, 2016.

[10] X. Li, J. L. Li, Z. X. Liao, F. Jin, F. S. Zhang, and J. T. Xiong, "Microstructure evolution and mechanical properties of rotary friction welded TC4/SUS321 joints at various rotation speeds," *Materials and Design*, vol. 99, pp. 26–36, 2016.

[11] P. Li, J. L. Li, H. G. Dong, and C. Z. Ji, "Metallurgical and mechanical properties of continuous drive friction welded copper/alumina dissimilar joints," *Materials and Design*, vol. 127, pp. 311-319, 2017.

[12] M. Kimura, M. Kusaka, K. Kaizu, and A. Fuji, "Effect of friction welding condition on joining phenomena and tensile strength of friction welded joint between pure copper and low carbon steel," *Journal of Solid Mechanics and Materials Engineering*, vol. 3, no. 2, pp. 187–198, 2009.

[13] R. Palanivel, I. Dinaharan, and R. F. Laubscher, "Assessment of microstructure and tensile behavior of continuous drive friction welded titanium tubes," *Materials Science and Engineering A*, vol. 687, pp. 249–258, 2017.

[14] Y. Morisada, T. Imaizumi, and H. Fujii, "Determination of strain rate in friction stir welding by three-dimensional visualization of material flow using X-ray radiography," *Scripta Materialia*, vol. 106, pp. 57–60, 2015.

[15] Y. N. Wei, J. T. Xiong, J. L. Li, F. S. Zhang, and S. H. Liang, "Microstructure and enhanced atomic diffusion of friction stir welding aluminium/steel joints," *Materials Science and Technology*, vol. 33, no. 10, pp. 1208–1214, 2017.

[16] M. Pourali, A. Abdollah-zadeh, T. Saeid, and F. Kargar, "Influence of welding parameters on intermetallic compounds formation in dissimilar steel/aluminum friction stir welds," *Journal of Alloys and Compounds*, vol. 715, pp. 1–8, 2017.

10

Effect of the Noble Metals Addition on the Oxidation Behavior of Ni_3Al

J. J. Ramos-Hernandez,[1] J. Porcayo-Calderon,[2] J. G. Gonzalez-Rodriguez,[2] Jan Mayén,[3] G. Lara-Rodriguez,[4] V. M. Salinas-Bravo,[5] and L. Martinez-Gomez[6,7]

[1]*Facultad de Química, Universidad Nacional Autónoma de Mexico, Edificio B, Cd. Universitaria, 04510 Ciudad de Mexico, Mexico*
[2]*CIICAp, Universidad Autónoma del Estado de Morelos, Avenida Universidad 1001, 62209 Cuernavaca, MOR, Mexico*
[3]*CIATEQ A.C., Eje 126 No. 225, Zona Industrial, 78395 San Luis Potosí, SLP, Mexico*
[4]*Instituto de Investigaciones en Materiales, Universidad Nacional Autónoma de Mexico, Cd. Universitaria, 04510 Ciudad de Mexico, Mexico*
[5]*Instituto Nacional de Electricidad y Energías Limpias, Avenida Reforma 113, Colonia Palmira, 62490 Cuernavaca, MOR, Mexico*
[6]*Instituto de Ciencias Físicas, Universidad Nacional Autónoma de Mexico, Avenida Universidad s/n, 62210 Cuernavaca, MOR, Mexico*
[7]*Corrosion y Protección (CyP), Buffon 46, 11590 Mexico City, Mexico*

Correspondence should be addressed to J. Porcayo-Calderon; jporcayoc@gmail.com

Academic Editor: Amit Bandyopadhyay

This paper discusses the effect of the addition of noble metals on the microstructure, mechanism, and oxidation kinetics of the Ni_3Al intermetallic alloy. Ni_3Al was doped with 1% (atomic percent) of Au, Ag, Pd, and Pt. Oxidation behavior of the alloys was evaluated at 900, 1000, and 1100°C in O_2 for 24 hours. XRD analysis showed that the addition of noble metals favored the oxide growth on preferential crystallographic planes. In addition, the preferential substitution of the noble metals in the Ni_3Al structure modifies the surface composition by increasing the Al/Ni ratio. It was observed that most of the alloys showed a subparabolic behavior, and only the intermetallic base and the alloy doped with Ag show a parabolic behavior at 900°C. The developed oxides were analyzed both superficially and in cross section by scanning electron microscopy (SEM) and energy dispersive X-ray analysis (EDXA). It was evident that only the intermetallic base showed the formation of a duplex oxide scale (Al_2O_3/NiO). The alloys doped with noble metals showed the oxide growth practically of pure Al_2O_3. This was due to a decrease in the diffusion of the Ni cations because of the presence of the noble metals in the crystalline structure.

1. Introduction

Ni_3Al intermetallic compounds are excellent candidates for high-temperature structural applications due to their thermochemical and mechanical properties. They can perform beyond conventional temperatures because of their excellent resistance to corrosion and oxidation [1]. During solidification of an alloy with stoichiometry close to that of Ni_3Al, there is in the liquid phase a competition for phase stability between β_{NiAl}, γ'_{Ni3Al}, and γ_{Ni} with stable eutectic (β-γ'), metastable (β-γ), and peritectic (γ'-γ) phases which have common borders in the Ni-Al binary diagram. Because the region of phase stability γ' is very narrow, any deviation in the Ni/Al ratio favors the stability of any of the aforementioned phases; that is, an increase in the Ni content tends to favor the stability of γ'-γ phase. According to Verin [2], the microstructure of the Ni-23Al-0.001Ce-0.004Co (at %) is composed of a combination of dendritic phases γ' and γ, where the matrix is the phase γ' and the interdendritic phase is the phase γ; similar results were obtained by Chiba and Hanada [3] for the alloy Ni-23Al (at %) and Jozwik et al. [4] for the alloy Ni-22Al-0.002Zr-0.001B (at %). Moreover, Hunziker and Kurz [5] mentioned that the directional solidification of intermetallic Ni-25Al (at %) shows a microstructure where the phases β, γ, and γ' are found. In this case, the γ' phase is formed during cooling of metastable eutectic β-γ. It has been shown that there are subsequent transformations after the solidification process, which complicates the interpretation of the phases produced.

On the other hand, the solidification rate is a factor that influences the microstructure of the alloy. When the solidification is slow, the convection in the liquid phase favors formation of a single γ'-phase. However, in a rapid solidification process, there is little convection in the liquid phase and it promotes the formation of a structure with phases γ'' and β [6]. The addition of a third element to the Ni-Al system may stabilize or destabilize a given phase. According to Inoue et al. [7], the addition of Cr, Mn, Fe, Co, or Si destabilizes the phase γ', and according to Goetzinger et al. [8], the addition of Ti stabilizes the phase. Sun et al. [9] indicate that Pd, Ag, and Cu are elements forming the γ phase, while Pint et al. [10] show that Pt promotes the reduction of the temperature of the phase transformation from β to γ' during the process of cooling and phase transformation from γ' to β during heating.

During the oxidation process, β and γ' phases play an important role due to their content of Al [11]. According to Haerig and Hofmann [12], in the early stages of the oxidation process, NiO and Al_2O_3 are formed simultaneously due to the diffusion of cations towards the outside of the alloy. Bardi et al. [13] show that the growth of the oxide is in the form of islands. Al consumption owing to the formation of Al_2O_3 diminishes the aluminum content below 17% (at %), which is the content needed to form the outer protective layer [14]. This favors the growth of NiO onto the Al_2O_3 layer. This pattern occurs because of the conducting properties of the formed oxides: Al_2O_3 is an amphoteric oxide (movement of cations and anions is possible) allowing outward diffusion of Ni^+ cations. Al_2O_3 forms internally due to inward diffusion of O^{2-} anions, while NiO is a p-type conductor, which preferably promotes conduction of cations [15–17]. Al_2O_3 growth onto alloys such as Ni_3Al can adopt different crystallographic arrangements: between 427 and 927°C, the γ-Al_2O_3 phase is stable; until 1077°C, the θ-Al_2O_3 phase is stable; and above 1077°C, the α-Al_2O_3 phase is stable [18]. The θ → α transformation is accompanied by a change in the transport properties of the scale. The θ-Al_2O_3 scale mainly grows by outward diffusion of cations, forming cavities in the scale/alloy interface, resulting in the deterioration of the adhesion layer on the substrate. In contrast, the growth of α scale is dependent on the diffusion of the O^{2-} anion inwards, so that the oxide fills the preformed cavities. This improves the adhesion of the layer on the substrate. Therefore, it is desirable for the growth of an α-Al_2O_3 layer [1, 14]. It has been observed that the only growth of an outer scale of Al_2O_3 is possible under the following conditions, by increasing both the temperature and concentration of Al [12], where the oxidation process is carried out at oxygen partial pressures below the equilibrium of NiO formation [13, 19–21] and also in alloys which were implanted with B^+ and Cr^+ ions [14]. Increasing the oxygen partial pressure and temperature, it is possible to form the spinel $NiAl_2O_4$ at the NiO/Al_2O_3 interface [22]. Garza et al. [23] attributed this behavior to the difference between the free energies of formation of Al_2O_3 and NiO which is four times larger than that between Al_2O_3 and $NiAl_2O_4$. Velon and Olefjord [15] mentioned that the spinel could also be formed in the initial stages of oxidation at low temperatures (300°C).

A key aspect to increase the oxidation resistance of an alloy is to improve the adherence of the protective oxide mainly because most applications implicate high-temperature thermal cycles [13, 24]. The most effective method is the addition of elements or compounds such as Hf, Y, Zr, Cr, Pd, Pt, and SiC [25–31], which can change the localized concentration of aluminum and reduce its growth rate. Gleeson et al. [32] indicate that the addition of Pt has beneficial effects in promoting the formation of a stable Al_2O_3 scale. Pt is an unreactive element whose addition decreases the chemical activity of Al in the γ' phase, and it increases the Al/Ni ratio, which promotes the formation of Al_2O_3 in certain crystallographic planes and increases the Al diffusion into the formed scale. It has been calculated that the preferential site of substitution of noble metals like Ag, Au, Pd, and Pt in the Ni_3Al structure occurs in locations where there are normally nickel atoms [33–37]; therefore, the addition of these elements can modify the oxidation behavior of the alloy. Therefore, this paper evaluates the effect of the addition of these elements in low concentration (1 at.%) on the oxidation behavior of the Ni_3Al intermetallic alloy.

2. Experimental Procedure

The intermetallic alloys with nominal composition $Ni_{75}Al_{25}$ and $Ni_{75}Al_{24}M_1$ (where M = Ag, Au, Pd, and Pt) were manufactured in quartz crucibles in an induction furnace under inert atmosphere using elements with purity higher than 99.99%. The ingots were prepared by the fusion of stoichiometric amounts of the high-purity elements (solid rods), and the melt was homogenized by the convective currents generated by the induction furnace. After melting the metal elements, the ingots were allowed to cool to room temperature inside the induction furnace. From the ingots, samples were cut with dimensions of 20 × 5 × 2 mm to carry out the tests.

XRD analysis of alloys was carried with a Bruker D8 Advance diffractometer equipped with a Cu Kα (k = 1.5418 Å; 40 kV) monochromatic source and a LYNXEYE™ 1D positional detector. Scans were performed from $2\theta = 5°$ to 100° with a 0.02° step and an irradiation time of 0.2 seconds.

In order to reveal the microstructure of the intermetallic alloys, they were etched by immersing in HF-HNO_3 solution (1 : 1 molar ratio) for 5 to 10 seconds at room temperature.

For the oxidation test, samples were grinded with a silicon carbide paper to 600-grit size, washed with water, and ultrasonically cleaned with acetone. In order to have a larger reaction area, two samples were employed in each test: one was used for surface analysis and the other for cross-sectional analysis.

Oxidation tests were carried out in pure oxygen at 900, 1000, and 1100°C. This range was chosen since at these temperatures, there is a change in the stability of the different

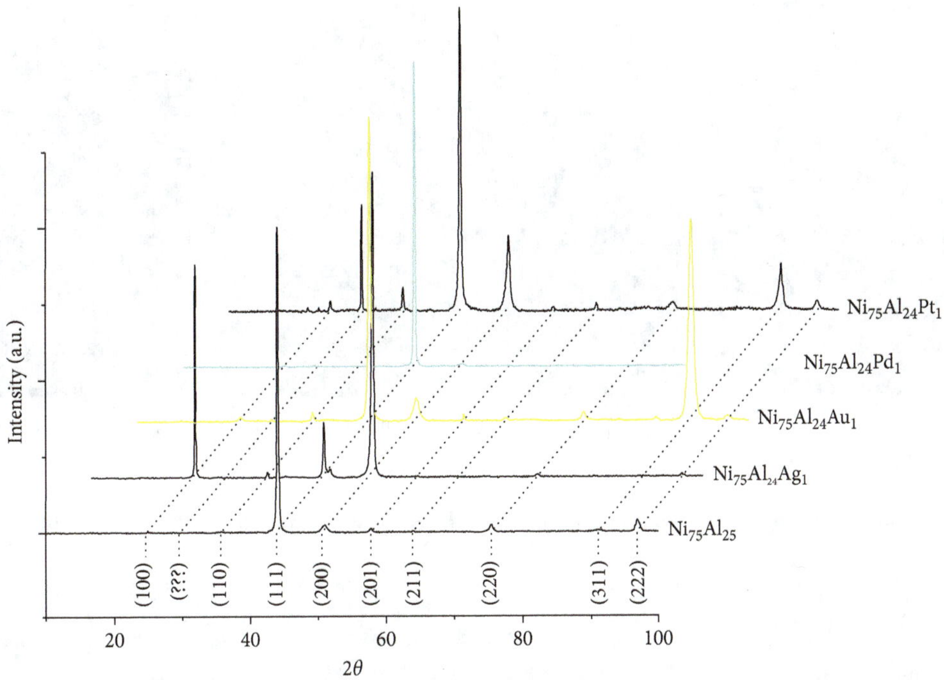

FIGURE 1: The XRD patterns of Ni_3Al and Ni_3Al doped with 1 at.% of noble metals.

phases of Al_2O_3; at 900°C, transition between γ-phase and θ-phase begins; at 1000°C, θ-phase is stable; and at 1100°C, α-phase is stable. Before each test, the electric furnace was preheated to the experimental temperature. The record of the weight gain curves as a function of exposure time was done with a thermobalance CI MK2-M5 Head LM 2-01 and the LabWeigh software. Analyses were carried out by means of a scanning electron microscope (Zeiss model DSM 960) and the energy dispersive X-ray analysis (EDXA).

3. Results and Discussion

3.1. XRD Analysis of Alloys. The XRD profiles of the alloys with and without noble metal additions are shown in Figure 1. The XRD profile of the Ni_3Al alloy shows the characteristic peaks (111), (200), (220), (311), and (222) corresponding to the L_{12} superstructure (γ'-phase), and their relative intensities are consistent with previous reports [38–40], which indicate a complete transformation from metals to the Ni_3Al intermetallic alloy.

The peaks observed in the XRD profiles of the doped alloys are wide, suggesting a small crystallite size. In addition, the intensities of some peaks increase and decrease depending on the added noble metal. The $Ni_{75}Al_{24}Ag_1$ profile shows highest intensities in (100) and (200), and the intensity of the peak decreased in (111); this suggests the crystal growth in two preferential planes. This preferential growth was also observed in the $Ni_{75}Al_{24}Pd_1$ alloy, where the profile shows a slight increase in the (222) peak, indicating a preferential unidirectional growth [41]. The XRD profile for the $Ni_{75}Al_{24}Au_1$ alloy shows a notable increase in the intensity of the (311) peak and a slight increase in the intensity of the (200) peak, and the characteristic peak (111) shows no change. On the other hand, the XRD profile of the $Ni_{75}Al_{24}Pt_1$ alloy shows an increase in the intensity of all its peaks, and an unidentified peak at $2\theta = 29.56$ is observed. Considering that the positions of the atoms in the unit cell affect the intensities but not the directions of the diffracted beams [42], the change in peak intensities can be completely attributable to the addition of noble metals.

Properties such as hardness show a significant dependence on crystallographic planes: for Ni_3Al, the highest value occurs in (100), followed by (111) and (110) [43]. However, several studies about the Ni_3Al oxidation indicate that there are no significant differences in the oxidation behavior between planes (100), (110), and (111) [44–48]. Taking into account that the planes (100) and (110) consist of a series of two layers: the most stable having a composition of 50% Ni and 50% Al and the other with 100% Ni and the plane (111) having a composition of 75% Ni-25% Al [49, 50]; then due to it, it is possible the presence of zones with different chemical compositions onto alloy surface (i.e., change in the Al/Ni ratio). In addition, substitution at preferential sites by the addition of noble metals can further change the surface chemical composition, thereby greatly influencing its oxidation behavior.

3.2. Microstructural Analysis. Figure 2 shows the microstructure of the intermetallic $Ni_{75}Al_{25}$ alloy, where the presence of two phases is observed. The EDXA showed that the dendrite is a phase with high concentration of Al, so that the (γ-γ') phase is in the matrix and the (β-γ') phase is in the dendrite core as indicated by Hunziker and Kurz [51] and Rosas et al. [52], respectively. During the solidification process, there is no cooling rate sufficiently slow to maintain equilibrium in the melt; that is, there is insufficient time for the composition gradients to be diffusion removed and the dendrites retained [53]. Its formation also may be

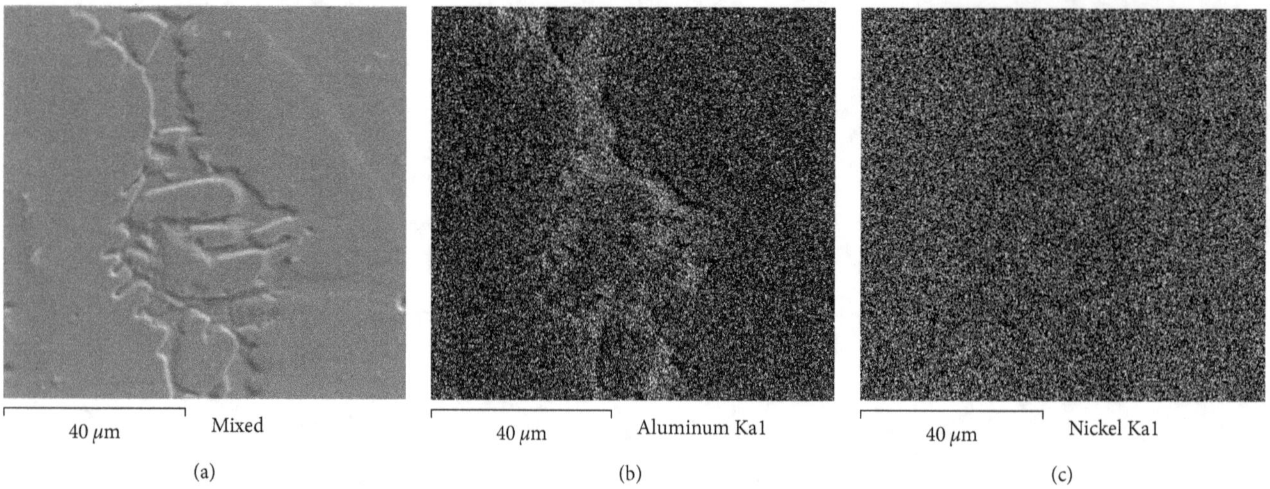

FIGURE 2: Microstructure of the $Ni_{75}Al_{25}$ intermetallic alloy (a) and EDXA mapping for Al (b) and Ni (c) of the dendrite structure.

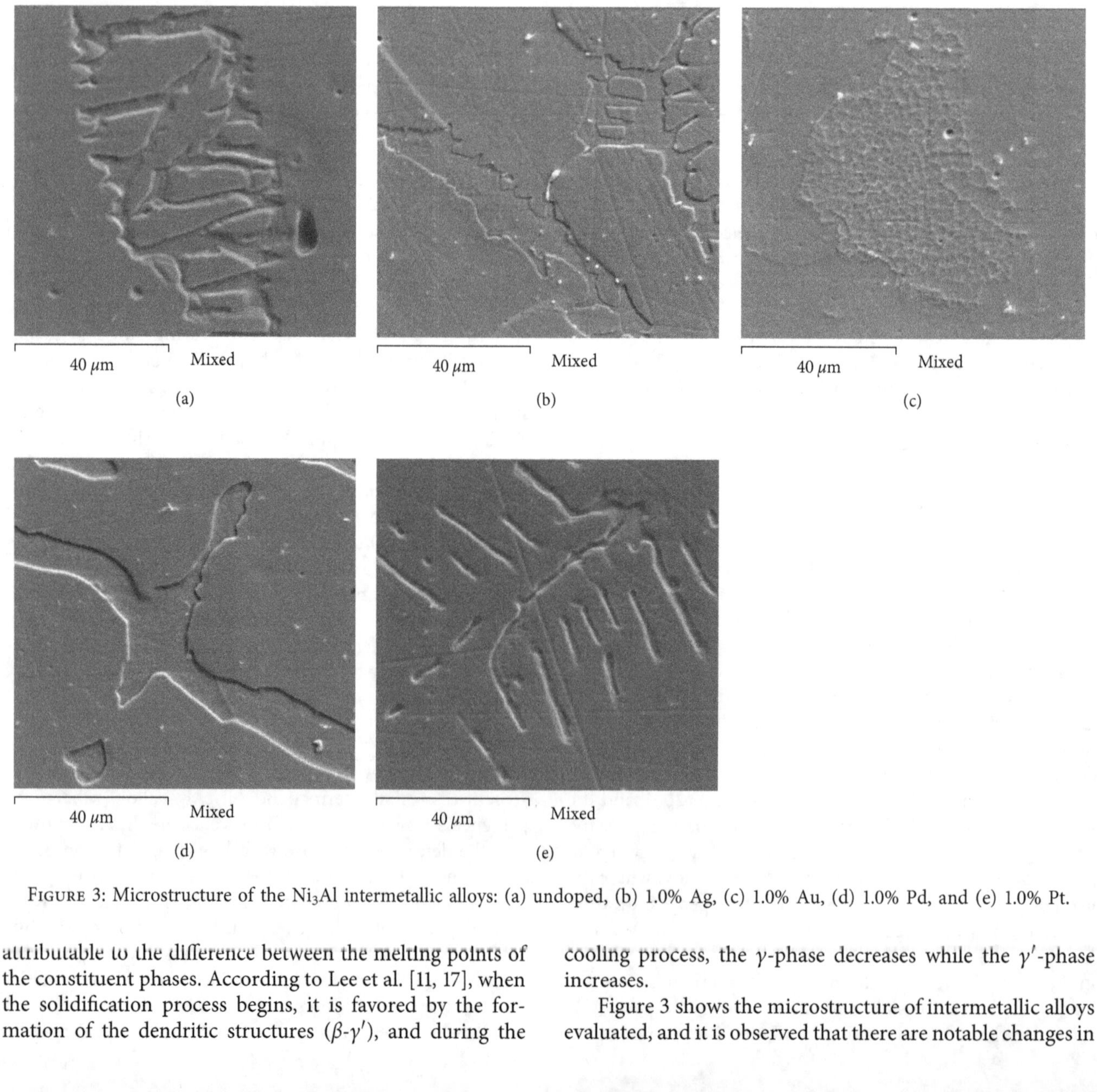

FIGURE 3: Microstructure of the Ni_3Al intermetallic alloys: (a) undoped, (b) 1.0% Ag, (c) 1.0% Au, (d) 1.0% Pd, and (e) 1.0% Pt.

attributable to the difference between the melting points of the constituent phases. According to Lee et al. [11, 17], when the solidification process begins, it is favored by the formation of the dendritic structures (β-γ'), and during the cooling process, the γ-phase decreases while the γ'-phase increases.

Figure 3 shows the microstructure of intermetallic alloys evaluated, and it is observed that there are notable changes in

TABLE 1: Spaces between secondary arms and space ratio, taking as reference the undoped alloy.

Alloy	Space between secondary arms (μm)	Space ratio between secondary arms with respect to Ni_3Al
Ni_3Al	47.00	1.00
Ni_3Al + 1% Ag	23.30	0.49
Ni_3Al + 1% Au	73.04	1.55
Ni_3Al + 1% Pd	17.07	0.36
Ni_3Al + 1% Pt	10.24	0.22

the microstructure of the doped alloys, which can be explained according to the preferential growth plane of each alloy (Figure 1) and an increase in the solidification rate. The dendrite structure of $Ni_{75}Al_{25}$ (Figure 3(a)) has a spacing between secondary arms of 47 μm, and this value is typical in a conventional solidification process [54]. The space between the secondary arms of the dendrites is inversely related to the cooling rate during the solidification process, which has been correlated according to the following equation [55]:

$$d_2 = kv^{-1/3}, \quad (1)$$

where d_2 is the secondary dendrite arm spacing, v is the solidification cooling rate, and k is a constant of the alloy.

Since the solidification cooling rate is inversely proportional to the space between the secondary arms of the dendrites, from Table 1, it can be noticed that the solidification cooling rate in the $Ni_{75}Al_{24}Pd_1$, $Ni_{75}Al_{24}Pt_1$, and $Ni_{75}Al_{24}Ag_1$ alloys was faster than in the $Ni_{75}Al_{25}$ alloy, so that there is a lower space between the secondary dendrite arms; moreover, several directions of dendrites in the $Ni_{75}Al_{24}Pt_1$ alloy are congruent with its preferential growth plane. This same affirmation is consistent with the $Ni_{75}Al_{24}Pd_1$ alloy with almost only one preferential growth plane. The solidification cooling rate in the $Ni_{75}Al_{24}Au_1$ alloy was slower than the undoped alloy such that the dendrites almost disappeared.

It is observed that the addition of Ag favors an increase in the ratio γ-γ'/β due to the reduction in the number and volume of the interdendritic phase (Figure 3(b)). Because of the low solubility of Ag in Ni [56], it is possible to observe the presence of Ag-rich phases. According to Zhou and Guo [57, 58], the presence of Ag at grain boundaries begins during the solidification process since this element segregates at the solid-liquid interface, and at the same time, the Ag solubility decreases causing their precipitation into the grains. The microstructure of the Au-doped intermetallic alloy showed a honeycomb appearance (Figure 3(c)); that is, there is not a tendency to form precipitates as in the case of Ag-doped intermetallic alloy. Sheng et al. [59] found that the effect of addition of 1% Au on the microstructure of the NiAl intermetallic alloy causes a grain size reduction and the formation of α-$AlAu_2$ precipitates. It can be assumed that the absence of the α-$AlAu_2$ phase is due to the lower content of Al in Ni_3Al compared to that in NiAl. Figures 3(d) and 3(e) show the microstructure of the intermetallic alloy doped with Pd and Pt, respectively. The Pd-doped alloy has dendritic structures of bigger area; however, the relationship γ-γ'/β is much greater than in the alloy doped with Ag and Au, indicating that, at room temperature, the γ-γ' phase is stable. The Pt-doped alloy has very fine dendritic formations in which there is no enrichment of Al. In both cases, a uniform distribution of Pd and Pt is observed as in the case of the Au-doped intermetallic alloy. Meng et al. [60] indicate that the addition of Pd into Ni_3Al intermetallic alloy generates an increase of delocalized bonding electrons, and this causes a strengthening of the grain boundaries and improves the ductility of the alloy. Furthermore, Li et al. [61, 62], Oquab and Monceau [63], and Zhang et al. [64] mention that Pd plays an important role in stabilizing the β-phase in NiAl at high temperatures; the same effect is attributed to the addition of Pt [10]. No precipitates were observed in the interdendritic region of the alloys, indicating that the phases found there are results of a eutectic reaction, contrasting with the work of Lee et al. [11], which presents cell interdendritic microstructures that Hunziker and Kurz [5] defined as peritectic structures.

3.3. Oxidation Kinetics. Figure 4 shows the curves (solid lines) of weight gain per unit area as a function of exposure time during the process of oxidation of the intermetallic alloys between 900 and 1100°C. The curves were fitted (dashed lines) to an equation of the form [24, 65]:

$$\Delta m = kt^n, \quad (2)$$

where Δm is the weight gain per unit area, k is the oxidation rate constant, n is the exponent of velocity, and t is the time of exposure. The values of k and n to fit the experimental results are shown in Table 2. Adjustments were done using a nonlinear least-squares regression method (Gauss–Newton method) [66]. According to the n values obtained, it can be observed that only at 900°C, the oxidation kinetics of the Ni_3Al alloy and the Ag-doped alloy follows a parabolic behavior, all other n values correspond to a subparabolic behavior. The oxidation mechanism of a protective oxide that grows with parabolic kinetics mainly consists of the diffusion of cations from the metal-oxide interface to the oxide-gas interface; however, when the oxidation mechanism approaches to cubic kinetics, some authors attribute this behavior to the oxide grain growth [67, 68], although other authors claim that the formation of a protective layer like α-Al_2O_3 is more a rule than an exception [69]. An additional characteristic of the subparabolic kinetics is that at the beginning of the oxidation process, a great weight gain is observed with respect to that observed in a parabolic kinetic, being this a characteristic of the transient stage of the oxidation process.

At 900°C, Ag- and Au-doped alloys had the highest weight gain throughout the exposure period with respect to the undoped alloy, whereas Pd- and Pt-doped alloys had the lower weight gain. During first two hours, the Au-doped alloy has a significantly greater weight gain, while Ag-, Pd-, and Pt-doped alloys have a similar and slightly higher weight gain than the undoped alloy. At 1000°C, all doped alloys (except Pd-doped alloy) have the greater

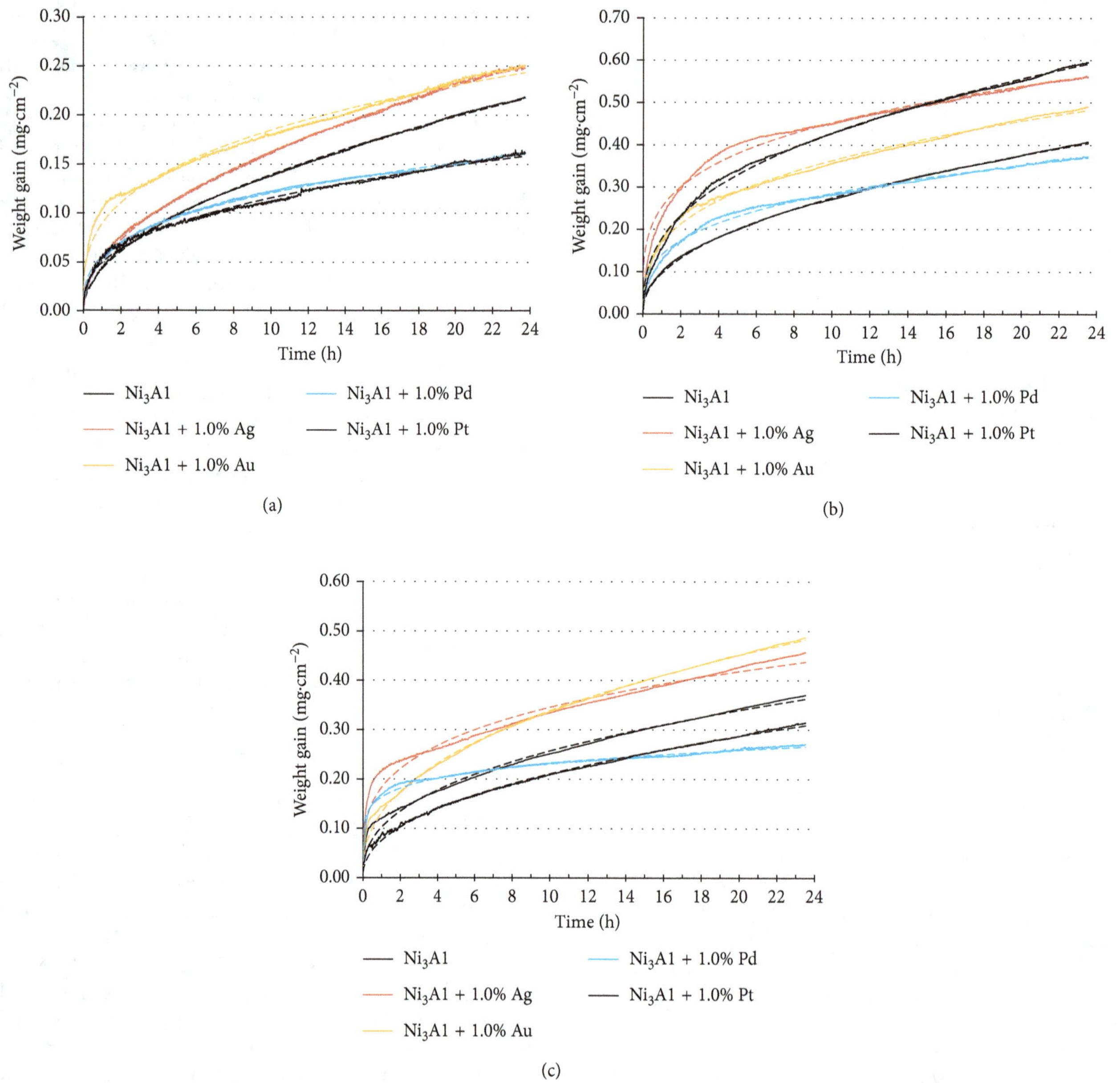

Figure 4: Weight gain of intermetallic alloys oxidized in O_2 at (a) 900°C, (b) 1000°C, and (c) 1100°C.

Table 2: Values of k and n obtained from the curves of weight gain.

Temperature (°C)		Ni_3Al	$Ni_3Al + 1.0\%$ Ag	$Ni_3Al + 1.0\%$ Au	$Ni_3Al + 1.0\%$ Pd	$Ni_3Al + 1.0\%$ Pt
900	k	0.0430	0.0513	0.0879	0.0553	0.0497
	n	0.5106	0.4998	0.3199	0.3350	0.3653
1000	k	0.0971	0.2515	0.1684	0.1402	0.1799
	n	0.4517	0.2541	0.3323	0.3070	0.3769
1100	k	0.1022	0.1814	0.1297	0.1604	0.0755
	n	0.4009	0.2785	0.4156	0.1602	0.4460

weight gain with respect to the undoped alloy, in a decreasing order of Au-, Ag-, and Pt-doped alloys. This behavior may be because the doped alloys stabilize the θ-Al_2O_3 phase and therefore reduce the α-Al_2O_3 phase transformation as Pt does according to Cadoret et al. [70] and Hayashi and Gleeson [71]. The θ-Al_2O_3 phase generates more pathways for diffusion because of its smaller grain size compared to α-Al_2O_3 phase. The k values at 1000°C show magnitudes that explain the high increase in weight gain for doped alloys. At 1100°C, there is a clear grouping in

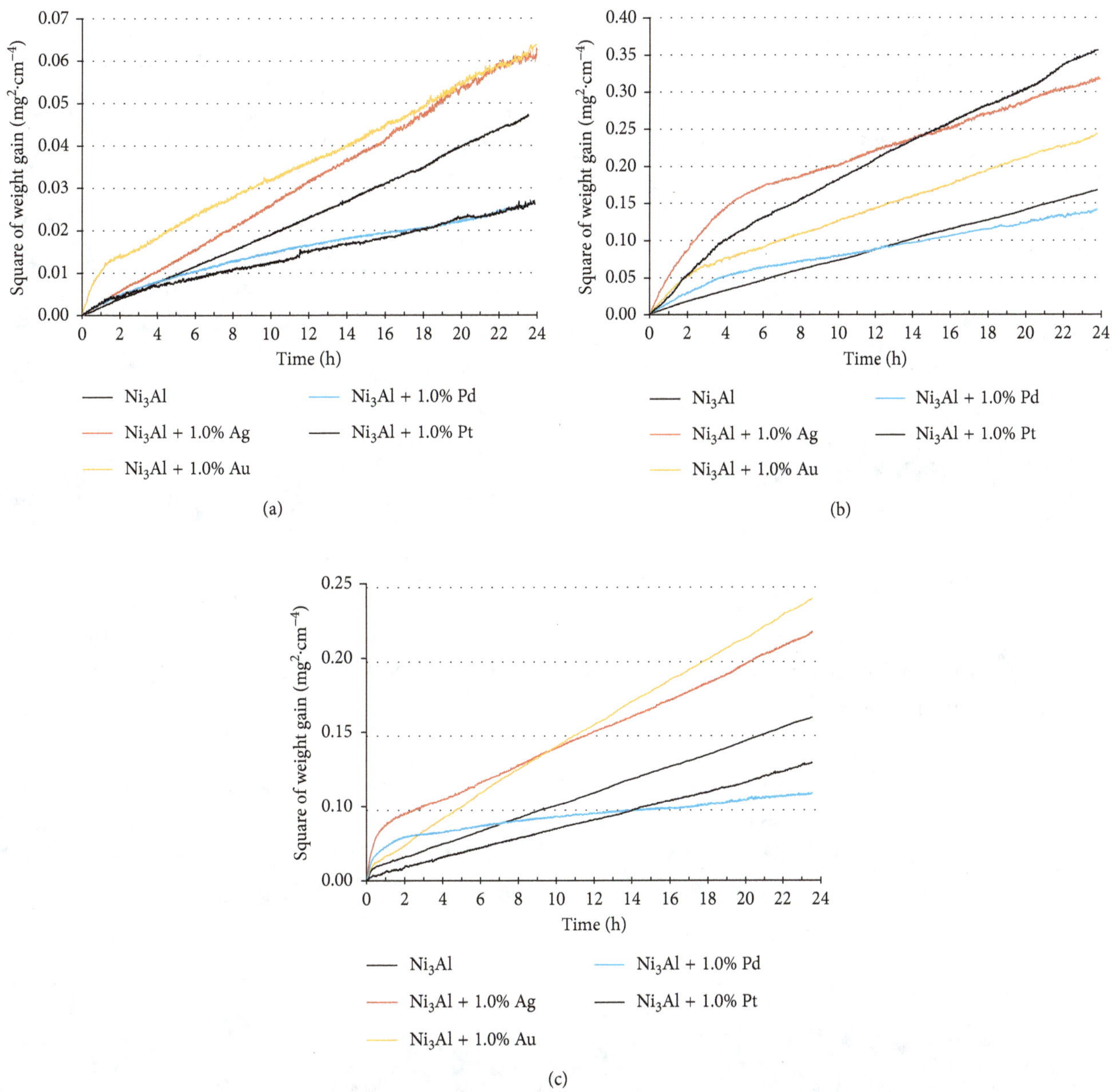

FIGURE 5: Square weight gain versus time of exposure of intermetallic alloys oxidized in O_2 at (a) 900°C, (b) 1000°C, and (c) 1100°C.

terms of weight gain between the Pd- and Pt-doped alloys (lowest) and the Ag- and Au-doped alloys (higher) compared to undoped alloy, similar as at 900°C. The Pt-doped alloy has the lower weight gain than the undoped alloy throughout the exposure period, while the Pd-doped alloy has a higher weight gain than the undoped alloy only in the first six hours. All alloys except Pd-doped alloy have a remarkably high weight gain in the first hour.

The weight gain curves of Figure 4 give an impression about the effect of the addition of noble metals in the oxidation process, but it is also necessary to know the rate of oxidation. Considering strictly a parabolic behavior $n = 0.5$, for comparative purposes (2) can be rewritten as

$$\Delta m^2 = k_p t, \tag{3}$$

where k_p is the parabolic rate constant. The linearity of (3) allows graphically displaying and estimating the k_p value considering the slope of the $\Delta m^2 - t$ relationship. In Figure 5, changes in the parabolic rate constant showing two k_p values can be identified, the first k_p value being greater than the second one; this is due to the fact that it is mathematically forcing a curve with a subparabolic behavior to be adjusted to a parabolic behavior [69], so the k_p value that will be taken into account to characterize the oxidation process will be that of the second slope which is presented in Table 3.

At 900°C, undoped and Ag-doped alloys show a constant k_p value throughout the exposure period. The k_p value of Pd- and Pt-doped alloys is similar, and it is lower than that for the undoped alloy. At 1000°C, the k_p value of the Au-doped

Table 3: Values of k_p obtained from slopes of Figure 5.

Temperature (°C)	Ni_3Al	$Ni_3Al + 1.0\%$ Ag	$Ni_3Al + 1.0\%$ Au	$Ni_3Al + 1.0\%$ Pd	$Ni_3Al + 1.0\%$ Pt
900	0.0019	0.0026	0.0022	0.0010	0.0010
1000	0.0072	0.0082	0.0085	0.0045	0.0136
1100	0.0054	0.0071	0.0095	0.0016	0.0040

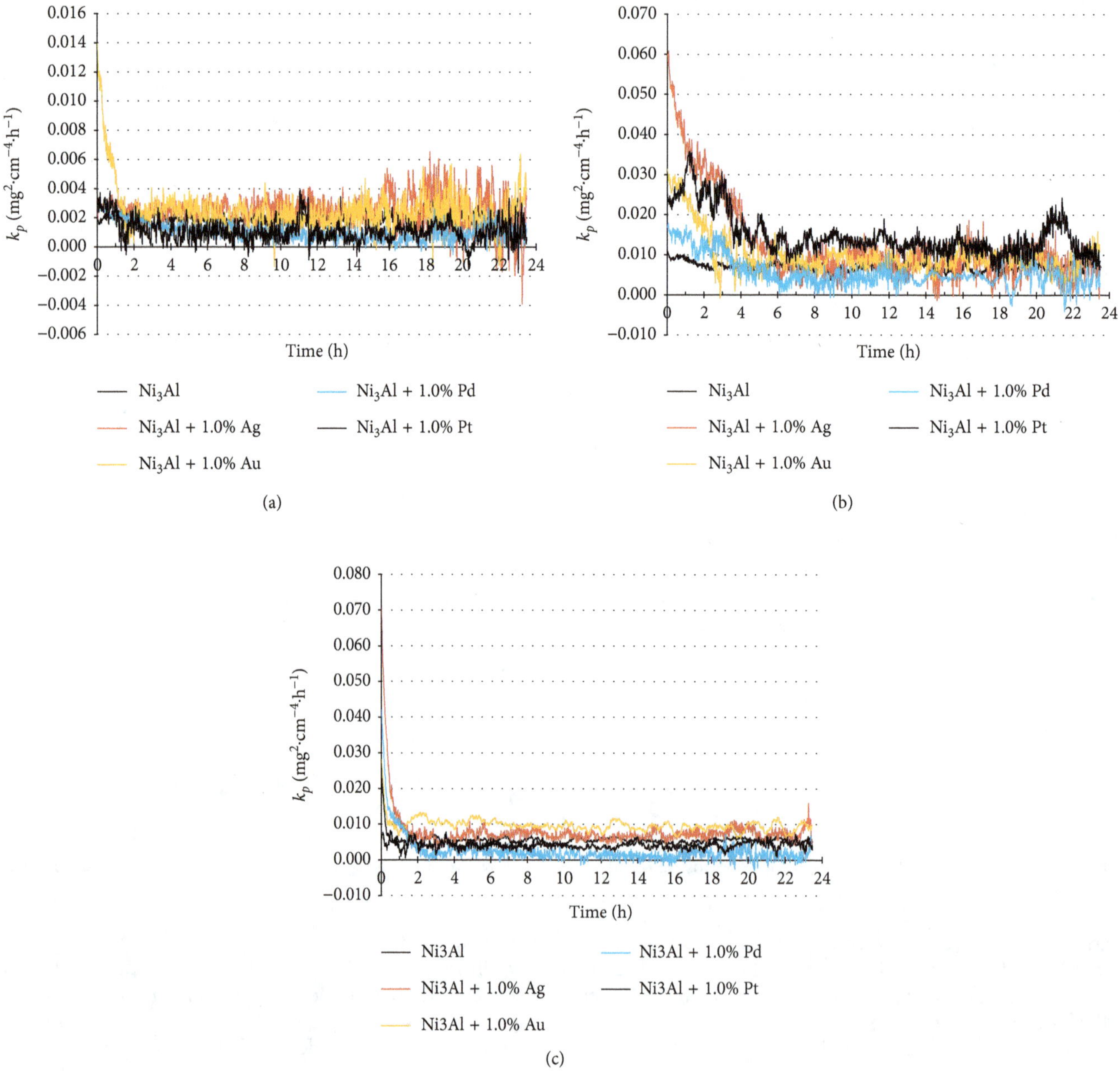

Figure 6: Instantaneous k_p versus time for the intermetallic alloys oxidized at (a) 900°C, (b) 1000°C, and (c) 1100°C.

Table 4: Average values of instantaneous k_p from Figure 6.

T (°C)	Ni_3Al	$Ni_3Al + 1\%$ Ag	$Ni_3Al + 1\%$ Au	$Ni_3Al + 1\%$ Pd	$Ni_3Al + 1\%$ Pt
900	0.0020	0.0026	0.0027	0.0011	0.0011
1000	0.0070	0.0132	0.0102	0.0059	0.0149
1100	0.0058	0.0088	0.0101	0.0032	0.0041

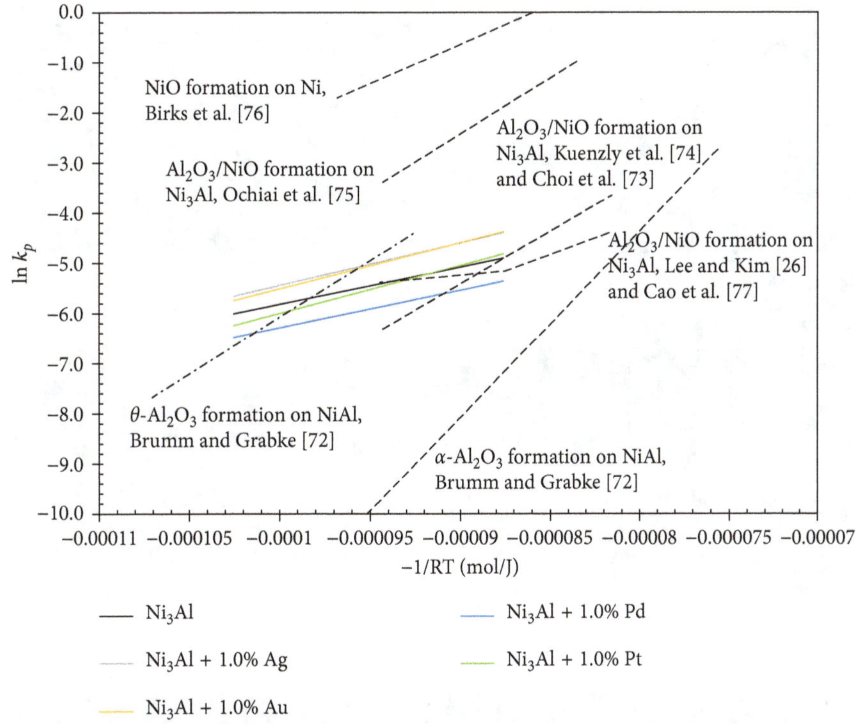

FIGURE 7: Arrhenius graph showing the k_p variation in function of temperature.

TABLE 5: Values of k_0 and Q obtained from the Arrhenius graph.

	Ni_3Al	Ni_3Al + 1.0% Ag	Ni_3Al + 1.0% Au	Ni_3Al + 1.0% Pd	Ni_3Al + 1.0% Pt
ln (k_0)	1.6031	3.0341	3.6400	1.1934	3.5693
Q (kJ/mol)	74.2443	84.7525	91.4718	74.7778	95.6762

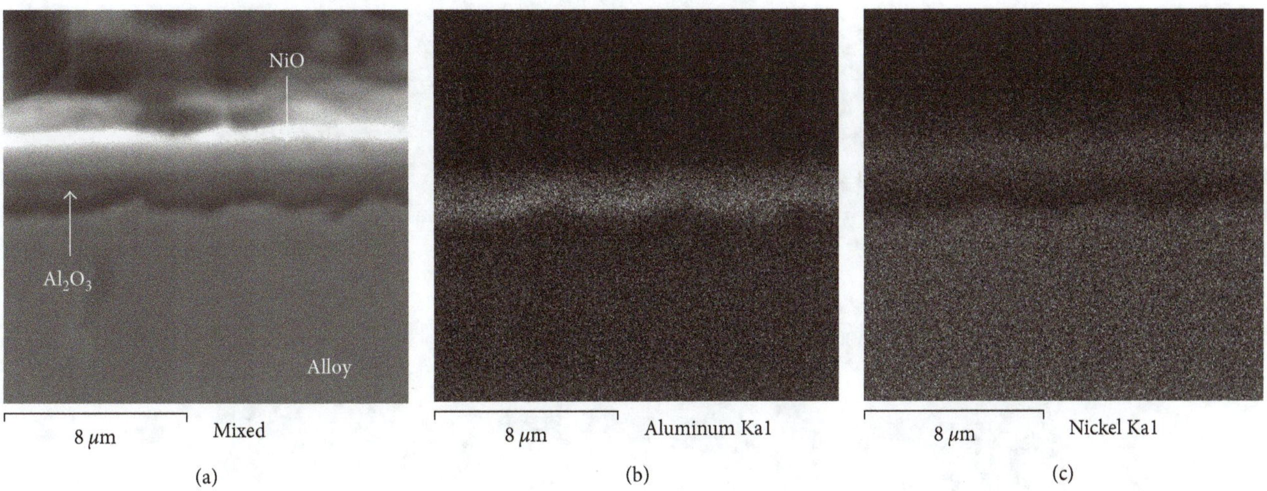

FIGURE 8: SEM micrograph (a) and mapping elements (b, c) of the scale formed onto Ni_3Al at 900°C.

alloy decreases at three hours after starting the test, the k_p value of Ag, Pd, and Pt doped alloys decreases at four hours after starting the test, while for the undoped alloy, its k_p value remains constant. At 1100°C, the k_p value of the undoped and doped alloys decrease at the first half an hour after starting the test, except the Pt-doped alloy, for which the k_p value remains constant. In most alloys, at all temperatures, it is observed at most only one change in k_p, except the Pd-doped alloy at 1100°C with two changes: half an hour and two hours after starting the test. At 1000°C, the k_p value of the

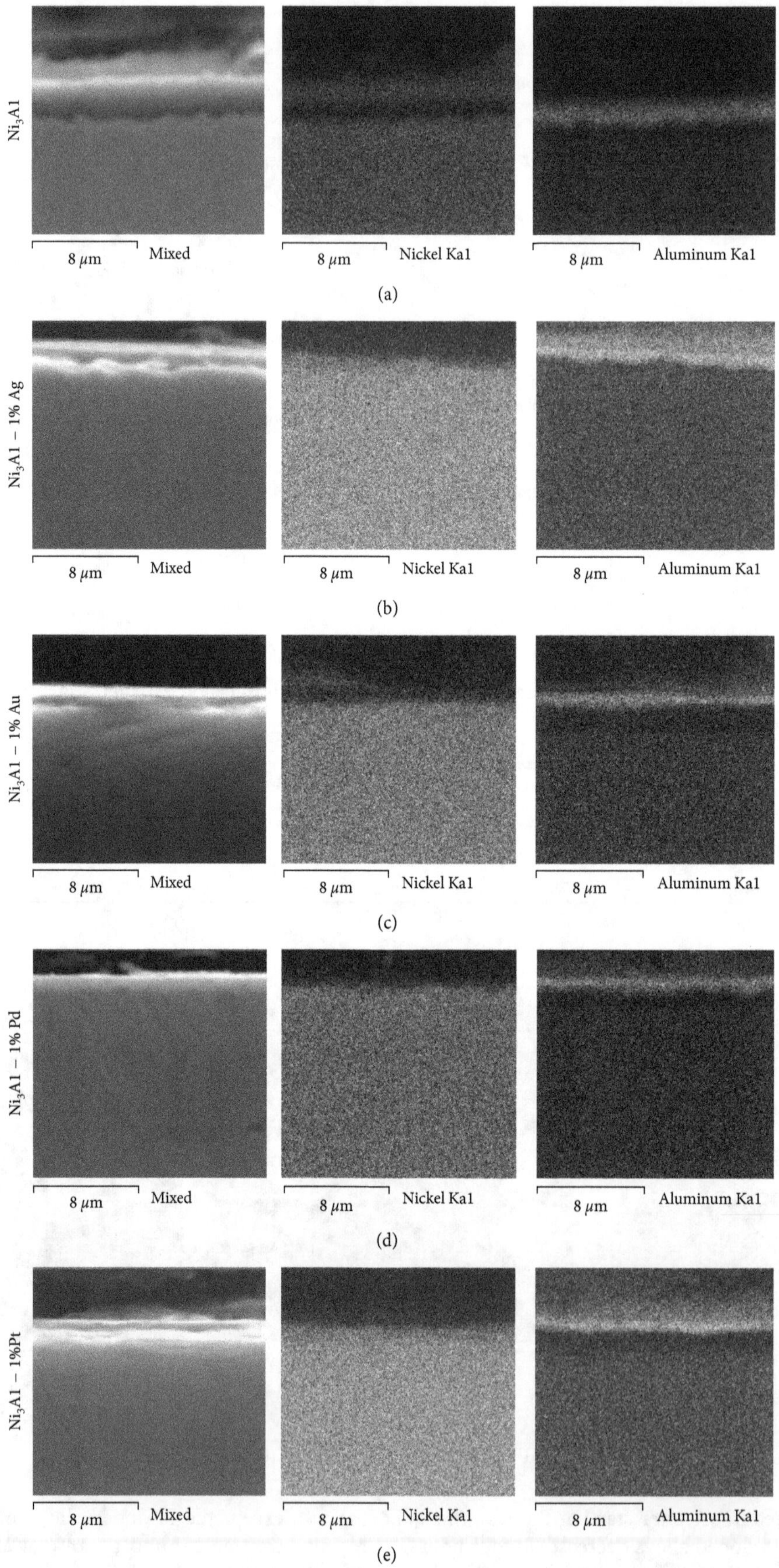

FIGURE 9: Cross section of intermetallic alloys and mapping elements after oxidation at 900°C: (a) Ni_3Al, (b) Ag-doped alloy, (c) Au-doped alloy, (d) Pd-doped alloy, (e) Pt-doped alloy.

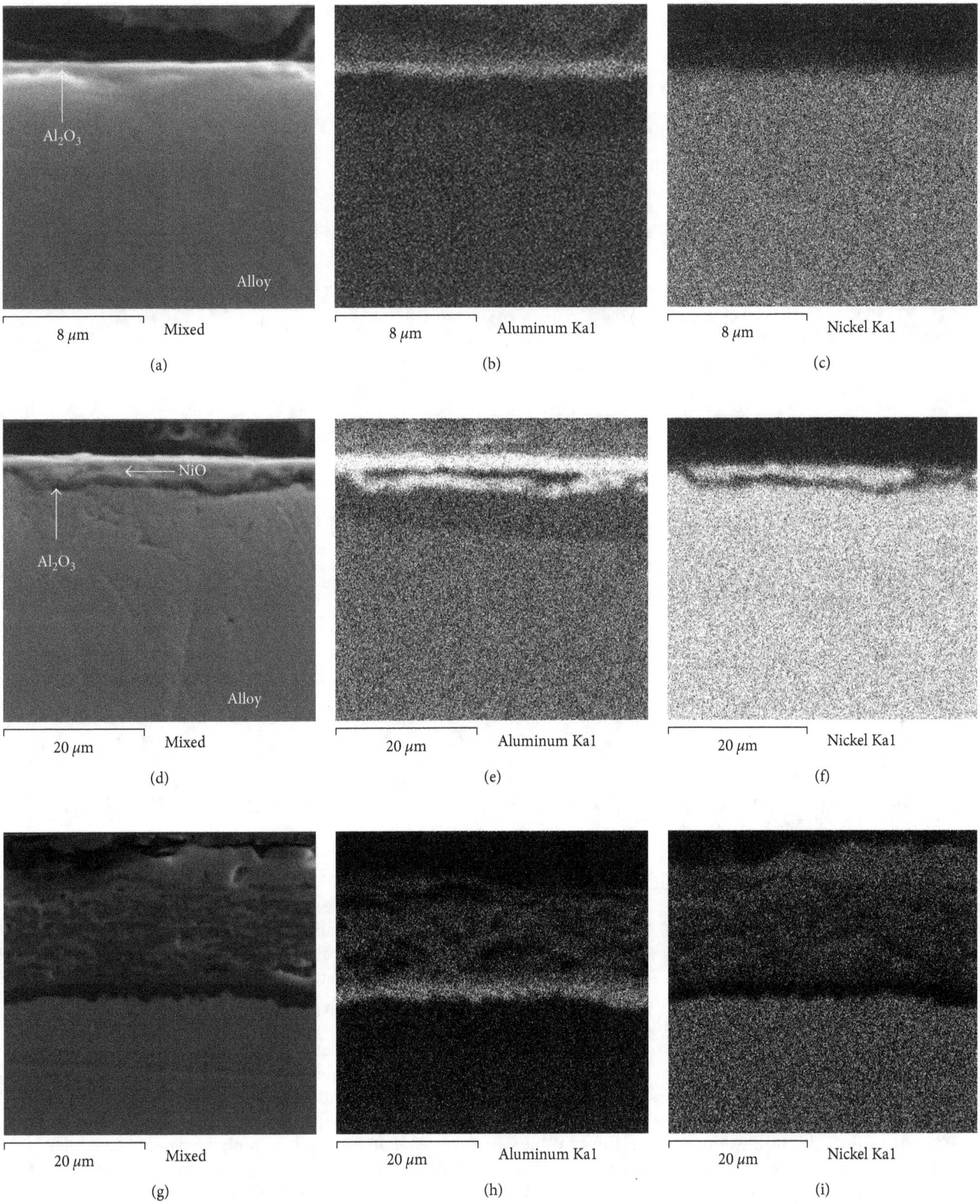

FIGURE 10: SEM micrograph and mapping elements: (a–c) Pd-doped alloy (900°C), (d–f) Pt-doped alloy (1000°C), and (g–i) Ag-doped alloy (1100°C).

Au-doped alloy decreases at three hours after starting the test, and for Ag-, Pd-, and Pt-doped alloys, the k_p value decreases at four hours after starting the test, while for the undoped alloy, its k_p value remains constant. At 1100°C, the k_p value of all alloys (except Pt) decreases after 30 min, and for Pt-doped alloy, the k_p value remains constant.

When it is not possible to obtain accurate measurement from the slopes, it is necessary to consider the average of the

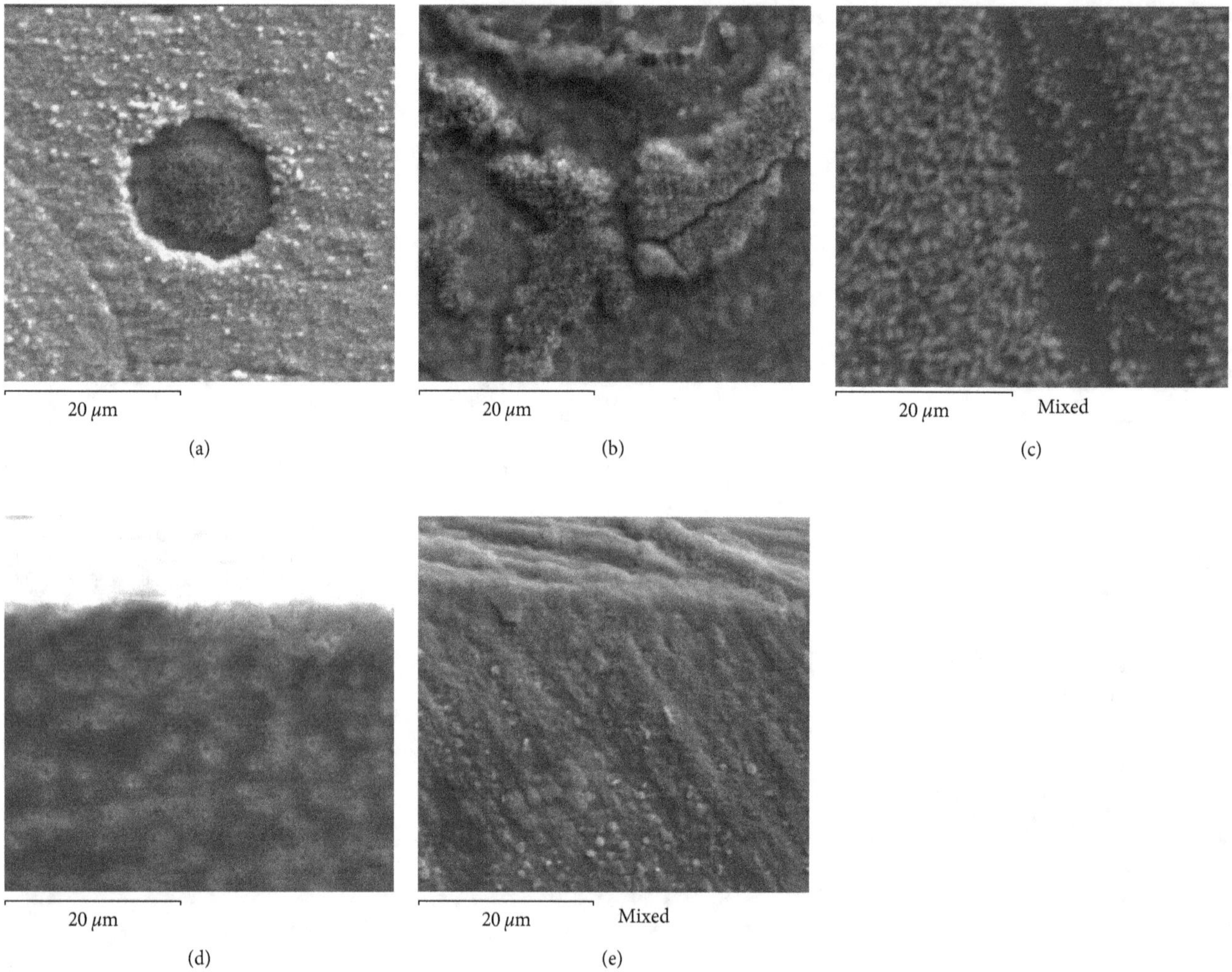

FIGURE 11: SEM micrographs showing the surface morphology of the oxide scales at 900°C: (a) Ni_3Al, (b) Ni_3Al + 1.0% Ag, (c) Ni_3Al + 1.0% Au, (d) Ni_3Al + 1.0% Pd, and (e) Ni_3Al + 1.0% Pt.

instantaneous parabolic constants, which results in differentiating (3) versus time [65],

$$\frac{d(\Delta m^2)}{dt} = k_p. \tag{4}$$

The instantaneous parabolic constant (k_p) of the oxidation tests is shown in Figure 6. This type of representation of the experimental data provides more information regarding the evolution of oxidation rates. At 900°C (Figure 6(a)), in the first two hours, the k_p values of the Au-doped alloy decrease seven times its equilibrium value, while in the Pd- and Pt-doped alloys, the k_p values decrease two times their equilibrium value; on the other hand, the Ag-doped alloy and undoped alloy maintain a constant rate. At 1000°C (Figure 6(b)), it is observed that the values of the instantaneous parabolic constant k_p of doped alloys tend to decrease in the first 6 hours, and then they remain constant, and for the undoped alloy, the k_p values remain constant. For the Ag-doped alloy, its k_p value decreases six times its equilibrium value, and for other doped-alloys, their k_p values decrease two times their equilibrium value. Finally, at 1100°C (Figure 6(c)), the Pt-doped alloy shows a stable behavior, and the other alloys show a fast decrease in their k_p values in the first hour.

In order to visualize the effect of the temperature on the rate constants, it is necessary to determine the activation energy of the oxidation process from the Arrhenius equation.

$$\ln k_p = \ln k_0 - Q \cdot (RT)^{-1}, \tag{5}$$

where Q is the activation energy, R is the universal gas constant, T is the temperature, k_p is the rate constant, and k_0 is the preexponential factor. Plotting ln k_p versus $-(RT)^{-1}$ from the slope of the lines, the Q value can be determined. Perez et al. [65] suggest assuming a parabolic kinetics in the whole temperature range; therefore, the average value of instantaneous k_p from Figure 6 was calculated. Table 4 shows the average values of instantaneous k_p.

In order to compare the results of this work, Figure 7 shows the instantaneous k_p values obtained along with those reported in other studies. At 900 and 1000°C, the k_p values of both doped and undoped alloys tend to k_p values where the formation of the θ-Al_2O_3 phase may be possible in NiAl as reported by Brumm and Grabke [72]. At 1100°C, the k_p values are concentrated at a point between the formation of

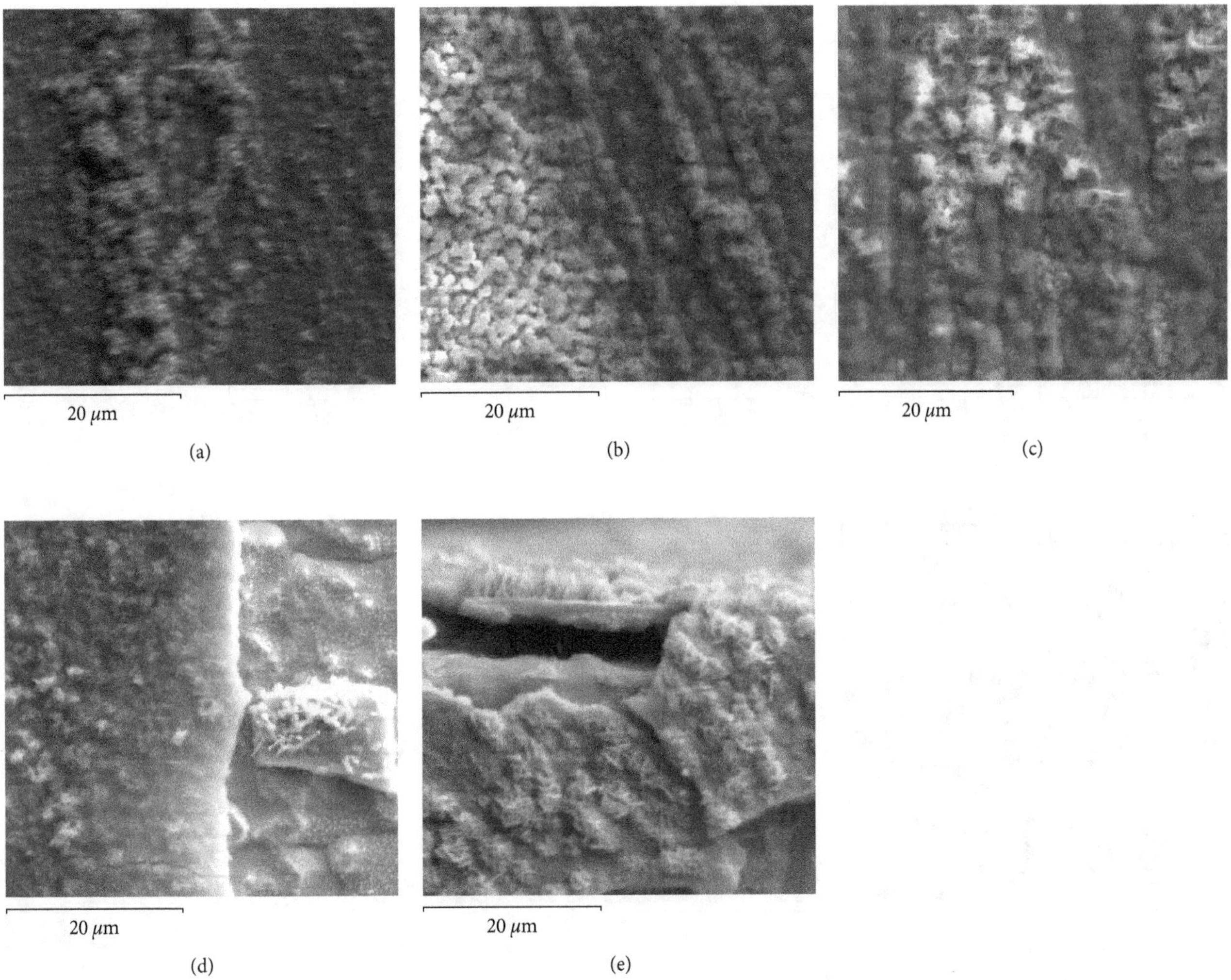

FIGURE 12: SEM micrographs showing the surface morphology of the oxide scales at 1000°C: (a) Ni_3Al, (b) $Ni_3Al + 1.0\%$ Ag, (c) $Ni_3Al + 1.0\%$ Au, (d) $Ni_3Al + 1.0\%$ Pd, and (e) $Ni_3Al + 1.0\%$ Pt.

θ-Al_2O_3 and α-Al_2O_3; an explanation for this behavior could be that, at this temperature, the stability of the α-Al_2O_3 begins [18] and this tends to decrease the weight gain [70, 71]. At 900°C and 1000°C, the k_p value of the undoped alloy is similar to results reported by Lee and Kim [26] and Choi et al. [73], and at 1100°C, it is similar to results of Kuenzly et al. [74]. The Pt-doped alloy has k_p values similar to the undoped alloy, Ag- and Au-doped alloys have higher k_p values, and the Pd-doped alloy has lower k_p values.

Activation energy is an important parameter that characterizes the predominant oxidation mechanism [75]; for example, the results presented by Ochiai et al. [75] and Kuenzly et al. [74] suggest that, in both cases, the oxidation mechanism is similar, although the experimental conditions were different, that is, cyclic oxidation tests and continuous oxidation tests. On the other hand, analyzing the results for the formation of α-Al_2O_3 scale [72] and NiO scale [76], it is observed that the preexponential factor is inversely proportional to the protective capacity of the protective oxide. Table 5 shows the activation energies and preexponential factors obtained from the Arrhenius graph (Figure 7). The activation energy value of Ni_3Al is similar to those values reported by Lee and Kim [26] and Cao et al. [77], but almost three times lower than that reported by other authors [73, 74]. This difference is because the intermetallic alloy used in this study was not thermally treated. Activation energy values of the doped alloys show that the additions of the noble elements affect the Ni_3Al oxidation mechanism in the following order: Pt > Au > Ag > Pd. On the other hand, the Pd-doped alloy showed the lowest value of the pre-exponential factor.

3.4. SEM Analysis. SEM analysis showed that the oxide scale formed on all alloys had nonuniform thicknesses. In the case of Ni_3Al, the average thicknesses observed were between 2.44 and 2.86 μm in the temperature range studied. At 900°C (Figure 8), a double-layer oxide was developed, an outer Ni-rich layer and an internal Al-rich layer, with similar thicknesses. Other authors have described similar results [78]. However, at higher temperatures, only the presence of an Al-rich oxide was detected.

In general, the thickness of the oxide layer developed onto doped alloys was lower than that observed onto Ni_3Al.

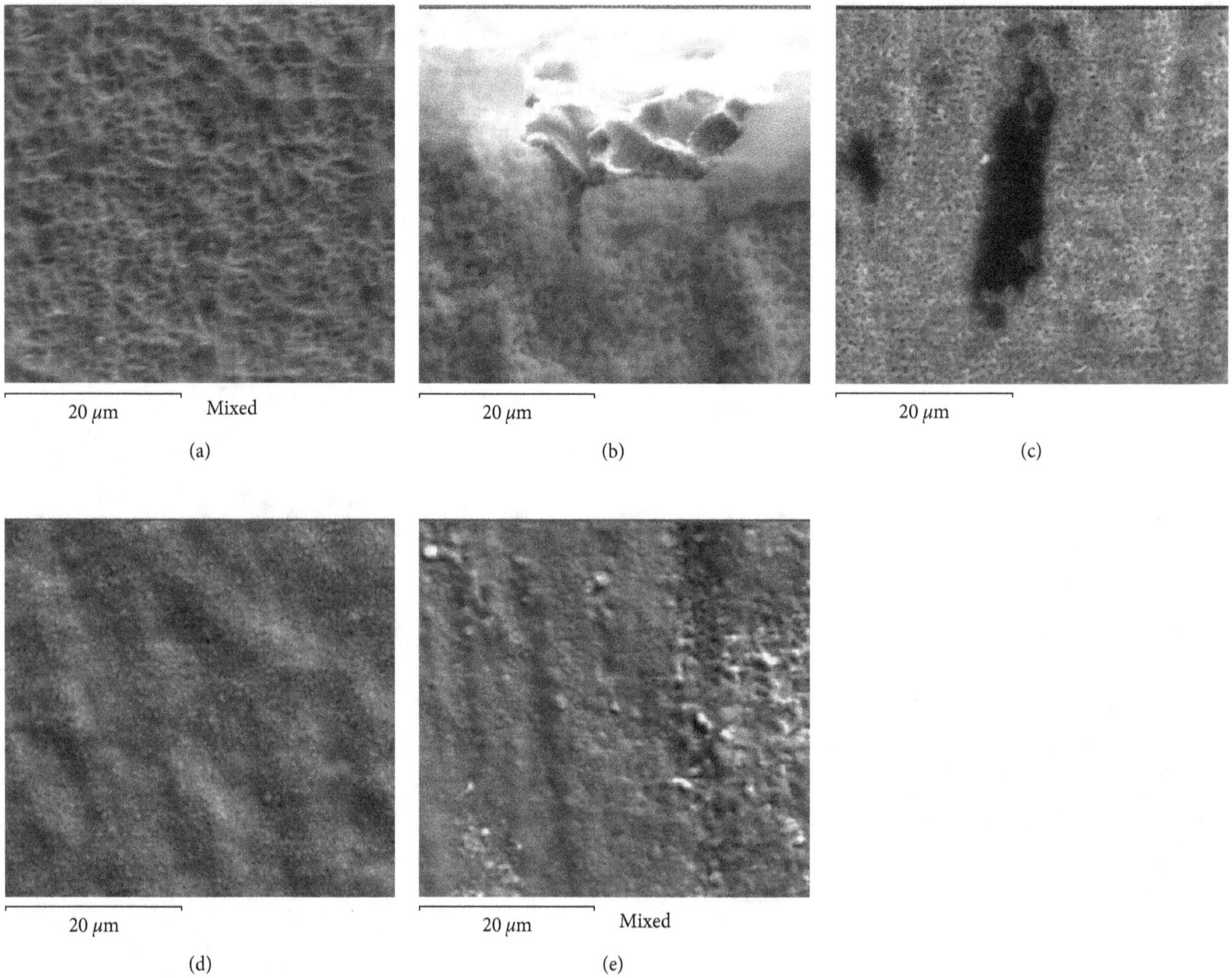

FIGURE 13: SEM micrographs showing the surface morphology of the oxide scales at 1100°C: (a)Ni_3Al, (b)$Ni_3Al + 1.0\%$ Ag, (c) $Ni_3Al + 1.0\%$ Au, (d) $Ni_3Al + 1.0\%$ Pd, and (e) $Ni_3Al + 1.0\%$ Pt.

At 900°C, the average thicknesses were between 0.98 and 1.5 μm, at 1000°C, between 2.13 and 2.20 μm (except the Pt-doped alloy, its average thickness was 4.61 μm), and at 1100°C, between 1.43 and 2.55 μm (except the Ag-doped alloy, its average thickness was 4.52 μm). Figure 9 shows the thickness of the oxide formed on the intermetallic alloys oxidized to 900°C; similar aspects were observed at higher temperatures. At all test temperatures, the Pd-doped alloy showed the oxide layer with the lowest thickness (Figure 10(a)). In general, all doped alloys developed an Al-rich monolayer oxide (Figures 10(a)–10(c)), as has been reported in other studies [79]. In the case of alloys doped with Pt (at 1000°C) and with Ag (at 1100°C), zones with multilayer oxides and thicknesses up to 20 μm were observed (Figures 10(d) and 10(g)). In all cases, an Al-depleted zone was observed at the interface alloy oxide (Figures 8 and 10). This is due to the consumption of this element for the formation of protective oxide, Al_2O_3 [80]. Similar depletion processes have been reported in studies of corrosion by molten salts [81–85].

Figure 11 shows the surface morphology of the oxides formed at 900°C. The formation of oxides in the form of needles can be observed only onto nondoped alloy and onto those alloys doped with Ag and Au. This type of morphology is associated with the presence of the metastable phases of alumina [86–88], whose transformation from γ-Al_2O_3 to θ-Al_2O_3 occurs between 800 and 950°C [72]. It is worth mentioning that the alloys that presented this type of morphology had a greater weight gain (Figure 4(a)).

Figure 12 shows the surface morphology of the oxide scale at 1000°C. It is observed that, at this temperature, there is a greater presence of oxides in the form of needles (θ-Al_2O_3) both in the nondoped alloy and in those alloys doped with Ag, Au, and Pt. The alloy doped with Pd showed a lower presence of this phase, possibly due to a decrease in the growth rate of the θ-Al_2O_3 phase. Again, the surface morphology is correlated with the observed weight gain (Figure 4(b)); that is, the greater the presence of the θ-Al_2O_3 phase, the greater the mass gain.

On the other hand, the morphology of the oxides formed at 1100°C was completely different (Figure 13). In this case, the oxides formed show wrinkle shaped, which is characteristic of the α-Al_2O_3 phase [72]. This morphology is more evident in the nondoped alloy and in those doped with Ag

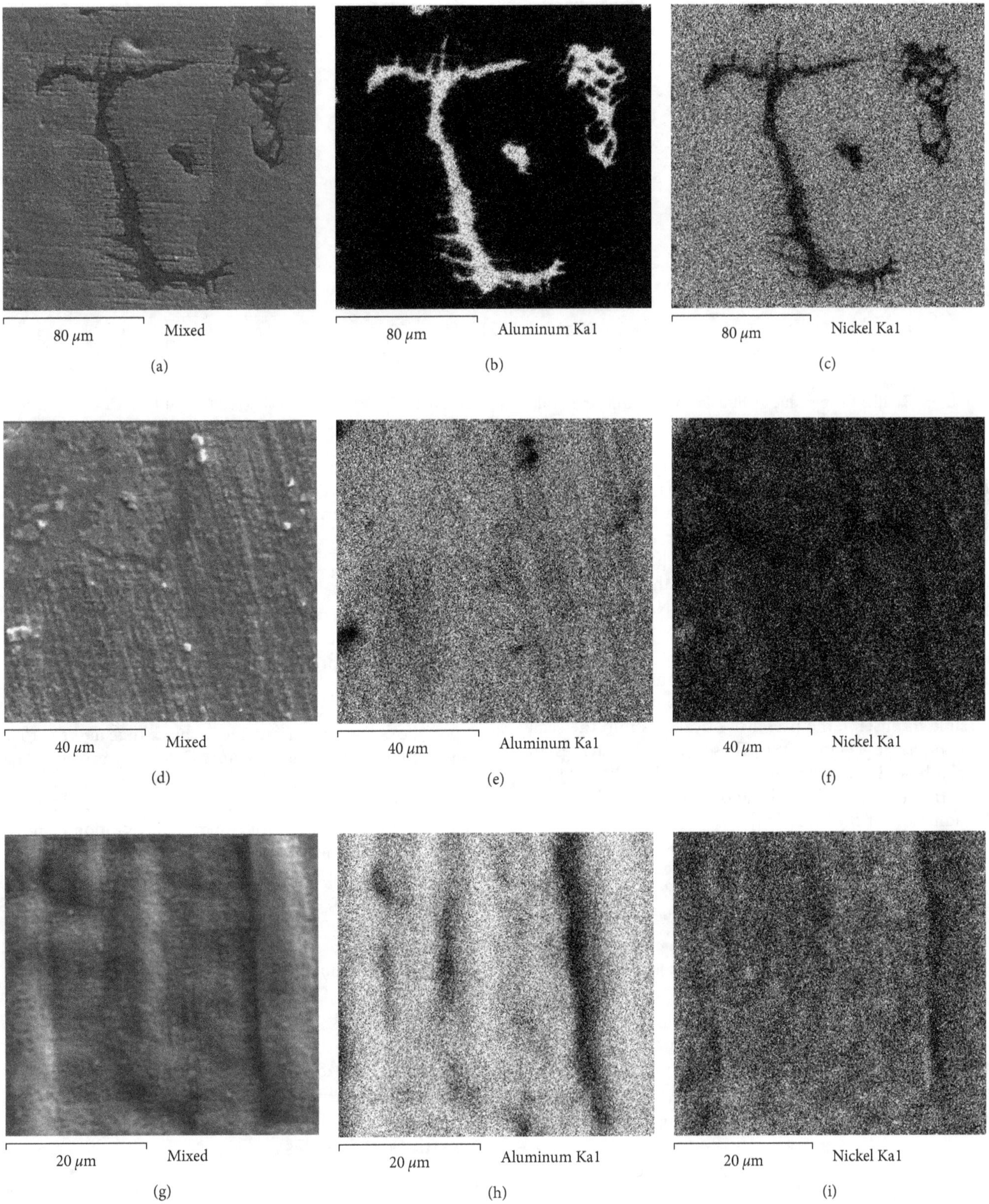

FIGURE 14: Superficial aspect and mapping elements of the oxide scale formed onto Ni_3Al at (a) 900°C, (d) 1000°C, and (g) 1100°C.

and Au, and in the alloys doped with Pd and Pt, a more compact and dense morphology was observed. These morphological differences indicate a difference in the growth rates of the oxide layers. In all cases, the weight gain at 1100°C was lower than that observed at 1000°C (Figure 4). This suggests that oxide layers rich in α-Al_2O_3 are more protective than those rich in θ-Al_2O_3.

Figure 14 shows the typical morphology of the Ni_3Al oxidized in oxygen for 24 hours. In general, a granular scale involving both heterogeneous Ni and Al oxides, mainly Ni

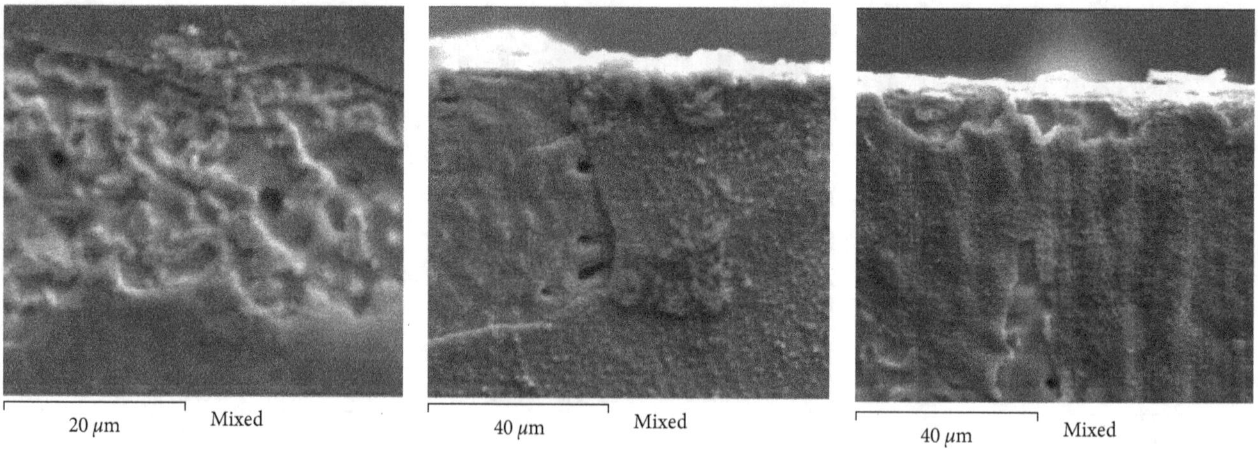

(a) (b) (c)

FIGURE 15: Cavities formed due to diffusion of Al during the oxidation process of the Ni_3Al in oxygen at (a) 900°C, (b) 1000°C, and (c) 1100°C.

oxides, as reported by Cao et al. [77] were observed. At 900°C, there are well-defined regions rich in Al_2O_3 which largely coincide with the interdendritic regions (Figure 14(a)). This suggests that the lateral growth of NiO could not fully cover the Al_2O_3, and the initial stage of oxidation occurred preferentially in the interdendritic phases [11]. At 1000 and 1100°C, the Ni_3Al surface was oxidized uniformly in comparison with the oxidized alloy at 900°C, but the surface mainly contains Al_2O_3 (Figures 14(d) and 14(g)). In some regions, the oxide layer has been cracked, probably during cooling, and it was possible to observe the presence of cavities into the scale-alloy interface (Figure 15). These sorts of cavities are formed by the consumption of Al and Ni in the course of formation of the oxide scale, and they have an impact on the adhesion of the scale [70].

The difference in the oxidation process between the dendritic and the interdendritic phases lies in the thermochemistry of alumina formation between the phases; that is, in the NiAl-rich dendritic phase, the following reaction may occur:

$$4NiAl + 3O_2 \leftrightarrows 2Al_2O_3 + 4Ni \ (\Delta G = -940.52\,kJ\,@\,900^{\circ}C) \tag{6}$$

On the other hand, in the interdendritic phase where the Ni3Al phase predominates, the following reaction occurs:

$$4Ni_3Al + 3O_2 \leftrightarrows 2Al_2O_3 + 12Ni \ (\Delta G = -895.40\,kJ\,@\,900^{\circ}C) \tag{7}$$

Thermochemically, the dendritic phase is more favored than the interdendritic phase. Considering also that in the doped alloys there is a considerable decrease in the dendritic phase (Section 3.2), it can be established that there exists a relation between the microstructural change and the thermochemical equilibrium conditions for the formation of an alumina layer. On the other hand, the absence of NiO in the surface of the doped alloys can also be attributed to the increase in the grain size of the Al_2O_3; that is, to say, a decrease in the diffusion of Ni through the grain boundaries of the Al_2O_3 layer, which is related to the kinetics of subparabolic growth.

4. Conclusions

The XRD analysis of the alloys showed that the addition of 1% (atomic percent) of noble metals to the Ni_3Al intermetallic alloy promotes the preferential growth planes and increases the solidification rate by changing the microstructure of the base alloy. The oxidation behavior of doped alloys may be attributable to the change in the surface composition, taking into account that every growth plane has different compositions and that the preferential substitution site of the noble metals in the structure of Ni_3Al increases the Al/Ni ratio. In general, the addition of noble metals promotes the formation of oxide layers rich in Al_2O_3 and is evident that the diffusion of the Ni cations through Al_2O_3 scale is decreased, since the formation of NiO during the test time was not evident. Ag and Au additions increase the value of the parabolic rate constant k_p, while Pd and Pt decrease, the k_p values tend to values where the formation of θ-Al_2O_3 and α-Al_2O_3 is possible. The noble metal addition increased the activation energy value of the oxidation process compared to the nondoped Ni_3Al, except Pd. The low solubility of Ag and its low melting point caused microstructural localized failures into Ag-doped alloy.

Conflicts of Interest

The authors declare that they have no conflicts of interest regarding the publication of this paper.

Acknowledgments

Financial support from Consejo Nacional de Ciencia y Tecnología (CONACYT, México) (Project no. 159898 and Ph.D. scholarship to J. J. Ramos-Hernandez, Registration no. 202898) is gratefully acknowledged.

References

[1] S. C. Deevi and V. K. Sikka, "Nickel and iron aluminides: an overview on properties, processing, and applications," *Intermetallics*, vol. 4, no. 5, pp. 357–375, 1996.

[2] A. S. Verin, "Certain features of the structure and anisotropy in Ni_3Al single crystals," *Metal Science and Heat Treatment*, vol. 36, no. 2, pp. 99–102, 1994.

[3] A. Chiba and S. Hanada, "Relationship between the APD structure and ductility of rapidly solidified Ni_3Al alloys," *Intermetallics*, vol. 4, no. 1, pp. 55–59, 1996.

[4] P. Jozwik, E. Jezierska, and Z. Bojar, "TEM investigations of Ni_3Al-based intermetallic polycrystals," *Materials Chemistry and Physics*, vol. 81, no. 2-3, pp. 448–451, 2003.

[5] O. Hunziker and W. Kurz, "Solidification microstructure maps in Ni-Al alloys," *Acta Materialia*, vol. 45, no. 12, pp. 4981–4992, 1997.

[6] T. Mawari and T. Hirano, "Effects of unidirectional solidification conditions on the microstructure and tensile properties of Ni_3Al," *Intermetallics*, vol. 3, no. 1, pp. 23–33, 1995.

[7] A. Inoue, H. Tomioka, and T. Masumoto, "An improvement of ductility and strength of the L12 compounds in Ni-Al-X (X = Cr, Mn, Fe, Co or Si) systems by melt quenching," *Journal of Materials Science Letters*, vol. 1, no. 9, pp. 377–380, 1982.

[8] R. Goetzinger, M. Barth, D. M. Herlach, O. Hunziker, and W. Kurz, "Disorder trapping in Ni_3(Al, Ti) by solidification from the undercooled melt," *Materials Science and Engineering: A*, vol. 226–228, pp. 415–419, 1997.

[9] J. Sun, C. S. Lee, and J. K. L. Lai, "Deformation microstructure of Ni_3Al intermetallic compound macroalloyed with Pd," *Materials Letters*, vol. 31, no. 1-2, pp. 1–4, 1997.

[10] B. A. Pint, S. A. Speakman, C. J. Rawn, and Y. Zhang, "Deformation and phase transformations during the cyclic oxidation of Ni-Al and Ni-Pt-Al," *The Journal of the Minerals, Metals & Materials Society*, vol. 58, no. 1, pp. 47–52, 2006.

[11] D. Lee, M. L. Santella, I. M. Anderson, and G. M. Pharr, "Thermal aging effects on the microstructure and short-term oxidation behavior of a cast Ni_3Al alloy," *Intermetallics*, vol. 13, no. 2, pp. 187–196, 2005.

[12] M. Haerig and S. Hofmann, "Mechanisms of Ni_3Al oxidation between 500°C and 700°C," *Applied Surface Science*, vol. 125, no. 1, pp. 99–114, 1998.

[13] U. Bardi, A. Atrei, and G. Rovida, "Initial stages of oxidation of the Ni_3Al alloy: a study by X-ray photoelectron spectroscopy and low energy He+ scattering," *Surface Science*, vol. 239, no. 1-2, pp. L511–L516, 1990.

[14] C. Mei, S. Patu, S. Changxu, and S. Jianian, "Effects of B+ and Cr+ ion implantation on the oxidation of Ni_3Al," *Journal of Materials Science*, vol. 28, no. 20, pp. 5508–5513, 1993.

[15] A. Velon and I. Olefjord, "Oxidation behavior of Ni_3Al and Fe_3Al: II. Early stage of oxide growth," *Oxidation of Metals*, vol. 56, no. 5-6, pp. 425–452, 2001.

[16] T. Norby and P. Kofstad, "Electronic, ionic, and protonic conductivities of a commercial, polycrystalline α-Al_2O_3," *High Temperatures-High Pressures*, vol. 20, pp. 345–360, 1988.

[17] D. B. Lee and M. L. Santella, "High temperature oxidation of Ni_3Al alloy containing Cr, Zr, Mo, and B," *Materials Science and Engineering: A*, vol. 374, no. 1-2, pp. 217–223, 2004.

[18] S. G. Addepalli, B. Ekstrom, N. P. Magtoto, J. Lin, and J. A. Kelber, "STM atomic-scale characterization of the γ'-Al_2O_3 film on Ni_3Al(111)," *Surface Science*, vol. 442, no. 3, pp. 385–389, 1999.

[19] A. M. Venezia and C. M. Loxton, "Low pressure oxidation of Ni_3Al alloys at elevated temperatures as studied by X-ray photoelectron spectroscopy and Auger spectroscopy," *Surface Science*, vol. 194, no. 1-2, pp. 136–148, 1988.

[20] A. M. Venezia, C. M. Loxton, and J. A. Horton, "Temperature and pressure effects for oxygen adsorption and oxidation of Ni_3Al alloyed with chromium and zirconium," *Surface Science*, vol. 225, no. 1-2, pp. 195–205, 1990.

[21] U. Bardi, A. Atrei, and G. Rovida, "Initial stages of oxidation of the Ni_3Al alloy: structure and composition of the aluminum oxide overlayer studied by XPS, LEIS and LEED," *Surface Science*, vol. 268, no. 1–3, pp. 87–97, 1992.

[22] P. Pérez, J. L. González-Carrasco, P. Adeva, V. A. C. Haanappel, and M. F. Stroosnijder, "Effect of Cr- and Y-implantation on oxidation behaviour of powder metallurgical Ni_3Al," *Materials Science and Technology*, vol. 15, no. 3, pp. 345–351, 1999.

[23] M. Garza, N. P. Magtoto, and J. A. Kelber, "Characterization of oxidized Ni_3Al(110) and interaction of the oxide film with water vapor," *Surface Science*, vol. 519, no. 3, pp. 259–268, 2002.

[24] V. A. C. Haanappel, P. Pérez, J. L. González-Carrasco, and M. F. Stroosnijder, "The cyclic oxidation behaviour of PM Ni_3Al in air at elevated temperatures: the effect of Cr- and Y-implantation," *Intermetallics*, vol. 6, no. 5, pp. 347–356, 1998.

[25] T. T. Chang, Y. C. Pan, and T. H. Chuang, "The oxidation behavior of Ni_3Al-Zr alloys with various zirconium contents," *Journal of Alloys and Compounds*, vol. 243, no. 1-2, pp. 126–132, 1996.

[26] D. B. Lee and D. J. Kim, "The oxidation of Ni_3Al containing decomposed SiC-particles," *Intermetallics*, vol. 9, no. 1, pp. 51–56, 2001.

[27] W. W. Lee, D. B. Lee, M. H. Kim, and S. C. Ur, "High temperature oxidation of an oxide-dispersion strengthened NiAl," *Intermetallics*, vol. 7, no. 12, pp. 1361–1366, 1999.

[28] A. Velon and D. Q. Yi, "Influence of Cr on the Oxidation of Fe_3Al and Ni_3Al at 500°C," *Oxidation of Metals*, vol. 57, pp. 13–31, 2002.

[29] V. Deodeshmukh, N. Mu, B. Li, and B. Gleeson, "Hot corrosion and oxidation behavior of a novel Pt+ Hf-modified γ'-Ni_3Al+γ-Ni-based coating," *Surface and Coatings Technology*, vol. 201, no. 1, pp. 3836–3840, 2006.

[30] F. Qin, J. W. Anderegg, C. J. Jenks, B. Gleeson, D. J. Sordelet, and P. A. Thiel, "The effect of Pt on Ni_3Al surface oxidation at low-pressures," *Surface Science*, vol. 601, no. 1, pp. 146–154, 2007.

[31] S. J. Hong, G. H. Hwang, W. K. Han, and S. G. Kang, "Cyclic oxidation of Pt/Pd-modified aluminide coating on a nickel-based superalloy at 1150°C," *Intermetallics*, vol. 17, no. 6, pp. 381–386, 2009.

[32] B. Gleeson, N. Mu, and S. Hayashi, "Compositional factors affecting the establishment and maintenance of Al_2O_3 scales on Ni-Al-Pt systems," *Journal of Materials Science*, vol. 44, no. 7, pp. 1704–1710, 2009.

[33] A. V. Ruban and H. L. Skriver, "Calculated site substitution in ternary γ'-Ni_3Al: temperature and composition effects," *Physical Review B*, vol. 55, no. 2, pp. 856–874, 1997.

[34] C. Wolverton and D. De Fontaine, "Site substitution of ternary additions to Ni_3Al (γ') from electronic-structure calculations," *Physical Review B*, vol. 49, no. 17, pp. 12351–12354, 1994.

[35] C. Y. Geng, C. Y. Wang, and T. Yu, "Site preference and alloying effect of platinum group metals in γ'-Ni_3Al," *Acta Materialia*, vol. 52, no. 18, pp. 5427–5433, 2004.

[36] C. Jiang and B. Gleeson, "Site preference of transition metal elements in Ni_3Al," *Scripta Materialia*, vol. 55, no. 5, pp. 433–436, 2006.

[37] C. Jiang, D. J. Sordelet, and B. Gleeson, "Site preference of ternary alloying elements in Ni_3Al: a first-principles study," *Acta Materialia*, vol. 54, no. 4, pp. 1147–1154, 2006.

[38] B. Bhattacharya and R. K. Ray, "Deformation behavior of a Ni_3Al (B, Zr) alloy during cold rolling: part I. Changes in order and structure," *Metallurgical and Materials Transactions A*, vol. 31, no. 12, pp. 3001–3010, 2000.

[39] Y. Xu, S. Kameoka, K. Kishida, M. Demura, A. P. Tsai, and T. Hirano, "Catalytic properties of Ni_3Al intermetallics for methanol decomposition," *Materials Transactions*, vol. 45, no. 11, pp. 3177–3179, 2004.

[40] L. Lu, M. O. Lai, and S. Zhang, "Evolution and characterization of a Ni_3Al intermetallic compound during mechanical alloying," *Materials & Design*, vol. 15, no. 2, pp. 79–86, 1994.

[41] T. Hirano, "Improvement of room temperature ductility of stoichiometric Ni_3Al by unidirectional solidification," *Acta metallurgica et materialia*, vol. 38, no. 12, pp. 2667–2671, 1990.

[42] B. D. Cullity and S. R. Stock, *Elements of X-ray Diffraction*, , pp. 125–170, Prentice Hall, Pearson New Iinternational Edition, Upper Saddle River, NJ, USA, 2014.

[43] R. Seymour, A. Hemeryck, K. I. Nomura et al., "Nanoindentation of NiAl and Ni_3Al crystals on (100), (110), and (111) surfaces: a molecular dynamics study," *Applied Physics Letters*, vol. 104, no. 14, p. 141904, 2014.

[44] G. F. Cotterill, H. Niehus, and D. J. O'Connor, "An STM study of the initial stages of oxidation of Ni_3Al (110)," *Surface Review and Letters*, vol. 3, no. 3, pp. 1355–1363, 1996.

[45] Y. G. Shen, D. J. O'Connor, and R. J. MacDonald, "Studies of surface composition, structure and oxygen adsorption of Ni_3Al (110) by Li+ and He+ ion scattering," *Surface and Interface Analysis*, vol. 18, no. 10, pp. 729–739, 1992.

[46] A. Rosenhahn, J. Schneider, C. Becker, and K. Wandelt, "Oxidation of Ni_3Al (111) at 600, 800, and 1050 K investigated by scanning tunneling microscopy," *Journal of Vacuum Science & Technology A: Vacuum, Surfaces, and Films*, vol. 18, no. 4, pp. 1923–1927, 2000.

[47] S. Degen, A. Krupski, M. Kralj et al., "Determination of the coincidence lattice of an ultra thin Al_2O_3 film on Ni_3Al (111)," *Surface science*, vol. 576, no. 1–3, pp. L57–L64, 2005.

[48] A. Rosenhahn, J. Schneider, C. Becker, and K. Wandelt, "The formation of Al_2O_3-layers on Ni_3Al (111)," *Applied Surface Science*, vol. 142, no. 1–4, pp. 169–173, 1999.

[49] S. P. Chen, A. F. Voter, and D. J. Srolovitz, "Oscillatory surface relaxations in Ni, Al, and their ordered alloys," *Physical Review Letters*, vol. 57, no. 11, pp. 1308–1311, 1986.

[50] W. J. Chia and Y. W Chung, "Surface spectroscopy and temperature-programmed desorption studies of the interaction of water vapor with Ni_3(Al, Ti) (100) and (111) surfaces as a function of temperature," *Intermetallics*, vol. 3, no. 6, pp. 505–509, 1995.

[51] O. Hunziker and W. Kurz, "Directional solidification and phase equilibria in the Ni-Al system," *Metallurgical and Materials Transactions A*, vol. 30, no. 12, pp. 3167–3175, 1999.

[52] G. Rosas, R. Esparza, A. Bedolla, O. Flores, and R. Perez, "Effect of Li alloying on the microstructure and mechanical properties of Ni_3Al," *Advances in Technology of Materials and Materials Processing Journal*, vol. 9, pp. 99–102, 2007.

[53] M. H. Jacobs, *TALAT Lecture 1203: Phase Diagrams*, European Aluminium Association, Bruxelles, Belgium, 2010.

[54] E. J. Lavernia and T. S. Srivatsan, "The rapid solidification processing of materials: science, principles, technology, advances, and applications," *Journal of Materials Science*, vol. 45, no. 2, pp. 287–325, 2010.

[55] Z. Jingchen and Y. Ping, *The Effect of Cooling Rate of Solidification on Microstructure and Alloy Element Segregation of as Cast Alloy 718*, pp. 133–140, Central Iron and Steel Research Institute, Beijing, China, 2015.

[56] J. Xu, U. Herr, T. Klassen, and R. S. Averback, "Formation of supersaturated solid solutions in the immiscible Ni-Ag system by mechanical alloying," *Journal of Applied Physics*, vol. 79, no. 8, pp. 3935–3945, 1996.

[57] J. Zhou and J. T. Guo, "Effect of Ag alloying on the microstructure of NiAl," *Materials Letters*, vol. 56, no. 3, pp. 178–182, 2002.

[58] J. Zhou and J. T. Guo, "Effect of Ag alloying on microstructure, mechanical and electrical properties of NiAl intermetallic compound," *Materials Science and Engineering: A*, vol. 339, no. 1-2, pp. 166–174, 2003.

[59] L. Sheng, W. Zhang, J. Guo, F. Yang, Y. Liang, and H. Ye, "Effect of Au addition on the microstructure and mechanical properties of NiAl intermetallic compound," *Intermetallics*, vol. 18, no. 4, pp. 740–744, 2010.

[60] C. Meng, J. Guo, and Z. Hu, "Mechanism of macroalloying-induced ductility in Ni_3Al," *Journal of Materials Science & Technology*, vol. 10, pp. 279–284, 1994.

[61] M. J. Li, X. F. Sun, H. R. Guan, X. X. Jiang, and Z. Q. Hu, "The degradation of (Ni,Pd)Al coatings on superalloy IN738 during isothermal oxidation," *Surface and Coatings Technology*, vol. 185, no. 2-3, pp. 172–177, 2004.

[62] M. J. Li, X. F. Sun, H. R. Guan, X. X. Jiang, and Z. Q. Hu, "Cyclic oxidation behavior of palladium-modified aluminide coating," *Surface and Coatings Technology*, vol. 167, no. 1, pp. 106–111, 2003.

[63] D. Oquab and D. Monceau, "In-situ SEM study of cavity growth during high temperature oxidation of β-(Ni,Pd)Al," *Scripta Materialia*, vol. 44, no. 12, pp. 2741–2746, 2001.

[64] Y. Zhang, J. A. Haynes, W. Y. Lee et al., "Effects of Pt incorporation on the isothermal oxidation behavior of chemical vapor deposition aluminide coatings," *Metallurgical and Materials Transactions A*, vol. 32, no. 7, pp. 1727–1741, 2001.

[65] P. Peréz, M. Eddahbi, J. L. González-Carrasco, and P. Adeva, "Effect of the processing route on the oxidation behaviour of a Ni_3Al powder metallurgy alloy," *Intermetallics*, vol. 7, no. 6, pp. 679–687, 1999.

[66] J. J. Ramos-Hernandez, J. Porcayo-Calderon, V. M. Salinas-Bravo, C. D. Arrieta-Gonzalez, J. G. Gonzalez-Rodriguez, and L. Martinez-Gomez, "Phase stability diagrams for high temperature corrosion processes," *Mathematical Problems in Engineering*, vol. 2013, pp. 1–7, 2013.

[67] A. Atkinson, R. I. Taylor, and A. E. Hughes, "A quantitative demonstration of the grain boundary diffusion mechanism for the oxidation of metals," *Philosophical Magazine A*, vol. 45, no. 5, pp. 823–833, 1982.

[68] D. Naumenko, B. Gleeson, E. Wessel, L. Singheiser, and W. J. Quadakkers, "Correlation between the microstructure, growth mechanism, and growth kinetics of alumina scales on a FeCrAlY alloy," *Metallurgical and Materials Transactions A*, vol. 38, no. 12, pp. 2974–2983, 2007.

[69] W. J. Quadakkers, D. Naumenko, E. Wessel, V. Kochubey, and L. Singheiser, "Growth rates of alumina scales on Fe-Cr–Al alloys," *Oxidation of Metals*, vol. 61, no. 1-2, pp. 17–37, 2004.

[70] Y. Cadoret, D. Monceau, M. Bacos, P. Josso, V. Maurice, and P. Marcus, "Effect of platinum on the growth rate of the oxide scale formed on cast nickel aluminide intermetallic alloys," *Oxidation of Metals*, vol. 64, no. 3-4, pp. 185–205, 2005.

[71] S. Hayashi and B. Gleeson, "Early-stage oxidation behavior of Pt-modified γ'-Ni_3Al-based alloys with and without Hf addition," *Oxidation of Metals*, vol. 71, no. 1-2, pp. 5–19, 2009.

[72] M. W. Brumm and H. J Grabke, "The oxidation behaviour of NiAl-I. Phase transformations in the alumina scale during oxidation of NiAl and NiAl-Cr alloys," *Corrosion Science*, vol. 33, no. 11, pp. 1677–1690, 1992.

[73] S. C. Choi, H. J. Cho, Y. J. Kim, and D. B. Lee, "High-temperature oxidation behavior of pure Ni_3Al," *Oxidation of Metals*, vol. 46, no. 1-2, pp. 51–72, 1996.

[74] J. D. Kuenzly and D. L. Douglass, "The oxidation mechanism of Ni_3Al containing Yttrium," *Oxidation of Metals*, vol. 8, no. 3, pp. 139–178, 1974.

[75] S. Ochiai, T. Suzuki, Y. Kojima, M. Kobayashi, and Y. Torisaka, "Effect of yttrium addition on the high-temperature oxidation resistance of Ni_3Al alloy," *Journal of the Japan Institute of Metals and Materials*, vol. 53, no. 6, pp. 585–592, 1989.

[76] N. Birks, G. H. Meier, and F. S. Pettit, "High-temperature corrosion resistance," *Journal of the Minerals, Metals & Materials Society*, vol. 39, no. 12, pp. 28–31, 1987.

[77] G. Cao, L. Geng, Z. Zheng, and M. Naka, "The oxidation of nanocrystalline Ni_3Al fabricated by mechanical alloying and spark plasma sintering," *Intermetallics*, vol. 15, no. 12, pp. 1672–1677, 2007.

[78] N. Mu, T. Izumi, L. Zhang, and B. Gleeson, "The development and performance of novel Pt+Hf-modified γ'-Ni_3Al +γ-Ni bond coatings for advanced thermal barrier coatings systems," in *Superalloys 2008*, R. C. Reed, K. A. Green, P. Caron et al, Eds., TMS (The Minerals, Metals & Materials Society), 2008.

[79] C. Jiang and B. Gleeson, "Surface segregation of Pt in γ'-Ni_3Al: a first-principles study," *Acta Materialia*, vol. 55, no. 5, pp. 1641–1647, 2007.

[80] Y. Niu, X. J. Zhang, Y. Wu, and F. Gesmundo, "The third-element effect in the oxidation of Ni–xCr–7Al (x=0, 5, 10, 15at.%) alloys in 1atm O_2 at 900–1000°C," *Corrosion Science*, vol. 48, no. 12, pp. 4020–4036, 2006.

[81] J. Porcayo-Calderon, O. Sotelo-Mazon, V. M. Salinas-Bravo, C. D. Arrieta-Gonzalez, J. J. Ramos-Hernandez, and C. Cuevas-Arteaga, "Electrochemical performance of Ni_20Cr coatings applied by combustion powder spray in $ZnCl_2$-KCl molten salts," *International Journal of Electrochemical Science*, vol. 7, pp. 1134–1148, 2012.

[82] A. Luna-Ramirez, J. Porcayo-Calderon, G. Salinas-Solano, C. D. Arrieta-Gonzalez, V. M. Salinas-Bravo, and L. Martinez, "Electrochemical performance of Fe-Al intermetallic alloys with addition of Li, Ni and Ce in $NaVO_3$ at 700°C," *International Journal of Electrochemical Science*, vol. 8, pp. 4641–4653, 2013.

[83] J. Porcayo-Calderon, O. Sotelo-Mazon, M. Casales-Diaz, J. A. Ascencio-Gutierrez, V. M. Salinas-Bravo, and L. Martinez-Gomez, "Electrochemical study of Ni_20Cr coatings applied by HVOF process in $ZnCl_2$-KCl at high temperatures," *Journal of Analytical Methods in Chemistry*, vol. 2014, pp. 1–11, 2014.

[84] G. Salinas-Solano, J. Porcayo-Calderon, J. G. Gonzalez-Rodriguez, V. M. Salinas-Bravo, J. A. Ascencio-Gutierrez, and L. Martinez-Gomez, "High temperature corrosion of inconel 600 in NaCl-KCl molten salts," *Advances in Materials Science and Engineering*, vol. 2014, pp. 1–9, 2014.

[85] O. Sotelo-Mazon, J. Porcayo-Calderon, C. Cuevas-Arteaga et al., "Corrosion performance of Ni-based alloys in sodium metavanadate," *International Journal of Electrochemical Science*, vol. 11, pp. 1868–1882, 2016.

[86] J. Doychak, J. L. Smialek, and C. A. Barrett, "The oxidation of Ni-rich Ni-Al intermetallics," in *Oxidation of High Temperature Intermetallics*, T. Grobstein and J. Doychak, Eds., p. 41, Minerals, Metals & Material Society, 1988.

[87] K. Zhao, Y. B. Zhou, Y. H. Ma, L. H. Lou, and Z. Q. Hu, "Effect of microstructure on the oxidation behavior of two nickel-based superalloys," *Corrosion*, vol. 61, no. 10, pp. 961–967, 2005.

[88] G. Liu, M. Li, M. Zhu, and Y. Zhou, "Transient of alumina oxide scale on β-NiAl coated on M38G alloy at 950°C," *Intermetallics*, vol. 15, no. 10, pp. 1285–1290, 2007.

11

Study on the High Temperature Friction and Wear Behaviors of Cu-Based Friction Pairs in Wet Clutches by Pin-on-Disc Tests

Er-Hui Zhao,[1] Biao Ma,[1,2] and He-Yan Li[1,2]

[1] *School of Mechanical Engineering, Beijing Institute of Technology, Beijing, China*
[2] *Collaborative Innovation Center of Electric Vehicles in Beijing, Beijing, China*

Correspondence should be addressed to He-Yan Li; lovheyan@bit.edu.cn

Academic Editor: Carlos Navarro

This work is devoted to the study of the high temperature friction and wear behaviors of Cu-based friction pairs in wet clutches under different temperatures, rotation speeds, and loads. Pin-on-disc tests are carried out on the UMT-3. The friction coefficient, wear factor, and high temperature wear mechanism are primarily analyzed. The results show that as the temperature rises from 120°C to 420°C, the friction coefficient increases from 0.28 to 0.35 at first and then decreases to 0.30, when the vibration of friction coefficient is significantly identified. Meanwhile, the wear factor grows gradually from $K = 7.9\times10^{-8}$ g/Nm to $K = 41.8\times10^{-8}$ g/Nm at first and then grows sharply to $K = 112.2\times10^{-8}$ g/Nm. The main wear mechanisms are abrasive wear and ploughing wear when the temperature is below 345°C, and the wear seriously deteriorates when the temperature exceeds 345°C, when the wear mechanism changes to adhesive wear and delamination wear.

1. Introduction

Cu-based powder metallurgy friction pairs are widely used in the wet multidisc clutch, which is one of the main parts in tracked-vehicle integrated transmissions. As power switching and torque transfer devices, wet clutches often work under extremely atrocious conditions, such as high initial sliding velocity, rapid temperature rise, and high surface pressure. The failures of wet clutches due to the friction and wear problems have become a limiting factor for the use of integrated transmissions. Therefore, the friction and wear behaviors of Cu-based friction pairs in wet clutches have to be thoroughly investigated.

The friction and wear behaviors of Cu-based friction pairs in wet clutches crucially depend on the operating conditions. However, the operating conditions in wet clutches are complex and unstable. Gao et al. [1], Deur et al. [2], Ompusunggu et al. [3], and Iqbal et al. [4] proposed numerical models and tests to investigate the engagement of wet clutches. At the initial stage of engagement, friction plates are separated from the mating discs and rotate with a high relative speed. Then, asperity contacts emerge and the relative speed decreases sharply. During the engagement of wet clutches, a large quantity of heat is generated intensively in 1 second, and the temperature rises rapidly at the interfaces of friction pairs. Researchers, such as Mansouri et al. [5], Jen and Nemecek [6], Ingram et al. [7], Seo et al. [8], and Wenbin et al. [9], investigated the heat transfer, temperature distribution, and thermal stresses of wet clutches by numerical models and experiments. The temperature distribution at the contact surfaces and the thermal stresses in the friction pairs are nonhomogeneous. The friction pairs can buckle due to the elevated temperatures and thermal stresses that occur during clutch engagements. Therefore, the friction and wear behaviors of Cu-based friction pairs will be influenced by the deterioration of operating conditions.

The friction and wear behaviors of wet clutches have long been investigated by numerical and experimental methods. Xiong et al. [10] investigated the effects of Fe and SiO_2 friction components on the friction and wear behaviors of the Cu-based friction materials by an experimental method, and the wear mechanisms were analyzed. Ost et al. [11] investigated the friction and wear behaviors of paper-based wet clutch friction pairs by both SAE#II and pin-on-disc tests, and the

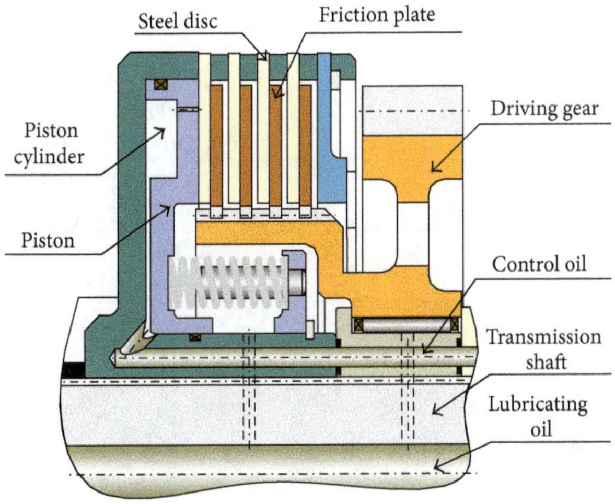

FIGURE 1: Structure diagram of wet multidisc clutch.

influences of material parameters and operating conditions on friction coefficient and wear rate were analyzed. Nyman et al. [12] investigated the influence of changes in the topography of the sintered friction material on the friction characteristics of wet clutches by an experimental method. Yao et al. [13] investigated the characteristics of a worn surface of Cu-based powder metallurgy brake materials after working under service condition, and the main mechanisms were discussed. Zhou et al. [14] investigated the drag torque in a two-speed dual clutch transmission by numerical and experimental methods. Li et al. [15] presented a methodology for prediction of wear in the friction lining of a wet clutch subjected to repeated engagement cycles. Pica et al. [16] provided a temperature and slip speed dependent model to investigate the torque characteristics of dry dual clutches. Gong et al. [17] investigated the wear behaviors of Cu-based friction wet clutches by ring-on-ring test, and the dominant wear mechanisms were discussed. Hoic et al. [18] investigated the wear behaviors of dry dual clutch by experimental characterization and mathematical model. However, the investigations presented above mainly focused on the friction and wear behaviors of wet clutch friction pairs under normal operating conditions, and few of them investigated the friction and wear behaviors of Cu-based friction pairs in wet clutches under severe operating conditions, especially the high temperature friction and wear behaviors.

In this paper, based on the Cu-based wet multidisc clutches used in tracked-vehicle integrated transmissions, as shown in Figure 1, pin-on-disc tests are conducted on the UMT-3 to investigate the friction and wear behaviors of Cu-based friction pairs under severe operating conditions, especially the high temperature friction and wear behaviors. The friction coefficient and wear factor under different operating conditions and the high temperature wear mechanisms are primarily analyzed.

2. Experimental Details

The friction and wear behaviors of Cu-based friction pairs in wet clutches crucially depend on the operating conditions. However, the local temperature, the local contact pressure, and the local relative sliding velocity are not constant on the contact surfaces of the friction pairs. Therefore, to better study the local friction and wear behaviors of Cu-based friction pairs under different operating conditions, pin-on-disc tests are conducted on a professional test apparatus, Universal Material Tester (UMT), the model of which used in this paper is UMT-3, provided by the Bruker Corporation in the United States. Generally, during the engagement of wet clutches, a large quantity of heat is generated intensively in 1 second, and the lubricant film can fail due to the rapid temperature rise. Therefore, in this paper, the high temperature friction and wear behaviors of Cu-based friction pairs are investigated by dry pin-on-disc tests, and there is no lubricant present during these experiments.

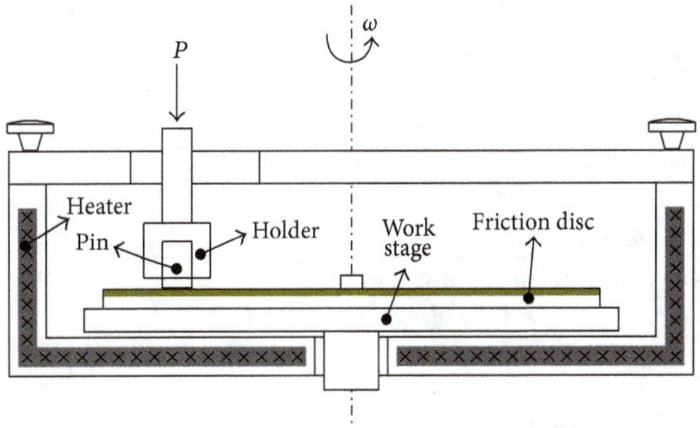

FIGURE 2: Elevated temperature chamber for rotary drives.

The structure of elevated temperature chamber for rotary drives is shown in Figure 2. The heating range of the elevated temperature chamber is 0~1000°C. The diameter of the steel pin is $d = 6$ mm, and the rotation radius of the sliding track on the friction disc is $r = 25$ mm. In the process of testing, the pin was fixed with the sensor module, and the friction disc rotated with the rotary chamber.

The steel pin was made of 65Mn steel, and the friction disc was made of Cu-based powder metallurgy material. Copper was the matrix material, and the iron powder (5% in weight), feldspar powder (5% in weight), carbon (6% in weight), and so forth were the additive materials. The RMS surface roughness of the friction disc was 0.61 μm. When the friction coefficient was stable after running in, we started the tests. When the temperature stabilized at the target temperature, we carried out the test procedures: (1) pressure the pin against the friction disc; (2) bring the electromotor to the target rotation speed; (3) record the rotation speed, load, temperature, and friction coefficient; (4) stop the electromotor and separate the pin and friction disc; (5) change the pin and disc with brand new ones. Operating parameter settings in pin-on-disc tests and corresponding values in wet clutches are shown in Table 1.

3. Results and Discussion

In order to investigate the high temperature friction and wear behaviors of Cu-based friction pairs, the test results presented below are carried out with five different temperature levels,

TABLE 1: Parameter settings in pin-on-disc tests and corresponding values in wet clutches.

Parameters	Values
Rotation speed in pin-on-disc tests (rpm)	400, 500, 600, 700, 800
Relative sliding velocity in wet clutches (mm/s)	1047, 1309, 1571, 1833, 2095
Load in pin-on-disc tests (N)	60, 80, 100, 120, 140
Average surface pressure in wet clutches (MPa)	2.1, 2.8, 3.5, 4.2, 4.9
Temperature in pin-on-disc tests (°C)	120, 195, 270, 345, 420
Temperature in wet clutches (°C)	120, 195, 270, 345, 420

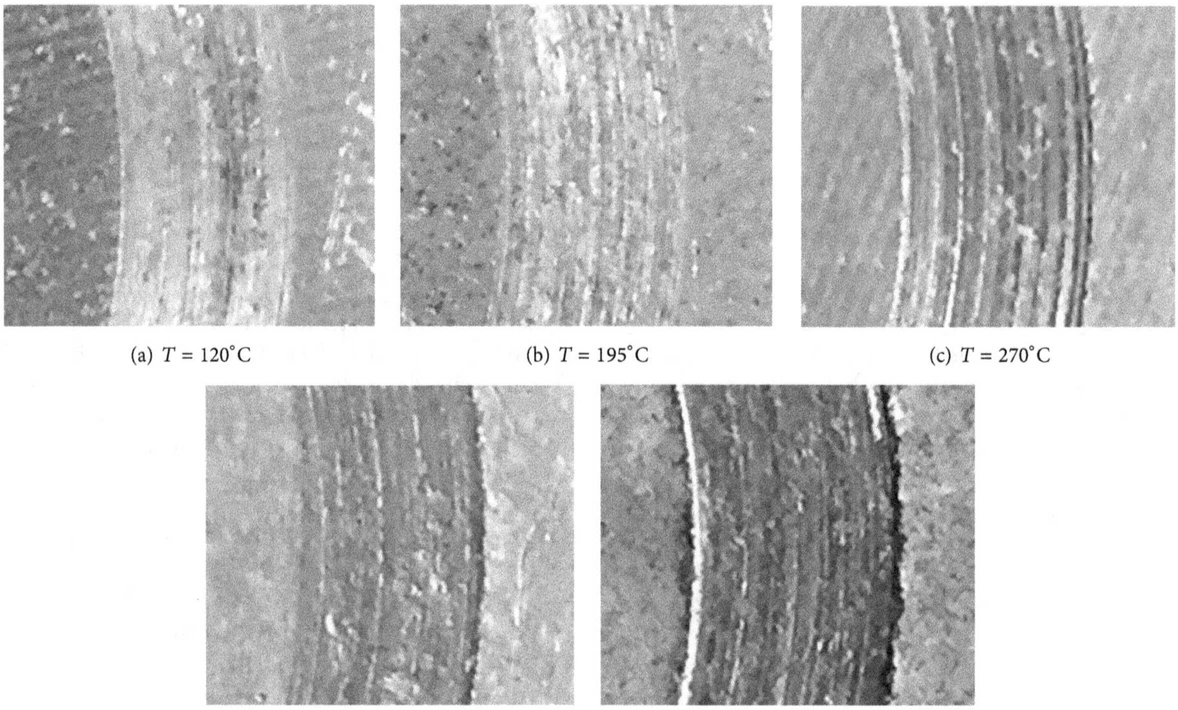

(a) $T = 120°C$ (b) $T = 195°C$ (c) $T = 270°C$

(d) $T = 345°C$ (e) $T = 420°C$

FIGURE 3: Friction discs after tests with different levels of temperature.

five different rotation speeds, and five different load levels. The test duration is 30 minutes. The friction coefficient, wear factor, and high temperature wear mechanism will be primarily analyzed in this part. The pin and the disc used in each experiment are brand new.

Figures 3(a)–3(e) show the surface state of the friction discs after tests with different temperatures. It is clear that the depth of wear groove increases significantly as the temperature rises. Furthermore, the friction condition and the wear mechanism of Cu-based friction pairs change in the process of temperature rising.

Figure 4(a) presents the test results of friction coefficient and average friction coefficient under different temperatures. It is clear that temperature rise has a significant influence on the friction coefficient of Cu-based friction pairs. When the temperature is $T = 120°C$, the friction coefficient is stable and the mean value of friction coefficient is about 0.28. As the temperature rises to $T = 195°C$, the friction coefficient has slight fluctuation and the mean value increases to about 0.32. With the temperature $T = 270°C$, the friction coefficient is the biggest, about 0.35, and begins to vibrate obviously. Then, as the temperature rises to $T = 345°C$, the friction coefficient deceases to about 0.33 and fluctuates strongly. When the temperature rises to $T = 420°C$, the vibration of friction coefficient is significantly identified, and the mean value of the friction coefficient decreases to about 0.30.

Figures 4(b) and 4(c) present the test results of friction coefficient and average friction coefficient under different rotation speeds and different loads. We can see that the effects of rotation speed and load on the friction coefficient of Cu-based friction pairs are also obvious. As the rotation speed increases from 400 rpm to 800 rpm, the average friction coefficient decreases gradually from about 0.38 to 0.32, and the vibration of the friction coefficient decreases at the same time. Similarly, as the load increases from 60 N to 140 N, the average friction coefficient decreases gradually from about 0.39 to 0.32, and there is no significant change in the vibration of the friction coefficient.

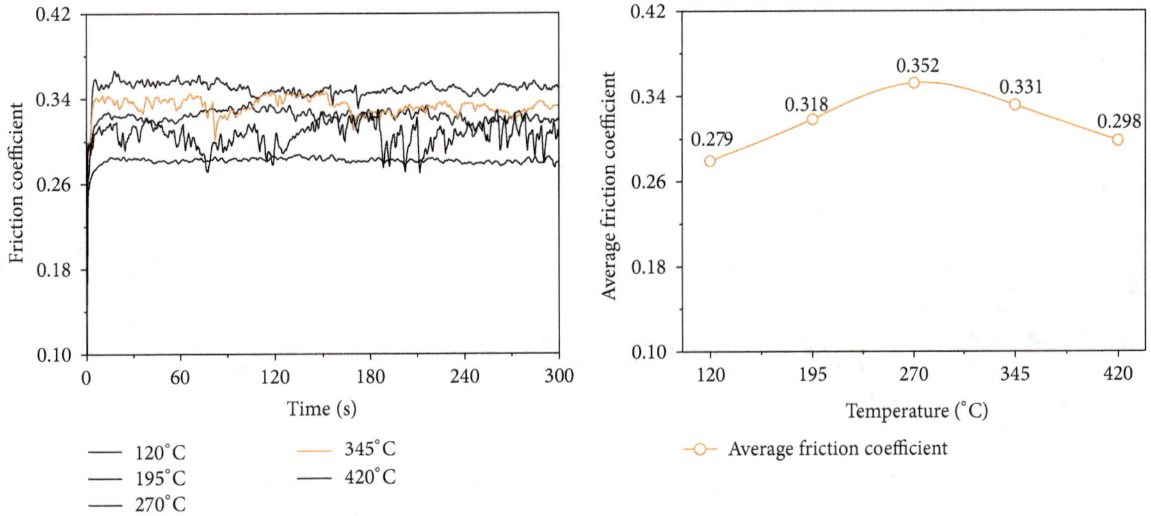

(a) Test results of friction coefficient under different temperatures

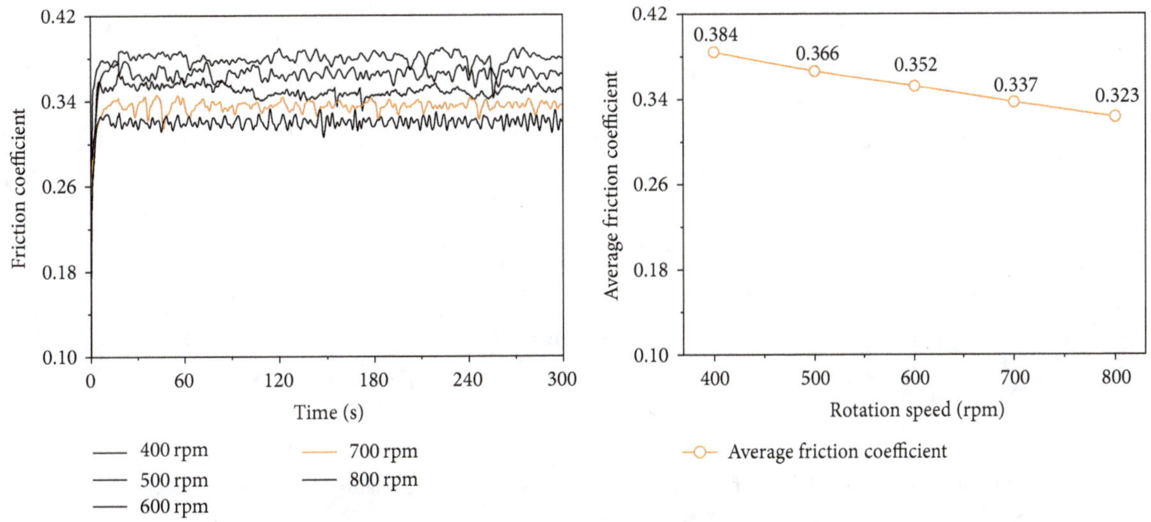

(b) Test results of friction coefficient under different rotation speeds

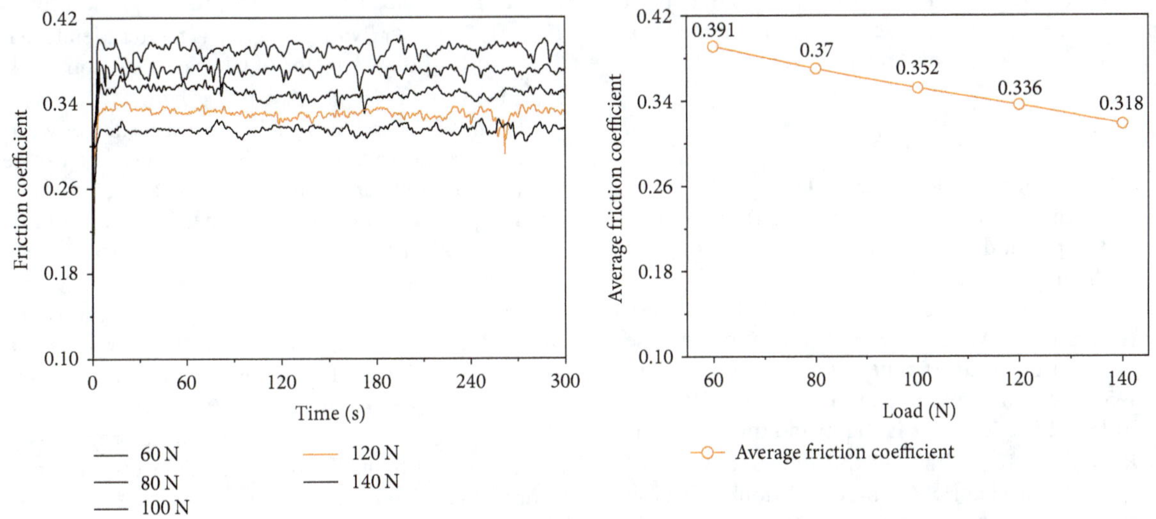

(c) Test results of friction coefficient under different loads

FIGURE 4: Test results of friction coefficient under different temperatures, rotation speeds, and loads.

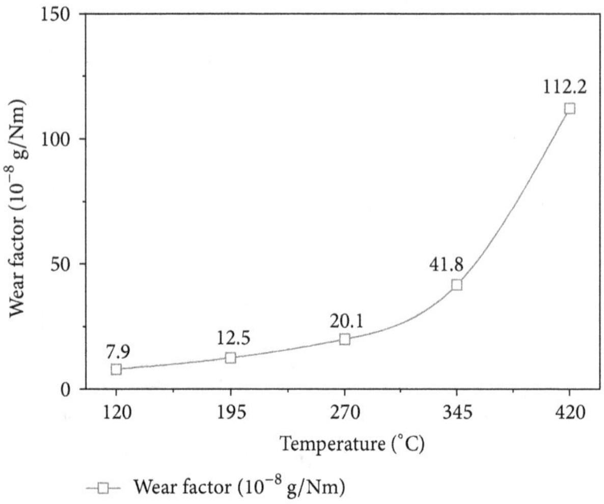

(a) Test results of wear factor under different temperatures

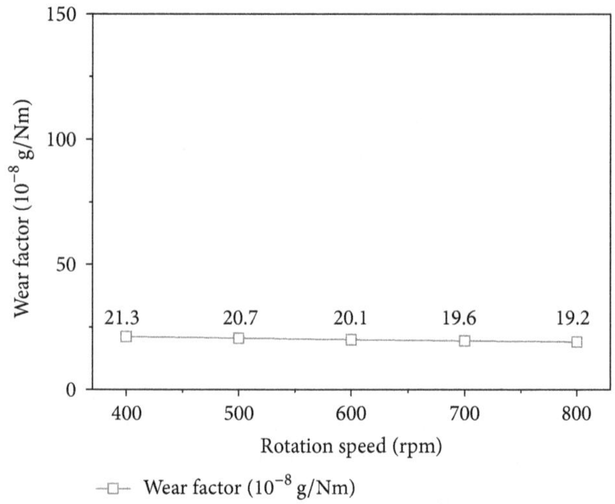

(b) Test results of wear factor under different rotation speeds

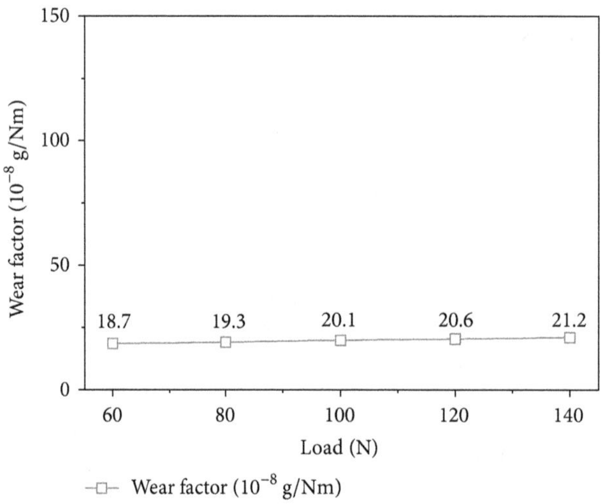

(c) Test results of wear factor under different loads

FIGURE 5: Test results of wear factor under different temperatures, rotation speeds, and loads.

The wear loss of the friction disc is measured by an electronic balance, and the wear factor can be calculated from the test result of weight loss by Archard's wear model:

$$K = \frac{W_{\text{test}}}{F_{\text{test}} S}, \quad (1)$$

where W_{test} is the test result of weight loss (g); F_{test} is the load of pin-on-disc test (N); and S is the total relative sliding distance of the pin and the disc (m). The unit of the wear factor K is g/Nm.

Figure 5(a) shows the test results of wear factor under different temperatures. We can see that the wear of Cu-based friction pairs seriously deteriorates when the temperature exceeds 345°C. When the temperature is $T = 120$°C, the wear of Cu-based friction pairs is slight, and the wear factor is about $K = 7.9 \times 10^{-8}$ g/Nm. As the temperature rises to $T = 270$°C, the wear factor of Cu-based friction pairs grows gradually to about $K = 20.1 \times 10^{-8}$ g/Nm. When the temperature is higher than $T = 300$°C, the growth of the wear factor becomes faster. When the temperature is $T = 345$°C, the wear factor of Cu-based friction pairs grows significantly to about $K = 41.8 \times 10^{-8}$ g/Nm. However, as the temperature rises to $T = 420$°C, the wear factor of Cu-based friction pairs grows sharply to about $K = 112.2 \times 10^{-8}$ g/Nm.

Figures 5(b) and 5(c) show the test results of wear factor under different rotation speeds and loads. We can see that there is no significant change in the wear factor of Cu-based friction pairs when the rotation speed or the load increases. As the rotation speed increases from 400 rpm to 800 rpm, the wear factor decreases slightly from about $K = 21.3 \times 10^{-8}$ g/Nm to $K = 19.2 \times 10^{-8}$ g/Nm. On the contrary, as the load increases from 60 N to 140 N, the wear factor of the Cu-based friction pairs increases slightly from about $K = 18.7 \times 10^{-8}$ g/Nm to $K = 21.2 \times 10^{-8}$ g/Nm.

Figures 6(a)–6(e) provide the SEM micrographs of worn surfaces of Cu-based friction pairs after test under different temperatures. It is observed that the wear mechanism changes obviously as the temperature rises from $T = 120$°C to $T = 420$°C. When the temperature is $T = 120$°C, the wear of the friction surface is slight, and the wear mechanism is abrasive

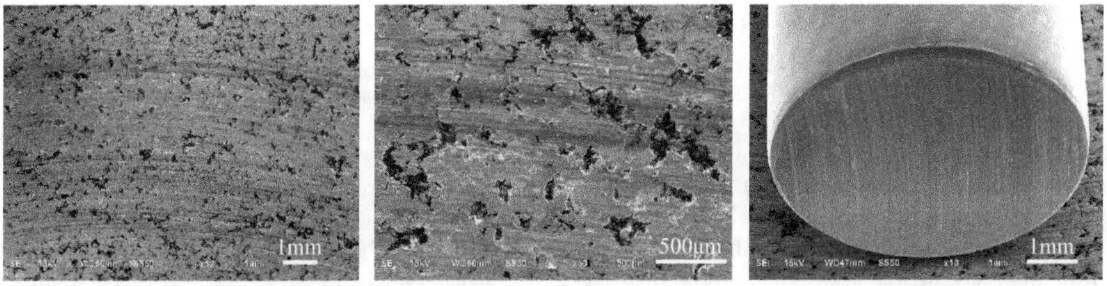

(a) $T = 120^{\circ}\mathrm{C}$

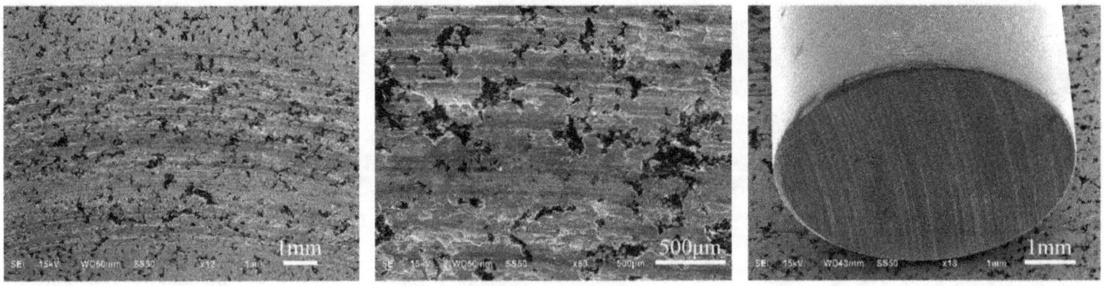

(b) $T = 195^{\circ}\mathrm{C}$

(c) $T = 270^{\circ}\mathrm{C}$

(d) $T = 345^{\circ}\mathrm{C}$

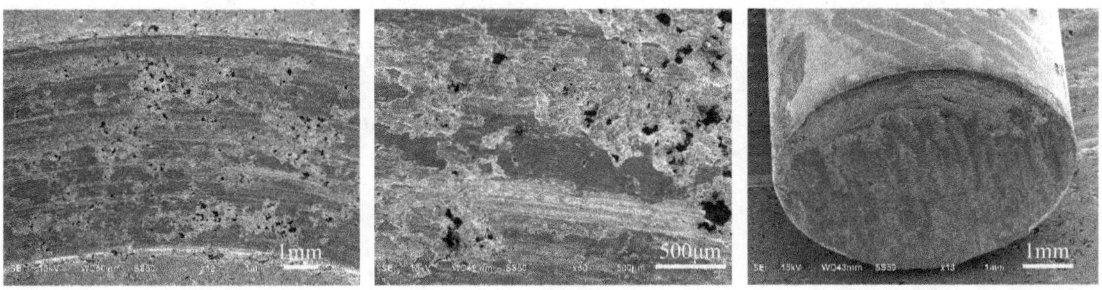

(e) $T = 420^{\circ}\mathrm{C}$

FIGURE 6: SEM micrographs of worn surfaces of Cu-based friction pairs after the test.

wear. As the temperature rises to $T = 195^\circ$C, the wear of the friction surface becomes more obvious, and the wear mechanism is still abrasive wear. When the temperature rises to $T = 270^\circ$C, some furrows emerge, and the wear mechanism becomes ploughing wear. As the temperature rises to $T = 345^\circ$C, the Cu-based powder metallurgy material begins to soften, and the main wear mechanism changes to adhesive wear. When the temperature rises continuously to extremely high $T = 420^\circ$C, the wear of Cu-based friction pairs deteriorates seriously, and there are massive exfoliations on the friction surface, when the wear mechanism changes to delamination wear.

4. Conclusions

The high temperature friction and wear behaviors of Cu-based friction pairs in wet clutches have been investigated based on pin-on-disc tests. The friction coefficient, wear factor, and high temperature wear mechanism have been primarily analyzed. The main conclusions are summarized as follows:

(1) When the temperature is 120°C, the friction coefficient of Cu-based friction pairs is stable. As the temperature rises to 420°C, the friction coefficient begins to vibrate significantly. The friction coefficient increases from 0.28 to 0.35 when the temperature rises from 120°C to 270°C and decreases to 0.30 when the temperature continuously rises to 420°C.

(2) When the temperature rises from 120°C to 270°C, the wear of Cu-based friction pairs is slight, and the wear factor grows gradually from $K = 7.9 \times 10^{-8}$ g/Nm to $K = 20.1 \times 10^{-8}$ g/Nm. However, as the temperature rises from 345°C to 420°C, the wear factor grows sharply from $K = 41.8 \times 10^{-8}$ g/Nm to $K = 112.2 \times 10^{-8}$ g/Nm.

(3) When the temperature is lower than 345°C, the main wear mechanisms are abrasive wear and ploughing wear. As the temperature rises to higher than 345°C, the wear of Cu-based friction pairs seriously deteriorates, and the wear mechanism changes to adhesive wear and delamination wear.

Conflicts of Interest

The authors declare that they have no conflicts of interest regarding the publication of this paper.

Acknowledgments

The authors would like to express their appreciation for the continuous support from Professor Wen-zhong Wang at Beijing Institute of Technology. Furthermore, the authors acknowledge the financial support from the National Natural Science Foundation of China (no. 51575042).

References

[1] H. Gao, G. C. Barber, and M. Shillor, "Numerical simulation of engagement of a wet clutch with skewed surface roughness," *Journal of Tribology*, vol. 124, no. 2, pp. 305–312, 2002.

[2] J. Deur, J. Petric, J. Asgari, and D. Hrovat, "Modeling of wet clutch engagement including a thorough experimental validation," *SAE Technical Papers*, 2005.

[3] A. P. Ompusunggu, P. Sas, and H. Van Brussel, "Modeling and simulation of the engagement dynamics of a wet friction clutch system subjected to degradation: An application to condition monitoring and prognostics," *Mechatronics*, vol. 23, no. 6, pp. 700–712, 2013.

[4] S. Iqbal, F. Al-Bender, A. P. Ompusunggu, B. Pluymers, and W. Desmet, "Modeling and analysis of wet friction clutch engagement dynamics," *Mechanical Systems and Signal Processing*, vol. 60-61, pp. 420–436, 2015.

[5] M. Mansouri, M. Holgerson, M. M. Khonsari, and W. Aung, "Thermal and dynamic characterization of wet clutch engagement with provision for drive Torque," *Journal of Tribology*, vol. 123, no. 2, pp. 313–323, 2001.

[6] T.-C. Jen and D. J. Nemecek, "Thermal analysis of a wet-disk clutch subjected to a constant energy engagement," *International Journal of Heat and Mass Transfer*, vol. 51, no. 7-8, pp. 1757–1769, 2008.

[7] M. Ingram, T. Reddyhoff, and H. A. Spikes, "Thermal behaviour of a slipping wet clutch contact," *Tribology Letters*, vol. 41, no. 1, pp. 23–32, 2011.

[8] H. Seo, C. Zheng, W. Lim, S. W. Cha, and S. Han, "Temperature prediction model of wet clutch in coupling," in *Proceedings of the 7th IEEE Vehicle Power and Propulsion Conference, VPPC 2011*, pp. 1–4, IEEE, Chicago, IL, USA, September 2011.

[9] L. Wenbin, H. Jianfeng, F. Jie, C. Liyun, and Y. Chunyan, "Simulation and application of temperature field of carbon fabric wet clutch during engagement based on finite element analysis," *International Communications in Heat and Mass Transfer*, vol. 71, pp. 180–187, 2016.

[10] X. Xiong, J. Chen, P. Yao, S. Li, and B. Huang, "Friction and wear behaviors and mechanisms of Fe and SiO_2 in Cu-based P/M friction materials," *Wear*, vol. 262, no. 9-10, pp. 1182–1186, 2007.

[11] W. Ost, P. De Baets, and J. Degrieck, "The tribological behaviour of paper friction plates for wet clutch application investigated on SAE#II and pin-on-disk test rigs," *Wear*, vol. 249, no. 5-6, pp. 361–371, 2001.

[12] P. Nyman, R. Mäki, R. Olsson, and B. Ganemi, "Influence of surface topography on friction characteristics in wet clutch applications," *Wear*, vol. 261, no. 1, pp. 46–52, 2006.

[13] P.-P. Yao, H.-C. Sheng, X. Xiong, and B.-Y. Huang, "Worn surface characteristics of Cu-based powder metallurgy bake materials for aircraft," *Transactions of Nonferrous Metals Society of China*, vol. 17, no. 1, pp. 99–103, 2007.

[14] X. Zhou, P. Walker, N. Zhang, B. Zhu, and J. Ruan, "Numerical and experimental investigation of drag torque in a two-speed dual clutch transmission," *Mechanism and Machine Theory*, vol. 79, pp. 46–63, 2014.

[15] M. Li, M. M. Khonsari, D. M. C. McCarthy, and J. Lundin, "On the wear prediction of the paper-based friction materialin a wet clutch," *Wear*, vol. 334-335, pp. 56–66, 2015.

[16] G. Pica, C. Cervone, A. Senatore, M. Lupo, and F. Vasca, "Dry dual clutch torque model with temperature and slip speed

effects," *Intelligent Industrial Systems*, vol. 2, no. 2, pp. 133–147, 2016.

[17] T. Gong, P. Yao, Y. Xiao et al., "Wear map for a copper-based friction clutch material under oil lubrication," *Wear*, vol. 328-329, pp. 270–276, 2015.

[18] M. Hoic, N. Kranjcevic, Z. Herold, J. Deur, and V. Ivanovic, "Experimental characterization and modeling of dry dual clutch wear," *SAE International Journal of Passenger Cars-Mechanical Systems*, vol. 7, no. 1, pp. 326–337, 2014.

12

Study on Thickness Thinning Ratio of the Forming Parts in Single Point Incremental Forming Process

Mingshun Yang,[1] Zimeng Yao,[1] Yan Li,[1] Pengyang Li,[1] Fengkui Cui,[2] and Lang Bai[1]

[1]*Faculty of Mechanical and Precision Instrument Engineering, Xi'an University of Technology, Xi'an 710048, China*
[2]*School of Mechatronics Engineering, Henan University of Science and Technology, Luoyang, China*

Correspondence should be addressed to Zimeng Yao; 15829585702@163.com

Academic Editor: Laszlo Toth

An excessive thickness-reducing ratio of the deformation zone in single point incremental forming of the metal sheet process has an important influence on the forming limit. Prediction of the deformation zone thickness is an important approach to control the thinning ratio. Taking the 1060 aluminum as the research object, the principle of thickness deformation in the single point incremental forming process was analyzed; the finite element model was established using ABAQUS. A formula with high accuracy to predict the deformation zone thickness was fitted with the simulation results, and the influences of process parameters, such as tool diameter, step down, feeding speed, sheet thickness, and forming angle, on thinning ratio were analyzed. The accuracy of the finite element simulation was verified by experiment. A method to control the thinning rate by changing the forming trajectory was proposed. The results showed that the obtained value by using the fitted formula is closer to the experimental results than that obtained by the sine theorem. The thinning rate of the deformation zone increases with the increase of tool diameter, forming angle, and sheet thickness and decreases with the increase of step down, while the feeding speed had no significant effect on the thinning ratio. The most important factor of the thinning ratio is the forming angle, and the thinning ratio can be effectively reduced by using the forming trajectory with a uniformly distributed pressing point.

1. Introduction

Single point incremental forming (SPIF) is a new type of sheet metal-forming technology integrating numerical control (NC) technology, computer technology, and plastic-forming technology [1]. In the forming process, according to the shape of the target part, CAM software is used to generate a toolpath with a series of high-profile NC codes, and the tool subsequently follows the toolpath generated and deforms the sheet in small increments, accumulating plastic deformation between the tool and material that can result in obtaining the target shape [2]. Better flexibility, lower cost, high degree of automation, and enhanced formability are special features of this process in the customized and small batch production of metal sheet parts, especially in artificial, medical appliance, aerospace, and household products [3, 4].

The forming quality, including the excessive thinning rate of wall thickness, is still one of the greatest challenges for the development of single point incremental forming [5–7]. The thickness thinning ratio is directly related to the formability of the forming parts; thus, the study of the thickness law of forming parts is of great significance [8]. There are many studies from researchers all over the world on the thickness rules of forming parts. McAnulty et al. [9] analyzed the experimental results from 35 papers that reported the study of the influence of process parameters (tool diameter, tool shape, step down, feed rate, rotation direction, and spindle speed) on the formability in SPIF. Malhotra et al. [10] used a fracture model and finite element analyses to predict the fracture in SPIF of the cone and the funnel parts; the forming forces, fracture depths, and thinning from finite element analyses were used to validate predictions, and the results showed that the local nature of deformation was a primary factor for the increased formability in SPIF. Hussain et al. [11] obtained the final thickness of the forming parts with varying wall angle by the sine law; on this basis, the maximum wall angle was obtained from four conical frustums of varying wall angle. Jackson and Allwood [12] analyzed the wall thickness using the deformation mechanism of SPIF. Kim and Park [13] analyzed the influence of processing parameters on the wall thickness. Bouffioux et al. [14] simulated the forming process and provided better forecasting of the wall thickness. Silva et al. [15] investigated the failure mechanisms in SPIF with different tool radii and the evolution of the wall thickness of truncated pyramids with five different tool radii; the results showed that a critical threshold exists for the ratio between the sheet thickness and the tool radius that distinguishes between fracture with and without previous necking. Young and Jeswiet [16] found that the thickness of the deforming region only depends on the forming angle by theoretical analysis and used the cosine theorem to predict the sheet thickness value. Hirt et al. [17] verified the cosine theorem through experiments and finite element simulations. Al-Ghamdi and Hussain [18] found that the ratio of the tool radius to the sheet thickness could control the formability in SPIF; the ratio was a threshold value, and the formability was noticed to increase with decrease in the tool radius above the threshold value and decrease with the reduction in tool radius below the threshold value. Martins et al. [19] proposed that the cosine theorem is suitable for deformation thickness prediction and that only shear deformation occurred in the sheet-forming process. Zhou [20] studied the influence of the forming angle on the thickness value of the deformation area and analyzed the thickness of deformation region in a straight wall. Xiao and Gao [21] adopted the method of numerical simulation and experiment to study the critical forming angle of uniform thickness in the single point incremental forming of a conical part process. Li [22] found that the minimum sheet thickness is closely related to the tool diameter and that work hardening can lead to sheet material strength enhanced significantly. Yu and Gao [23] proposed a theoretical calculation formula for the minimum wall thickness and reduced the sheet thinning successfully by multistep forming. Cao and Gao [24] found a method to produce parts with equal wall thickness by means of increasing processing variables after analyzing the relationships among feed, plate material thickness, forming angle, and maximum stable area. Wei and Gao [25] developed thickness prediction software based on the law of sines that could accurately predict the part's thickness. Zhao and Zhan [26] obtained thinning distribution and sheet distortion pattern after studying the metal flawing law in the forming process.

The current study on thickness thinning is mainly based on the shear deformation that occurs under the action of the forming tool on sheet material, for which the sine law is not adequate in predicting and controlling the wall thickness. In this paper, the single point incremental forming process of a 1060 aluminum sheet is simulated with finite element simulation software ABAQUS to study the influences of different process parameters, including tool diameter, step down, feeding speed, sheet thickness, and forming angle, on the wall thinning. The experimental results obtained are fitted by MATLAB to generate an empirical formula that could better predict the wall thickness, and the formula is verified by experiments on HASS milling machine; the formula predicting the wall thickness in the forming area and method to control the thinning rate are obtained.

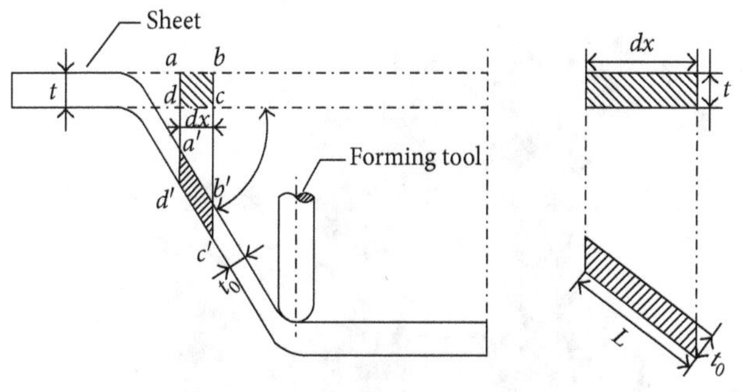

FIGURE 1: Illustration of the shear deformation in the SPIF process. α is the forming angle, t is the initial sheet thickness, t_0 is the thickness of deformation, $abcd$ with length of dx is the unit to be deformed, and $a'b'c'd'$ is the deformed unit.

2. Theoretical Background

Single point incremental forming process is a nonlinear large displacement process. The plastic deformation of metal material results in the continuous squeezing of the material as the tool is moved. Because the axial force is greater than the radial and tangential forces in the forming process, it is usually assumed that shear deformation occurs only in the axial direction, as shown in Figure 1.

According to the constant volume principle in plastic deformation, the following formula can be obtained:

$$t \cdot dx = t_0 \cdot L, \qquad dx = L \cdot \cos\alpha, \tag{1}$$

From the above, t_0 can be calculated as $t_0 = t\cos\alpha$.

The wall thinning rate is an important indicator of formability that can reflect the thickness variation law and changing scope of the metal sheet. When the wall thinning rate exceeds a certain range, the part will have lower strength, thereby greatly affecting the precision of the formed part. The wall thinning rate can be defined as follows:

$$\varphi = \frac{t - t_0}{t} \times 100\%. \tag{2}$$

In the forming process, both radial movement and tangential movement occur while the tool moves along the axial direction, and the sheet will generate radial deformation and tangential deformation via the squeezing force and friction force of the forming tool. Although tangential deformation and radial deformation are smaller than the axial deformation, they still have effects on the wall thickness distribution. Therefore, the thinning rate obtained from formulas (1) and (2) is not accurate enough, and the thinning rate of the forming part is in relation to not only the initial thickness and forming angle, but also the tool forming diameter, step down, and feeding speed. Based on formula (1), the systematically comprehensive wall thickness calculation formula is given by

$$t_0 = a \cdot D + b \cdot Z + c \cdot F + d \cdot \alpha + e \cdot t + f, \tag{3}$$

where a, b, c, d, e, and f are undetermined coefficients, D is the tool diameter, Z is the step down, and F is the feed speed.

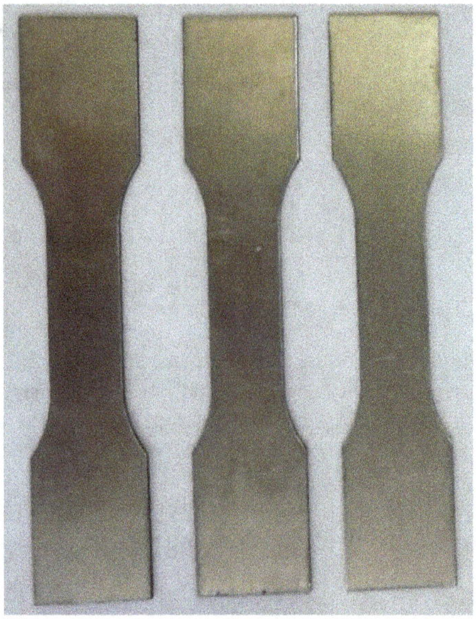

Figure 2: Tension test specimen.

Figure 3: Tension test machine.

In this paper, the relationship between the wall thinning rate and the processing parameters are established by simulation results and the undetermined coefficients are obtained; the formula and the simulation results are verified experimentally.

3. FEM Simulation

3.1. Simulation Model. The commercial finite element software ABAQUS can solve the problems of nonlinear materials and geometrically nonlinear problems, as well as accurately simulating the contact process with long contact time and large plastic deformation. ABAQUS is used as the finite element simulation software combined with the characteristics of SPIF. The sheet material is 1060 aluminum; the specimens are shown in Figure 2. The tensile test is conducted on the HT2402 tensile test machine, as shown in Figure 3; the loading speed is 1.5 mm/min and the temperature is room temperature (20°C), with each specimen being subjected to three repeated tests, and the average value is taken as the final tensile test result of the specimen. The material properties and the stress-strain properties are shown in Tables 1 and 2, respectively. In performing the simulation, because the experiment device of SPIF is a relatively complex experimental system, to reduce the difficulty of modeling and shorten the time of simulation, some simplifications are used under the unchanged relative motion relationship. The model consists of a spherical forming tool, a fixture, and a metal sheet; the fixture is simplified as two rings of the same size. The model can satisfy the relative motion relationship in the forming process by applying the boundary conditions. Forming tools and fixtures are considered as rigid bodies in the model; a rigid body does not participate in the calculation of stress and strain, thereby, saving CPU computing resources and reducing the computation time. The sheet size is 140 mm × 140 mm; the outer radius of the fixture is 70 mm, the inner radius of the fixture is 55 mm, and the upper radius of the forming part is 45 mm.

Table 1: Materials properties of 1060 aluminum.

Material	Density ρ(g/mm^3)	Elastic modulus (GPa)	Poisson's ratio	Yield strength (Mpa)
Al 1060	2.71	70	0.3	30

Table 2: Stress-strain properties of 1060 aluminum material.

Stress (MPa)	20	44	64	77	87	91	98
Strain	0.002	0.0169	0.0507	0.0845	0.1183	0.1352	0.169

The process parameters depend on the experimental conditions. SPIF is a complex plastic-forming process and a dynamic contact problem that requires high efficiency and a stable dynamic property algorithm. Taking A1 in Table 3 as an example, the simulation time is 121381 s, and the numerical simulation is conducted on a computer equipped with an Intel Core i7-6700 (3.4 GHz) and 32 GB memory.

Unit selection and grid division should follow accurate and efficient principles; thus, the S4R shell element was selected to simulate the forming process, and the mesh size is 1 mm × 1 mm. The forming tool and fixture do not need to be divided because they are rigid bodies. The surface-surface contact model was used, the dynamic explicit algorithm was selected, and a penalty function was employed. Coulomb friction was chosen as the friction type because the tangential force is smaller than the axial and radial forces.

Because the trajectory of forming tool is complex in the forming process, it is necessary to simulate the trajectory by periodic amplitude curves when simulation-forming cone parts. The forming tool will follow $X = A_1 \cos(2\pi t_n)$ in the X direction and follow $Y = B_1 \sin(2\pi t_n)$ in the Y direction, where A_1 and B_1 are the radius and t_n is the analysis step time; the motions along the X and Y directions can be synthesized

Table 3: Experimental parameter.

Number	Tool diameter (mm)	Step down (mm)	Feed speed (mm/min)	Forming angle (°)	Sheet metal thickness (mm)
A1	6	1	200	45	0.8
A2	10	1	200	45	0.8
A3	14	1	200	45	0.8
B1	10	0.6	200	45	0.8
B2	10	0.8	200	45	0.8
B3	10	1	200	45	0.8
C1	10	1	100	45	0.8
C2	10	1	200	45	0.8
C3	10	1	300	45	0.8
D1	10	1	200	45	0.8
D2	10	1	200	53	0.8
D3	10	1	200	60	0.8
E1	10	1	200	45	0.6
E2	10	1	200	45	0.8
E3	10	1	200	45	1

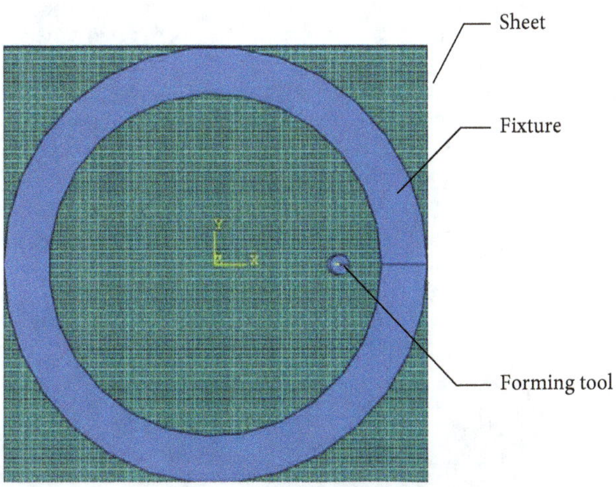

Figure 4: Finite element model.

Figure 5: Target shape.

to generate the trajectory for forming tool. The FE model and the target shape are shown in Figures 4 and 5, respectively.

3.2. Simulation Scheme. There are many parameters affecting the thickness thinning ratio in the single point incremental forming process. The simulation scheme is designed to include five forming parameters (forming tool diameter, step down, feed speed, sheet thickness, and forming angle), as shown in Table 3.

3.3. Simulation Results and Analysis. By using the view section function, the part is symmetrically cut along the X–Z section; "STH" is selected, which means sheet thickness; the sheet thickness is output to display the wall thickness, and the inquiry function is used to obtain the thickness values of five cells in the stable zone, respectively, located at depths of 10 mm, 13 mm, 15 mm, 17 mm, and 20 mm. The simulation results of the thickness distribution are shown in Figure 6, and the thickness values of different parameters are shown in Table 4.

Using multilinear regression to fit the above data using MATLAB, the undetermined coefficients of (3) could be obtained. Thus, the wall thickness can be calculated as follows:

$$t_0 = 0.59 - 0.0016 \cdot D + 0.0065 \cdot Z + 0 \cdot F - 0.0127\alpha + 0.6755 \cdot t. \quad (4)$$

The initial thickness t in Table 3 and the average thickness t_0 of the forming area in Table 4 were substituted in (2). The relationship between the thickness thinning ratio and the processing parameters obtained is shown in Figure 7.

The wall thinning rate clearly increases from 31.17% to 32.85% when the tool diameter changes from 6 mm to 14 mm (A1 to A3). The wall thinning rate decreases from 31.7% to 31.32% when the step down changes from 0.6 mm to 1.0 mm (B1 to B3), but the variation is not obvious. Based on the data of experiment nos. C1~C3, the wall thinning rate remains approximately 31.30% with the feed speed changing; that is, the feed speed has little effect on the thinning rate. According to the data of experiment nos. D1~D3, the wall thinning rate increases from 31.32% to 55.91% when the forming angle changes from 45° to 60°; thus, the forming angle has a great influence on the thinning rate. From the data of experiment nos. E1~E3, the wall thinning rate increases from 31.03% to 31.60% when the sheet thickness

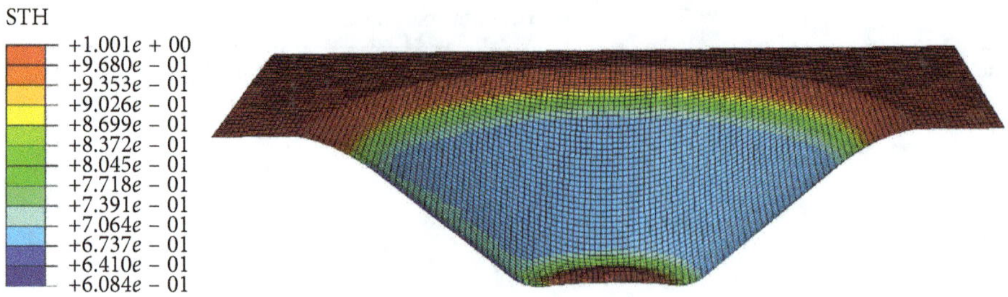

FIGURE 6: Simulation results.

TABLE 4: Thickness value.

Thickness number	10 mm	13 mm	15 mm	17 mm	20 mm	Average (mm)
A1	0.5540	0.5447	0.5529	0.5554	0.5434	0.5501
A2	0.5491	0.5390	0.5414	0.5399	0.5433	0.5425
A3	0.5396	0.5350	0.5241	0.5450	0.5424	0.5372
B1	0.5448	0.5409	0.5487	0.5508	0.5438	0.5458
B2	0.5506	0.5389	0.5485	0.5517	0.5550	0.5489
B3	0.5471	0.5530	0.5448	0.5536	0.5487	0.5494
C1	0.5647	0.5388	0.5515	0.5452	0.5482	0.5497
C2	0.5471	0.5530	0.5448	0.5536	0.5487	0.5494
C3	0.5494	0.5471	0.5518	0.5538	0.5469	0.5498
D1	0.5471	0.5530	0.5448	0.5536	0.5487	0.5494
D2	0.4491	0.4420	0.4628	0.4612	0.4562	0.4543
D3	0.3637	0.3365	0.3411	0.3639	0.3583	0.3527
E1	0.4193	0.4061	0.4128	0.4175	0.4134	0.4138
E2	0.5471	0.5530	0.5448	0.5536	0.5487	0.5494
E3	0.6905	0.6804	0.6828	0.6813	0.6847	0.6840

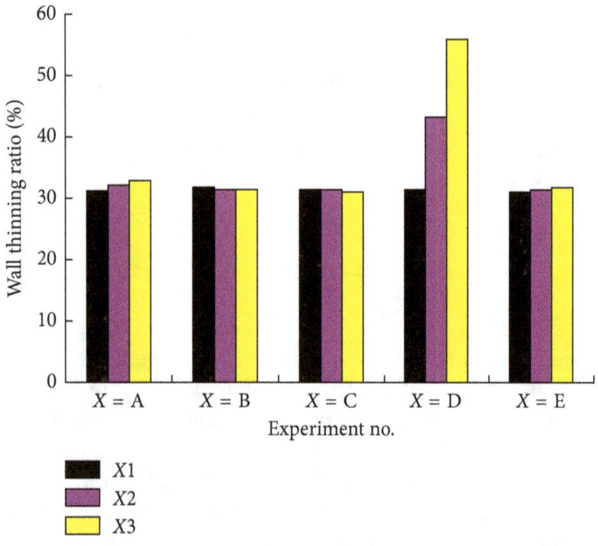

FIGURE 7: Relationship between thickness thinning ratio and process parameters.

increases from 0.6 mm to 1.0 mm, that is, the variation is not obvious.

4. Experimental Results

As shown in Figure 8, the HASS high-speed NC milling machine was used. The size of fixture is the same as that in the simulation model, and the experiment parameters are

FIGURE 8: SPIF experiment unit.

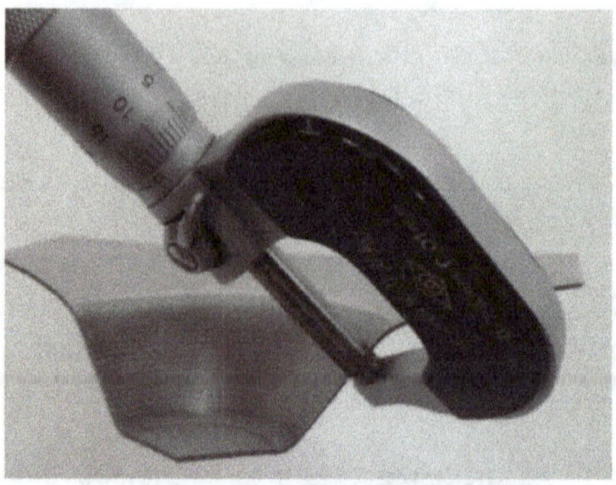

FIGURE 9: The measuring method.

Table 5: Comparison between the theoretical and experimental results of the average thickness value.

Process parameters	Tool diameter (mm)			Step down (mm)			Feed speed (mm/min)			Forming angle (°)			Feed speed (mm/min)		
	6	10	14	0.6	0.8	1.0	100	200	300	45	53	60	0.62	0.81	1.08
$t_{experiment}$	0.630	0.616	0.616	0.607	0.622	0.616	0.606	0.616	0.603	0.616	0.492	0.381	0.458	0.616	0.845
$t_{fitting}$	0.625	0.612	0.615	0.608	0.610	0.612	0.603	0.612	0.584	0.612	0.475	0.421	0.450	0.612	0.802
$t_{initial}$	0.638	0.631	0.631	0.630	0.630	0.631	0.622	0.631	0.587	0.631	0.506	0.446	0.453	0.631	0.834
$p_{fitting}$	0.79	0.65	0.16	0.16	1.92	0.65	0.49	0.65	3.15	0.65	3.46	10.50	1.70	0.65	5.09
$p_{fitting}$	1.27	2.43	2.43	3.79	1.29	2.44	2.64	2.44	2.65	2.44	2.85	17.06	5.59	2.44	1.30

the same as those in Table 3. The measured sheet thickness values are shown in Figure 9; the formed part was cut along the wall thickness direction with the help of wire-electrode cutting, and then the thickness value was obtained using a micrometer. To minimize error, an average of five readings was calculated to represent the thickness for each forming part. The thickness values of the formed part at depths of 10 mm, 13 mm, 15 mm, 17 mm, and 20 mm were measured.

The theoretical wall average thickness value $t_{initial}$ (mm) in the forming area can be calculated using (1), and the error value of the experiment wall average thickness value $t_{experiment}$ (mm) is $p_{initial}$ (%). The fitting wall average thickness value $t_{fitting}$ in the forming area can be calculated using (4), and the error value with the experiment wall average thickness value $t_{experiment}$ (mm) is $p_{fitting}$. The comparison results are shown in Table 5 and Figure 10.

Table 5 and Figure 10 show that the fitting formula has better accuracy than the initial formula, demonstrating that the fitting formula based on the simulation results can better predict the thickness than the cosine calculation formula. According to (2), the relationship between the thickness thinning ratio and the processing parameters obtained is shown in Figure 11.

Li et al. [8] studied the effect of SPIF on the thickness distribution through experimental investigations and found that the tool diameter and step down was nearly independent of the location of thickness; however, the thickness thinning ratio was found to increase continuously as the tool diameter grows. The same variation is clearly observed in Figure 11: the wall thinning ratio clearly increases from 30.15% to 32.11% when the tool diameter changes from 6 mm to 14 mm (A1 to A3) but no obvious change occurs in the thinning rate (decreasing from 31.87% to 31.02) with the step down changing from 0.6 mm to 1.0 mm (B1 to B3). Bagudanch et al. [27, 28] obtained the maximum temperature, forming force, formability, and surface roughness in the SPIF by using the design of experiments approach; the results showed that the feed speed has a significant influence on the maximum temperature and the forming force, whereas it has little effect on the formability. Therefore, from the data of experiment nos. C1~C3, the feed speed has little effect on the thickness thinning ratio. According to the cosine calculation formula, the forming angle has the most significant effect on the thickness; from the data of experiment nos. D1~D3, the wall thinning rate increases from 31.02% to 57.29%, with the forming angle changing from 45° to 60°. From the data of experiment nos. E1~E3, the wall thinning rate increases from 29.66% to 32%, with the sheet thickness changing from 0.6 mm to 1.0 mm.

Taking A1 as an example, the results of the thickness ratio obtained from the simulation and experiment are 31.17% and 30.15%, showing that the results of the simulation are correct and reliable. The forming tool is assumed to be a rigid body in the simulation process, and the wastage of the sheet and tool caused by friction is not considered; a degree of deviation is found between the simulation and experiment results, but the deviation is notably small. Therefore, the simulation results can be used to guide the experiment to reduce the thinning rate.

5. Method of Controlling the Thinning Ratio

According to the previous analysis, the thickness thinning ratio of the deformation zone affects the strength and formability of the forming parts, and the tool diameter, the step down, the forming angle, and other processing parameters affect the thinning ratio. On this basis, the forming trajectory has a great influence on the forming process [29]; therefore, it is necessary to study the influence of the forming trajectory on the thickness thinning ratio. It is expected that the thinning rate can be controlled by controlling the forming trajectory. In previous simulations and experimental processing, the tool was found to always move along only one direction and the pressure point was found to be concentrated at the same location; thus, the material cannot flow uniformly in forming process, and the thinning ratio of the forming region will become notably large when the impresses are produced, as shown in Figure 12.

To ensure the material flows more uniformly, it is best to control the sheet thinning rate and increase the thickness of the forming region. The alternating inversion trajectory and uniformly press point trajectory are employed to simulate the forming process. A sheet with a thickness value of 1 mm and three different forming trajectories are used to simulate the process for the same processing parameters. The three different forming trajectories are shown in Figure 13.

The experimental forming parts and the simulation results under three forming trajectories are shown in Figures 14 and 15, respectively; the thickness distribution is found to be the most uniform under the uniformly press point trajectory.

The simulation results and experimental results of the wall thickness for the three different forming trajectories are shown in Figure 16; the simulation results are found to be close to the experimental results. The simulation results are as follows: the average thickness is 0.6840 mm and the thinning rate is 31.6%, for the single forming trajectory; the average thickness is 0.6920 mm and the thinning rate is 30.8%, for the

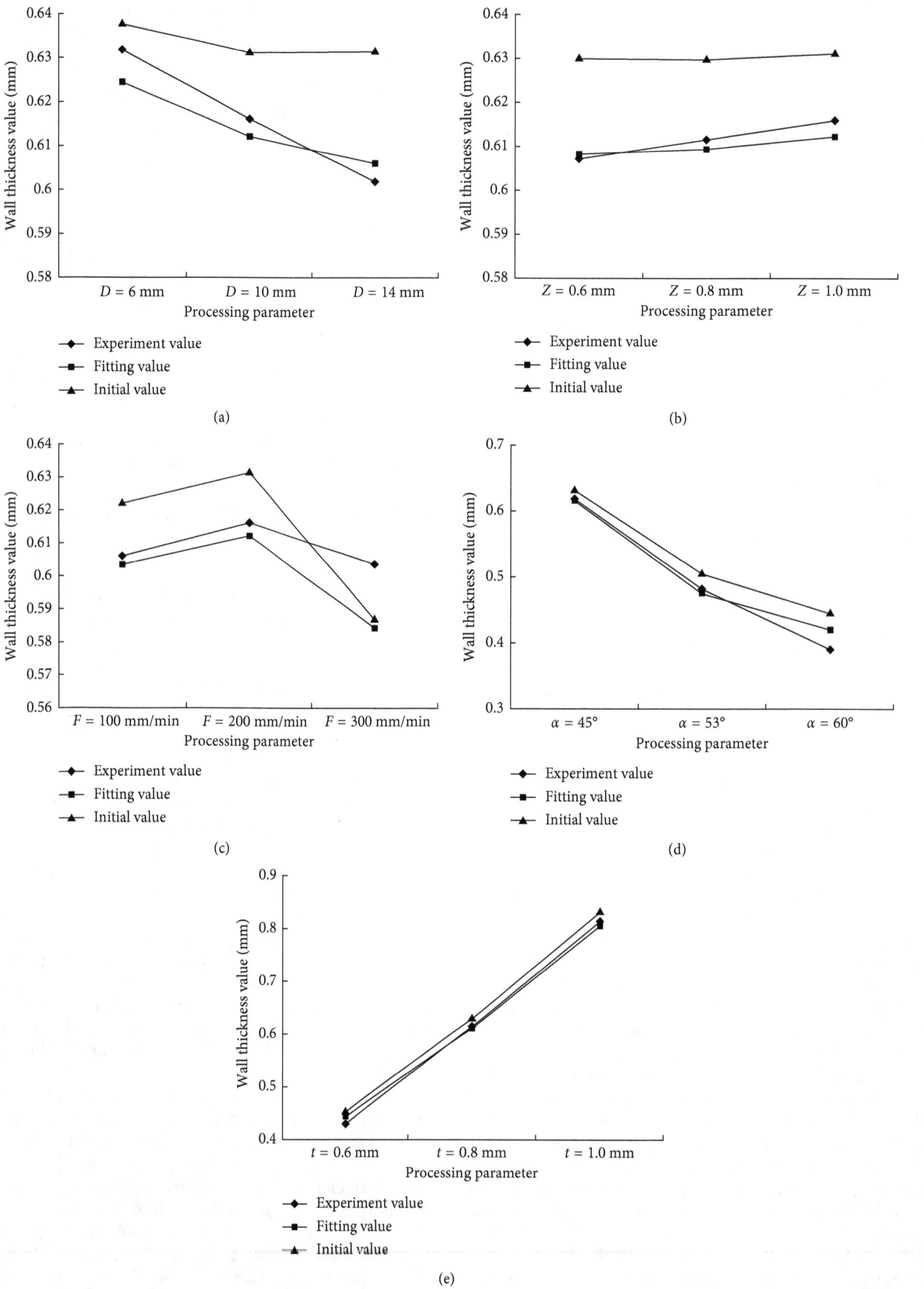

FIGURE 10: Comparisons between the theoretical and experimental results. (a) The wall thickness values under different tool diameters, (b) the wall thickness values under different step down values, (c) the wall thickness values under different feed speeds, (d) the wall thickness values under different forming angles, and (e) the wall thickness values under different initial thickness values.

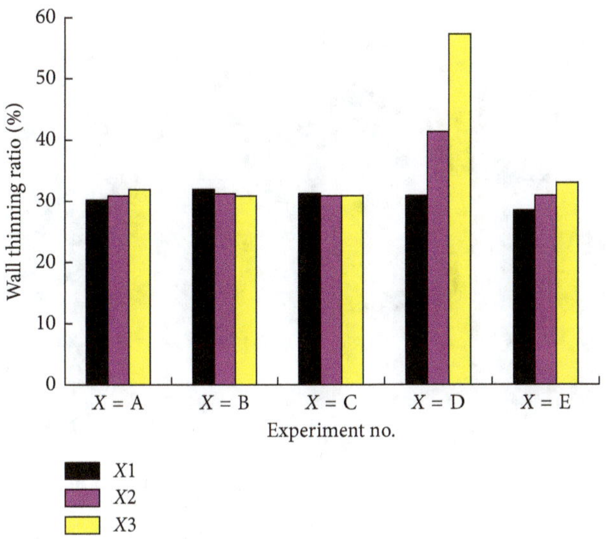

FIGURE 11: Relationship between the experimental thickness thinning ratio and the process parameters.

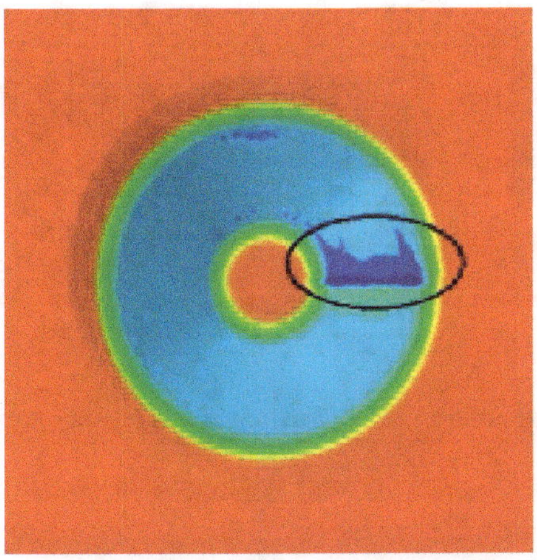

(a)

(b)

FIGURE 12: Effect of a single forming trajectory.

alternating inversion trajectory; and the average thickness is 0.7114 mm and the thinning rate is 28.86%, for the uniformly press point trajectory. From these results, the alternating

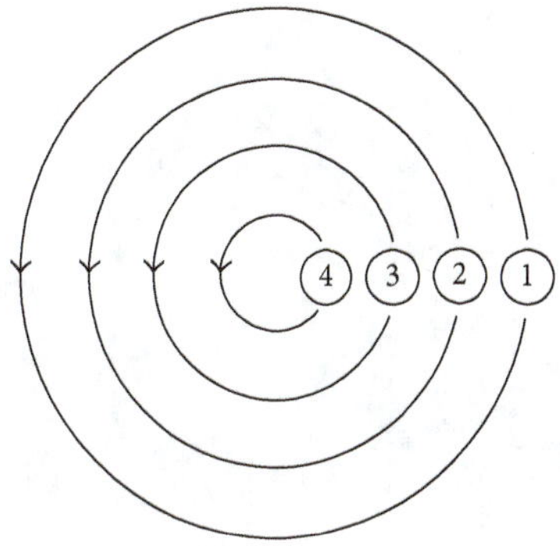

(a)

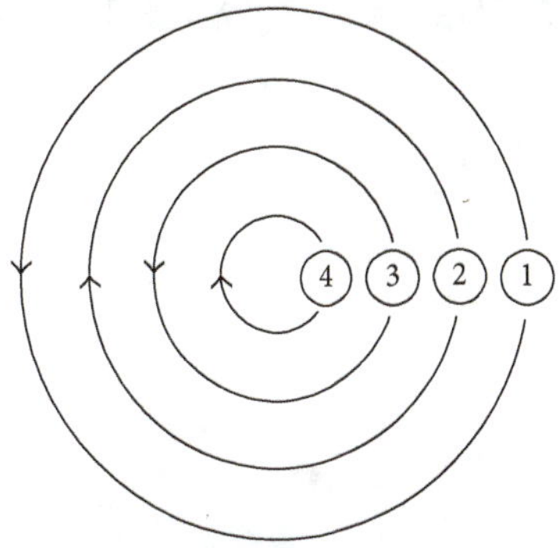

(b)

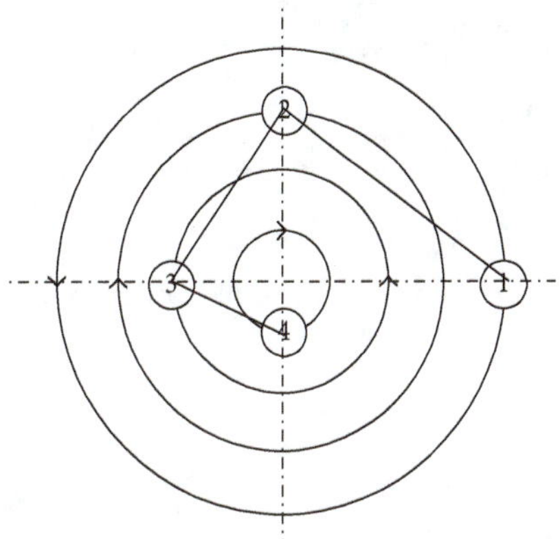

(c)

FIGURE 13: Three different forming trajectories: (a) the single forming trajectory, (b) the alternating inversion trajectory, and (c) the uniform press point trajectory.

inversion trajectory could slightly decrease the thinning rate and the uniformly press point trajectory could obviously decrease the thinning rate; thus, the uniformly press point trajectory can be used to control the sheet-thinning ratio.

6. Conclusions

Because the thickness thinning ratio of the deformation zone affects the strength and formability of the forming parts, the influence of different process parameters on the thinning mechanism of the wall thickness and the changing law of the thinning ratio of the forming region was studied and analyzed by using both finite element simulation and experiments. The following conclusions were drawn:

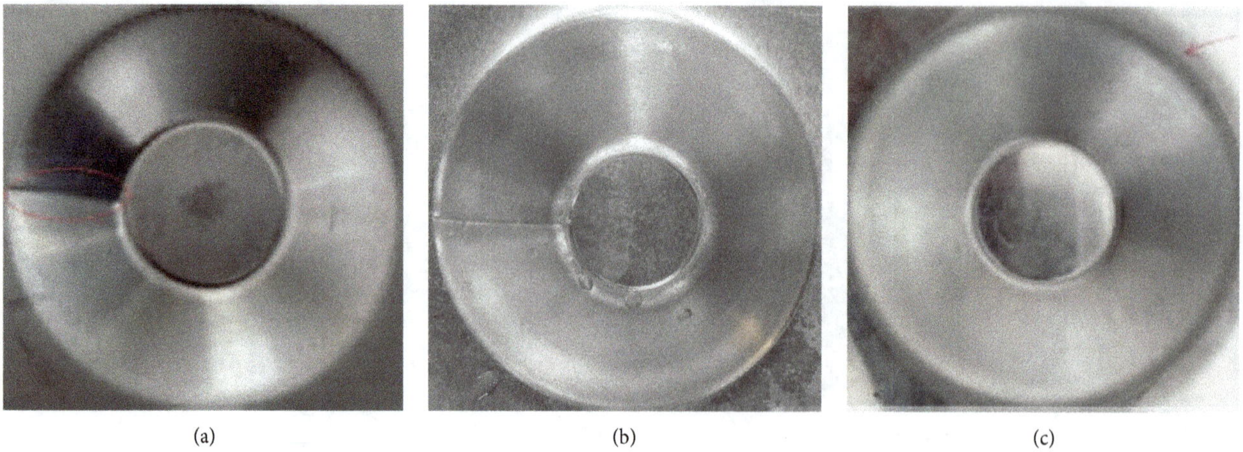

FIGURE 14: Forming parts of three forming trajectories: (a) single forming trajectory, (b) alternating inversion trajectory, and (c) uniformly press point trajectory.

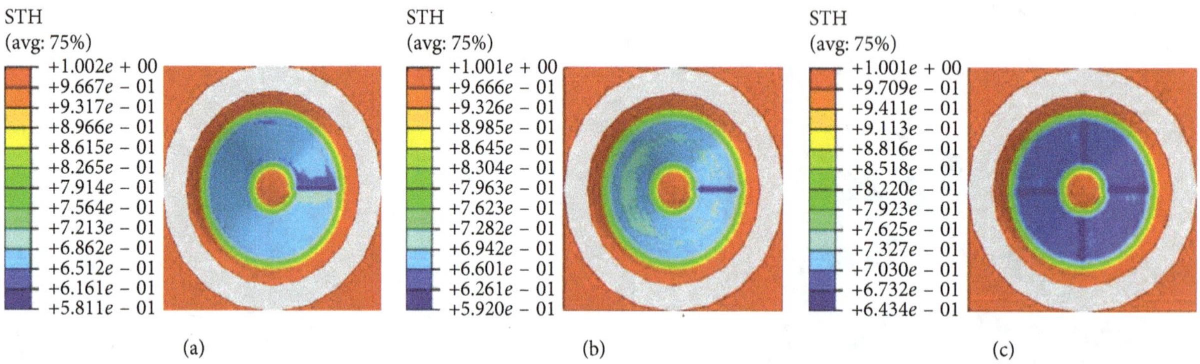

FIGURE 15: Simulation results of three forming trajectories: (a) single forming trajectory, (b) alternating inversion trajectory, and (c) uniformly press point trajectory.

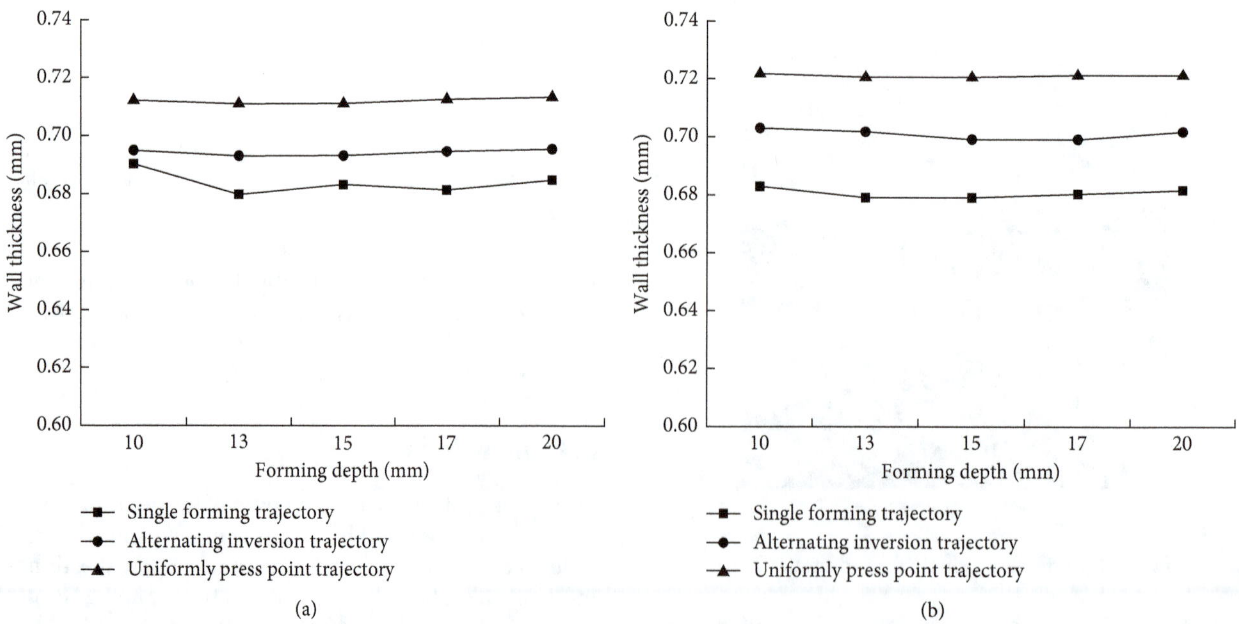

FIGURE 16: Thickness under three different forming trajectories. (a) Simulation results of the wall thickness and (b) experimental results of the wall thickness.

(1) Based on the finite element model established and the SPIF process simulated, the average wall thickness values of forming area under different processing parameters were obtained. The high-precision prediction formula of wall thickness obtained was based on the average wall thickness data of the deformation area under different process parameters. The influences of tool head diameter, step down, feed speed, sheet thickness, and forming angle on the change of thinning rate were analyzed, and the laws of change were analyzed and verified by experiments.

(2) In the process of single point incremental forming, the diameter of the tool head has a significant effect on the thinning ratio and the thinning ratio increases with the increase of the diameter of the tool head. The thinning ratio is reduced when the step down increases, but the influence of feed rate on the thinning ratio is notably small. The thinning ratio is increased rapidly when the forming angle increases; thus, the forming angle is the most important factor affecting the thinning ratio. The thinning ratio increases with the thickness of the material, showing that the thickness of the sheet has a significant effect on the thinning ratio of the wall thickness. The uniform press point trajectory was found to reduce the thinning ratio.

(3) In the experiment, the forming parameters were selected with thin sheet material, smaller diameter tool head, smaller step down, larger feed speed, and smaller forming angle to obtain a lower thinning ratio. This method was found to improve the forming limit of sheet and the forming precision.

(4) The three forming trajectories of single forming trajectory, alternating inversion trajectory, and uniformly press point trajectory were analyzed by FEM or experiment; the uniformly press point trajectory was found to make the thickness distribution uniform and to increase the thickness thinning ratio.

(5) The influences of different process parameters on the thinning rate of wall thickness were obtained by both simulation and experimental study of cone parts. No clear mechanism was found from analysis of the data. The formula fitting algorithm is slightly simplistic, and the accuracy is not adequate. Further analysis of the thinning rate mechanism will be conducted, and a more comprehensive and accurate algorithm will be employed to optimize the thickness formulate to better control the wall thinning rate.

Conflicts of Interest

The authors declare that they have no conflicts of interest.

Acknowledgments

The authors would like to acknowledge the National Natural Science Foundation of China (51475366 and 51475146), the Science and Technology Planning Project of Shaanxi Province (Grant no. 2016JM5074), and the Key Laboratory of Scientific Research Projects of Shaan'xi Educational Committee (Grant no. 12JS072).

References

[1] S. B. M. Echrif and M. Hrairi, "Research and progress in incremental sheet forming processes," *Materials and Manufacturing Processing*, vol. 26, no. 11, pp. 1404–1414, 2011.

[2] V. C. Do, D. T. Nguyen, J. H. Cho, and Y. S. Kim, "Incremental forming of 3D structured aluminum sheet," *International Journal of Precision Engineering and Manufacturing*, vol. 17, no. 2, pp. 217–223, 2016.

[3] J. Jeswiet, F. Micari, G. Hirt et al., "Asymmetric single point incremental forming of sheet metal," *CIRP Annals-Manufacturing Technology*, vol. 54, no. 2, pp. 88–114, 2005.

[4] K. R. Jackson, J. M. Allwood, and M. Landert, "Incremental forming of sandwich panels," *Journal of Materials Processing Technology*, vol. 204, no. 1–3, pp. 290–303, 2008.

[5] R. Bahloul, H. Arfa, and H. Belhadjsalah, "A study on optimal design of process parameters in single point incremental forming of sheet metal by combining Box–Behnken design of experiments, response surface methods and genetic algorithms," *International Journal of Advanced Manufacturing technology*, vol. 74, no. 1–4, pp. 163–185, 2014.

[6] G. Hussain, K. A. Al-Ghamdi, H. Khalatbari et al., "Forming parameters and forming defects in incremental forming process," *Materials and Manufacturing Processes*, vol. 29, no. 4, pp. 454–460, 2014.

[7] Z. Yao, Y. Li, M. Yang, Q. Yuan, and P. Shi, "Parameter optimization for deformation energy and forming quality in single point incremental forming process using response surface methodology," *Advances in Mechanical Engineering*, vol. 9, no. 7, pp. 1–15, 2017.

[8] J.-C. Li, C. Li, and T.-G. Zhou, "Thickness distribution and mechanical property of sheet metal incremental forming based on numerical simulation," *Transactions of Nonferrous Metals Society of China*, vol. 22, no. 1, pp. s58–s64, 2012.

[9] T. McAnulty, J. Jeswiet, and M. Doolan, "Formability in single point incremental forming: a comparative analysis of the state of the art," *CIRP Journal of Manufacturing Science and Technology*, vol. 16, pp. 43–54, 2017.

[10] R. Malhotra, L. Xue, T. Belytschko, and J. Cao, "Mechanics of fracture in single point incremental forming," *Journal of Materials Processing Technology*, vol. 212, no. 7, pp. 1573–1590, 2012.

[11] G. Hussain, L. Gao, and N. U. Dar, "An experimental study on some formability evaluation methods in negative incremental forming," *Journal of Materials Processing Technology*, vol. 186, no. 1–3, pp. 45–53, 2007.

[12] K. Jackson and J. Allwood, "The mechanics of incremental sheet forming," *Journal of Materials Processing Technology*, vol. 209, no. 3, pp. 1158–1174, 2008.

[13] Y. H. Kim and J. J. Park, "Effect of process parameters on formability in incremental forming of sheet metal," *Journal of Materials Processing Technology*, vol. 130-131, pp. 42–46, 2002.

[14] C. Bouffioux, C. Lequesne, H. Vanhove et al., "Experimental and numerical study of an AlMgSc sheet formed by an

incremental process," *Journal of Materials Processing Technology*, vol. 211, no. 11, pp. 1684–1693, 2011.

[15] M. B. Silva, P. S. Nielsen, N. Bay, and P. A. F. Martins, "Failure mechanisms in single-point incremental forming of metals," *International Journal of Advanced Manufacturing Technology*, vol. 56, no. 9–12, pp. 893–903, 2011.

[16] D. Young and J. Jeswiet, "Wall thickness variations in single-point incremental forming," *Proceedings of the Institution of Mechanical Engineers*, vol. 218, no. 11, pp. 1453–1459, 2004.

[17] G. Hirt, S. Junk, M. Bambach et al., "Process limits and material behaviour in incremental sheet forming with CNC-tools," *Materials Science Forum*, vol. 426–432, pp. 3825–3830, 2003.

[18] K. A. Al-Ghamdi and G. Hussain, "Threshold tool-radius condition maximizing the formability in SPIF considering a variety of materials: experimental and Fe investigations," *International Journal of Machine Tools and Manufacture*, vol. 88, pp. 82–94, 2014.

[19] P. Martins, N. Bay, M. Skjoedt et al., "Theory of single point incremental forming," *CIRP Annals-Manufacturing Technology*, vol. 57, no. 1, pp. 247–252, 2008.

[20] L. Zhou, "Research on the thickness change laws in numerical control incremental sheet forming," *Journal of Mechanical Engineering*, vol. 47, no. 18, pp. 50–54, 2011.

[21] S. Xiao and J.-Z. Gao, "Research on critical forming angle of uniform thickness in single-path incremental forming for conical part," *Forging & Stamping Technology*, vol. 1, no. 37, pp. 49–54, 2012.

[22] J. Li, "Experimental and numerical research on thickness variation dieless sheet mental incremental forming," *Hot Working Technology*, vol. 7, no. 40, pp. 1–4, 2011.

[23] C. Yu and J. Gao, "Numerical simulation for thickness thinning of deformation zone on hole-flanging by incremental forming," *Forging & Stamping Technology*, vol. 2, no. 40, pp. 53–59, 2015.

[24] H.-Z. Cao and J.-Z. Gao, "Research on influencing factors of region with stable thickness of conical part in incremental forming process," *Forging & Stamping Technology*, vol. 6, no. 37, pp. 17–23, 2012.

[25] H.-Y. Wei and L. Gao, "Investigation of thickness distribution of incrementally formed sheet mental parts," *Journal of Shandong University*, vol. 6, no. 39, pp. 82–87, 2009.

[26] W. Zhao and Y. Zhan, "Study on law of incremental forming square cone-shaped parts based on numerical simulation," *Hot Working Technology*, vol. 1, no. 42, pp. 103–105, 2013.

[27] I. Bagudanch, M. L. Garcia-Romeu, G. Centeno, A. Elías-Zúñiga, and J. Ciurana, "Forming force and temperature effects on single point incremental forming of polyvinylchloride," *Journal of Materials Processing Technology*, vol. 219, pp. 221–229, 2015.

[28] I. Bagudanch, M. Vivesmestres, M. Sabater et al., "Polymer incremental sheet forming process: temperature analysis using response surface methodology," *Materials & Manufacturing Processes*, vol. 32, no. 1, pp. 44–53, 2016.

[29] L. Filice, G. Ambrogio, and M. Gaudioso, "Optimised tool-path design to reduce thinning in incremental sheet forming process," *International journal of Material Forming*, vol. 6, no. 1, pp. 173–178, 2013.

13

Optimal Modes for the Fabrication of Aluminum Nanopowders by the Electrical Explosion of Wires

Alexei Pustovalov,[1] Vladimir An,[1] and Jin-Chun Kim[2]

[1] *Institute of High Technology Physics, National Research Tomsk Polytechnic University, 30 Lenin Ave., Tomsk 634050, Russia*
[2] *School of Materials Science and Engineering, University of Ulsan, 93 Daehak-ro, Nam-Ku, Ulsan 44610, Republic of Korea*

Correspondence should be addressed to Vladimir An; an_vladimir@tpu.ru

Academic Editor: Antonio Riveiro

The paper is aimed at studying the impact of initial conditions of electrical explosion of wires on energy characteristics of the explosion and some other properties of the obtained aluminum powders. Explosion modes where the energy input into the wire has the maximal level were found. These modes are optimal for fabrication of powders with the best properties. The powders have the highest value of the specific surface of 14.5 m^2/g, a narrow histogram of the particle size distribution, and a narrow distribution histogram with a high polydispersity coefficient of 0.7.

1. Introduction

Metal nanopowders are used in the development of high-quality construction materials. Because of their structural peculiarities, the products of powder metallurgy are more thermally stable and resist better cyclic temperature gradients and deformation stress [1]. The transition from bulk metals to highly dispersed powders and nanoparticles is characterized by a change in a number of fundamental material properties, such as melting temperature, electron work function, and chemical reactivity [2]. This allows us to prepare novel metal and composite materials with improved mechanical, electrophysical, magnetic, and physicochemical characteristics. In addition, these nanopowders are used as highly active catalysts [3, 4], lubricants [5, 6], and activators of combustion and sintering [7, 8].

One of the fabrication methods for metal nanopowders is the electrical explosion of wires (EEW) [8]. EEW is characterized by low energy consumption related to the pulsed supply of energy to the wire, technology simplicity, and the availability of raw materials. EEW is carried out in an inert gas and experiment parameters are empirically found in order to produce pure metal nanopowders. So far, the level of the energy input into the wire during EEW has been seen to exercise the most significant influence over the obtained powder dispersivity [9]. However, a great number of variables such as the capacity of capacitor battery, the initial voltage of its charge, the diameter and length of the exploded wire, circuit inductance, and gas pressure complicate the calculation of EEW in ensuring the desired mode of explosion for fabrication of powders with chosen wires.

The objective of the present work is to study the influence of the energy input into the wire and the initial parameters of the electrical explosion of aluminum wires on the properties of produced aluminum nanopowders.

2. Methods and Approach

Experiments on electrical explosion of aluminum wires were carried out in the setup described in [10]. To simplify the experiments, EEW was performed at constant capacity of 1.08 μF, circuit inductance of 0.82 μH, and argon pressure of $3 \cdot 10^5$ Pa. Aluminum wire of 0.25 mm in diameter was used for the experiments.

The starting voltage was varied from 20 to 30 kV, and the exploded wire length was varied from 30 to 180 mm. The response on a change of parameters was controlled with waveforms recorded using a current shunt and an oscilloscope Tektronix 2014B. Additionally, the EEW current and voltage waveforms were treated in order to calculate the

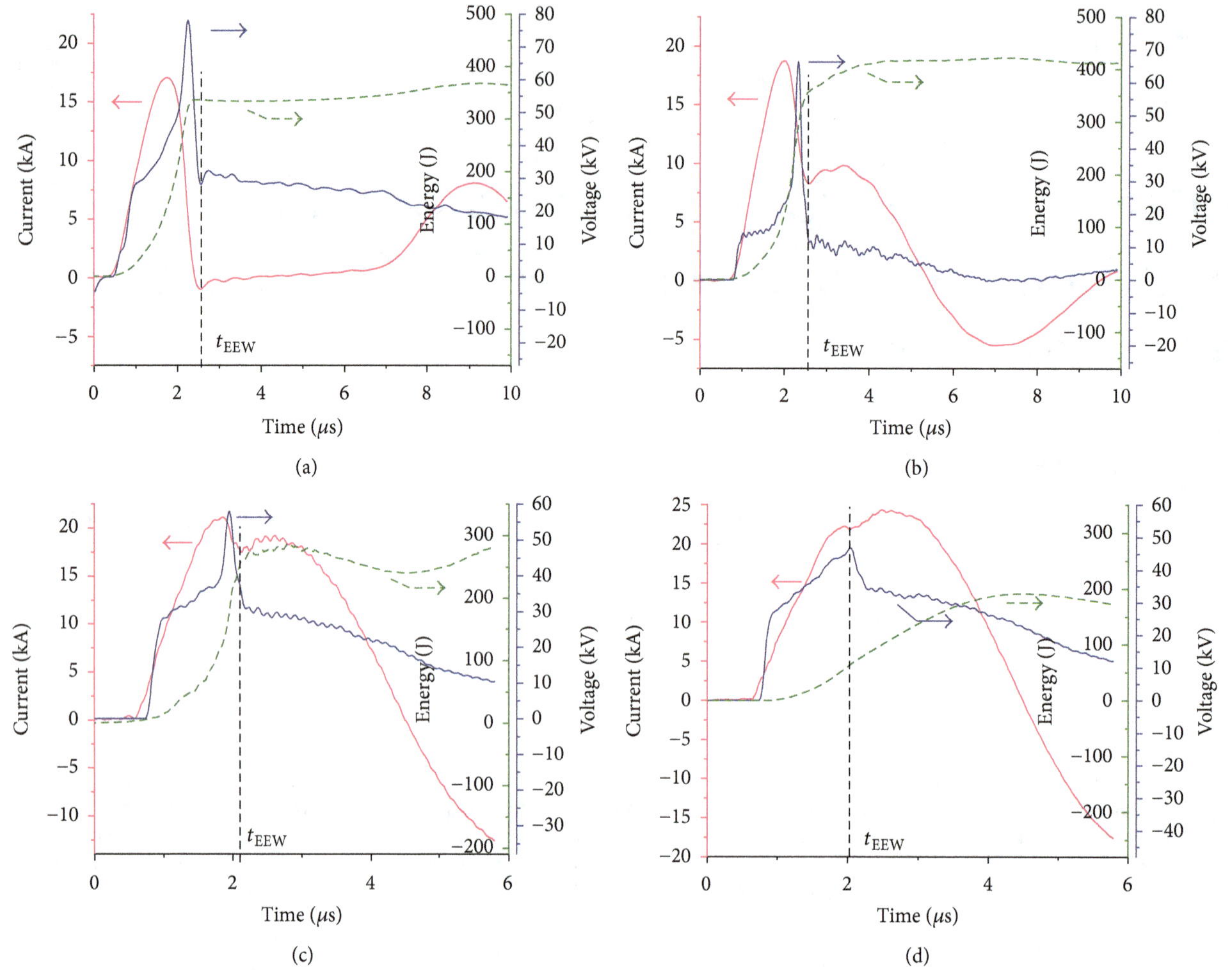

FIGURE 1: The EEW current and voltage waveforms: $U_0 = 30$ kV. (a) $l = 175$ mm, $e/e_S = 1.1$, and $e_{arc}/e_S = 0.2$; (b) $l = 135$ mm, $e/e_S = 1.6$, and $e_{arc}/e_S = 0.4$; (c) $l = 85$ mm, $e/e_S = 2.0$, and $e_{arc}/e_S = 1.2$; (d) $l = 60$ mm, $e/e_S = 1.7$, and $e_{arc}/e_S = 2.8$.

energy input into the wire and also the energy released during the charge arc stage. The energy released in the discharge gap in the range of time from the beginning of the current rise until the inflection moment of the current waveform (t_{EEW} in Figure 1) was considered as the energy input into the wire and the rest of the released energy as that released during the arc discharge stage.

The energy input into the wire was calculated using (1) offered by the authors of [11]:

$$W_{EEW} = \frac{CU_0^2}{2} - \frac{C\left(U_0 - (1/C)\int_0^{t_{EEW}} i(t)\,dt\right)^2}{2} - \frac{L \cdot i(t_{EEW})^2}{2} - R\int_0^{t_{EEW}} i(t)^2 dt, \quad (1)$$

where C is the capacitor capacity, U_0 is the voltage of capacitor charge, i is the EEW current, L is the inductance of setup circuit, R is the circuit resistance, and t_{EEW} is the duration of an EEW current pulse. According to the authors of [11], the accuracy of the calculations of the energy input into the wire is not less than 4–6%. The calculations did not take into account the inductance of the exploded wire which was not less than 10%. In order to compare the various modes of electrical explosion of wires, the energy input into the wire was expressed in the form of a dimensionless value:

$$\frac{e}{e_s} = \frac{W_{EEW}}{V_W \cdot e_S}, \quad (2)$$

where e is the specific energy, V_W is the volume of the exploded wire, and e_S is the energy of aluminum sublimation (32.9 J/mm^3). The energy released during the charge arc stage was also converted into a dimensionless value e_{arc}/e_S.

The obtained aluminum powders were passivated in air using the method described in [12]. The shape and the average size of the particles were studied using a transmission electron microscopy JEM-2100 F (JEOL, Japan). A particle size distribution histogram was additionally built. For this purpose, more than 15 microphotographs obtained at different magnitudes (totally more than 3000 particles) were treated. This histogram served for the calculations of the number-average (a_n), surface-average (a_S), and mass-average (a_m) particle size. According to a_n/a_m ratio, the particle polydispersion coefficient (k) characterizing the particle size "dispersion" was calculated.

The specific surface area (S_{sp}) was measured according to the BET method using a Sorbtometer-M device. The specific surface area was used for the calculations of the surface-average particle diameter according to the following equation:

$$d_S = \frac{6}{\rho S_{sp}} \cdot 1000 \text{ [nm]}, \quad (3)$$

where ρ is the density of aluminum, 2.7 g/cm^3.

The phase composition was determined using an X-ray diffractometer, Shimadzu XRD-7000 (Japan). According to the data of X-ray peak maximum enlargement, the size of coherent scattering regions (average size of crystallites) was calculated using the Debye-Scherrer equation:

$$d_{CSR} = \frac{k \cdot \lambda}{\beta \cdot \cos(2\theta/2)}, \quad (4)$$

where k is the dimensionless coefficient (for spherical particles it is 0,9), λ is the X-ray wavelength (0.1540593 nm), β is the full width at half maximum, and 2θ is the diffraction angle.

The content of metallic aluminum (Al^0) in samples was measured using the volumetric method during interaction of the metal with a 10% NaOH solution.

3. Results and Discussion

Figure 1 shows the EEW current and voltage waveforms. This figure also illustrates the dependence of the energy input during the discharge period versus time at a fixed voltage of 30 kV. In this experiment, the only exploded wire length was varied from 175 to 30 mm.

Variation in the exploded wire length leads to a rather high change in the EEW waveforms. At the longest length (Figure 1(a)), the EEW mode is characterized by two separately occurring processes. The first process is electrical explosion itself. The second one is the EEW arc stage. This explosion mode is characterized by a high coefficient of energy transfer from the capacitor to the wire: more than 80%. In the literature, this mode is called mode with a «current pause» [9, 13]. Nevertheless, the value of the energy input into the wire is not high and is equal to 330 J ($1.1e_S$). The decrease of the exploded wire length up to 85 mm leads to fusion of electrical explosion itself and the discharge arc stage: a transition to the mode without current pause [9, 13, 14]. The level of the energy input into the wire decreases down to 270 J. However, an increase in the specific energy (up to $2e_S$) into the wire $2eS$ is observed due to an increase in the exploded wire volume. Moreover, the energy released during the arc discharge stage increases up to $1.2e_S$ (Figure 1(c)). Then, when decreasing the length (Figure 1(d)), a bigger part of energy is released during the arc stage, whereas the specific energy input into the wire decreases down to $1.7e_S$ (130 J). Probably, it is due to a rather low electrical strength of argon. When the wire length is less than 80 mm, an earlier start of the arc discharge stage is observed. It leads to shunting of the energy input into the wire (Figure 1(d)).

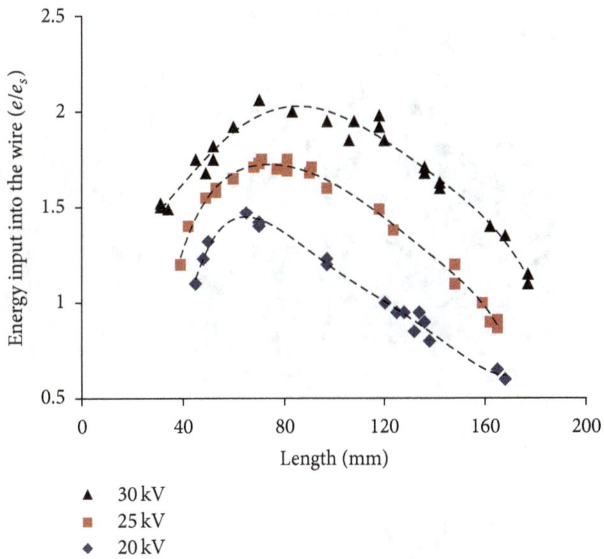

FIGURE 2: Energy input into the wire during explosion versus the exploded wire length and the voltage.

Figure 2 depicts the dependence of the specific energy input into the wire during the explosion on the exploded wire length for a series of experiments carried out at a voltage of 20, 25, and 30 kV.

The regularities of the change of e/e_S from l at different U_0 reveal similar character. The minimal e/e_S value is observed during electrical explosion of wires with the maximal length. A decrease in the exploded wire length leads to an increase in e/e_S. However, an increase in the specific energy input into the wire is observed up to a definite value of the working voltage. For the series of experiments at $U = 20$ kV, the maximal e/e_S value (1.5) is observed at $l = 65$ mm. For the series of experiments at $U = 30$ kV, the maximal e/e_S value is 2 at $l \approx 80$ mm. The further increase of the wire length leads to a fast decrease of the energy input into the wire.

Initial EEW conditions for fabrication of nanopowders were chosen taking into account the obtained data. The initial EEW conditions and properties of the powders produced under these conditions are given in Table 1.

Sample Al-1, obtained at the lowest level of the specific energy input into the wire during the explosion ($0.9e_S$), has the lowest specific surface area that equals 5.6 m^2/g. The particles have a spherical shape; there are both big particles of 4.5 μm and small particles of 25 to 400 nm. According to the literature data [15], the big particles form during the metal melting and sputtering, whereas the origin of the small particles is related to the metal evaporation during the explosion and the further condensation of the vapors. On the particles surface, we observe an approximately 5 nm thick well-defined layer which consists of aluminum oxide-hydroxide phases according to the literature data [2] formed during the passivation of the powders in the air. The presence of these phases is confirmed by the volumetric analysis,

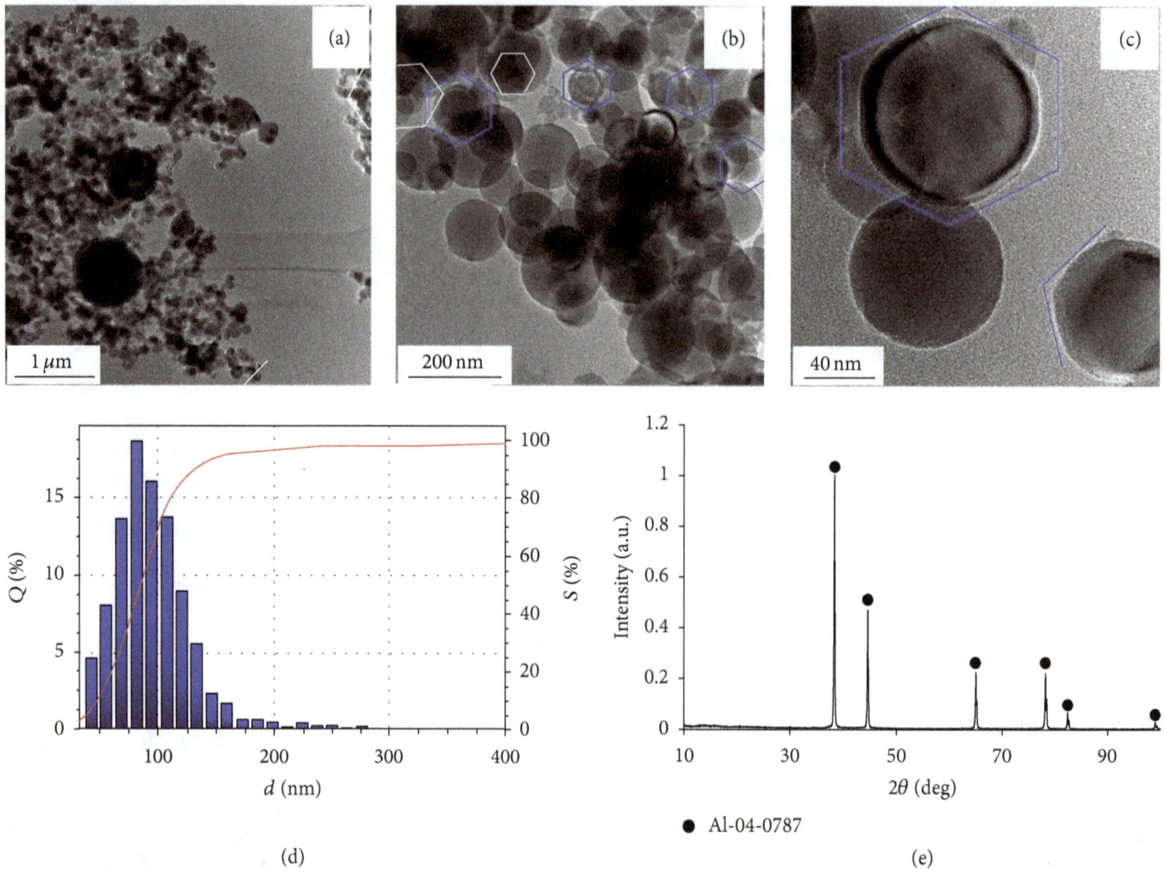

FIGURE 3: Microphotographs (a, b, c), histogram of the particle size distribution (d), and XRD pattern (e): sample Al-5: a_n = 130 nm; a_S = 150 nm; a_m = 180 nm; k = 0.7.

TABLE 1: Impact of EEW energetic parameters on the characteristics of Al powders.

Sample	U_0, kV	l, mm	e/e_S	e_{ars}/e_S	S_{sp}, m^2/g	d_s, nm	Al^0, wt.%
Al-1	20	120	0.9	0.1	5.6	400	96
Al-2	20	75	1.4	0.2	8.1	275	93
Al-3	25	120	1.5	0.1	9.2	240	92
Al-4	25	75	1.7	0.8	11.2	200	91
Al-5	30	85	2.0	1.2	14.5	150	89
Al-6	30	60	1.7	2.8	13.9	160	90
Al-7	30	30	1.5	7.5	13.1	170	90

according to which the content of metallic aluminum in the sample is 96 wt.%.

The size of 95% of the particles ranges from 25 to 400 nm; their maximum distribution corresponds to 115 nm. The presence of 5% of particles with a size from 400 nm to 4.5 μm leads to an increase in the number of particles of average size up to 160 nm and up to 350 nm in the weight-average.

The sample is characterized by a rather low polydispersity coefficient: 0.4. When the particle size distribution is large, the surface-average particle diameter, calculated from the distribution histogram (a_S), is practically 2 times lower than the analogous value calculated from the specific surface area (d_S). The difference between a_S and d_S can be explained by the difference in techniques of sample preparation for analysis. The BET measurements are more accurate because the surface of all particles is measured in contrast of TEM, where most of big particles, sedimented in the solution, are not analyzed.

The increase in e/e_S leads to an increase in the specific surface area of the obtained particles. Highest S_{sp} was found for the samples prepared under the modes with the maximal level e/e_S equal to 2.0 (Al-5, Table 1). The microphotographs of sample Al-5 are given in Figure 3. The sample is characterized by a narrow range of the particle size distribution and a rather high polydispersity coefficient: 0.7. As a result, the surface average particle diameter, calculated from the distribution histogram, corresponds to the particle size calculated using the powder specific surface area. The size of most particles is in the range of 25–200 nm, whereas particles of approx. 1 μm are observed but their part is not high.

Most particles of nanometer size have a spherical shape. However, in contrast to the previous sample, we can observe particles having the shape of a hexagon (Figures 3(b) and 3(c)), where the diameter of such particles is always less than 100 nm. According to the authors [2], the presence of the faceted particles (white hexagonal or polygonal lines in Figures 3(b) and 3(c)) is related to the mechanism of their crystallization. Droplets with the size of 100 or more nanometers crystallize from a large number of nuclei of which the mutual growth limitations contribute to the formation of particles with a spherical shape. Crystallization of relatively small droplets can occur during the formation and growth of one or a few nuclei when the particle forms in the form of a hexagon.

The increase in the specific surface area of the studied samples from 5.6 m^2/g to 14.5 m^2/g leads to a twofold decrease in the sizes of coherent scattering regions from 150 nm to 70 nm which can be explained by a decrease in crystallite sizes. However, the crystallite size for the studied samples calculated using the Debye-Scherrer formula is lower than the average particle size calculated using the powder specific surface area (Table 1). This allows us to assume that the powder particle produced at $2.0e_S$ has a polycrystalline structure.

The study of the modes with an increase of e/e_S and fast rise of e_{arc}/e_S showed that the powder specific surface area under these explosion modes decreases with a decrease in e/e_S (samples Al-6 and Al-7, Table 1). The powder particles obtained at the highest energy released during the arc discharge stage (sample Al-7) have both a spherical shape and polygon shape. The difference between this sample and the one considered above consists in the appearance of particles sintered with each other. Probably it is due to the release of large amount of discharge arc stage heat resulting in the sintering of particles.

4. Conclusions

An increase in the specific energy input into the wire is observed when decreasing the length of the exploded aluminum wire and keeping constant the other EEW conditions. However, this increase occurs until a definite moment. The dependence of e/e_S on U_0 and l has a maximum; the further decrease in the length leads to either significant increase in the working voltage or increase in the energy released at the discharge arc stage and decrease in the specific energy input into the wire during the explosion.

The optimal EEW modes for the fabrication of metal Al powders in the studied range are the explosion modes with the maximal level of the energy input into the wire (U = 30 kV; l = 85 mm). The powders have the highest value of the specific surface of 14.5 m^2/g, a narrow histogram of the particle size distribution, and a high polydispersity coefficient.

Under the modes with an increase in the specific energy input into the wire during the arc discharge stage, the specific surface area decreases; the samples were found to have a high amount of sintered particles which can be provoked by the high thermal action of the energy released during the discharge arc stage.

Competing Interests

The authors declare that there are no competing interests regarding the publication of this paper.

Acknowledgments

This study was supported by the Russian Foundation for Basic Research (Grant RFBR-mol-a 16-38-00051). This work was also supported under the state assignment of the Ministry of Education and Science of Russia for 2014–2016 (Research Work no. 361). The authors would like to thank the Nano-Center and Scientific Analytical Centre at Tomsk Polytechnic University for the XRD, TEM, and SEM analyses. It is their pleasure to express their thanks to Ms. Mary Page for her invaluable help in correcting the manuscript.

References

[1] R. E. Smallman and A. H. W. Ngan, *Physical Metallurgy and Advanced Materials*, Butterworth-Heinemann, 2011.

[2] A. V. Korshunov, "Influence of dispersion aluminum powders on the regularities of their interaction with nitrogen," *Russian Journal of Physical Chemistry A*, vol. 85, no. 7, pp. 1202–1210, 2011.

[3] P. Rauwel, S. Küünal, S. Ferdov, and E. Rauwel, "A review on the green synthesis of silver nanoparticles and their morphologies studied via TEM," *Advances in Materials Science and Engineering*, vol. 2015, Article ID 682749, 9 pages, 2015.

[4] S. L. Suib, Ed., *New and Future Developments in Catalysis: Batteries, Hydrogen Storage and Fuel Cells*, Newnes, 2013.

[5] V. An, Y. Irtegov, and C. De Izarra, "Study of tribological properties of nanolamellar WS_2 and MoS_2 as additives to lubricants," *Journal of Nanomaterials*, vol. 2014, Article ID 865839, 8 pages, 2014.

[6] V. An and Y. Irtegov, "Tribological properties of nanolamellar MoS_2 doped with copper nanoparticles," *Journal of Nanomaterials*, vol. 2014, Article ID 731073, 7 pages, 2014.

[7] G. V. Sakovich, V. A. Arkhipov, A. B. Vorozhtsov, S. S. Bondarchuk, and B. V. Pevchenko, "Investigation of combustion of HEM with aluminum nanopowders," *Nanotechnologies in Russia*, vol. 5, no. 1, pp. 91–107, 2010.

[8] Y.-S. Kwon, Y.-H. Jung, N. A. Yavorovsky, A. P. Illyn, and J.-S. Kim, "Ultra-fine powder by wire explosion method," *Scripta Materialia*, vol. 44, no. 8-9, pp. 2247–2251, 2001.

[9] Y. A. Kotov, "The electrical explosion of wire: a method for the synthesis of weakly aggregated nanopowders," *Nanotechnologies in Russia*, vol. 4, no. 7-8, pp. 415–424, 2009.

[10] A. V. Pustovalov, "Study of products of electrical explosion of iron wires in argon-oxygen mixture," *Advanced Materials Research*, vol. 872, pp. 206–213, 2014.

[11] I. F. Kvartskhava, V. V. Bondarenko, A. A. Pliutto, and A. A. Chernov, "Oscillographic determination of energy of electric explosion of wires," *Journal of Experimental and Theoretical Physics*, vol. 31, pp. 745–775, 1956.

[12] A. A. Gromov, U. Förter-Barth, and U. Teipel, "Aluminum nanopowders produced by electrical explosion of wires and passivated by non-inert coatings: characterisation and reactivity with

air and water," *Powder Technology*, vol. 164, no. 2, pp. 111–115, 2006.

[13] C. S. Wong, B. Bora, S. L. Yap, Y. S. Lee, H. Bhuyan, and M. Favre, "Effect of ambient gas species on the formation of Cu nanoparticles in wire explosion process," *Current Applied Physics*, vol. 12, no. 5, pp. 1345–1348, 2012.

[14] Y. S. Lee, B. Bora, S. L. Yap, and C. S. Wong, "Effect of ambient air pressure on synthesis of copper and copper oxide nanoparticles by wire explosion process," *Current Applied Physics*, vol. 12, no. 1, pp. 199–203, 2012.

[15] Y. Kinemuchi, K. Murai, C. Sangurai et al., "Nanosize powders of aluminum nitride synthesized by pulsed wire discharge," *Journal of the American Ceramic Society*, vol. 86, no. 3, pp. 420–424, 2003.

14

Computer Simulation of Metal Ions Transport to Uneven Substrates during Ionized Plasma Vapour Deposition

Tomáš Ibehej, Jakub Hromádka, and Rudolf Hrach

Department of Surface and Plasma Science, Faculty of Mathematics and Physics, Charles University, V Holesovickach 2, 180 00 Prague 8, Czech Republic

Correspondence should be addressed to Rudolf Hrach; rudolf.hrach@mff.cuni.cz

Academic Editor: Peter Majewski

We present a computational study of processes taking place in a sheath region formed near a negatively biased uneven substrate during ionized plasma vapour deposition. The sputtered metal atoms are ionized on their way to substrate and they are accelerated in the sheath near the substrate. They are able to penetrate to high-aspect-ratio structures, for example, trenches, which can be, therefore, effectively coated. The main technique used was a two-dimensional particle simulation. The results of our model predict the energy and angular distributions of impinging ions in low-pressure conditions which are characteristic for this method and where typical continuous models fail due to unfulfilled assumptions. Input bulk plasma properties were computed by a "zero dimensional" global model which took into account more physical processes important on a scale of the whole magnetron chamber. Output parameters, such as electrostatic potential, energy of ions, and ion fluxes, were computed for wide range of conditions (electron density and substrate bias) to show the influence of these conditions on observed phenomena, penetration of sheath inside the trench, deceleration of argon and copper ions inside the trench, and local maxima of ion fluxes near the trench opening.

1. Introduction

When modifying metal or semiconductor solids during various material-treatment technologies, there is an important group of methods connected with both inert gas and chemically active plasmas, oxidation, sputtering, plasma deposition, plasma-enhanced chemical vapour deposition, plasma immersion ion implantation, and so on. In all these cases the detailed understanding of plasma-solid interaction is of a great importance. One of these techniques is the ionized plasma vapour deposition (IPVD) or more specifically Inductively Coupled Plasma Magnetron Sputtering (ICP-MS).

The deposition technique IPVD was presented in [1] as an improvement in filling of high-aspect-ratio features in microelectronics fabrication. Metal atoms sputtered from a target move through inductively coupled plasma of low pressure but high electron density where they are ionized. Ionization degree of metal atoms and flux fraction of metal ions can reach very high values during IPVD. Paper [2] described ionization up to 80% for aluminium atoms. Flux fraction of copper ions exceeding 40% was shown in [3]. Afterwards, the metal ions together with ions of the carrier gas move towards the negatively biased substrate. In the vicinity of the substrate, that is, in the presheath and sheath, the ions are accelerated and their angular distribution gets narrower. Ion energy can be regulated by the substrate bias. The flux of metal ions depends on power of the coil which generates the secondary plasma. The flux of ions can be regulated independently on their energy. Accelerated ions can reach the bottom of high-aspect-ratio trenches which are hence coated all over their surface [4]. Furthermore, the high-energy metal ions and ions of carrier gas are able to resputter the overhanging deposit which is often created at the opening of the trench and which makes further coating impossible [5, 6].

2. Description of the Simulation

In our contribution we studied the physical and chemical processes on surfaces, both flat and uneven, of solids immersed into plasma. The method of solution was the

TABLE 1: A list of processes included in global model and their parameters.

Process	Parameter	Ref.	Value
Electron impact ionization of Cu	Cross section	[9]	$f(E)^{b}$
Penning ionization of Cu	Cross section	[10]	4.9×10^{-15} cm^2
Ambipolar diffusion loss of Cu^+	Thermal mobility of Cu^+ in Ar	[11]	2.3 cm^{2}V^{-1}s^{-1} [c]
Ambipolar diffusion loss of Ar^+	Thermal mobility of Ar^+ in Ar	[11]	1.8 cm^{2}V^{-1}s^{-1} [c]
Electron impact Ar excitation	Cross section	[12]	$f(E)^{b}$
Diffusion loss of Ar^* [a]	Diffusion coefficient	[13]	54 cm^{2}s^{-1}
Collision of Ar^* [a] and electron	Collision frequency	[14]	$5 \times 10^{-8} n_e \sqrt{T_e}$ [d]
Decay of trapped resonant radiation of Ar^{**} [a]	Radiative lifetime of the resonance transition	[15]	1×10^{-8} s
Electron impact ionization of Ar^* [a]	Cross section	[16]	$f(E)^{b}$
Electron impact ionization of Ar	Cross section	[12]	$f(E)^{b}$

[a] Ar^* denotes the lowest metastable argon state and Ar^{**} a nearby radiative level; [b] A is function of electron energy. [c] The value is valid for neutral gas temperature 300 K and number density 2.69×10^{19} cm^{-3}. [d] Frequency is in s^{-1}, electron temperature in eV, and electron number density in cm^{-3}.

computer simulation as an increase in performance of computers in recent years allowed computer modelling to be included among other research tools as a valuable source of information. Plasma is an environment where phenomena over multiple length and time scales are coupled. There exists two basic modelling techniques, the particle models which are able to capture microscopic effects but they demand huge computational resources, while the macroscopic fluid models which are not so computationally demanding however give results with only limited accuracy caused by the lack of microscopic information. To solve specific problems the third way can be used, the hybrid models that take advantage of both fluid and particle modelling approach [7, 8].

Particle models, or more precisely PIC-MCC (i.e., Particle-in-Cell, Monte Carlo collisions) models, are quite frequently used techniques to simulate plasma behaviour near immersed solids. For example, similar models were used to simulate conditions during plasma immersion ion implantation process [17, 18]. Simulation [19] is an example of a common hybrid approach, electrons are treated continuously and ions as particles. Another approach, an iterative hybrid model [20], was used to study problem of plasma diagnostics. We also tried to develop a fully 3-dimensional self-consistent particle model [21] which has no symmetry requirements. However, its computational requirements are very high and hence the size of the mesh and the computational domain are limited. More complex model of IPVD [22] includes processes on the surface, the deposition itself including resputtering and movement of physisorbed particles on the surface. However, the plasma-related part is a continuous model. Our model is intended for precise determination of plasma behaviour near the substrate and it can be further supplemented with the surface-related phase of IPVD using Monte Carlo-based methods. Here, the microscopic nature of particle model results can be of great use.

2.1. Obtaining Input Plasma Parameters. A simple volume-averaged model was adopted from [23] to estimate the main plasma parameters of argon plasma with addition of sputtered and partially ionized metal, in our case copper. The main processes taken into account are summarized in Table 1 along with the list of their parameters and references from which they have been taken. The rate constants of electron collisions were numerically derived at runtime using cross sections from the literature. The processes mentioned in Table 1 were connected with each other in generation-loss balance equations which were solved as a system together with the conservation equation for metal atoms and quasineutrality condition. The equations for individual processes can be obtained from article [10] which deals with aluminium deposition rather than copper. The solution of Ar/Al plasma was also repeated to test the model.

2.2. Particle Model. Dimensionality of the particle model is $2d3v$, that is, two spatial dimensions and three dimensions of velocity. Particles follow two-dimensional trajectories and the electrostatic interaction (Poisson's equation) is solved in 2D. The third dimension of velocity follows Gaussian distribution and plays a part during collisions.

The particle model follows trajectories of three types of charged particles, electrons, argon ions, and copper ions. Their trajectories are influenced by Monte Carlo collisions with neutral argon atoms and an interaction with electrostatic forces originating from the biased substrate and from other charged particles in the computational domain. Copper neutral atoms do not participate in particle model collisions because of their negligible density. In this study we assumed total averaged copper density in the chamber 2×10^{12} cm^{-3} which is nearly three orders of magnitude lower than the density of argon.

Instead of random free path, our method of collision simulation generates random time to next collision. The cross section σ_0 of the imaginary "null collision" satisfies

$$\left(\sum_{i=1,2,\ldots} \sigma_i(g) + \sigma_0(g)\right) gn = \nu_{\max}, \quad (1)$$

where $\nu_{\max}$ is the lowest possible constant collision frequency and g is the relative velocity. Unlike the widely used null collision method, this approach does not neglect the velocity of scattering centres. This is important especially for the movement of ions whose velocity is comparable to the velocity of neutral collision centres, at least outside the sheath. This method is derived and described in detail in [24]. We included following scattering processes of charged species with argon neutrals: Ar^+: elastic scattering and charge transfer [25], electrons: elastic scattering, excitation to the first excited level, and ionization [12], and Cu^+: elastic scattering [26].

The electrostatic force was computed in the usual manner using the Particle-in-Cell method. Poisson's equation was solved numerically at each time-step using the sparse system solver UMFPACK [27]. For the integration of Newton's equation of motion we used Verlet velocity algorithm [28]. The computational domain was square-shaped with side of 5 mm. On one side, there was substrate with the two-dimensional trench with size of 2 mm (height) and 0.2 mm (width). The substrate was biased and the given potential, with respect to the plasma potential, was used as Dirichlet boundary condition along the substrate surface. The opposite side of the domain acted as a connection with bulk plasma. Particles could leave the domain through this side and other particles came through it to the domain from a source where we assumed Maxwell velocity distribution of charged particles with corresponding temperatures. The condition $U = 0$ volts was applied on this side. Remaining sides had periodic boundary conditions. The Cartesian mesh size was set to 1000×1000 and on average there were 1×10^7 particles in the domain. Figure 1 shows a diagram of the computational domain.

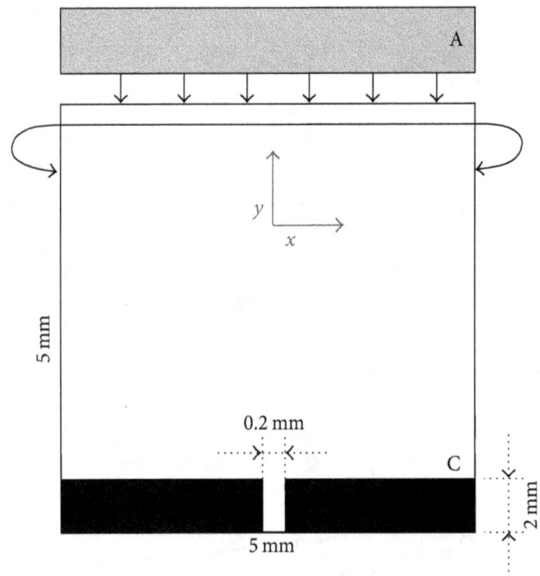

FIGURE 1: Scheme of the computational domain. A: source of particles, Dirichlet boundary condition $U = 0$ V was applied on this side. B: trenched substrate, Dirichlet boundary condition $U = U_b$. C: computational domain, period boundary condition was applied on remaining sides of the domain.

3. Results

3.1. Global Model Results. For the validation of the volume-averaged model we used the same input conditions as in [23]. Figure 2(a) shows a good agreement of results, electron temperature as a function of total amount of metal in the chamber which contains aluminium. Figure 2(a) also shows an interesting comparison of Al and Cu behaviour in the magnetron chamber. Increasing amount of sputtered metal (caused, e.g., by increasing magnetron power) results in cooling of electrons. This effect is much stronger for Al. Ionization energies are $E_{i,\mathrm{Al}} = 5.99$ eV and $E_{i,\mathrm{Cu}} = 7.73$ eV. There are certainly more electrons in the chamber capable of ionizing Al then Cu and hence they lose ionization energy more easily. It is also in agreement with the deceleration of the temperature decrease. Figure 2(b) shows the development of ionization degree. As long as the electron temperatures are similar for both cases (small amount of metal), aluminium is significantly more ionized. But faster decrease of electron temperature in chamber with aluminium causes also faster decrease of ionization degree.

However, for our simulations we used conditions typical for deposition of copper which were used in the experiment in [3]. One of the main variables in this study was RF power. It was shown that the RF power can effectively regulate electron density and consequently the ion-to-neutral flux fraction which is one of the crucial parameters in IPVD process. Therefore, as one of our studied parameters we chose RF power which corresponds linearly to electron density. Other parameters remained the same as in [21]. Results of the global model, some of which were used in the particle model, are shown in Table 2.

3.2. Particle Model Results. The particle model was run 40x; for each of these four values of RF power (100 W, 400 W, 700 W, and 1000 W) we simulated 10 different values of substrate bias (−20 V, −40 V,..., and −200 V). At first, we checked the shape and size of the sheath. Figure 3 shows two contour lines of electrostatic potential, $0.975 \times U_b$ and $0.875 \times U_b$. The substrate bias U_b was −40 V in panel (a) and −200 V in panel (b). In both panels two different RF powers are compared. As expected, the sheath is thicker for lower RF power (electron density) and for higher substrate bias. The penetration of sheath into the high-aspect-ratio trench obviously depends on the width of the sheath. The furthermost penetration of sheath was observed in the case $P = 1000$ W, $U_b = -20$ V (not displayed in Figure 3). First interesting result is that an increase of substrate bias can suppress the relative deformation of sheath near the trench. The sheath deformation affects directionality of ions passing through the sheath. Therefore, it is useful to estimate the deformation of sheath before the deposition.

Ions inside the trench were our next subject of interest. The most interesting results included mean energies of ions and the difference of Ar^+ and Cu^+ behaviour. Even the case of planar substrate showed difference in energy of impinging ions. Under our range of conditions, Ar^+ ions lost in average from 10% to 40% of mean energy gained during their way through sheath and presheath. On the other hand, Cu^+ ions lost maximally 11% of gained energy. A closer examination of the situation inside the trench can explain more. As was

TABLE 2: Parameters and results of global model.

Parameter	Values			
$[Cu^0] + [Cu^+]^b$	2×10^{12} cm^{-3}			
$T_g^{a,b}$	400 K			
$p^{a,b}$	5.3 Pa (40 mTorr)			
RF power[b]	100 W	400 W	700 W	1000 W
$n_e^{a,c}$	1×10^{11} cm^{-3}	3.5×10^{11} cm^{-3}	6×10^{11} cm^{-3}	8.5×10^{11} cm^{-3}
$T_e^{a,d}$	1.47 eV	1.46 eV	1.44 eV	1.42 eV
$[Cu^+]^{a,d}$	0.38×10^{11} cm^{-3}	0.94×10^{11} cm^{-3}	1.38×10^{11} cm^{-3}	1.75×10^{11} cm^{-3}
$[Ar^+]^{a,d}$	0.62×10^{11} cm^{-3}	2.56×10^{11} cm^{-3}	4.62×10^{11} cm^{-3}	6.75×10^{11} cm^{-3}

[a]Input of particle model; [b]free parameter; [c]obtained from [23]; [d]output of global model.

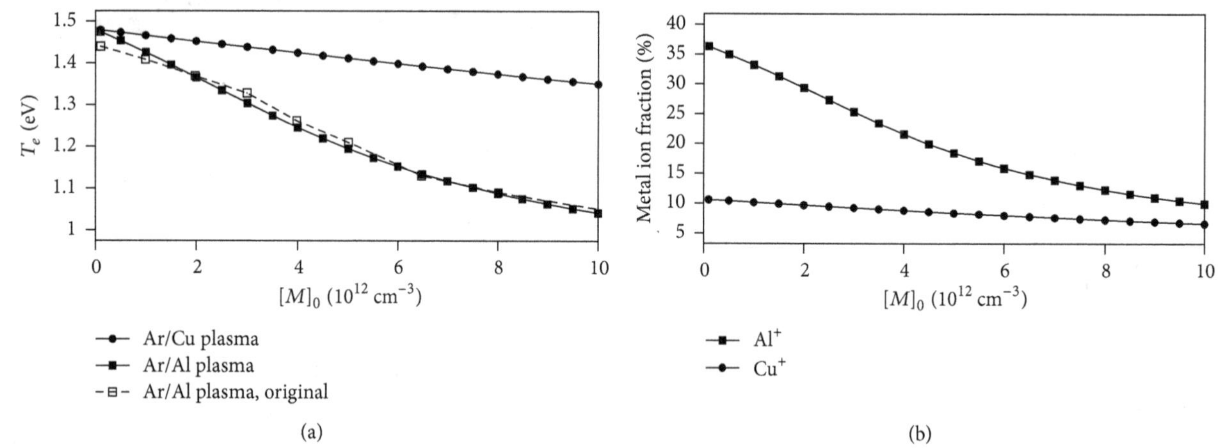

FIGURE 2: Validation of the global model and comparison of results for Al and Cu. (a) Electron temperature and (b) metal atoms ionization degree versus total average density of metal atoms in the chamber.

shown before, even in the worst case, most of the remaining electric field was located in the first quarter of the trench and still it was much lower than the field in the sheath in front of the substrate. Particles which enter the trench suddenly stop their acceleration and they should move with constant speed through the trench. Figure 4 shows the mean energies of Cu^+ (a) and Ar^+ (b) inside the trench. The conditions are $U_b = -100$ V and $P = 400$ W. Both species are decelerating but the effect is stronger for Ar^+ ions. The explanation is in the included collision processes. While only elastic collision between Cu^+ and Ar^0 is included, Ar^+ can also transfer its charge to a slow neutral Ar^0 which in fact causes significant deceleration of the ion.

Ar^+ ions do not only lose more energy during collision but, due to higher cross section, also collide more often. The collision can change the direction of their flight and the walls catch these ions more often. This results in lower fluxes of Ar^+ ions to the bottom of the trench Φ_{Ar+} with respect to their flux to planar substrate Φ^0_{Ar+}. Figure 5 shows, for example, that for U_b = 200 V and $P = 100$ W the flux of Ar^+ to the bottom of the trench is only approx. 3% of the flux of Ar^+ to planar substrate. On the contrary, 35% of Cu^+ flux lasts down to the bottom under the same conditions. The relative flux to the bottom decreases with increasing RF power and with increasing negative bias. This corresponds to the situation of thinner sheath which is more deformed by the trench and the directionality of ions is disrupted.

Finally, we observed the fluxes of ions along the boundary of the trench. Figure 6 shows the fluxes of (a) Cu^+ and (b) Ar^+ for different biases under the condition $P = 1000$ W. The coordinate s follows the boundary of the trench. That is, interval (0, 2.4) represents the flat surface "to the left" from the trench; interval (2.4, 4.4) represents "left wall" of the trench; interval (4.4, 4.6) is the bottom; (4.6, 6.6) is the "right wall" and (6.6, 9) is the flat surface "to the right" from the trench. Apart from the fluxes to the bottom which were discussed previously, the local maxima near the opening of the trench deserve attention. Due to high electric field, higher fluxes of ions were observed in the vicinity of the opening corners. It is interesting that the maxima are not located directly on the corner but right next to it. This effect was already observed in [17] and explained using the law of conservation of angular momentum. In addition, our observation showed that the effect of flux peaks near the trench corner is stronger in the case of thinner and hence more deformed sheaths.

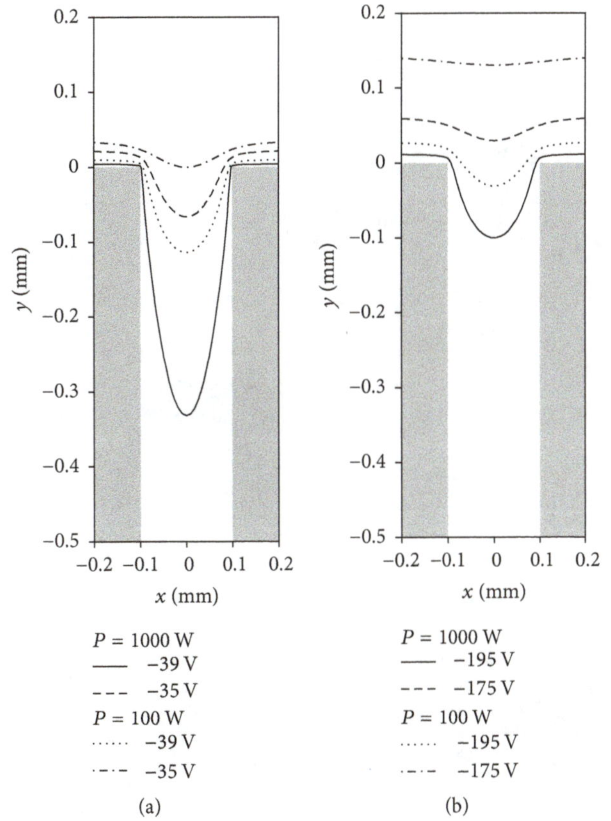

FIGURE 3: Penetration of sheath into the trench for different values of RF power and substrate bias. For each case two contours are plotted: $0.975 \times U_b$ and $0.875 \times U_b$. (a) $U_b = -40$ V, (b) $U_b = -200$ V.

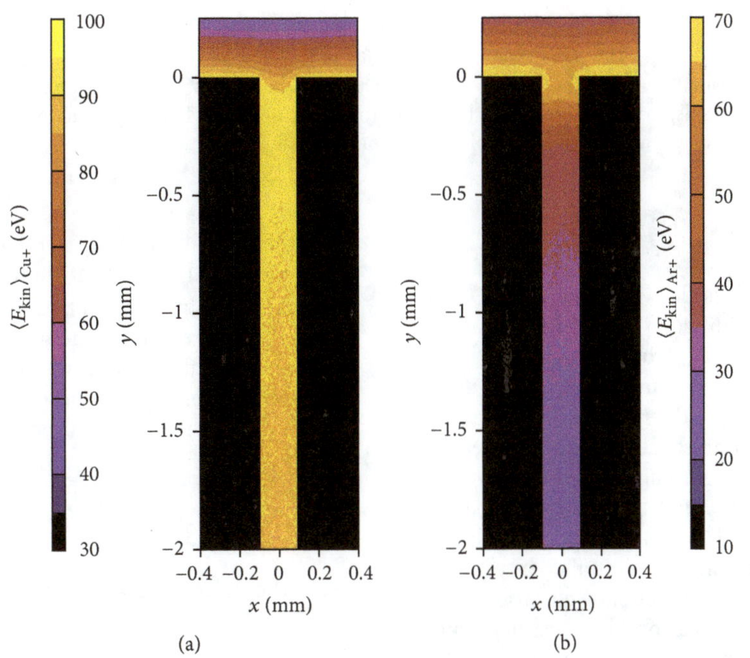

FIGURE 4: Mean energy of (a) Cu^+ and (b) Ar^+ ions inside the trench.

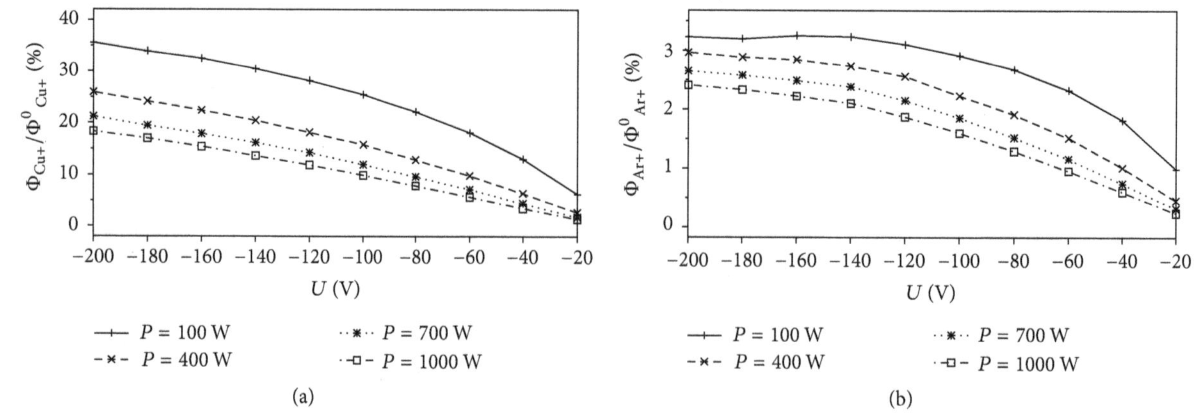

FIGURE 5: Ratio of flux to the bottom and flux to planar substrate of (a) Cu^+ and (b) Ar^+ ions under the whole range of studied conditions. Φ_X means flux of particles X to the bottom of the trench; $\Phi^0{}_X$ means flux of particles X to a planar substrate without trench.

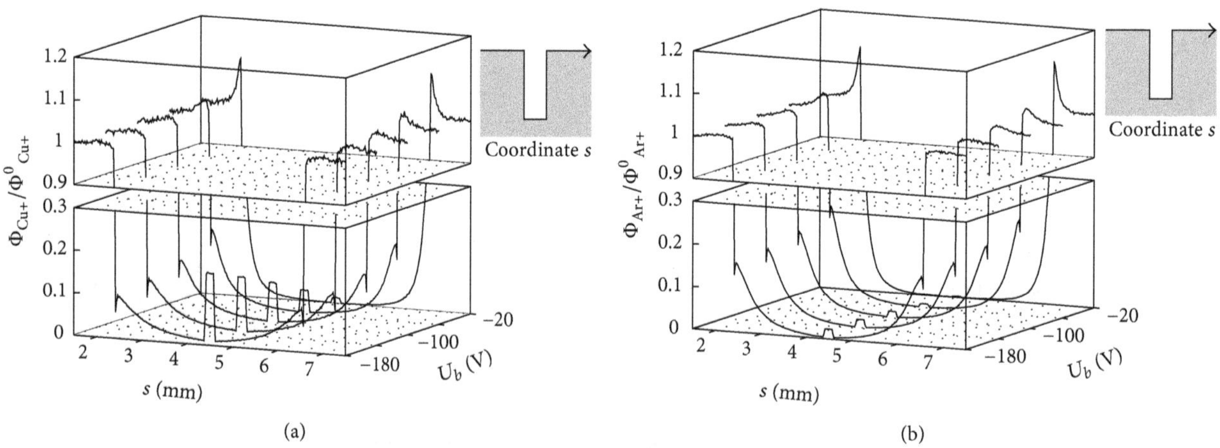

FIGURE 6: Flux of (a) Cu^+ and (b) Ar^+ ions along the trench boundary for different biases. RF power P = 1000 W. Φ_X means flux of particles X along the trench; $\Phi^0{}_X$ is a constant that describes flux of particles X to a planar substrate without trench. Coordinate s starts on the edge of the domain (0 mm), reaches the beginning of the trench (2.4 mm), then the bottom (4.4 mm), goes along the bottom (4.4–4.6 mm), then along the side (4.6–6.6 mm), and continues on remaining planar part of substrate (6.6–9 mm).

4. Conclusions

We presented a PIC-MCC model which simulated Ar/Cu plasma behaviour in the vicinity of a two-dimensional high-aspect-ratio trench. Our model provided us with detailed information about ion fluxes, number densities, energy and angular distributions, and distribution of electrostatic potential in the sheath and presheath region. The test case was IPVD of copper in argon atmosphere. A volume-averaged model obtained the input parameters and showed the difference between Al and Cu in the magnetron chamber which was caused by different ionization energy of both species. Afterwards, the particle model showed various effects of changing RF power and bias on the sheath thickness and shape and on the amount of ions capable of reaching the bottom of the trench. Thicker sheaths, which can be achieved by higher bias or lower RF power, proved to be more suitable for depositions on the bottom of the trench. The differences between the copper ions and ions of carrier gas were described and attributed to high cross section of charge transfer interaction between Ar^+ and Ar^0. The results of our model predict the energy and angular distributions of impinging metal ions in low-pressure conditions which are characteristic for ionized plasma vapour deposition method and therefore they can be used for the optimization of this technology.

Conflicts of Interest

The authors declare that they have no conflicts of interest.

Acknowledgments

The authors gratefully acknowledge the support of the Grant Agency of Charles University, Project no. 220215.

References

[1] S. M. Rossnagel and J. Hopwood, "Magnetron sputter deposition with high levels of metal ionization," *Applied Physics Letters*, vol. 63, no. 24, pp. 3285–3287, 1993.

[2] S. M. Rossnagel and J. Hopwood, "Metal ion deposition from ionized mangetron sputtering discharge," *Journal of Vacuum Science & Technology B: Microelectronics and Nanometer Structures Processing, Measurement, and Phenomena*, vol. 12, pp. 449–453, 1994.

[3] W. Wang, J. Foster, T. Snodgrass, A. E. Wendt, and J. H. Booske, "An rf sustained argon and copper plasma for ionized physical vapor deposition of copper," *Journal of Applied Physics*, vol. 85, no. 11, pp. 7556–7561, 1999.

[4] U. Helmersson, M. Lattemann, J. Bohlmark, A. P. Ehiasarian, and J. T. Gudmundsson, "Ionized physical vapor deposition (IPVD): a review of technology and applications," *Thin Solid Films*, vol. 513, no. 1-2, pp. 1–24, 2006.

[5] U. Hansen, S. Rodgers, and K. F. Jensen, "Modeling of metal thin film growth: Linking angstrom-scale molecular dynamics results to micron-scale film topographies," *Physical Review B*, vol. 62, no. 4, pp. 2869–2878, 2000.

[6] S. Hamaguchi and S. M. Rossnagel, "Journal of Vacuum Science & Technology B, Nanotechnology and Microelectronics: Materials, Processing, Measurement, and Phenomena," *Simulations of trench-filling profiles under ionized magnetron sputter metal deposition*, vol. 13, 183, 1995.

[7] A. Bogaerts, R. Gijbels, and W. Goedheer, "Hybrid Modeling of a Capacitively Coupled Radio Frequency Glow Discharge in Argon: Combined Monte Carlo and Fluid Model," *Japanese Journal of Applied Physics*, vol. 38, no. Part 1, No. 7B, pp. 4404–4415, 1999.

[8] V. Hruby and R. Hrach, "Three-dimensional Hybrid computer modeling of langmuir probes of finite dimensions in medium pressure plasmas," *IEEE Transactions on Plasma Sciences*, vol. 38, no. 9, pp. 2328–2331, 2010.

[9] M. A. Bolorizadeh, C. J. Patton, M. B. Shah, and H. B. Gilbody, "Multiple ionization of copper by electron impact," *Journal of Physics B: Atomic, Molecular and Optical Physics*, vol. 27, no. 1, p. 175, 1994.

[10] A. Bogaerts, R. Gijbels, and R. J. Carmand, "Collisional—radiative model for the sputtered copper atoms and ions in a direct current argon glow discharge," *Spectrochimica Acta Part B: Atomic Spectroscopy*, vol. 53, no. 12, pp. 1679–1703, 1998.

[11] L. M. Chanin and M. A. Biondi, "Mobilities of mercury ions in helium, neon, and argon," *Physical Review A: Atomic, Molecular and Optical Physics*, vol. 107, no. 5, 1219 pages, 1957.

[12] A. V. Phelps, "Compilation of Electron Cross Sections [online]," (accessed January 7, 2015), http://jila.colorado.edu/~avp/collision_data/electronneutral/ELECTRON.TXT.

[13] A. H. Futch and F. A. Grant, "Mean life of the $^3P_2$ metastable argon level," *Physical Review A: Atomic, Molecular and Optical Physics*, vol. 104, no. 2, pp. 356–361, 1956.

[14] S. A. Cohen, *Plasma Etching: An Introduction*, D. M. Manos and D. L. Flamm, Eds., p. 230, Academic, San Diego, 1989.

[15] T. Holstein, "Imprisonment of resonance radiation in gases," *Physical Review*, vol. 72, no. 12, 1212 pages, 1947.

[16] L. Vriens, "Calculation of absolute ionisation cross sections of He, He^*, He^+, Ne, Ne^*, Ne^+, Ar, Ar^*, Hg and Hg^*," *Physics Letters*, vol. 8, no. 4, pp. 260-261, 1964.

[17] G. Keller, U. Rüde, L. Stals, S. Mändl, and B. Rauschenbach, "Simulation of trench homogeneity in plasma immersion ion implantation," *Journal of Applied Physics*, vol. 88, no. 2, pp. 1111–1117, 2000.

[18] Y. Miyagawa, M. Ikeyama, S. Miyagawa, M. Tanaka, and H. Nakadate, "Plasma analysis for the plasma immersion ion implantation processing by a PIC-MCC simulation," *Computer Physics Communications*, vol. 177, no. 1-2, pp. 84–87, 2007.

[19] B. Briehl and H. M. Urbassek, "Simulation of sheath dynamics and current nonuniformity in plasma-immersion ion implantation of a patterned surface," *Journal of Applied Physics*, vol. 93, no. 8, pp. 4420–4431, 2003.

[20] V. Hruby and R. Hrach, "Computational simulation of metal ion propagation from plasma to substrates with uneven surfaces," *Vacuum*, vol. 90, no. 1, pp. 109–113, 2013.

[21] Z. Pekárek and R. Hrach, "Influence of non-axial magnetic field in a 3D3V Particle-In-Cell plasma model," *Vacuum*, vol. 82, pp. 244–247, 2008.

[22] J. Lu and M. J. Kushner, "Trench filling by ionized metal physical vapor deposition," *Journal of Vacuum Science and Technology A*, vol. 19, no. 5, pp. 2652–2663, 2001.

[23] J. Hopwood and F. Qian, "Mechanisms for highly ionized magnetron sputtering," *Journal of Applied Physics*, vol. 78, no. 2, pp. 758–765, 1995.

[24] S. Roucka and R. Hrach, "Extending PIC models to higher pressures—enhanced model of collisions," *IEEE Transactions on Plasma Science*, vol. 39, no. 11, pp. 3244–3250, 2011.

[25] A. V. Phelps, "The application of scattering cross sections to ion flux models in discharge sheaths," *Journal of Applied Physics*, vol. 76, no. 2, pp. 747–753, 1994.

[26] A. Bogaerts, M. Van Straaten, and R. Gijbels, "Description of the thermalization process of the sputtered atoms in a glow discharge using a three-dimensional Monte Carlo method," *Journal of Applied Physics*, vol. 77, no. 5, pp. 1868–1874, 1995.

[27] T. A. Davis, "Algorithm 832: UMFPACK V4.3—an unsymmetric-pattern multifrontal method," *ACM Transactions on Mathematical Software*, vol. 30, no. 2, pp. 196–199, 2004.

[28] L. Verlet, "Computer 'experiments' on classical fluids. I. Thermodynamical properties of Lennard-Jones molecules," *Physical Review*, vol. 159, no. 1, pp. 98–103, 1967.

15

Fabrication of Metal Nanoparticle Arrays in the ZrO_2(Y), HfO_2(Y), and GeO_x Films by Magnetron Sputtering

Oleg Gorshkov,[1] Ivan Antonov,[2] Dmitry Filatov,[1] Maria Shenina,[1] Alexander Kasatkin,[2] Alexander Bobrov,[1] Maria Koryazhkina,[2] Irina Korotaeva,[3] and Mikhail Kudryashov[3]

[1] *Research and Educational Center for Physics of Solid State Nanostructures, Lobachevsky State University of Nizhny Novgorod, 23 Gagarin Ave., Nizhny Novgorod 603950, Russia*
[2] *Research Institute for Physics and Technology, Lobachevsky State University of Nizhny Novgorod, 23 Gagarin Ave., Nizhny Novgorod 603950, Russia*
[3] *Department of Physics, Lobachevsky State University of Nizhny Novgorod, 23 Gagarin Ave., Nizhny Novgorod 603950, Russia*

Correspondence should be addressed to Dmitry Filatov; dmitry_filatov@inbox.ru

Academic Editor: Ilia Ivanov

The single sheet arrays of Au nanoparticles (NPs) embedded into the ZrO_2(Y), HfO_2(Y), and GeO_x ($x \approx 2$) films have been fabricated by the alternating deposition of the nanometer-thick dielectric and metal films using Magnetron Sputtering followed by annealing. The structure and optical properties of the NP arrays have been studied, subject to the fabrication technology parameters. The possibility of fabricating dense single sheet Au NP arrays in the matrices listed above with controlled NP sizes (within 1 to 3 nm) and surface density has been demonstrated. A red shift of the plasmonic optical absorption peak in the optical transmission spectra of the nanocomposite films (in the wavelength band of 500 to 650 nm) has been observed. The effect was attributed to the excitation of the collective surface plasmon-polaritons in the dense Au NP arrays. The nanocomposite films fabricated in the present study can find various applications in nanoelectronics (e.g., single electronics, nonvolatile memory devices), integrated optics, and plasmonics.

1. Introduction

In recent years, the investigations of the optical and electronic properties of the metal nanoparticles (MNPs) embedded into various dielectric matrices have attracted much attention [1]. These nanocomposite materials are promising for the applications in nonvolatile memory [2–5], single electronics [6], integrated optics [7], plasmonics [8], and so forth. For instance, the MNPs are known to play a role of the electromagnetic field concentrators in the resistive random access memory (RRAM) cells and, therefore, to improve the stability of these ones [4] and to increase the yield of the RRAM devices [2]. In the last decade, a new direction in the studies of the dielectric films with embedded MNPs has emerged, namely, the investigations of the photoconductivity in the MNP arrays in the dielectric films due to the optically excited plasmon resonance (PR) in the MNP arrays. The motivation for this activity was to control the conductivity of the nanocomposite materials by means of the optical pumping in the visible wavelength band [9].

Ion implantation with subsequent annealing is used widely to fabricate the MNPs in the dielectric matrices [10]. This method allows forming the MNPs with controlled sizes and densities within a wide range. However, the MNPs formed by ion implantation are featured by an essential scatter in sizes originating from the stochastic nature of the ion interaction with solids. The alternating deposition of the metal and dielectric films (e.g., by Electron Beam Evaporation or by Magnetron Sputtering) with subsequent annealing is another method applied to fabricate the MNP arrays in the dielectric matrices [11]. The metal film material coagulates into the MNPs during annealing thus forming the single sheet MNP arrays separated by the dielectric spacers with the predefined thickness. The Au nanoparticle (NP) arrays in the SiO_2 matrix have been studied most comprehensively to date, mainly, due to poor solubility and diffusivity of Au in SiO_2 [12] that favors the coagulation of the Au films into the NPs. On the other hand, yttria-stabilized zirconia ZrO_2(Y) and hafnia HfO_2(Y) [4, 13] as well as germanium dioxide GeO_2 [14] are promising oxide matrices for various applications. However, the structure and the optical properties of the MNP arrays in the nanocomposite films formed by Magnetron Sputtering from the above materials have been studied insufficiently to date.

In [14, 15], we have reported the fabrication of dense Au NP arrays inside the ZrO_2(Y) films using layer-by-layer Magnetron Sputtering followed by annealing. The uniform Au NPs of 1 to 3 nm in diameter (subject to the initial Au film thickness) were arranged in a single sheet inside the ZrO_2(Y) films. The optical transmission spectra measurements revealed a red shift of the plasmon absorption peaks in the Au NPs (down to ~650 nm) attributed to the collective plasmon excitation in the dense NP arrays.

It should be noted that dense enough Au NP arrays in the oxide matrices specified above manifesting the collective plasmonic excitations could be fabricated by ion implantation as well [16]. However, forming the high density NP arrays requires high enough ion doses that in turn leads to high concentration of the radiation defects. From this viewpoint, Magnetron Sputtering has a considerable advantage as compared to ion implantation in forming the dense NP arrays because Magnetron Sputtering provides the formation of more uniform NPs in the defect-free dielectric matrix.

In the present paper, the results of comprehensive studies of the structure and optical properties of the Au NP arrays in the ZrO_2(Y), HfO_2(Y), and GeO_x ($x \approx 2$) films formed with layer-by-layer Magnetron Sputtering subject to the conditions of the deposition process are reported. The purpose of the present study was to investigate the capabilities of Magnetron Sputtering in forming the MNP arrays in the above dielectric matrices with controlled morphology and optical properties necessary for various nanoelectronic and photonic applications.

2. Materials and Methods

The samples for investigations were fabricated by Magnetron Sputtering using Torr International 2gl-1g2-eb4-thl vacuum setup for thin film deposition on the fused silica and Si(100) substrates. The underlying oxide layers 20 nm in thickness were deposited on the substrate surface by radio frequency (RF) Magnetron Sputtering in the Ar-O_2 gas mixture (50-50% mol.) at the substrate temperature $T_g = 300 \pm 10$°C. Then, the islanded Au films with nominal thickness $d_{Au} = 0.5$, 1.0, and 2.0 nm were deposited by direct current (dc) Magnetron Sputtering in Ar ambient at $T_g = 200 \pm 10$°C. Finally, the Au films were capped by the cladding layers of the same dielectric with the thickness of 20 nm. So far, the overall thickness of the nanocomposite films was ≈ 40 nm. The oxide/Au/oxide stacks were annealed in the Ar flow at the temperature T_A = 450 ± 10°C in order to make the islanded Au films coagulate into the NPs. The annealing times t_A were 2 min, 1 hr, and 2 hrs. Also, the 40 nm thick dielectric films without the NPs were fabricated to serve as the reference samples.

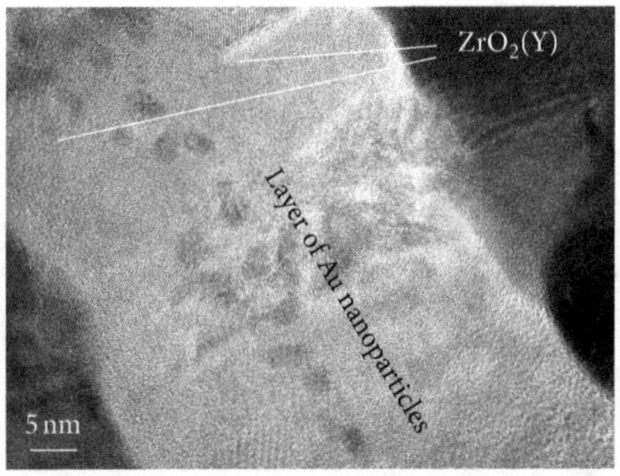

FIGURE 1: High resolution X-TEM image of a ZrO_2(Y)/Au(1 nm)/ZrO_2(Y)/Si stack after annealing at 450°C for 2 min.

The investigations of the structure of the nanocomposite films deposited onto the Si substrates were carried out by High Resolution Cross-Sectional Transmission Electron Microscopy (HR X-TEM) using Jeol® JEM 2100/F electron microscope. The single crystal Si(100) substrates served as a reference in the HR TEM measurements with atomic resolution. The optical properties of the nanocomposite films formed on the fused silica substrates were investigated by optical transmission spectroscopy at 300 K using Varian® Cary™ 6000i spectrophotometer. The refractive indices of the reference dielectric films were measured by spectrum-resolved ellipsometry using MicroPhotonic® PhE-102 instrument.

3. Results and Discussion

The results of the structural investigations of the nanocomposite films by HR X-TEM have demonstrated that during annealing the islanded Au films transform into the two-dimensional arrays of almost round NPs. For example, Figure 1 presents a HR X-TEM image of a ZrO_2(Y)/Au(1 nm)/ZrO_2(Y)/Si stack after annealing at 450°C for 2 min. The TEM image shows the Au NPs with the diameter $D = 2.3 \pm 0.4$ nm and with the average surface density $N_s \approx 7.3 \cdot 10^{12}$ cm^{-2} to be confined almost in a single sheet in the middle of the dielectric film within a thin layer of ~10 nm thickness. The location of the NP layer inside the dielectric film is determined by the thicknesses of the underlying layer and of the cladding one. In the case of the stack, the TEM image of which is shown in Figure 1, the underlying layer and the cladding one had the same thickness ≈ 20 nm. The HR X-TEM images have shown also the Au NPs to have a single crystal structure whereas ZrO_2(Y) layers had a nanocrystalline structure with the grain sizes of 5 to 15 nm.

Figures 2(a), 2(b), and 2(c) present typical optical transmission spectra of the ZrO_2(Y)/Au/ZrO_2(Y)/SiO_2, HfO_2(Y)/Au/HfO_2(Y)/SiO_2, and GeO_2/Au/GeO_2/SiO_2 stacks for d_{Au} = 0.5 nm, 1.0 nm, and 2.0 nm, respectively, after annealing for

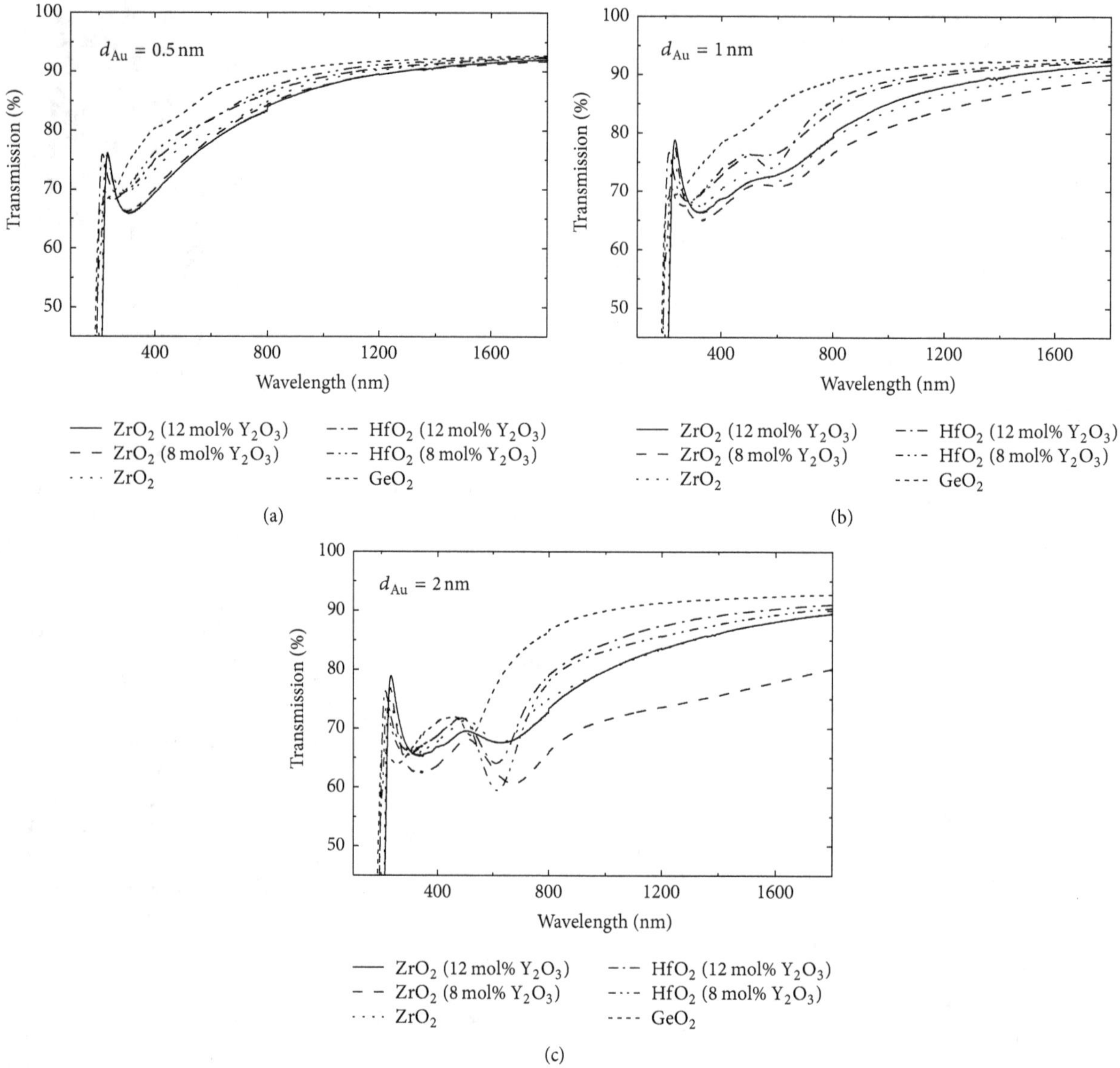

FIGURE 2: Typical optical transmission spectra (300 K) of ZrO_2(Y)/Au/ZrO_2(Y)/SiO_2, HfO_2(Y)/Au/HfO_2(Y)/SiO_2, and GeO_2/Au/GeO_2/SiO_2 stacks with different values of the nominal Au film thickness d_{Au} after annealing at 450°C for 1 hr.

1 hr. Besides a background due to the impurity absorption in the dielectric films and the interference fringes typical for the thin films, the plasmon absorption peaks in the Au NPs were observed in the optical transmission spectra.

In the ZrO_2(Y)- and HfO_2(Y)-based films, the PR peaks were observed at the wavelengths $\lambda \approx 623$ nm and ≈ 607 nm, respectively, while in the GeO_2-based film the PR peak was observed at $\lambda \approx 529$ nm. Different values of the spectral positions of the PR peaks observed for the Au NP arrays formed in the same conditions coincide with the differences in the refractive index n of the respective matrix materials (2.15 to 2.20 for ZrO_2(Y), 1.95 to 2.05 for HfO_2(Y), and ≈ 1.6 for GeO_2, according to the spectrum-resolved ellipsometry data for the probing light wavelength 632.8 nm), in full agreement with Mie scattering theory [17]. The scatter in the values of n obtained for the ZrO_2(Y) and HfO_2(Y) films was due to different molar fraction of Y_2O_3 in different samples [18]. It should be noted that a nonuniform distribution of Y in the host oxide films could introduce an additional scatter in the values of n measured in these ones as well [18].

It is worth noting that the expected PR peak wavelengths calculated for the separate Au NPs with the same sizes in the ZrO_2(Y) and HfO_2(Y) matrices according to Mie scattering theory [17] fall into the range from 590 to 600 nm. For the GeO_2 matrix, the expected PR peak wavelength was ≈ 470 nm. The red shift of the PR peak observed in the experiment relative to the calculated values for the noninteracting NPs was attributed to the collective plasmon excitation in dense Au NP arrays [19]. The parameter $a = D/l$, where l is the average spacing between the NPs, can serve as a criterion of the MNP array density [19]. For example, for the ZrO_2(Y)/Au/ZrO_2(Y)/Si stack, the TEM image of which is shown in Figure 1, $a \sim 0.62$.

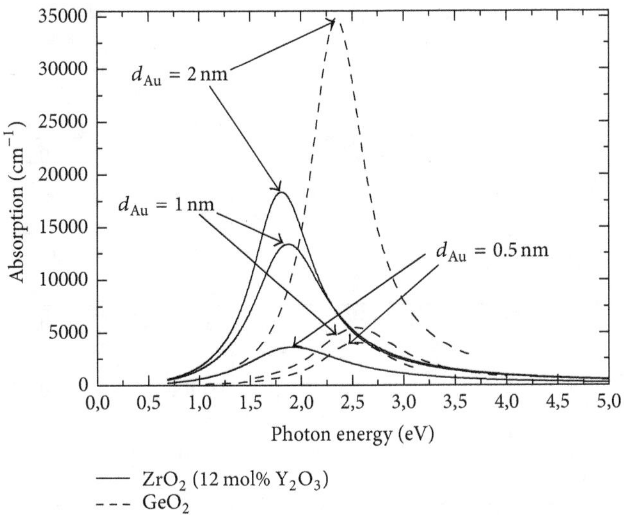

FIGURE 3: Plasmonic optical absorption spectra (300 K) of the $ZrO_2(Y)/Au/ZrO_2(Y)/SiO_2$ and $GeO_2/Au/GeO_2/SiO_2$ stacks after annealing at 450°C for 1 hr. The optical absorption spectra were calculated from the respective optical transmission ones presented in Figures 2(a) and 2(c).

It is worth noting that usually the oxide films deposited by RF Magnetron Sputtering are featured by a considerable excess surface charge density, which also could potentially induce a shift in the frequency of the plasmon resonance in the MNPs due to the charging of these ones [20]. Indeed, the *C-V* measurements performed on the *n*-Si(001)/ZrO_2(Y)/Au metal-oxide-semiconductor (MOS) stacks with the 40 nm thick ZrO_2(Y) layers deposited by RF Magnetron Sputtering in the same conditions as the ones studied in the present paper have revealed a built-in negative charge with the effective surface density ~ 10^{12} cm^{-2} [21]. However, the excess charging of the NPs with electrons would result in a blue shift of the plasmon resonance in these ones [20] whereas red shift of the plasmon resonance with respect to the bulk has been observed in the present paper. In any case, the effect of the excess charge in the MNPs on the absorption signatures of these ones is quite small and limited in very small NPs [22]. Note that just such NPs were the objects of investigations in the present study. So far, one can conclude that the collectivization of the plasmon excitations in the dense NP arrays rather than the charging of the NPs was the primary factor governing the optical properties of the NPs studied in the present work.

From the plasmonic optical absorption spectra (see, e.g., Figure 3) obtained from the transmission ones by standard technique, the average diameter D, surface density of NPs N_s, the volume fraction of the NPs N_v calculated for the 10 nm thick layer, within which the NPs are confined, and the array density parameter a have been evaluated using Mie scattering theory [17]. The results of the optical transmission data analysis for the annealing time t_A = 2 min, 1 hr, and 2 hrs are presented in Table 1. As shown in Table 1, the averaged NP diameter D and the array density parameter a as well as the relative volume fraction of NPs N_v increased with increasing nominal thickness of the initial Au film d_{Au} for all dielectric film materials. The values of a for the NPs arrays obtained from the TEM data and from the optical absorption spectra demonstrate the applicability of Magnetron Sputtering to fabricate the dense Au NP arrays in the ZrO_2(Y), HfO_2(Y), and GeO_2 thin films, in which the conditions for the manifestation of the collective plasmon excitations are provided.

The maximum values of a have been observed in ZrO_2(Y) and HfO_2(Y) with the Y_2O_3 molar fraction ≈ 0.12 and d_{Au} ≈ 2 nm that could be attributed to increased diffusivity of Au in the Y-stabilized zirconia and hafnia due to large concentration of the oxygen vacancies in these materials (the latter is equal to 1/2 of the one of Y [18]). So far, the greater the concentration of the stabilizing oxide Y_2O_3 in the film material, the better the conditions for the diffusion of Au, which promote the growth of the Au NPs during annealing. This conclusion was supported also by the fact that in the samples based on ZrO_2(Y) with the Y_2O_3 molar fraction ≈ 0.12 the values of the array density parameter a (as well as the ones of the surface density of the NPs N_s) were the largest as compared to other matrices as well as to other values of the Y_2O_3 molar fraction in ZrO_2(Y) and HfO_2(Y) for every value of d_{Au}. The smallest value of a has been observed in the sample based on amorphous GeO_2 (see Table 1). In the HfO_2(Y)-based samples, the doping by the stabilizing oxide Y_2O_3 affected the Au NP array density parameter a weaker than in the ZrO_2(Y)-based ones.

On the other hand, the concentration of Y_2O_3 affected the size of the NPs D formed in HfO_2(Y) as well as in ZrO_2(Y). The increasing of the Y_2O_3 molar fraction resulted in the decreasing of the NP diameter D accompanied by the corresponding increasing of the NP surface density for each value of d_{Au}. The largest values of NP diameter D have been observed in the undoped ZrO_2 and HfO_2 matrices.

Earlier, the NP diameter D in the ZrO_2(Y)-based nanocomposite films was found to increase with increasing nominal Au film thickness d_{Au} for all values of the annealing time t_A within the range from 2 min to 2 hrs [15]. The results obtained within the present study have revealed, in general, the same behavior of D with increasing d_{Au} for the HfO_2(Y)-based films as well as for the GeO_2-bases ones. However, the smallest NPs with D ≈ 0.72 nm and N_s ≈ 3.4 · 10^{13} cm^{-2} have been observed in ZrO_2 with the highest concentration of Y_2O_3 and d_{Au} = 0.5 nm and t_A = 2 hrs (see Table 1). In this case, the decrease of the NP diameter with increasing t_A was attributed to the dissolving of the Au NPs during annealing because the detachment of the Au atoms from the NPs begins to prevail over the attachment at long enough annealing time t_A [15].

The variation of the stabilizing oxide Y_2O_3 molar fraction in ZrO_2(Y) and HfO_2(Y) has been found to affect the density parameter a weakly (see Table 1). This observation can be explained as follows. As one can see from Table 1, in general, in the HfO_2(Y) based films as well as in the ZrO_2(Y)-based ones the NP diameter D increased with decreasing Y_2O_3 molar fraction at given d_{Au} and t_A. Again, this effect may be ascribed to the increase in the concentration of the vacancies in the oxygen sublattice with increasing Y concentration that,

TABLE 1: The average diameter D, the surface density N_s, the volume fraction N_v, and the array density parameter a for the Au NPs in the ZrO_2(Y), HfO_2(Y), and GeO_2 based nanocomposite films fabricated by Magnetron Sputtering followed by annealing. Uncertainty assessment for all parameters is ~15%.

d_{Au}, nm	t_A	Film material	D, nm	N_s, cm^{-2}	N_v	a
0.5	2 min	ZrO_2 (12 mol.% Y_2O_3) [15]	1.1	$4.7 \cdot 10^{12}$	0.0028	0.23
		ZrO_2 (8 mol.% Y_2O_3)	—	—	—	—
		HfO_2 (12 mol.% Y_2O_3)	—	—	—	—
		HfO_2 (0 mol.% Y_2O_3)	—	—	—	—
		GeO_2	—	—	—	—
0.5	1 hr	ZrO_2 (12 mol.% Y_2O_3) [15]	1.2	$1.1 \cdot 10^{13}$	0.0097	0.40
		ZrO_2 (8 mol.% Y_2O_3)	—	—	—	—
		ZrO_2	1.9	$2.8 \cdot 10^{12}$	0.0106	0.27
		HfO_2 (12 mol.% Y_2O_3)	—	—	—	—
		HfO_2 (0 mol.% Y_2O_3)	1.9	$3.0 \cdot 10^{12}$	0.0108	0.28
		GeO_2	1.8	$2.3 \cdot 10^{12}$	0.0065	0.23
0.5	2 hrs	ZrO_2 (12 mol.% Y_2O_3) [15]	0.72	$3.4 \cdot 10^{13}$	0.0068	0.42
1.0	2 min	ZrO_2 (12 mol.% Y_2O_3) [15]	1.9	$4.0 \cdot 10^{12}$	0.0151	0.38
		ZrO_2 (8 mol.% Y_2O_3)	2.7	$1.2 \cdot 10^{12}$	0.0115	0.29
		HfO_2 (12 mol.% Y_2O_3)	—	—	—	—
		HfO_2 (0 mol.% Y_2O_3)	2.1	$2.9 \cdot 10^{12}$	0.0147	0.30
		GeO_2	—	—	—	—
1.0	1 hr	ZrO_2 (12 mol.% Y_2O_3) [15]	1.6	$5.0 \cdot 10^{12}$	0.0281	0.36
		ZrO_2 (8 mol.% Y_2O_3)	2.6	$1.6 \cdot 10^{12}$	0.0147	0.31
		ZrO_2	3.0	$1.4 \cdot 10^{12}$	0.0180	0.34
		HfO_2 (12 mol.% Y_2O_3)	2.2	$2.5 \cdot 10^{12}$	0.0205	0.30
		HfO_2 (0 mol.% Y_2O_3)	3.6	$6.8 \cdot 10^{11}$	0.0164	0.32
		GeO_2	1.5	$6.0 \cdot 10^{12}$	0.0103	0.27
1.0	2 hrs	ZrO_2 (12 mol.% Y_2O_3) [15]	1.7	$8.6 \cdot 10^{12}$	0.0223	0.50
2.0	2 min	ZrO_2 (12 mol.% Y_2O_3) [15]	1.9	$8.3 \cdot 10^{12}$	0.0274	0.54
		ZrO_2 (8 mol.% Y_2O_3)	2.4	$6.3 \cdot 10^{12}$	0.0482	0.44
		HfO_2 (12 mol.% Y_2O_3)	2.2	$7.6 \cdot 10^{12}$	0.0421	0.43
		HfO_2 (0 mol.% Y_2O_3)	3.5	$1.9 \cdot 10^{12}$	0.0404	0.43
		GeO_2	2.0	$9.1 \cdot 10^{12}$	0.0410	0.42
2.0	1 hr	ZrO_2 (12 mol.% Y_2O_3) [15]	1.8	$1.1 \cdot 10^{13}$	0.0346	0.60
		ZrO_2 (8 mol.% Y_2O_3)	2.6	$3.7 \cdot 10^{12}$	0.0370	0.40
		ZrO_2	3.4	$1.2 \cdot 10^{12}$	0.0230	0.36
		HfO_2 (12 mol.% Y_2O_3)	3.2	$2.8 \cdot 10^{12}$	0.0576	0.45
		HfO_2 (0 mol.% Y_2O_3)	3.4	$2.4 \cdot 10^{12}$	0.0528	0.45
		GeO_2	2.2	$4.4 \cdot 10^{12}$	0.0488	0.36
2.0	2 hrs	ZrO_2 (12 mol.% Y_2O_3) [15]	1.9	$7.9 \cdot 10^{12}$	0.0284	0.53

in turn, facilitates the diffusion of Au from the NPs into the rest of the volume of the ZrO_2(Y) and HfO_2(Y) films during annealing. On the other hand, the surface density of the NPs N_s decreased with decreasing Y_2O_3 content. For example, in the case of ZrO_2(Y) film with $d_{Au} \approx 0.5$ nm annealed during $t_A \approx 1$ hr, a notable increase of the NP size D by ≈58% (from 1.2 nm up to 1.9 nm) along with the reduction of the surface density of the NPs N_s by ≈75% (from $1.1 \cdot 10^{13}$ cm^{-2} down to $2.8 \cdot 10^{12}$ cm^{-2}) with decreasing the Y_2O_3 molar fraction from 12% down to 0% had been observed. For the HfO_2(Y) based film, this tendency was hard to trace in the case of thin initial Au film ($d_{Au} \approx 0.5$ nm) and short annealing time ($t_A \approx 2$ min) because the plasmonic absorption peaks were manifested poorly in the optical transmission spectra of this series of samples. However, in the cases of $d_{Au} \approx 1.0$ and 2.0 nm for $t_A \approx 2$ min and 1 hr, this tendency was expressed clearly. For instance, for $d_{Au} \approx 1.0$ nm and $t_A \approx 1$ hr the sizes of the Au NPs increased by 64% (from ≈2.2 nm up to ≈3.6 nm) whereas N_s reduced by ≈73% (from ≈$2.5 \cdot 10^{12}$ cm^{-2} down to ≈$6.8 \cdot 10^{11}$ cm^{-2}) with decreasing of the Y_2O_3 molar fraction from 12% down to 0%. Note that the differences in the relative increasing of the Au NP sizes D and the decreasing of N_s make 10 to 15% for each value of d_{Au} and each t_A, that is, fall into the uncertainties of the determination of the above parameters from the optical transmission spectra. So far, the variations of the NP volume fraction N_v as well as of the array density parameter a fall into the uncertainty limits. This tendency can also point to the redistribution of the Au atoms among the NPs in the array during annealing: the Au atoms detach from the smaller NPs and attach the bigger ones. The greater the stabilizing oxide Y_2O_3 molar fraction is (i.e., the greater the oxygen vacancy concentration), the more the redistribution is pronounced. Therefore, the decrease of the Y_2O_3 molar fraction complicates the diffusion of the Au atoms during annealing thus preventing the redistribution of these ones among the NPs and increasing the NP surface density.

4. Conclusion

The structural parameters and optical properties of the single sheet Au nanoparticle arrays formed inside the ZrO_2(Y), HfO_2(Y), and GeO_x ($x \approx 2$) nanometer-thick films with layer-by-layer Magnetron Sputtering deposition with subsequent annealing have been studied, subject to the technological parameters of the film fabrication process. The possibility of the fabrication of the single sheet Au nanoparticle arrays in the ZrO_2(Y), HfO_2(Y), and GeO_2 thin films positioned at a predefined distance from the substrate and from the film surface has been demonstrated. The technological regimes for forming the dense uniform Au nanoparticle arrays, in which the conditions for the collective plasmon excitations take place, were found. The fabricated nanocomposite films can find various applications used in single electronic devices, nonvolatile memory, integrated optics, and plasmonics.

Conflicts of Interest

The authors declare that there are no conflicts of interest regarding the publication of this paper.

Acknowledgments

The present work has been supported by Ministry of Education and Science, Russian Federation (Project no. 16.7864.2017). The TEM measurements were performed using the shared research facilities of Research and Educational Center for Physics of Solid State Nanostructures at Lobachevsky State University of Nizhny Novgorod.

References

[1] J. Yang and H. Liu, *Metal-Based Composite Nanomaterials*, Springer International Publishing, Switzerland, 2015.

[2] W. Guan, S. Long, R. Jia, and M. Liu, "Nonvolatile resistive switching memory utilizing gold nanocrystals embedded in zirconium oxide," *Applied Physics Letters*, vol. 91, no. 6, Article ID 062111, 2007.

[3] S. V. Tikhov, O. N. Gorshkov, I. N. Antonov, A. P. Kasatkin, and M. N. Koryazhkina, "The forming process in resistive-memory elements based on metal-insulator-semiconductor structures," *Technical Physics Letters*, vol. 40, no. 10, pp. 837–840, 2014.

[4] W.-Y. Chang, K.-J. Cheng, J.-M. Tsai et al., "Improvement of resistive switching characteristics in TiO_2 thin films with embedded Pt nanocrystals," *Applied Physics Letters*, vol. 95, no. 4, Article ID 042104, 2009.

[5] M. H. Lee and C. S. Hwang, "Resistive switching memory: Observations with scanning probe microscopy," *Nanoscale*, vol. 3, no. 2, pp. 490–502, 2011.

[6] M. H. Devoret, D. Esteve, and C. Urbina, "Single-electron transfer in metallic nanostructures," *Nature*, vol. 360, no. 6404, pp. 547-553, 1992.

[7] A. L. Stepanov and I. B. Khaibullin, "Fabrication of metal nanoparticles in sapphire by low-energy ion implantation," *Reviews on Advanced Materials Science*, vol. 9, no. 2, pp. 109–129, 2005.

[8] M. Pelton and G. W. Bryant, *Introduction to Metal-Nanoparticle Plasmonics*, Wiley & Sons, New York, NY, USA, 2013.

[9] P. Banerjee, D. Conklin, S. Nanayakkara, T.-H. Park, M. J. Therien, and D. A. Bonnell, "Plasmon-induced electrical conduction in molecular devices," *ACS Nano*, vol. 4, no. 2, pp. 1019–1025, 2010.

[10] A. Meldrum, R. Lopez, R. H. Magruder, L. A. Boatner, and C. W. White, "Structure and properties of nanoparticles formed by ion implantation," *Topics in Applied Physics*, vol. 116, pp. 255–285, 2010.

[11] S. H. Cho, S. Lee, D. Y. Ku et al., "Growth behavior and optical properties of metal-nanoparticle dispersed dielectric thin films formed by alternating sputtering," *Thin Solid Films*, vol. 447-448, pp. 68–73, 2004.

[12] J. D. McBrayer, R. M. Swanson, and T. W. Sigmon, "Diffusion of Metals in Silicon Dioxide," *Journal of the Electrochemical Society*, vol. 133, no. 6, pp. 1242–1246, 1986.

[13] U. S. Kuzminov, E. E. Lomonova, and V. V. Osiko, *Refractory Materials from the Cold Crucible*, Science, Moscow, Russia, 2004.

[14] S. V. Tikhov, O. N. Gorshkov, D. A. Pavlov et al., "Capacitors with nonlinear characteristics based on stabilized zirconia with built-in gold nanoparticles," *Technical Physics Letters*, vol. 40, no. 5, pp. 369–371, 2014.

[15] O. N. Gorshkov, I. N. Antonov, D. O. Filatov et al., "Forming dense arrays of gold nanoparticles in thin films of yttria stabilized zirconia by magnetron sputtering," *Technical Physics Letters*, vol. 42, no. 1, pp. 36–39, 2016.

[16] O. N. Gorshkov, D. A. Pavlov, V. N. Trushin et al., "Peculiarities in the formation of gold nanoparticles by ion implantation in stabilized zirconia," *Technical Physics Letters*, vol. 38, no. 2, pp. 185–187, 2012.

[17] G. Mie, "Beiträge zur Optik trüber Medien, speziell kolloidaler Metallösungen," *Annalen der Physik*, vol. 330, no. 3, pp. 377–445, 1908.

[18] V. I. Aleksandrov, V. V. Osikov, A. M. Prokhorov, and V. M. Tatarintsev, "The formation of high-temperature materials by direct high-frequency fusion in a cold container," *Russian Chemical Reviews*, vol. 47, no. 3, pp. 213–237, 1978.

[19] B. Auguié and W. L. Barnes, "Collective resonances in gold nanoparticle arrays," *Physical Review Letters*, vol. 101, no. 14, Article ID 143902, 2008.

[20] S. Arnon and E. Rosenkrantz, "Enhanced absorption of light by charged nanoparticles," *Optics Letters*, vol. 35, no. 8, pp. 1178–1180, 2010.

[21] S. V. Tikhov, O. N. Gorshkov, M. N. Koryazhkina, I. N. Antonov, and A. P. Kasatkin, "Specific features of nonequilibrium depletion accompanied by the trapping of minority carriers by surface states in metal–insulator–semiconductor structures," *Technical Physics Letters*, vol. 42, no. 2, pp. 138–142, 2016.

[22] T. Bian, R. Chang, and P. T. Leung, "Optical interactions with a charged metallic nanoshell," *Journal of the Optical Society of America B: Optical Physics*, vol. 33, no. 1, pp. 17–26, 2016.

16

Identifying Key Factors towards Highly Reflective Silver Coatings

Bo Zheng,[1,2] Lai Peng Wong,[1] Linda Y. L. Wu,[3] and Zhong Chen[2]

[1]*MacDermid Enthone, MacDermid Performance Solutions, 26 Tuas West Road, Singapore 638382*
[2]*School of Materials Science and Engineering, Nanyang Technological University, 50 Nanyang Avenue, Singapore 639798*
[3]*Singapore Institute of Manufacturing Technology, 2 Fusionopolis Way, No. 08-04, Innovis, Singapore 138634*

Correspondence should be addressed to Bo Zheng; bzheng002@ntu.edu.sg and Zhong Chen; aszchen@ntu.edu.sg

Academic Editor: Francesco Ruffino

This paper attempts to identify key factor(s) for highly reflective silver (Ag) coatings. Investigation was made over the crystal orientation and surface roughness, using several types of surfaces, including electroplated Ag polycrystal films, physical vapour deposited polycrystal Ag films, and single crystal Ag foils with different crystal orientations. In each type of the surfaces, surface roughness was varied so that, for different combinations of crystal orientation, roughness would elucidate the key factors towards highly reflective Ag coatings. It is found that surface roughness plays a critical role in determining the reflectance, while the crystal orientation has negligible effect. The mean reflectance and one-way ANOVA analysis indicate that the single crystal Ag foils with three orientations performed statistically the same in the same roughness group at significance level $\alpha = 0.05$. Moreover, correlation between the surface reflectance and surface roughness has been proposed for the benefit of coating design. Refection data obtained from the polycrystalline silver samples are used to verify the accuracy of the proposed correlations. It was observed that the development surface area ratio, S_{dr}, is a better roughness indicator in predicting the reflectance of polycrystalline silver films than the arithmetic average roughness, R_a.

1. Introduction

In recent years, high brightness light emitting diodes (HB LEDs) which emit white light have been expanding their markets to replace incandescent and halogen bulbs for general illuminations applications because of their higher efficiencies and are more pleasing to the eyes [1]. An LED consists of lead frame, in which a light emitting semiconductor chip is positioned, and epoxy encapsulation. Since the light emitting chip is placed on the lead frame, the lead frame must have excellent electrical, thermal, and reflective properties [2]. High light extraction efficiency is one of the critical performance indicators for HB LEDs. Due to the large contrast in the refractive indices among the different materials used, certain amount of the emitted photons within the LED chip is trapped by its internal reflection. As a result, the amount of light output is reduced. Improving the light extraction efficiency is vital for the development of HB LEDs. One of the methods is to reduce the light scattering by depositing a highly reflective silver (Ag) thin film on the lead frame surface [3–8]. The Ag layer also serves the purpose of enhancing the solderability and heat dissipation of the lead frame.

On the other hand, plasmon, a quantum of plasma oscillation, has also played more and more important role in studying the properties of metals and semiconductors. Surface plasmon resonance technology has been widely used in high resolution lithography, sensing, and microscopy [9–11]. Recently, this technology has also been used to improve the emission efficiency of organic LED for developing high brightness LEDs [12]. However, as surface plasmon resonance technology is still not widely accepted in the lead frame industries, this technology is under consideration of the current work.

Many metals have good reflection of light. The reason that Ag is selected among different metallic materials is due to its high theoretical reflectance of up to 95% in the visible light wavelength region and its ability to form a smooth coating. The excellent reflective characteristics of Ag can help improve the LED luminous efficiency [13, 14].

In lead frame industries, the lead frame substrates will be pretreated chemically to obtain a smoother surface. After that, highly reflective Ag coating will be electrodeposited on the substrate with the aid of suitable brighteners, levellers, and wetting agents. These additives are proprietary information by the suppliers. Besides the plating chemistry, the operating conditions such as pH, temperature, current density, and agitation will also greatly affect the quality of the Ag coating [15].

Ag is chosen because it has the highest theoretical reflectivity of visible light among all metals. However since the cost of Ag is high too, ways to save the cost of raw material or the production cost have been explored by the industry. For example, spot plating can selectively electrodeposit silver on specific functional areas instead of the entire lead frame surfaces. In this way, the amount of Ag deposited on the nonfunctional areas of the lead frame can be significantly reduced, resulting in lower production cost. In addition, high speed silver plating, which can help to improve the productivity, is preferred in order to reduce the mass production cost.

In general, the reflectance of a metal is governed by the amount of light reflected off the metal specularly [17]. The relationship between the microstructure and brightness of electroplated nickel has been reported few decades ago [18]. This study has shown the effect of grain size, crystal texture, and surface morphology of nickel surface on its light reflecting capability. The study has also confirmed that not all fine-grained deposits are bright.

On the other hand, the structural criteria for a metal surface to show mirror brightness were previously characterized by Nikolić et al. via their scanning tunneling microscopy (STM) as well as atomic force microscopy (AFM) studies [17, 19, 20]. First, mirror brightness is governed by flat and mutually parallel parts of the metal surface which exhibit smoothness at the atomic level; second, there are distances of about several atomic diameters between the adjacent flat parts of the surface [19]. However, the correlation between the crystal growth and crystal orientation of silver coating and its reflection was not examined in that study. Besides, polycrystalline films are normally used in the previous research.

The aim of the current study is to identify key factor(s) behind Ag reflection of visible light. Potential factors include crystal orientation, crystal size, and surface roughness. The correlation between crystal orientation and reflectance of silver film needs to be confirmed by separating the influence of surface roughness. This can be best achieved using single crystal Ag with different crystal orientation but with controllable surface roughness.

In this study, high purity single crystal Ag foils with specific orientations of the (100), (110), and (111) crystal planes will be used to study the relationship between the crystal orientation and the light reflectance. Attention will be paid to the individual effect of the crystal orientation and surface topological factor. The comprehensive studies of surface topography and surface texture on the optical properties of single crystal Ag surfaces can provide more affirmative guidance to the practical manufacturing of polycrystalline Ag films. In addition, polycrystalline Ag films were produced by physical vapour deposition and electroplating. Reflectance data from these polycrystalline Ag surfaces are used to verify the findings and correlation models derived from the single crystal study.

2. Experimental

2.1. Materials. High purity (99.999%) silver single crystals with specific orientations of the (100), (110), and (111) were purchased from Goodfellow Cambridge, Ltd., UK. These one-sided polished single crystals have diameter of 10 mm and thicknesses of 2 mm for the (100) and (111) samples and 1 mm for the (110) sample. The orientation tolerance of these single crystals was ±3°.

The relationship between the surface structural characteristics of single crystals and their optical properties was studied by carefully varying the surface roughness of single crystals. This was carried out by grinding/polishing machine (TegraPol-21, Struers) at room temperature.

Brass panels (Kocour) were used as substrates for preparing electroplated and physical vapour deposited Ag films. Before the film deposition, the brass panels were soak cleaned, electrolytic cleaned, and etched in Enprep 110 EC, Enprep 220 EC (MacDermid Enthone), and 10% H_2SO_4 in sequence prior to use. After the cleaning and etching step, a thin layer of Ni was electrodeposited on the cleaned brass panels using Lectro-Nic 10-03 (MacDermid Enthone) at 55°C, with a current density at 0.12 A/cm^2. The nickel thickness was measured by fluorescent X-ray coating thickness measurement gauge (Seiko SFT 9300) and was controlled in the range of 0.55–0.75 μm. The electrodeposited silver was prepared by a plating facility (XRD Ag-7000, MacDermid Enthone) at 57°C, under 0.3 A/cm^2 current density. The thickness of the Ag films was controlled within 2–2.5 μm while the different roughness of Ag films was obtained by controlling the brightener levels in the Ag electrolyte and current densities.

Physical vapour deposited (PVD) Ag films were prepared using UDP 550 - 4, TCL. Similar to the electrodeposition, a layer of Ni (0.55–0.75 μm) was electrodeposited on the brass panels following the same procedures. Before the deposition, middle frequency hydrogen plasma etching was used to remove native oxides so as to leave a clean and chemically balanced Ni surface. The etching time was 30 min at 8.0×10^{-4} torr. After that, sputtering was carried out using alternate current (AC) at 50 Hz. The chamber pressure was kept at 1.7×10^{-3} torr during sputtering, and the applied voltage and current were 60 V and 0.2 A, respectively. The thickness of each Ag film was controlled around 2.0–2.5 μm. As the PVD deposition follows the Ni surface morphology underneath the Ag film, the different roughness of Ag films was obtained by controlling the roughness of deposited Ni layer.

2.2. Optical Reflectivity. The specular reflectance (SR) was measured by spectroscopic reflectometer (Raditech, Taiwan) at a light incidence angle of 0°. A crystalline silicon wafer was used as a standard reference. Intensity change of the light reflected from the surface was measured as a function of wavelength (λ = 300 nm–1000 nm) at normal incidence. The specular reflectance of silver could be readily calculated based on the optical constant of silicon. Reflectance readings were taken at λ = 450 nm and 550 nm because these two wavelengths are critical in the LED application, accounting for the centre emission wavelength of, respectively, the blue and green light of a white light LED. The average reflectance of five locations on each Ag surface was reported.

2.3. Materials Characterization. X-ray diffraction was conducted for the crystal structure and orientation of the samples. LabX XRD (XRD-6000, Shimadzu) with Cu K_α radiation (40 kV, 30 mA) was used. The samples were scanned using the $\theta - 2\theta$ method at a scan rate of 2° min^{-1}.

The surface morphology was observed under scanning electron microscope (SEM, JSM - 6360A, JEOL). Elemental analysis was done using the energy-dispersive X-ray spectroscopy (EDX) attached to the SEM.

2.4. Surface Topology. The surface roughness was measured using AFM (DI-3100, Digital Instruments). The average of five surface roughness measurements was taken for each sample.

The quantitative surface roughness analysis was done reporting the arithmetic average roughness (R_a). In addition, S_{dr}, the developed surface area ratio, was also reported in this work with its correlation with the light reflection studied. S_{dr} is defined as the ratio of developed 3D surface area to the planar area [21, 22] and could potentially represent the surface feature better than R_a.

3. Results and Discussions

3.1. Microstructure and Surface Texture Analysis. The crystal structure of the three commercially available single crystals Ag, electrodeposited Ag, and physical vapour deposited Ag samples was examined using X-ray diffraction (XRD). Figure 1 shows that the three Ag crystal samples display only (200), (220), and (111) diffraction peaks, respectively, which indicates that each sample indeed exhibits single crystal orientation. On the other hand, the XRD patterns of an electrodeposited and the PVD Ag films show four diffraction peaks, namely, the (111), (200), (220), and (311) peaks. This indicates that the electrodeposited and PVD Ag films are polycrystalline. It was noted that the XRD peaks of the electroplated Ag film are broader than the ones of the PVD counterpart, indicating slightly poor crystallinity of the former.

In order to correlate the crystal orientation and reflectance of Ag crystals, the influence of their surface morphology (or roughness) on the reflectance must be eliminated. Hence, Ag crystals with three specific single crystal orientations but with controllable surface roughness parameter were utilized.

The three Ag (100), Ag (110), and Ag (111) crystals were either polished or roughened to different roughness, which could be quantitatively categorized into five groups, that is,

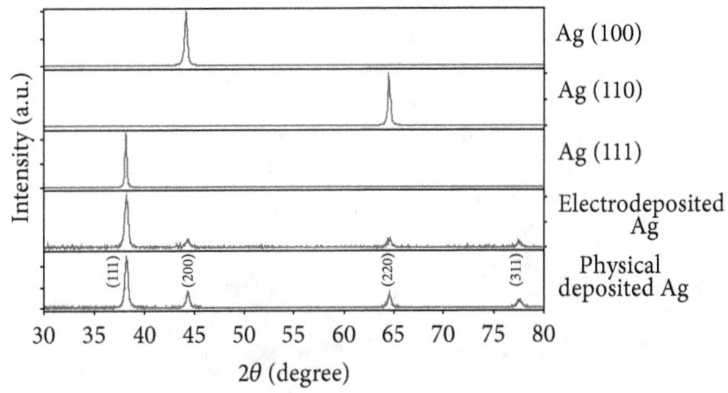

FIGURE 1: XRD patterns of Ag (100), Ag (110), and Ag (111) crystals as well as the electrodeposited and PVD polycrystalline silver films.

(A1) R_a = 0.00–5.00 nm; (B1) R_a = 5.01–10.00 nm; (C1) R_a = 10.01–20.00 nm; (D1) R_a = 20.01–30.00 nm; and (E1) R_a = 30.01–45.00 nm, as summarized in Table 1. The average R_a of the crystals were obtained by AFM analysis and they represent the arithmetic average of the variations in the height of the surface relative to a reference plane [23]. The average R_a for A1, B1, C1, D1, and E1 groups were determined to be 3.2 ± 1 nm, 6.6 ± 2 nm, 12.8 ± 2 nm, 26.5 ± 3 nm, and 36.2 ± 5 nm. The 3D AFM images in Table 1 also portray the surface topographies of the Ag crystals with different R_a roughness.

3.2. Reflection Characteristics. Specular reflectance is a numerical value of the visual brightness or mirror quality of a surface [24]. The reflectance of a sample can be determined by the ratio of reflected light intensity over incident light intensity. However, in practice, it is difficult to determine the incident light intensity before it strikes onto the sample surface. Hence, a relative intensity is first calculated whereby the reflected light intensity from the sample is divided by the incident light intensity from a standard. For the specular reflectance measurement, the change in reflected light intensity is detected while the incidence angle is 0°. Since the optical constants of silicon are well understood, the specular reflectance of the sample can be readily calculated from these known values [25].

From Figure 2, the Ag (100), Ag (110), and Ag (111) crystals with R_a = 0.00–5.00 nm show very high specular reflectance ~ 90% over λ = 400–800 nm. This result is quite consistent with the reflectance of silver metal obtained from the literature [16]. Depending on the radiation wavelength, metals respond differently; that is, their reflectivity varies with the wavelength of incident light. Even though silver has very high reflectivity over the visible light region, its reflectivity decreases significantly under the exposure of short optical wavelength radiation, particularly those below 500 nm. This is because, in this blue light region, the refractive index of silver is increased slightly which in turn affects the silver reflectivity [26, 27].

Previous studies [28, 29] have confirmed that perfectly smooth surfaces allow specular (mirror-like) reflection which obeys law of reflection; that is, the angle of incident beam is equal to the angle of reflected beam, whereas rough surfaces cause diffuse reflection as the incident beams are reflected

Table 1: The surface roughness of Ag (100), Ag (110), and Ag (111) crystals was quantitatively categorized into five different groups, that is, (A1) R_a = 0.00–5.00 nm; (B1) R_a = 5.01–10.00 nm; (C1) R_a = 10.01–20.00 nm; (D1) R_a = 20.01–30.00 nm; and (E1) R_a = 30.01–45.00 nm. Their average R_a were determined from the AFM analysis.

Group	Measured average R_a/nm	Ag (100)	Ag (110)	Ag (111)
A1	3.2 ± 1	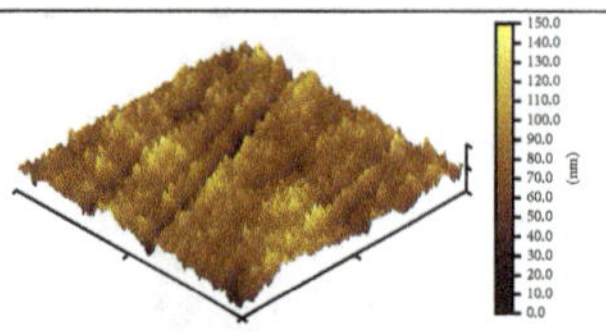		
B1	6.6 ± 2	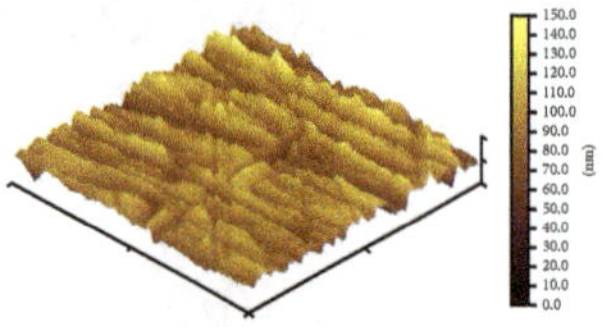		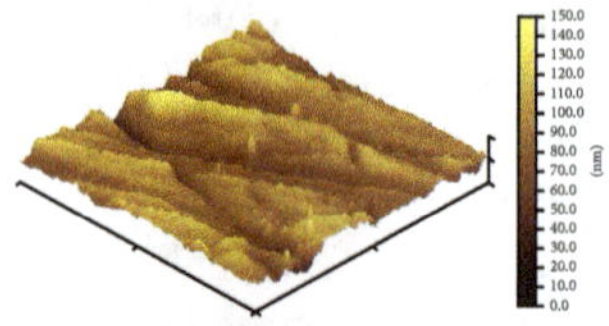
C1	12.8 ± 2			
D1	26.5 ± 3			
E1	36.2 ± 5			

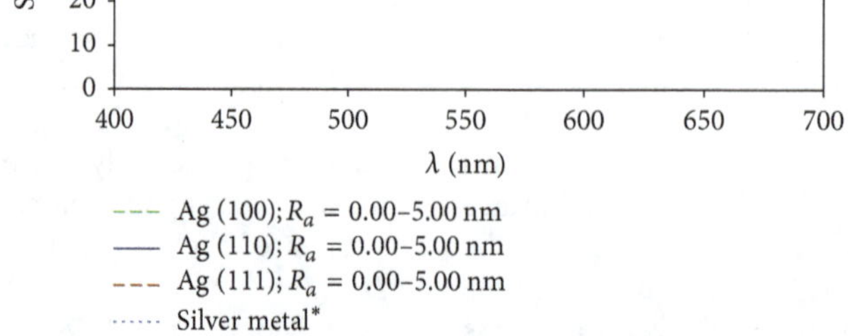

Figure 2: Specular reflectance (SR) versus wavelength spectrum of Ag (100), (110), (111) crystals with R_a = 0.00–5.00 nm and silver metal. *The reflectance of silver metal was obtained from the literature values of silver optical constants published by Palik [16].

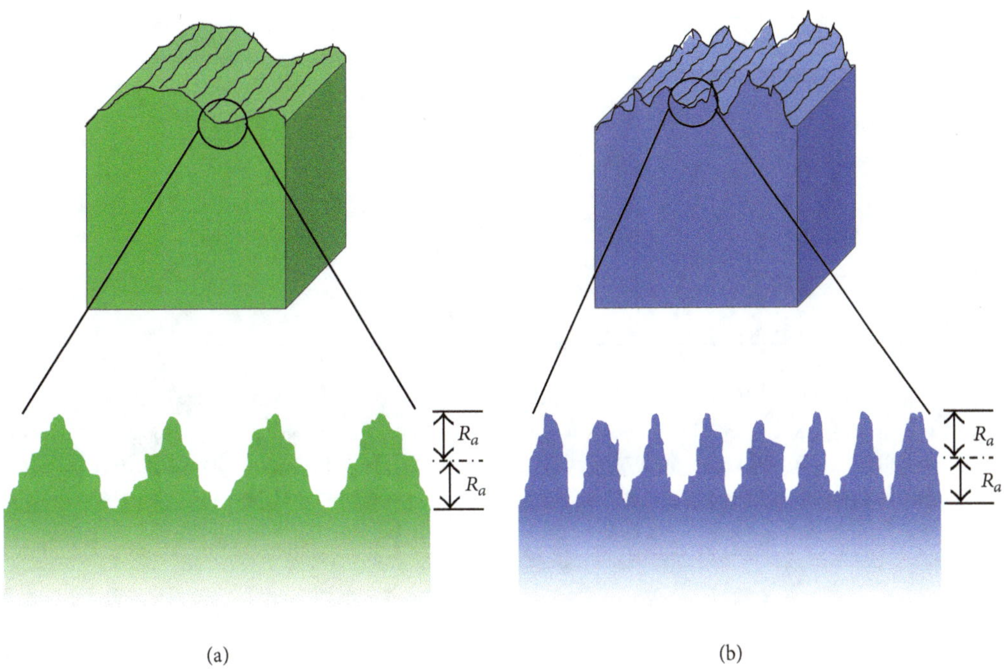

SCHEME 1: Schematic illustration of different surface profiles having the same R_a value.

in various directions. When the surface roughness increases, reflected light becomes more diffusive so the surface looks matt or dull. As a consequence, the surface roughness of the Ag sample could significantly affect its reflective characteristics.

3.3. Surface Topography Measurement and Reflectance. In most cases, surface roughness is often denoted by average R_a, which is the most commonly used roughness parameter in specifying the quality of a material surface. In addition to R_a, other roughness parameters, such as R_q (root-mean-square), R_{max} (difference in height between the highest and lowest points on the surface relative to the reference plane), and R_z (the average absolute value of the five highest and five lowest points), are also widely used to describe the vertical characteristics of a material surface.

Although R_a has been employed as a general guideline in characterizing surface texture in most industries for decades, R_a alone is not sufficient to provide a comprehensive surface profile of a material. This is because R_a result does not provide lateral information and R_a values cannot differentiate surfaces with different features between peaks and valleys, which are critical in determining the surface reflection of light. Surfaces with protruding peaks, cavity, and other irregularities may give the same average R_a value. For example, Schemes 1(a) and 1(b) depict different surface profiles having the same R_a value but the reflectance may be different from the two surfaces.

Hence, developed interfacial area ratio (S_{dr}), or more appropriately for the current context, the developed surface area ratio, is considered in this study. S_{dr} describes the change in 3D surface area relative to the 2D planar surface area. In contrast to R_a which does not provide detailed information on the 3D surfaces, S_{dr} is able to characterize the 3D surface topography of materials. The surface area includes the influence of variation in the height and number of peaks. S_{dr} is defined as follows:

$$S_{dr} = \frac{(\text{Texture surface area} - \text{Planar area}) * 100\%}{\text{Planar area}}. \quad (1)$$

Similar to R_a, the effect of S_{dr} on the reflectance is studied by categorizing the S_{dr} values obtained from the AFM measurement into 6 different groups, that is, (A2) S_{dr} = 0.00–0.50%; (B2) S_{dr} = 0.51–1.00%; (C2) S_{dr} = 1.01–2.00%; (D2) S_{dr} = 2.01–3.00% and (E2) S_{dr} = 3.01–4.00%; (F2) S_{dr} = 4.01–5.00%. The average S_{dr} for groups A2, B2, C2, D2, E2, and F2 were determined to be 0.3 ± 0.04%, 0.7 ± 0.05%, 1.5 ± 0.27%, 2.18 ± 0.07%, 3.18 ± 0.1%, and 4.48 ± 0.09%, respectively.

The reflectance at λ = 450 and 550 nm is used for comparison. A decrease in specular reflectance was observed and this is associated with an increase in surface roughness, regardless of the fact that it is expressed either in R_a (Figures 3(a) and 3(b)) or in S_{dr} (Figures 3(c) and 3(d)). The high R^2 of the regression equations in Figure 3 have indicated that the relationship between the surface roughness and reflectance can be well described as a linear relationship.

Although the roughness affects both the specular reflectance at 450 nm and at 550 nm, $R_{550\,nm}\%$ is always higher than $R_{450\,nm}\%$ because of the blue light effect explained above.

The SEM images of Ag (100), Ag (110), and Ag (111) crystals have also confirmed that the decrease in specular

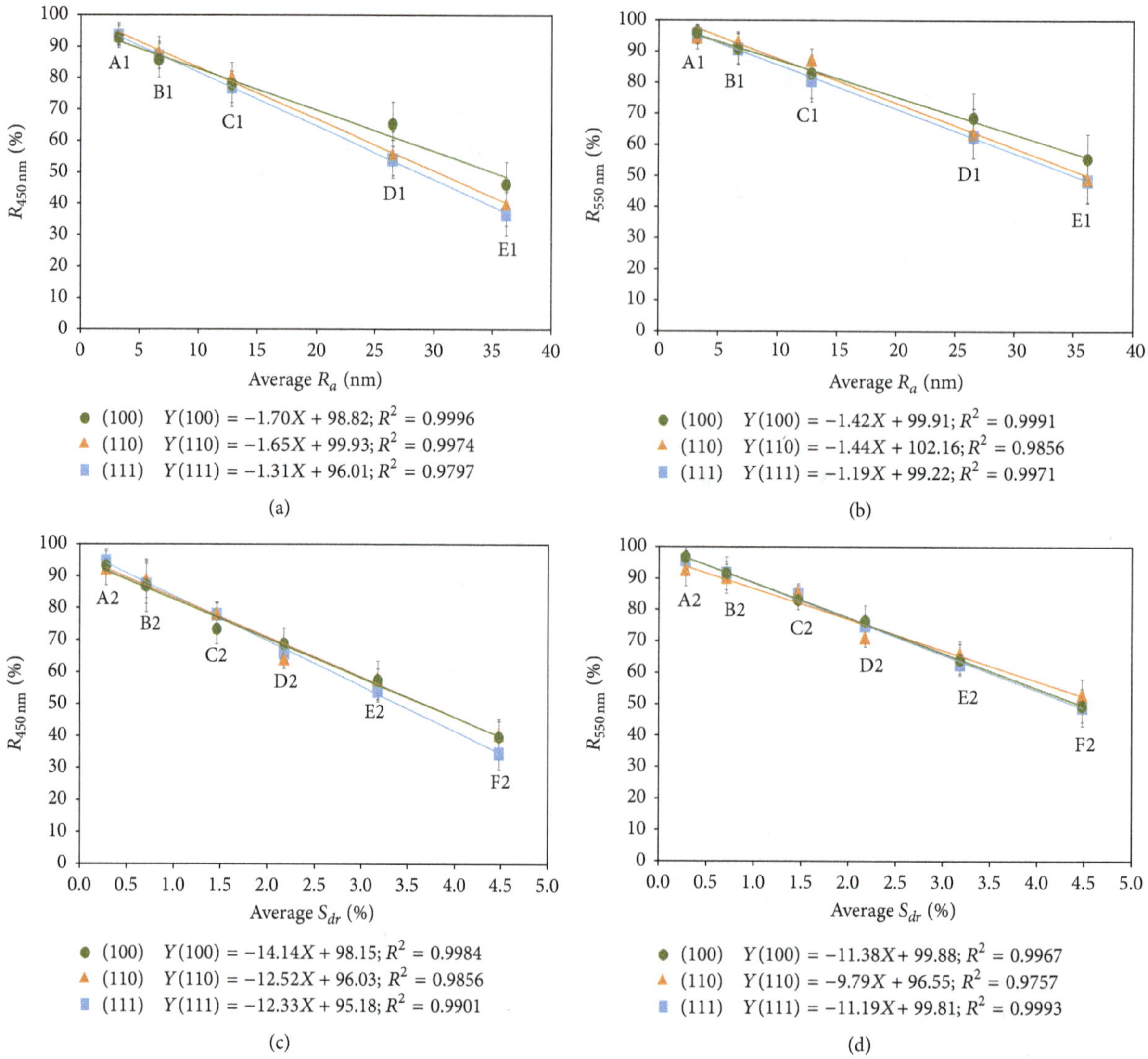

FIGURE 3: The dependence of specular reflectance on (a-b) the average R_a for Groups A1–E1; (c-d) the average developed interfacial area ratio (S_{dr}) values for Groups A2–F2 of Ag (100), Ag (110), and Ag (111) crystals at $\lambda = 450$ nm and 550 nm, respectively.

reflectance is greatly associated with an increase in irregularities in the surface morphologies of the single crystals as shown in Figure 4.

3.4. Correlation between Crystal Orientations and Their Reflection. Figure 3 has confirmed that both R_a and S_{dr} could be used as roughness indicators to represent the surface topology of single crystals. Hence, the effects of crystal orientations on the reflectance are also investigated based on these two indicators. From Figures 5(a) and 5(b), it seems that there is no significant difference in reflectance among the three single crystals at both 450 nm and 550 nm wavelengths for Groups A1–C1, while Ag (111) may exhibit slightly higher reflectance when compared to Ag (100) and Ag (110) in Groups D1–E1. Hence, the mean and one-way ANOVA of $R_{450\,nm}$% and $R_{550\,nm}$% among Groups A1–E1 are evaluated and their results are shown in Table 2. The one-way ANOVA has indicated that it is likely that there is no statistically significant difference in the means $R_{450\,nm}$% and $R_{550\,nm}$% as their P values > 0.05 (significance level set for one-way ANOVA) among Ag (100), Ag (110), and Ag (111) for all Groups A1–E1.

On the other hand, when S_{dr} is used as a roughness indicator to represent the surface topology of these Ag crystals, Figures 5(c) and 5(d) have also shown that there is not much difference in reflectance among Ag (100), Ag (110), and Ag (111) for similar S_{dr}. Table 3 shows the mean and one-way ANOVA of $R_{450\,nm}$% and $R_{550\,nm}$% among Groups A2–F2. The one-way ANOVA has also indicated that the means $R_{450\,nm}$% and $R_{550\,nm}$% of Ag (100), Ag (110), and Ag (111) are statistically the same ($P > 0.05$). Both R_a and S_{dr} indicators are independent of the wavelengths.

Tables 2 and 3 have clearly shown that there is no strong correlation between crystal orientation and reflectance. This

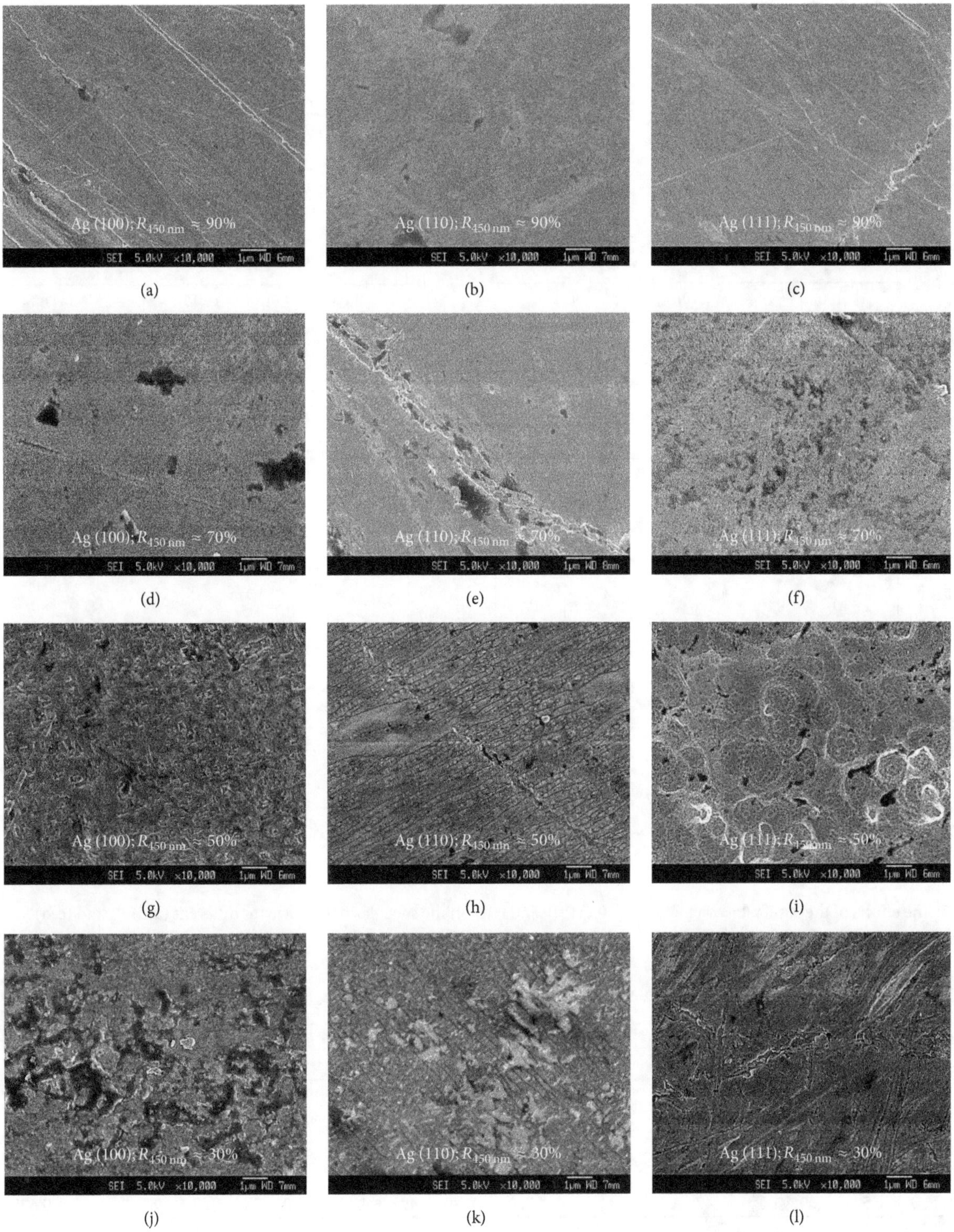

FIGURE 4: Surface morphologies of single crystal Ag (100), Ag (110), and Ag (111) with different specular reflectance (a)–(c) $R_{450\,nm} \approx 90\%$; (d)–(f) $R_{450\,nm} \approx 70\%$; (g)–(i) $R_{450\,nm} \approx 50\%$; (j)–(l) $R_{450\,nm} \approx 30\%$.

suggests that it is the surface roughness that has the main effect in controlling the reflectance of Ag rather than the way crystallites orient themselves at microscale. Assuming that the reflectance measurement of the (100), (110), and (111) surfaces at normal incidence is optically isotropic, the factor that influences the differences in reflectance should be from the irregularities of material surface. Moreover, although a single crystal with (100), (110), or (111) orientations exhibits

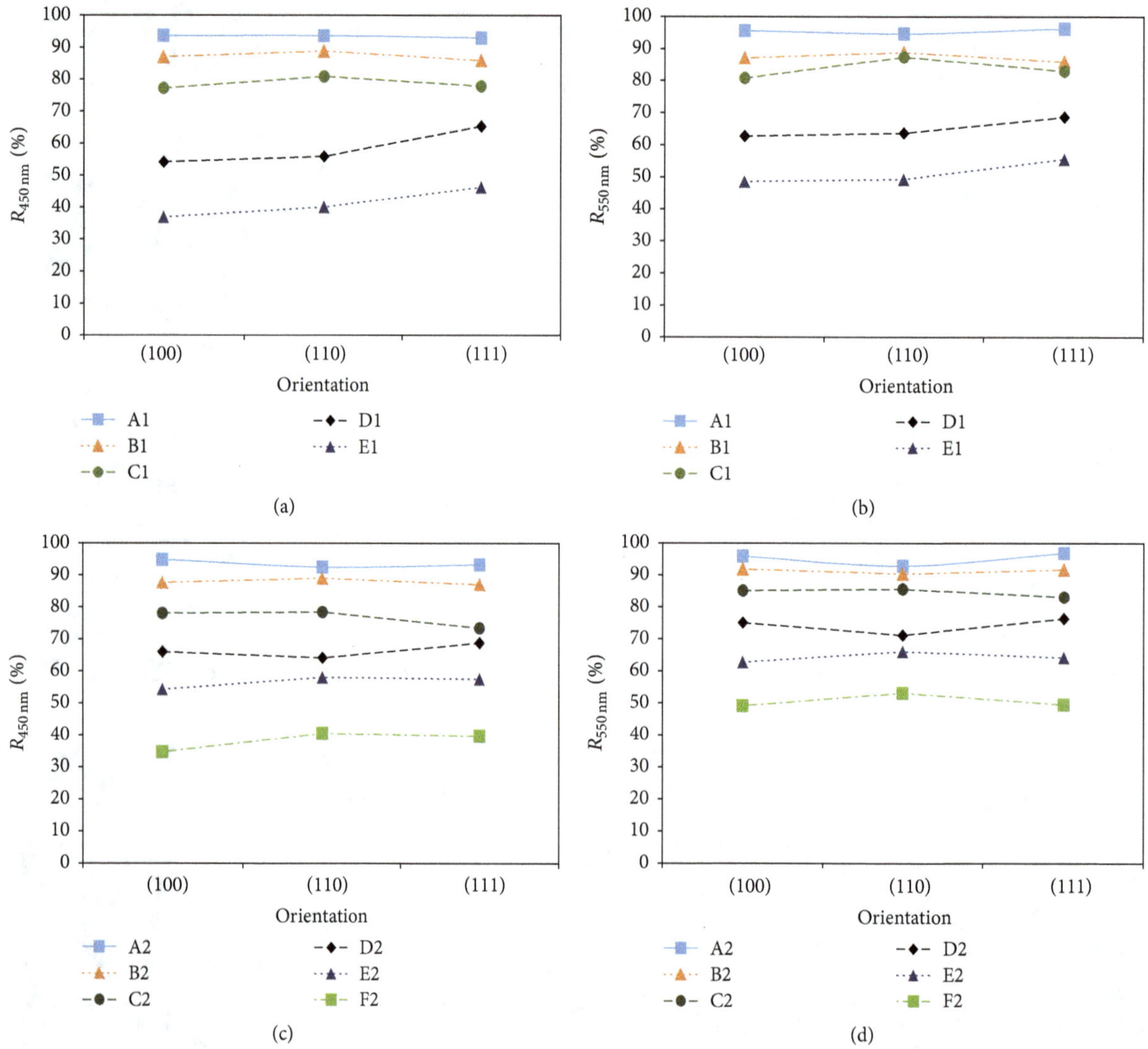

FIGURE 5: The effects of the Ag crystal orientations (100), (110), and (111) on the specular reflectance for different (a-b) R_a and (c-d) S_{dr} groups at $\lambda = 450$ nm and 550 nm, respectively.

TABLE 2: Mean and one-way ANOVA of $R_{450\,\text{nm}}$% and $R_{550\,\text{nm}}$% among Groups A1–E1. Significance level $\alpha = 0.05$.

Group	Mean $R_{450\,\text{nm}}$% ± SD			ANOVA ($R_{450\,\text{nm}}$%)	Mean $R_{550\,\text{nm}}$% ± SD			ANOVA ($R_{550\,\text{nm}}$%)
	(100)	(110)	(111)	P	(100)	(110)	(111)	P
A1	93.5 ± 3	93.5 ± 3	92.9 ± 2	0.745	94.5 ± 4	94.5 ± 4	96.1 ± 2	0.261
B1	86.9 ± 5	88.7 ± 4	85.8 ± 6	0.571	90.9 ± 4	93.2 ± 2	90.9 ± 5	0.463
C1	76.6 ± 3	79.7 ± 3	76.1 ± 4	0.332	81.0 ± 5	86.1 ± 2	84.2 ± 3	0.113
D1	54.8 ± 7	56.9 ± 8	63.3 ± 7	0.176	63.1 ± 6	64.5 ± 8	69.7 ± 8	0.179
E1	37.8 ± 5	39.4 ± 3	47.6 ± 7	0.251	47.3 ± 4	48.0 ± 7	53.6 ± 8	0.627

different surface symmetry and atoms arrangement if viewed from normal incidence, it is the microscopic roughness of the Ag surface that has caused the light scattering. When the microscopic roughness is much smaller than the wavelength, the light is reflected specularly from the surface.

3.5. Correlation Models. The individual effect of the three crystal orientations and their surface topological factors have been discussed above. Figure 5 shows that there is no significant and direct correlation between crystal orientation and reflectance; the orientation contribution towards the reflectance is neglected in developing the statistical model.

The statistical model is built based on the linear regression analysis of the surface roughness. Since R_a and S_{dr} are the results of 2D and 3D roughness measurements of the surface texture of Ag, respectively, two statistical models of the

TABLE 3: Mean and one-way ANOVA of $R_{450\,nm}\%$ and $R_{550\,nm}\%$ among Groups A2–F2. Significance level $\alpha = 0.05$.

Group	Mean $R_{450\,nm}\% \pm$ SD			ANOVA ($R_{450\,nm}\%$)	Mean $R_{550\,nm}\% \pm$ SD			ANOVA ($R_{550\,nm}\%$)
	(100)	(110)	(111)	P	(100)	(110)	(111)	P
A2	94.7 ± 3	92.4 ± 5	93.2 ± 2	0.363	95.8 ± 3	92.8 ± 5	96.7 ± 3	0.142
B2	87.6 ± 5	88.9 ± 5	87.0 ± 6	0.638	91.7 ± 2	90.3 ± 5	91.6 ± 4	0.699
C2	78.0 ± 3	78.4 ± 3	73.4 ± 3	0.213	85.1 ± 3	85.6 ± 3	83.1 ± 3	0.419
D2	66.0 ± 4	64.2 ± 4	68.8 ± 4	0.196	75.0 ± 3	71.2 ± 3	76.4 ± 5	0.154
E2	54.2 ± 4	58.0 ± 3	57.5 ± 4	0.225	62.7 ± 3	66.0 ± 4	64.1 ± 5	0.131
F2	34.6 ± 6	40.4 ± 5	39.7 ± 5	0.605	48.1 ± 6	53.1 ± 5	49.4 ± 5	0.625

reflectance are developed to distinguish the effect of each roughness parameter.

The experimental data are treated with linear regression analysis and the two statistical equations (for $\lambda = 450$ nm and 550 nm) are established for each roughness parameter.

$$R_{450\,nm}\% = 97.42 - 1.495R_a \quad \left(R^2 = 0.944\right) \tag{2}$$

$$R_{550\,nm}\% = 99.82 - 1.307R_a \quad \left(R^2 = 0.959\right) \tag{3}$$

$$R_{450\,nm}\% = 96.30 - 12.90S_{dr} \quad \left(R^2 = 0.975\right) \tag{4}$$

$$R_{550\,nm}\% = 98.63 - 10.73S_{dr} \quad \left(R^2 = 0.966\right). \tag{5}$$

3.6. Simulation of the Reflection from Polycrystalline Silver Films. From the comprehensive studies of morphology and crystal structure on the optical properties of single crystals, it was found that surface roughness is the most critical factor in determining the reflectance of Ag. 2D and 3D roughness parameters (R_a and S_{dr}) have portrayed similar linear relationship between roughness and reflectance and their resulting equations are generated in (2)–(5).

For practical applications, polycrystalline films are used. Therefore, it is important to verify the applicability and validity of the developed models based on single crystal foils using data obtained from polycrystalline films. To do so, we apply the above derived equations to two types of polycrystal Ag coatings: electroplated and physical vapour deposited Ag films.

The correlations between R_a and reflectance and between S_{dr} and reflectance of electroplated and physical vapour deposited Ag are shown in Figure 6. Expectedly, there are strong linear correlations between roughness and reflectance regardless of the types of polycrystalline Ag coatings. However, the same R_a values have given rise to two significantly different $R_{450\,nm}\%$ or $R_{550\,nm}\%$, depending on the types of Ag coatings as shown in Figures 6(a) and 6(b). Physical vapour deposited Ag generally show lower $R_{450\,nm}\%$ and $R_{550\,nm}\%$ than electroplated Ag if the roughness is perceived in terms of R_a and plotted against reflectance of the Ag.

On the other hand, if S_{dr} is used as the independent variable to represent the roughness of Ag coatings (Figures 6(c) and 6(d)), it correlates well with both types of Ag coatings. This strongly suggests that developed surface area ratio is a better indicator to correlate with reflectance as it captures the surface texture of the coatings.

The 3D AFM images of the electrodeposited Ag and physical vapour deposited Ag samples in Figure 7 have also portrayed that both surface topographies are significantly different even though they have similar R_a values.

In order to determine the percentage errors of the reflectance forecast in (2)–(5), the experimental $R_{450\,nm}\%$ and $R_{550\,nm}\%$ results of polycrystalline Ag coatings are compared with the model $R_{450\,nm}\%$ or $R_{550\,nm}\%$ values. The average percentage errors of the models are summarized in Table 4.

The total average errors % of (2) and (3) are found to be much higher than those of (4) and (5). The total average errors of using R_a as the variable in predicting $R_{450\,nm}\%$ and $R_{550\,nm}\%$ are about 14.5% and 8.9%, respectively. When the 3D parameter S_{dr} is used for prediction, the total average error for $R_{450\,nm}\%$ is 5.8% and for $R_{550\,nm}\%$ is 6.0%. This result has further suggested that S_{dr} is a better roughness indicator in predicting the reflectance of polycrystalline silver when compared to R_a.

4. Conclusions

The key factors on the reflectance of silver is studied using single crystal silver foils with specific crystal orientations of n (100), (110), and (111) with different surface roughness. Besides the most commonly used 2D parameter R_a, a hybrid parameter S_{dr} which describes the 3D surface topography is also employed in the surface roughness analysis. It is found that a decrease in specular reflectance is associated with an increase in the surface roughness of single crystals, regardless of the fact that it is expressed either in R_a or in S_{dr}. Moreover, after carefully separating the influence of surface roughness, it is observed that the surface roughness is the main factor in controlling the reflectance of Ag, while crystal orientation has a negligible effect on the light reflection. The mean and one-way ANOVA analysis have indicated that Ag (100), Ag (110), and Ag (111) with similar roughness are statistically the same ($P > 0.05$). When the microscopic roughness is much smaller than the wavelength, the light is reflected specularly from the surface. After treating the experimental data with linear regression analysis, two statistical equations (for $\lambda = 450$ nm and 550 nm) are established for R_a and S_{dr}. The accuracy of these models is verified via electrodeposited and physical vapour deposited silver films. The use of R_a as roughness indicator to predict the reflectance of silver is highly dependent on the types of coatings whereas S_{dr} is a better indicator to correlate with reflectance as it is

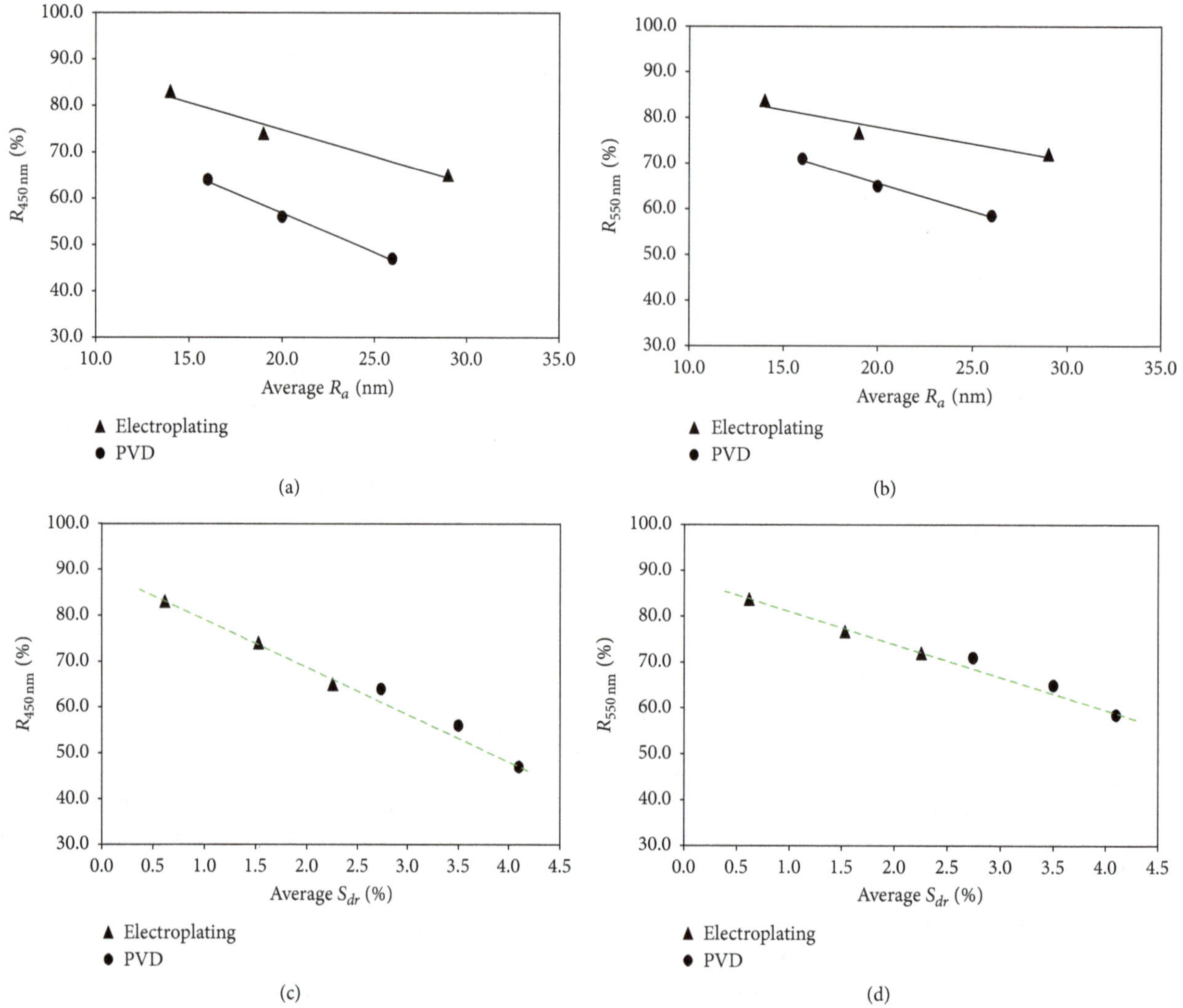

FIGURE 6: The dependence of specular reflectance of electroplated and physical vapour deposited Ag on the (a-b) average R_a; (c-d) average S_{dr} at $\lambda = 450$ nm and 550 nm, respectively.

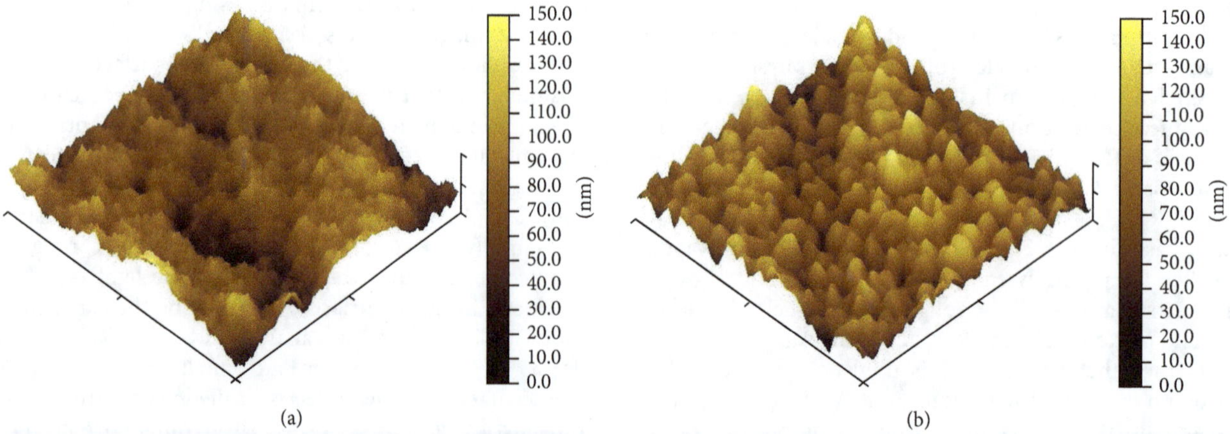

FIGURE 7: AFM images of (a) electrodeposited Ag, $R_a = 13.8$ nm, $S_{dr} = 2.49\%$; (b) physical vapour deposited Ag, $R_a = 14.5$ nm, $S_{dr} = 4.38\%$.

TABLE 4: The average errors of the statistical models for predicting the reflectance of electroplated and physical vapour deposited silvers.

Samples	R_a, nm	S_{dr}%	Experimental $R_{450\,nm}$%	Experimental $R_{550\,nm}$%	Model (2) $R_{450\,nm}$%	Error % (2)	Model (3) $R_{550\,nm}$%	Error % (3)	Model (4) $R_{450\,nm}$%	Error % (4)	Model (5) $R_{550\,nm}$%	Error % (5)
Electroplated Ag number 1	14.3	0.6	83.1	83.7	76.0	8.5	81.1	3.1	88.3	6.2	92.0	9.9
Electroplated Ag number 2	19.2	1.5	74.1	76.7	68.8	7.2	74.8	2.5	76.6	3.3	82.2	7.2
Electroplated Ag number 3	29.1	2.3	65.8	72.9	53.9	18.0	61.8	15.3	67.1	2.1	74.4	2.0
PVD-Ag number 1	17.0	2.7	64.1	71.4	72.0	12.3	77.6	8.7	61.0	4.9	69.2	3.0
PVD-Ag number 2	20.3	3.5	56.4	65.9	67.1	19.1	73.3	11.3	51.2	9.2	61.1	7.3
PVD-Ag number 3	26.1	4.1	47.8	58.5	58.4	22.0	65.7	12.3	43.4	9.2	54.6	6.6
Total average error						14.5		8.9		5.8		6.0

independent on the type of coatings. The reason is that S_{dr} is capable of capturing the surface features in 3D better.

Conflicts of Interest

The authors declare that there are no conflicts of interest regarding the publication of this paper.

Acknowledgments

The authors wish to acknowledge MacDermid Enthone for their equipment and financial support. They would also like to thank Mr. Zheng Wei Xia and Ms. Lin Ji for preparing the physical vapour deposited silver.

References

[1] J.-W. Park, Y.-B. Yoon, S.-H. Shin, and S.-H. Choi, "Joint structure in high brightness light emitting diode (HB LED) packages," *Materials Science and Engineering A*, vol. 441, no. 1-2, pp. 357–361, 2006.

[2] K.-C. Song-Zhu, O. Tohru, Y. Hitoshi et al., "Multilayered Sn/Ag_3Sn electroplating on Cu alloys for high reliable electronic/electric materials," in *Proceedings of the International Conference on Electronics Packaging, ICEP*, IEEE, Toyama, Japan, April 2014.

[3] S. H. Kee, W. J. Kim, and J. P. Jung, "Reflection characteristics of electroless deposited Sn-3.5Ag for LED lead frames," *Surface and Coatings Technology*, vol. 235, pp. 778–783, 2013.

[4] Y.-Y. Choi and T.-Y. Seong, "Morphological stability of Ag reflector for high-power GaN-based vertical light-emitting diode by addition of Ni layer," *Superlattices and Microstructures*, vol. 73, pp. 342–349, 2014.

[5] W.-S. Yum, J.-W. Jeon, J.-S. Sung, S. Jin, and T.-Y. Seong, "Ag/Ni/Ag multilayer reflector for GaN-based vertical light-emitting diode," *Japanese Journal of Applied Physics*, vol. 52, no. 5, Article ID 052101, 2013.

[6] J. H. Kim, J. H. Choi, and M. W. Shin, "Thermal and optical performance of a light-emitting-diode metal package with an integrated reflector and heat spreader structure," *IEEE/OSA Journal of Display Technology*, vol. 9, no. 10, pp. 794–799, 2013.

[7] S. Pimputkar, J. S. Speck, S. P. Denbaars, and S. Nakamura, "Prospects for LED lighting," *Nature Photonics*, vol. 3, no. 4, pp. 180–182, 2009.

[8] A. I. Zhmakin, "Enhancement of light extraction from light emitting diodes," *Physics Reports*, vol. 498, no. 4-5, pp. 189–241, 2011.

[9] R. Carles, C. Farcau, C. Bonafos et al., "Three dimensional design of silver nanoparticle assemblies embedded in dielectrics for raman spectroscopy enhancement and dark-field imaging," *ACS Nano*, vol. 5, no. 11, pp. 8774–8782, 2011.

[10] J. R. Guerreiro, N. Teixeira, V. De Freitas, M. G. Sales, and D. S. Sutherland, "A saliva molecular imprinted localized surface plasmon resonance biosensor for wine astringency estimation," *Food Chemistry*, vol. 233, pp. 457–466, 2017.

[11] K. Kim, W. Lee, K. Chung et al., "Molecular overlap with optical near-fields based on plasmonic nanolithography for ultrasensitive label-free detection by light-matter colocalization," *Biosensors and Bioelectronics*, vol. 96, pp. 89–98, 2017.

[12] K. Y. Yang, K. C. Choi, and C. W. Ahn, "Surface plasmon-enhanced spontaneous emission rate in an organic light-emitting device structure: Cathode structure for plasmonic application," *Applied Physics Letters*, vol. 94, no. 17, Article ID 173301, 2009.

[13] P. Mottier, *LEDs for Lighting Applications*, Wiley, 2010.

[14] Y. Arakawa, Y. Hirayama, K. Kishino, and H. Yamaguchi, *Compound Semiconductors 2001*, Taylor & Francis, 2002.

[15] B. Zheng, L. P. Wong, L. Y. Wu, and Z. Chen, "Cyclic voltammetric study of high speed silver electrodeposition and dissolution in low cyanide solutions," *International Journal of Electrochemistry*, vol. 2016, Article ID 4318178, 11 pages, 2016.

[16] E. D. Palik, *Handbook of Optical Constants of Solids*, vol. 3, Academic press, 1998.

[17] N. D. Nikolić, Z. Rakočević, and K. I. Popov, "Structural characteristics of bright copper surfaces," *Journal of Electroanalytical Chemistry*, vol. 514, no. 1-2, pp. 56–66, 2001.

[18] R. Weil and R. Paquin, "The relationship between brightness and structure in electroplated nickel," *Journal of the Electrochemical Society*, vol. 107, no. 2, pp. 87–91, 1960.

[19] N. D. Nikolić, Z. Rakočević, and K. I. Popov, "The STM analysis of a silver mirror surface," *Journal of the Serbian Chemical Society*, vol. 66, no. 10, pp. 723–727, 2001.

[20] N. D. Nikolić, Z. Rakočević, and K. I. Popov, "Reflection and structural analyses of mirror-bright metal coatings," *Journal of Solid State Electrochemistry*, vol. 8, no. 8, pp. 526–531, 2004.

[21] K. Stout, *Development of Methods for the Characterisation of Roughness in Three Dimensions*, Penton Press, London, Uk, 2000.

[22] K. J. Stout, W. P. Dong, E. Mainsah, N. Luo, T. Mathia, and H. Zahouani, *The Development of Methods for the Characterisation of Roughness in Three Dimensions*, EUR 15178, EN Commission of the European Communities, Brussels, Belgium, 1993.

[23] F. P. Bowden and D. Tabor, *The Friction and Lubrication of Solids*, Clarendon Press, 1950.

[24] E. Barkman, "Specular and diffuse reflectance measurements of aluminum surfaces," in *Appearance of Metallic Surfaces*, STP478, ASTM International, 1970.

[25] D. A. Miller, *Optical Properties of Solid Thin Films by Spectroscopic Reflectometry and Spectroscopic Ellipsometry*, City University of New York, 2008.

[26] O. P. Jaseentha and C. S. Menon, "Characterization of optical, electrical and structural properties of silverphthalocyanine thin films," *Journal of Materials Science: Materials in Electronics*, vol. 19, no. 7, pp. 602–606, 2008.

[27] M. Bass and Optical Society of America, *Handbook of Optics: Devices, Measurements, and Properties*, McGraw-Hill, 2 edition, 2001.

[28] V. Goossens, N. Gotzen, S. Van Gils et al., "Predicting reflections of thin coatings," *Surface and Coatings Technology*, vol. 204, no. 5, pp. 551–557, 2009.

[29] L. Shen, A. Ji, and J. Shen, "Silver mirror reaction as an approach to construct superhydrophobic surfaces with high reflectivity," *Langmuir*, vol. 24, no. 18, pp. 9962–9965, 2008.

17

Surface Organic Modification of TiO_2 Powder and Relevant Characterization

Hong Zhou, Sijia Sun, and Hao Ding

Beijing Key Laboratory of Materials Utilization of Nonmetallic Minerals and Solid Wastes, National Laboratory of Mineral Materials, School of Materials Science and Technology, China University of Geosciences, Beijing 100083, China

Correspondence should be addressed to Hao Ding; dinghao@cugb.edu.cn

Academic Editor: Michele Iafisco

Surface organic modification was conducted to TiO_2 with modifiers to improve the dispersity and comparability of pigment TiO_2 in application system by adjusting particle surface characteristics. Then, modification effects were characterized according to the changes in wetting contact angle and activation index of TiO_2 before and after modification. Moreover, the modification mechanisms of sodium stearate and sodium oleate were studied by analyzing the characteristics of TiO_2 surface functional groups in modification system and effects of modifiers. The results showed that, after being wet-processed with sodium stearate and sodium oleate, TiO_2 could turn from surface hydrophilic to inductive hydrophobic with controllable degree. The wetting contact angle of modified TiO_2 increased from 7° to 125.6° and 121.3°, respectively. The dispersity of TiO_2 in organic medium was stronger than that in inorganic medium. The modifiers formed absorption with chemical property on TiO_2 particle surface, so the inductive hydrophobicity of surface was stable.

1. Introduction

Titanium dioxide (TiO_2) is metallic oxide with excellent physical and chemical properties. Its particle in submicron scale (0.2–0.3 μm) is the component of the best white pigment; nanometer TiO_2 (less than 100 nm) is excellent photocatalytic material. Therefore, TiO_2 draws general attention. With properties of high refractive index, no toxicity, strong stability, high brightness, high covering power, and so forth, TiO_2 has been widely used in fields of coating, plastic, papermaking, printing ink, chemical fiber, and so forth [1–7]. As the photocatalyst, nanometer TiO_2 is applied in fields of water and air purification [8–11], breaking down procedures of organic pollutant degradation [12, 13], medical waste treatment [14, 15], and so forth. Actually, above properties and their roles are dependent on the control over the purity, crystal form, and size distribution of TiO_2 powder in synthesis and preparation. Moreover, the postprocessing is also important in optimizing the surface morphology and compatibility with application system. Therefore, the processing for TiO_2 powder is necessary when improving the application performance.

Micro-nano TiO_2 is poor in compatibility with high polymer material because of small grain size, high surface energy, and, especially, strong hydrophilia for surface hydroxylation. Actually, the bad dispersity and agglomeration of TiO_2 in organic system impede its pigment performance or photocatalytic performance. However, this problem can be solved by improving the dispersity of TiO_2 in application system. In the way of particle surface modification, the surface wettability can be adjusted and the morphology of surface functional groups be changed. Guosheng et al. conducted surface organic modification to nano-TiO_2 particle with silane coupling agent. The results show that TiO_2 turns from surface hydrophilic to surface hydrophobic [16]. In addition, TiO_2 particle was also modified with sodium stearate and polyethylene glycol, lauryl sodium sulfate and triethanolamine, and sodium hexametaphosphate by Gang et al. [17], Xihai et al. [18], Hui [19], Ding et al. [20–22], and Weng et al. [23], respectively. All of these studies show that the dispersity of TiO_2 improves in corresponding medium. This work also indicates that the application effects of modifiers with carboxylic in structure are more ideal among above modifiers [24–26]. Except for surface modifier, it is also important to make preassessment for TiO_2 modification effects. At present, the surface modification of powder is mainly evaluated by measurement with contact angle [27, 28] and activation index [29]. However, the modification effects and stability of changed surface wettability cannot be evaluated. Moreover, the modification mechanism cannot be reflected with the interaction between modifier and TiO_2 particle. This work conducted surface modification to TiO_2 particle in micrometer scale by wet-modification with modifiers, sodium stearate, and sodium oleate. Suspension was prepared with TiO_2 powder and water in the weight ratio of 1 : 1 and heated to and fixed at 65°C. Then, appropriate dosage of modifier was added to adjust pH value of modifier equal to or smaller than 7. The modified suspension was obtained after 30 minutes of heat preservation and constant stirring. After drying and dispersing the modified suspension, modified TiO_2 powder can be obtained. Then, the modification effects of TiO_2 were characterized according to the changes of wetting contact angle and activation index in different liquid mediums before and after modification. Finally, the modification mechanisms of TiO_2 were studied by testing the infrared spectrums and analyzing the morphology of surface functional groups before and after modification with sodium stearate and sodium oleate.

2. Experiments

2.1. Raw Materials and Reagent. TiO_2 used in tests was anatase-type titanium oxide powder manufactured by Henan Billions Chemicals Co., Ltd. In sample test, the particle size was median diameter (d_{50}) of 0.37 μm; specific surface area was 6.125 m^2/g^{-1}; whiteness was 98.80%; covering power was 12.99 g/m^2; oil absorption was 24.40 g/100 g; and water-wetting contact angle was 7°.

The organic modifiers in test, sodium stearate and sodium oleate, were all chemically pure reagents provided by Beijing JingweiHuabo Co., Ltd. The reagents should be disposed as 5% water solution by heating.

2.2. Process of TiO_2 Organic Modification. Surface modification for TiO_2 was conducted by water suspension wet surface modification. Suspension was prepared with TiO_2 powder and water in weight ratio of 1 : 1 (solid content 50%) and heated to constant temperature (65°C). Then, the modifier solution was added to TiO_2 water suspension by ratio to adjust the pH value equal to or less than 7. Modified suspension was obtained after thermal insulation and constant stirring for 30 minutes. TiO_2 powder can be obtained after drying and dispersing the modified suspension.

2.3. Performance and Structural Characterization

2.3.1. Wetting Contact Angle. Contact angle (θ) refers to the included angle between gas-liquid surface and solid-liquid surface (including liquid) in the junction of three phases, gas, liquid, and solid. As the physical quantity quantitatively reflecting the wetting degree on solid surface, contact angle can represent the change in surface characteristics of TiO_2 before and after modification. Therefore, contact angle can be used to determine the modification effects. When measuring the angle, the sample to be tested should be pressed as the solid slice with smooth surface. Then, contact angle gauge (JC2000D, Shanghai powereach digital technology equipment Co. Ltd.) should be used to measure and take photo of wetting contact angle.

2.3.2. Activation Index. Activation index refers to the percentage of powder floating on the water-gas interface accounting for the total volume after mixing the modified powder in water. Particles with surface hydrophobicity may float on the interface in the comprehensive action of adhesive force, gravity, buoyancy, water pressure, and so forth in moist periphery of three phases. The stronger the surface hydrophobicity of particle is, the greater the adhesive force in moist periphery and the larger the floating trend will be. Therefore, the floating powder can intuitively reflect the surface modification effects.

2 g modified sample was correctly weighed and placed in pear-shaped funnel with 200 mL water. After oscillation and stewing, the sedimentation at bottom in hierarchy was introduced to beaker for drying and weighing. The activation index of sample H could be calculated by following formula:

$$H = \frac{\text{Sedimentation quantity in total quality}}{\text{Total quality}} \times 100\%. \quad (1)$$

2.3.3. Infrared Spectroscopy. Infrared spectroscopy analysis can be used to detect various keys in molecule, structure of functional group, and binding characteristics of different parts in composite materials. After compressing the sample to be tested and KBr, Spectrum 100 infrared spectrometric analyzer manufactured by Shanghai PerkinElmer Co., Ltd. was used for infrared spectrometry and analysis. The scanning scope of instrument was 4000 cm^{-1}–400 cm^{-1}, and the resolution ratio was 3 cm^{-1}.

3. Results and Discussion

3.1. Change of Modified TiO_2 Wetting Contact Angle with Modifiers and Dosage. Figure 1 shows the influences of modifier dosage on the contact angle of TiO_2 wetted by water after TiO_2 particles were modified with sodium stearate and sodium oleate. Figure 2 shows the microscopic image of wetting contact angle.

According to Figure 1, the contact angle continuously enlarged with the increase of modifier dosage after TiO_2 is modified with sodium stearate and sodium oleate. The angles are 125.6° (modified with sodium stearate) and 121.3° (modified with sodium oleate), larger than 90°, critical between hydrophile and hydrophobicity, when modifier dosage is 1.5%. The contact angle of TiO_2 is 7° before modification. Obviously, TiO_2 surface has turned from strong hydrophilia to hydrophobicity after modification. However, the wetting contact angle on TiO_2 surface decreases in small range when the dosage further increases. The results show that both modifiers can change the wettability of TiO_2 surface. Moreover, the hydrophobicity formed in modification can be controlled with modifier dosage. In other words, the wettability on

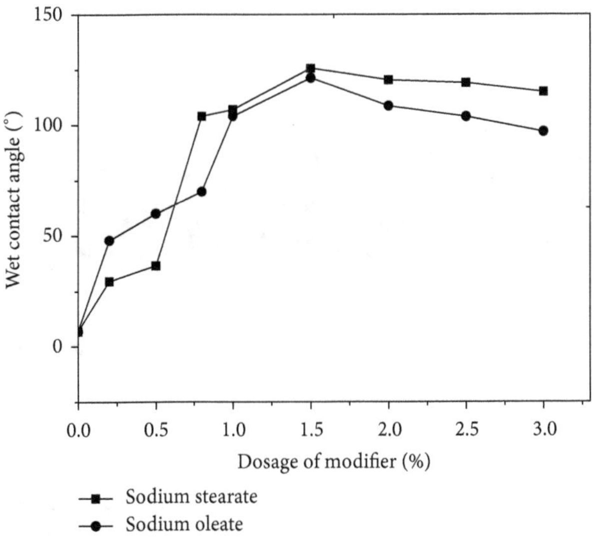

FIGURE 1: The wetting contact angle of modified TiO_2 in water.

TiO_2 particle surface and its degree are controllable. Figure 2 intuitively reflects the changes in Figure 1.

In this work, TiO_2 was modified with sodium stearate and sodium oleate (in dosage of 1.5%), respectively. Figure 3 shows the contact angles of modified product and unmodified TiO_2 wetted by water, absolute ethyl alcohol, and kerosene. It can be seen that the contact angle of modified TiO_2 wetted by absolute ethyl alcohol is about 21°, while the angle wetted by water is larger than 120° (see Figure 1). This means that the hydrophilia of TiO_2 with alcohol is stronger than that with water because alcohol has double characters of organics and water. Therefore, modified TiO_2 with surface strong hydrophobicity has stronger wetting action than water. The wetting contact angle of modified TiO_2 in kerosene (organic nonpolar solvent) is slightly smaller than that in ethyl alcohol (less than 20°). In other words, the hydrophilia of modified TiO_2 in kerosene is much stronger than that in alcohol. It means that the surface of modified TiO_2 particle has stronger organic nonpolarity. Moreover, the above results further specify that modifiers form the effective absorption on the surface of TiO_2 particle, so the particle turns from surface hydrophilia to hydrophobicity. Figure 3 also shows that the contact angle of unmodified TiO_2 wetted by alcohol and kerosene is about 20° larger than that wetted by water. In other words, the wetness degree of alcohol and kerosene is lower than water. This means that unmodified TiO_2 has strong surface hydrophilicity.

Before and after modification, wetting contact angle of TiO_2 particles changed in different mediums. This means that TiO_2 particle has extremely strong hydrophilia which may turn to strong hydrophobicity after organic modification. The wettability of modified TiO_2 particle in organic medium is stronger than that in inorganic medium, which is beneficial to the improvement of dispersity in organic medium.

3.2. Changes in Activation Index of Modified TiO_2. Activation index is an important index evaluating modification. The stronger the hydrophobicity TiO_2 particles are, the larger the activation index will be. Figure 4 shows the changes in activation index of TiO_2 powder modified by sodium stearate and sodium oleate. According to the figure, the activation index of modified TiO_2 greatly increases with the dosages of sodium stearate and sodium oleate increasing from 0% to 1.5% and from 0% to 1%, respectively. Moreover, the activation index reached the maximum, 98.6% and 89.7%, when the dosages of sodium stearate and sodium oleate were 1.5% and 1%, respectively. However, with the further increase of sodium stearate and sodium oleate, the activation index of TiO_2 may slightly decrease or keep stable. This is because the surface modification of TiO_2 is incomplete if the modifier dosage is little, so the activation index is small. With the increase of modifier dosage, however, the unimolecular coverage is realized with modifier on the surface of TiO_2 to make the modifier dosage saturated. At this time, the activation index is the maximum. If the modifier dosage further increases, multilayer coats will form on the surface of TiO_2 particles to make some types of group outward and reduce the hydrophobicity of TiO_2. Therefore, the activation index will decrease. The change of activation index shows that after modifying the sodium stearate and sodium oleate of TiO_2, the hydrophilia on surface will turn to strong hydrophobic effects. This is obviously the interaction between TiO_2 particle and modifier molecule, as well as the coating on the surface of TiO_2 [30, 31].

This means that TiO_2 surface turns from hydrophilia to strong hydrophobicity after being modified with sodium stearate and sodium oleate. This is obviously the result of TiO_2 particle and modifier molecule covering TiO_2 surface. Above results are consistent with the analysis on contact angle test.

3.3. Function Mechanism of Modifiers and TiO_2 Particle

3.3.1. IR Analysis on TiO_2 Modified with Sodium Stearate and Sodium Oleate. The mechanism of TiO_2 surface showing strong hydrophobicity can be further understood and hydrophobicity stability be determined by studying the properties of interaction between modifiers, sodium stearate and sodium oleate and TiO_2 surface. Therefore, the infrared spectroscopies were measured for sodium stearate and sodium oleate, as well as TiO_2 modified by these two modifiers. Figures 5 and 6 show the measurement results.

Figure 5 shows that there are symmetrical and dissymmetrical characteristic absorption peaks of -CH_2- at 2,917 cm^{-1} and 2,849 cm^{-1} in the infrared spectroscopy of sodium stearate (see Figure 5(a)). Moreover, there is characteristic absorption peak of -CH_3 at 2,956 cm^{-1} and stretching absorption peaks of carbonyl (C=O) and carboxyl (COO-) at 1,557 cm^{-1} and 1,470 cm^{-1}. All of these reflect the structural characteristics of sodium stearate. Compared with raw material of sodium stearate and TiO_2, there are absorption peaks at 2,996 cm^{-1} and 2,885 cm^{-1} in the infrared spectroscopy of modified TiO_2. This is obviously the characteristics of -CH_2- and -CH_3, showing that sodium stearate has attached to TiO_2 surface.

According to the comparison, the absorption peak of modified TiO_2 at 1,504 cm^{-1} resulted from the displacement

FIGURE 2: Microscopic image of wet contact angle of TiO_2. (a) Wet contact angle of TiO_2, (b) wet contact angle of modified TiO_2 with sodium stearate, and (c) wet contact angle of modified TiO_2 with sodium oleate.

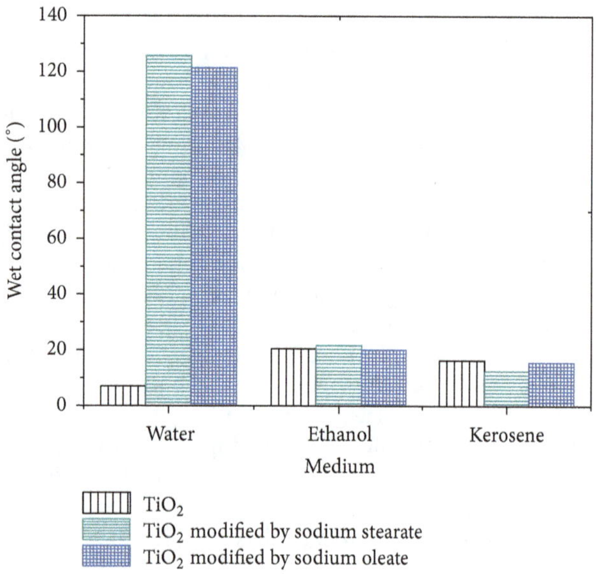

FIGURE 3: Change of wet contact angle of TiO_2.

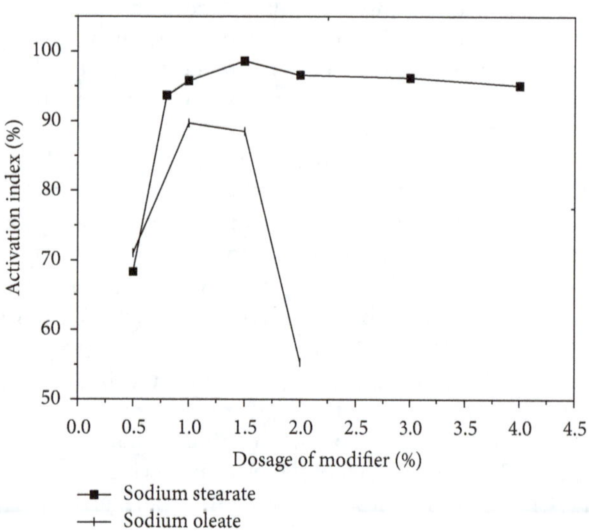

FIGURE 4: Activation index of modified TiO_2 with the amount of modifier.

of absorption peaks at 1557 cm^{-1} and 1470 cm^{-1} in the infrared spectrum of sodium stearate. It means that the chemical environment of C=O and -COO^- had significant change after sodium stearate reacted with TiO_2. This may result from the chemical action between functional group of sodium stearate and surface functional group of TiO_2 (the hydroxide radical produced in Ti or Ti hydrolysis). Therefore, it is believed that the attachment of sodium stearate to TiO_2 particle surface has chemical properties. Sodium stearate is closely and firmly combined with TiO_2 particle, so the changes in surface characteristics of TiO_2 for modification are stable.

As shown in Figure 6, the infrared spectral features of TiO_2 modified with sodium oleate are similar to those of TiO_2 modified with sodium stearate. In the infrared spectroscopy of modified TiO_2, there are characteristic absorption peaks representing methylene and methyl at 2,852 cm^{-1} and 2,922 cm^{-1}. This means that sodium oleate has attached to TiO_2 surface. The absorption peak at 1,465 cm^{-1} indicates that chemical action occurred between -COO^- and TiO_2 in sodium oleate. Therefore, sodium oleate is also closely combined with TiO_2 particle for stable modification of TiO_2.

3.3.2. Functional Group Morphology of TiO_2 Particle Surface in Water Medium. TiO_2 is one of the most typical oxides of surface hydroxylation, while the surface hydroxylation of TiO_2 as the semiconductor material is complex with various hydroxyl groups. Perrin made quantitative analysis on hydration reaction of titanium ion (Ti^{4+}) in water medium. The hydration reactions of Ti^{4+} in different levels and constants are as follows [32]:

$$Ti^{4+} + H_2O \rightleftharpoons Ti(OH)^{3+} + H^+ \quad pk_1 = 14.15 \quad (2)$$

$$Ti(OH)^{3+} + H_2O \rightleftharpoons Ti(OH)_2^{2+} + H^+ \quad pk_2 = 13.73 \quad (3)$$

$$Ti(OH)_2^{2+} + H_2O \rightleftharpoons Ti(OH)_3^{+} + H^+ \quad pk_3 = 13.39 \quad (4)$$

$$Ti(OH)_3^{+} + H_2O \rightleftharpoons Ti(OH)_4 + H^+ \quad pk_4 = 13.06 \quad (5)$$

$$pk_1 = \log \frac{[Ti(OH)^{3+}][H^+]}{[Ti^{4+}]} \quad (6)$$

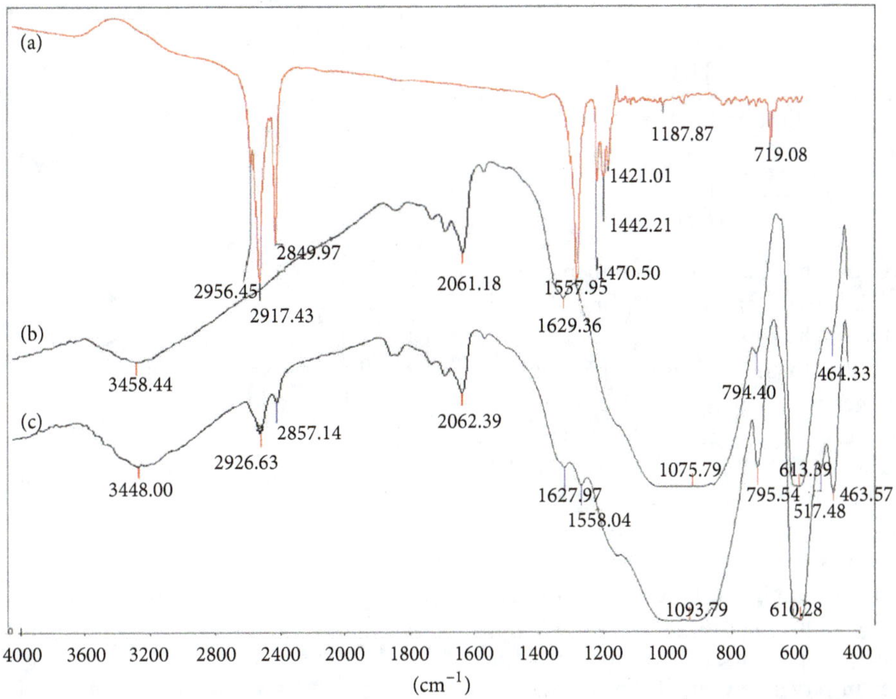

FIGURE 5: IR of TiO_2 modified by sodium stearate. (a) Sodium stearate; (b) TiO_2; (c) modified TiO_2.

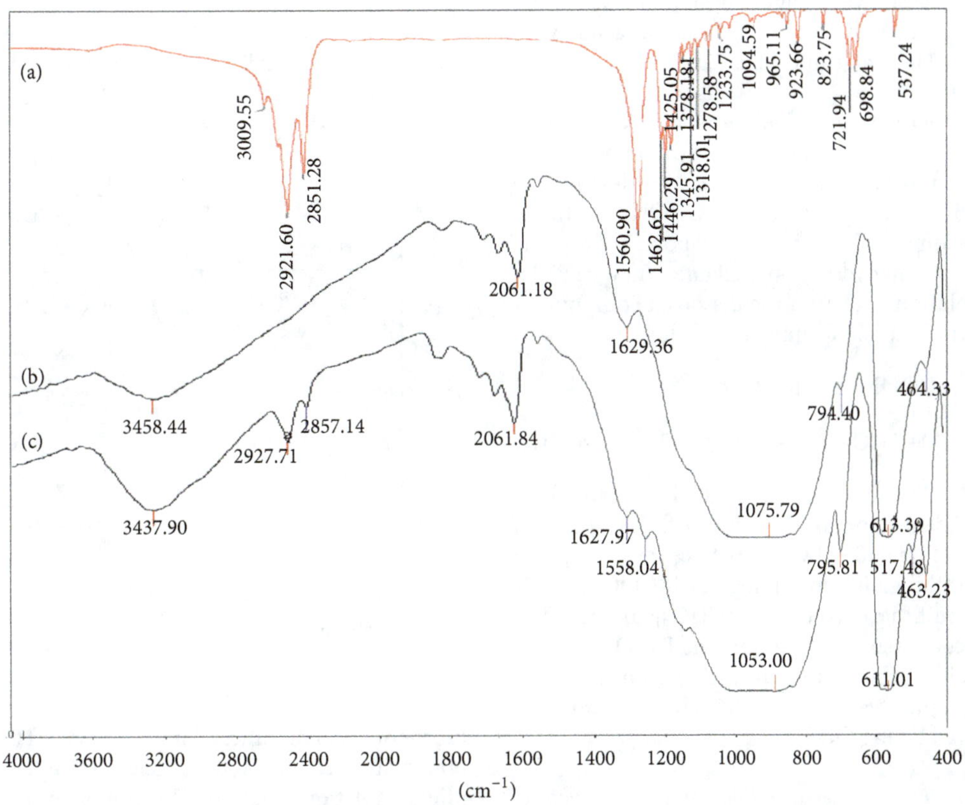

FIGURE 6: IR of TiO_2 modified by sodium oleate. (a) Sodium oleate; (b) TiO_2; (c) modified TiO_2.

$$pk_2 = \log\frac{\left[\text{Ti(OH)}_2^{\ 2+}\right][\text{H}^+]}{\left[\text{Ti (OH)}^{3+}\right]} \qquad (7)$$

$$pk_3 = \log\frac{\left[\text{Ti(OH)}_3^{\ +}\right][\text{H}^+]}{\left[\text{Ti (OH)}_2^{\ 2+}\right]} \qquad (8)$$

$$pk_4 = \log\frac{[\text{Ti (OH)}_4][\text{H}^+]}{[\text{Ti(OH)}_3^{\ +}]} \qquad (9)$$

Therefore, the relationship between components, Ti^{4+}, $Ti(OH)^{3+}$, $Ti(OH)_2^{\ 2+}$, $Ti(OH)_3^{\ +}$, and $Ti(OH)_4$, and pH value ($-\log[H^+]$) is shown in Figure 7 [32].

Figure 7 shows that with the increase of pH value in system, Ti^{4+} tends to generate polyhydroxy. When pH < 1, $Ti(OH)_2^{\ 2+}$ is taken as the principal; when pH = 2, $Ti(OH)_3^{\ +}$ is taken as the principal; when pH > 4, Ti^{4+} will turn to $Ti(OH)_4$ component. The result also shows that even when pH = 0~1, the concentration of OH^- in system is extremely small. If the external action of hydrolysis is weak, no Ti^{4+} and $Ti(OH)^{3+}$ component is represented, meaning that the capability of Ti^{4+} is strong in generating various hydroxylates. It is obvious because the constants (pk_1~pk_4) of Ti^{4+} hydrolysis reaction in different levels keep in very high value (13.06~14.15).

The hydroxylation characteristic of unsaturated Ti on TiO_2 surface can be deducted according to the hydrolysis behavior of Ti^{4+}. In other words, Ti is covered by a large amount of hydroxyl, so surface functional groups of TiO_2 are mainly OH-group as shown in Figure 8 [33]. This is similar with the molecule model of TiO_2 with surface attached with water proposed by Huaxiang and Xuxu [33].

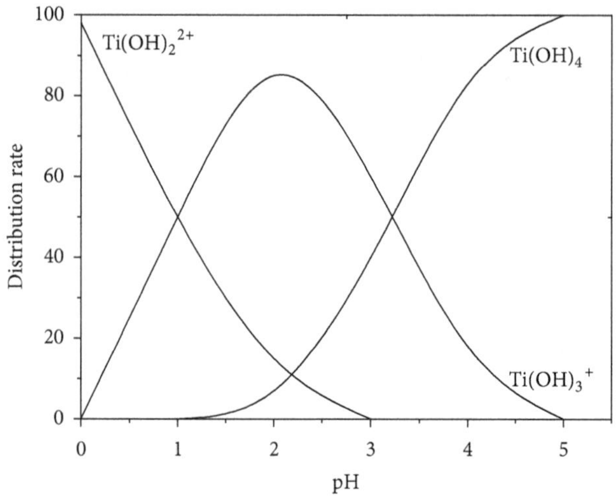

FIGURE 7: Distribution of Ti^{4+} hydrolysis component with pH value.

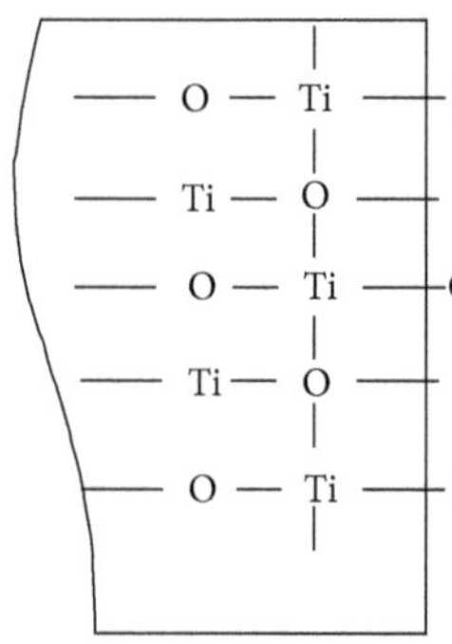

FIGURE 8: Schematic diagram of TiO_2 surface adsorbed hydroxyl.

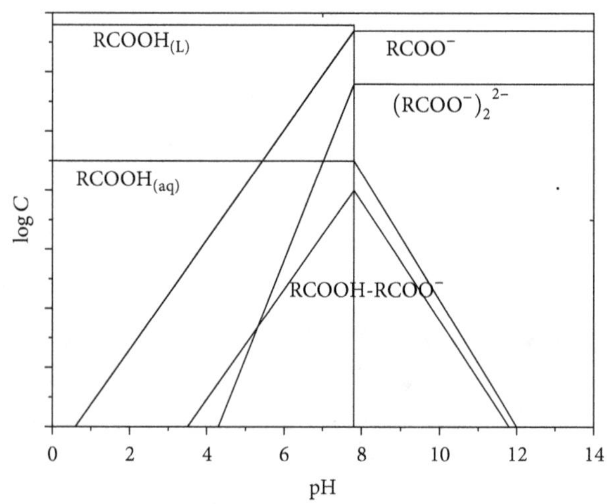

FIGURE 9: Distribution of sodium oleic acid in aqueous solution.

3.3.3. Dissociation Morphology of Sodium Stearate and Sodium Oleate in Water Solution. The molecular formula of sodium stearate is $CH_3(CH_2)_{16}COONa$, while that of sodium oleate is $CH_3(CH_2)_7CH{=}CH(CH_2)_7COONa$ (R-COONa for short). Both of them are composed of long chain of polar hydrophilic carboxyl and nonpolar hydrophobic alkane. They can dissolve in water solution, and then the dissolved components are rehydrated. The reaction formula is as follows:

$$\text{R-COONa} \longrightarrow \text{R-COO}^- + \text{Na}^+ \qquad (10)$$

$$\text{R-COO}^- + \text{H}_2\text{O} \longrightarrow \text{R-COOH} + \text{OH}^- \qquad (11)$$

Dianzuo and Yuehua [34] studied the relationship between the concentration of R-COO and R-COOH in sodium oleate solution and pH value (see Figure 9) [34]. The figure shows that the main morphology of R-COONa is R-COOH in the modification condition of TiO_2, and only few R-COONa have coexistence of R-COOH and $R\text{-}COO^-$ in this work. Therefore, it is believed that the action groups of TiO_2 modified with sodium stearate and sodium oleate are mainly R-COOH, followed by $R\text{-}COO^-$.

3.3.4. Model of Action between Modifier and TiO_2 Surface. The modification mechanism of sodium stearate and sodium oleate can be obtained according to above analysis on surface functional groups of TiO_2 particle and the dissociation morphology of modifiers. Firstly, R-COOH forms a hydrogen bond with OH^- on TiO_2 particle surface for combination by dihydroxylation reaction. The combination can be realized by forming hydrogen bond with C=O and OH^- on TiO_2 surface through $R\text{-}COO^-$. Therefore, the modifier, R-COONa firmly attaches to TiO_2 particle surface to make organic

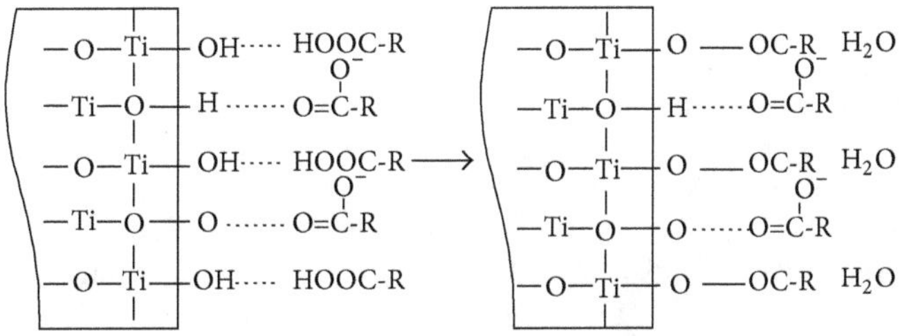

FIGURE 10: Schematic diagram of TiO_2 modified by sodium stearate and sodium oleate.

hydrocarbon chain distribute externally on particle surface. Therefore, TiO_2 has strong hydrophobicity. OH^- on TiO_2 surface is covered by R-COOH and R-COO^- attachment, so the absorption peaks reflect hydroxyl in infrared spectroscopy of TiO_2 after modification.

The reaction model of modifier on TiO_2 particle surface in wet process is established as according to above result analysis (See Figure 10).

4. Conclusions

Surface modification was conducted to TiO_2 particle with sodium stearate and sodium oleate. TiO_2 particle surface turns from hydrophilia to hydrophobicity, while the hydrophobicity degree can be controlled by changing modifier dosage.

In aqueous medium, wetting contact angle of TiO_2 particle greatly enlarges after modifying sodium stearate and sodium oleate. The increase of activation index shows that wettability in water medium becomes poor. The wetting contact angle of modified TiO_2 particle in absolute ethyl alcohol and kerosene is significantly lower than that in water medium, showing the good compatibility of organic matrix.

TiO_2 infrared spectroscopic analysis before and after modification shows that the modifier forms effective attachment on TiO_2 particle surface. With such attachment, inductive hydrophobicity forms on TiO_2 particle surface. It is relatively stable because attachment of modifier has certain chemical properties. The action model of TiO_2 modified with sodium stearate and sodium oleate was established based on the analysis on the functional groups of TiO_2 particle surface and the dissociation morphology of modifiers.

Disclosure

Zhou Hong is the first author.

Conflicts of Interest

The authors declare that they have no conflicts of interest.

Acknowledgments

The work was supported by National Natural Science Foundation of China "Fundamental Researches of Theory and Technology for Self-Assembly Compound of Calcium Carbonate and TiO_2 driven by Particle-Induced Hydrophobic Effect" (Grant no. 51474194).

References

[1] E. Reck and M. Richards, "TiO_2 manufacture and life cycle analysis," *Pigment and Resin Technology*, vol. 28, no. s3, pp. 149–157, 1999.

[2] S.-R. Lu, H.-L. Zhang, C.-X. Zhao et al., "Studies on the properties of a new hybrid materials containing chain-extended urea and SiO_2-TiO_2 particles," *Polymer*, vol. 46, no. 23, pp. 10484–10492, 2005.

[3] A. Chatterjee, "Effect of nanoTiO_2 addition on poly(methyl methacrylate): an exciting nanocomposite," *Journal of Applied Polymer Science*, vol. 116, no. 6, pp. 3396–3407, 2010.

[4] F. Loffredo, I. A. Grimaldi, A. De Girolamo Del Mauro et al., "Polyethylenimine/N-doped titanium dioxide nanoparticle based inks for ink-jet printing applications," *Journal of Applied Polymer Science*, vol. 122, no. 6, pp. 3630–3636, 2011.

[5] C. Xiaoquan and S. Wenhao, "Research progress on the preparation and application of nano titanium dioxide in papermaking," *Paper Science and technology*, vol. 6, pp. 109–114, 2010.

[6] A. Sotto, A. Boromand, R. Zhang et al., "Effect of nanoparticle aggregation at low concentrations of TiO_2 on the hydrophilicity, morphology, and fouling resistance of PES-TiO_2 membranes," *Journal of Colloid and Interface Science*, vol. 363, no. 2, pp. 540–550, 2011.

[7] J. C. Yu, J. Yu, W. Ho et al., "Light-induced super-hydrophilicity and photocatalytic activity of mesoporous TiO_2 thin films," *Journal of Photochemistry and Photobiology A: Chemistry*, vol. 148, no. 1, pp. 331–339, 2002.

[8] Y. Gao, G. Chen, Y. Oli et al., "Study on tribological properties of oleic acid-modified TiO_2 nanoparticle in water," *Wear*, vol. 252, no. 5, pp. 454–458, 2002.

[9] N. Maximous, G. Nakhla, W. Wan et al., "Performance of a novel ZrO_2/PES membrane for wastewater filtration," *Journal of Membrane Science*, vol. 352, no. 1, pp. 222–230, 2010.

[10] N. Maximous, G. Nakhla, W. Wan et al., "Preparation, characterization and performance of Al_2O_3/PES membrane for wastewater filtration," *Journal of Membrane Science*, vol. 341, no. 1, pp. 67–75, 2009.

[11] Z. Wanzhong, Q. Xueliang, Q. Lin et al., "Photocatalytic mechanism of nano titanium dioxide and its application in the treatment of organic wastewater," *Journal of Synthetic Crystals*, vol. 35, pp. 1026–1031, 2006.

[12] C. Hanlin, *Study on Photoelectrochemical degradation of organic pollutants and hydrogen production by modified titania nanotube array electrode*, South China University of Technology, 2015.

[13] L. Daohui, Ji. Jing, L. Tian et al., "Environmental behavior and biological toxicity of nanomaterials," *Science Bulletin*, vol. 23, pp. 3590–3604, 2009.

[14] Li. Shugan and J. Xiaoning, "Treatment method of medical polypropylene nonwoven fabric," *Industrial Textiles*, vol. 29, pp. 30–33, 2011.

[15] H. Park, Y. Park, W. Kim et al., "Surface modification of TiO_2 photocatalyst for environmental applications," *Journal of Photochemistry and Photobiology C: Photochemistry Reviews*, vol. 15, no. 1, pp. 1–20, 2013.

[16] C. Yao, G.-S. Gao, X.-P. Lin, X.-J. Yang, L.-D. Lu, and X. Wang, "Surface modification of nanosized TiO_2 with silane coupling reagent," *Journal of Inorganic Materials*, vol. 21, no. 2, pp. 315–321, 2006.

[17] X. Gang, Z. Huashan, Z. Haodong et al., "Study on surface modification of nano titanium dioxide," *Journal of Kunming University of Science and Technology (Science and Engineering)*, vol. 29, no. 4, pp. 39–42, 2004.

[18] H. Xihai, L. Huimin, L. Fei et al., "Study on surface modification of nano titanium dioxide," *inorganic salt industry*, vol. 44, no. 1, pp. 30–32, 2012.

[19] F. Hui, *Study on Surface Treatment and Application Performance of Rutile Titanium Dioxide*, Jiangsu University, 2007.

[20] H. Ding, S.-C. Lu, Y.-X. Deng, and G.-X. Du, "Mechano-activated surface modification of calcium carbonate in wet stirred mill and its properties," *Transactions of Nonferrous Metals Society of China (English Edition)*, vol. 17, no. 5, pp. 1100–1104, 2007.

[21] H. Ding, H. Yushu, and L. shouci, Ore sodium stearate modified heavy calcium carbonate of non metallic mines, wet ultrafine grinding 1997, (4): 37-40.

[22] H. Ding, H. Zhou, Y. X. Zheng, and M. M. Wang, "Effect of sodium stearate on grinding behaviour of calcium carbonate in wet stirred mill," *Materials Research Innovations*, vol. 17, no. 1, pp. s292–s296, 2013.

[23] Y.-X. Weng, L. Li, Y. Liu, L. Wang, and G.-Z. Yang, "Surface-binding forms of carboxylic groups on nanoparticulate TiO_2 surface studied by the interface-sensitive transient triplet-state molecular probe," *Journal of Physical Chemistry B*, vol. 107, no. 18, pp. 4356–4363, 2003.

[24] S. Mallakpour and E. Nikkhoo, "Surface modification of nano-TiO_2 with trimellitylimido-amino acid-based diacids for preventing aggregation of nanoparticles," *Advanced Powder Technology*, vol. 25, no. 1, pp. 348–353, 2014.

[25] D. Q. Vo, E.-J. Kim, and S. Kim, "Surface modification of hydrophobic nanocrystals using short-chain carboxylic acids," *Journal of Colloid and Interface Science*, vol. 337, no. 1, pp. 75–80, 2009.

[26] Y. Wang, W. Eli, L. Zhang, H. Gao, Y. Liu, and P. Li, "A new method for surface modification of nano-$CaCO_3$ and nano-Al_2O_3 at room temperature," *Advanced Powder Technology*, vol. 21, no. 2, pp. 203–205, 2010.

[27] T. Zhisong, L. Runjing, G. Fen et al., "The effect of coupling agent on the surface modification of nano $CaCO_3$," *Journal of Beijing University of Chemical Technology*, 2004.

[28] W. Shuzhong, M. Yixuan, H. Fuzeng et al., "Study on inorganic nano particles/styrene butadiene latex composite technology," *Encyclopedic Dictionary of Polymers*, vol. 15, no. 4, pp. 437–441, 2002.

[29] L. Qinfu and X. Xu, "Evaluation and Mechanism Study on the effect of surface modification of kaolin in coal series," *Journal of Materials Science*, vol. 14, no. 3, pp. 325–328, 2009.

[30] N. Li, G.-z. Ma, and B.-s. Xu, "Study on modification and properties of calcined kaolin with stearic acid," *Applied Chemical Industry*, vol. 38, no. 8, pp. 1136–1138, 2009.

[31] L. Lihua, "Study on surface modification of zinc borate," *Journal of Tangshan Normal University*, vol. 38, pp. 44–47, 2016.

[32] D. D. Perrin, "Stability constants of metal-ion-complexes," *IUPAC*, vol. 22, no. Chemical Data Series, 1979.

[33] L. huaxiang and W. Xuxu, "TiO_2 surface hydroxyl groups and their properties," *Progress in Chemistry*, vol. 19, no. 5, pp. 665–670, 2007.

[34] W. Dianzuo and H. Yuehua, *Flotation Solution Chemistry*, Hunan science and Technology Press, Changsha, Hunan, 1988.

18

Effects of Mechanical Alloying on Sintering Behavior of Tungsten Carbide-Cobalt Hard Metal System

Ahmet Ozan Gezerman and Burcu Didem Çorbacıoğlu

Department of Chemical Engineering, Faculty of Chemical and Metallurgical Engineering, Yıldız Technical University, Istanbul, Turkey

Correspondence should be addressed to Ahmet Ozan Gezerman; ahmet_ozan@yahoo.com

Academic Editor: Peter Majewski

During the last few years, efforts have been made to improve the properties of tungsten carbides (WCs) by preparing composite materials. In this study, we prepared WC particles by mechanical alloying and investigated the effects of mechanical alloying conditions, such as mechanical alloying time and mechanically alloyed powder ratio, on the properties of 94WC-6Co. According to experimental studies, increasing the mechanical alloying time causes an increase in the density of tungsten carbide samples and a decrease of crystal sizes and inner strength of the prepared materials. With the increase of mechanical alloying time, fine particle concentrations of tungsten carbide samples have increased. It is observed that increasing the mechanical alloying time caused a decrease of the particle surface area of tungsten carbide samples. Besides, the amount of specific phases such as Co_3W_3C and Co_6W_6C increases with increasing mechanical alloying time. As another subject of this study, increasing the concentration of mechanically alloyed tungsten carbides caused an increase in the densities of final tungsten carbide materials. With the concentrations of mechanically alloyed materials, the occurrence of Co_6W_6C and Co_3W_3C phases and the increase of crystallization are observed.

1. Introduction

Composite materials are composed of two or more materials. The properties of these materials, derived from those of the constituents, make them suitable for various applications [1, 2].

The properties of these materials, such as hardness and density, vary according to the industrial applications [3, 4]. In these studies, the density of tungsten carbide materials shows a change between 80% and 95%, while the hardness of these materials is between 580 and 1300 HV. Cemented carbides are hard and tough materials in which micrometer sized carbide particles are surrounded by a binder metal. Cemented carbides are produced by solid-state sintering of carbides and metal powders at temperatures in the range of 1300–1500°C. These materials are used under severe environmental conditions such as in the presence of abrasives (6000 revolutions at a load of 13.6 kg using a 229 mm diameter rubber wheel and dry sand) and under thermal shock (test samples were heated up to 1200°C with a heating speed of 400°C/min and cooled in water), high impact loads (3,000 kg load), and severe fatigue at temperatures as high as 1000°C. WC-Co is a hard metal cemented carbide that can be used under these conditions owing to its physical and mechanical properties. Besides the high abrasion and corrosion resistance, these materials show high temperature resistance, which is why they are used in several industries [5, 6].

In the last few years, in the electronic industry, the increased requirement of WC particles has resulted in the need for enhancing the properties of the product, such as 1500°C temperature, 500°C/min heating speed, and 4000 kg impact load [7–9]. Therefore, synthesis and sintering of WC particles have been adopted in several industries such as metal and building industries that use high temperature (>1500°C) and corrosion (95% salt spray conditions) and abrasive conditions (6000 revolutions at a load of 13.6 kg using a 229 mm diameter rubber wheel and dry sand) [10–12]. It is convenient to use WC particles with sizes in the nanometer range (<10 μm) in industrial processes such as metal industries [13–15]. Sintering below 500°C for 24 h can afford WC particles that have a maximum particle size of 0.2 μm as pointed out by Canakci et al. and Joo et al. [16, 17]. According to these studies, at 250–450°C, decreasing of the particle size ensured obtaining materials that have more densities (<92%) at sintering temperatures (500°C). Decreasing the WC particle size has an important advantage in that it yields a large particle surface area during high-temperature processes [18]. In those studies, increasing the surface area of particles depending on the decrease of particle sizes ensured that tungsten carbide materials have higher densities (85%) at high-temperature processes (<800°C).

An increase in the carbide amount during the sintering of WC particles has a negative impact on the abrasion resistance of the resulting material [15]. In that study, the concentration of the tungsten carbide materials is 10% maximum and abrasion resistance is measured up to 1000 kg vertical load. In addition, the C/W ratio in the WC particles plays an important role in the useful properties of the final material [19].

According to this, we obtained tungsten carbide materials that have density of 90%, at C/W ratio of 20–35% and 1000 HV hardness, with 5-year corrosion resistance in salt spray testing.

In this study, the effect of sintering on a 94WC-6Co system was investigated. For preparing the powder, mechanical alloying was used for grinding, and the effects of mechanical alloying on the powders sintered using different methods were investigated.

2. Experimental Procedure

In this study, WC powder materials that contain W/C ratio of 0–7% were supplied from Eurotungstene™.

In this study, the sintering behavior, microstructure, and mechanical properties of a 94WC-6Co hard metal system produced by using mechanical alloying were investigated. The powder samples used in this study were prepared by using two different methods.

In this study, two different methods were adopted to understand the behavior of mechanically alloyed 94WC-6Co samples at different concentrations and different lengths of time.

In the first method, the effect of alloying time on mechanically alloyed 94WC-6Co powders was investigated. The powder used in the second method was obtained by adding the 94WC-6Co powder mixture mechanically alloyed for 24 h to the 94WC-6Co powder mixture that was not mechanically alloyed at different concentrations (0, 25, 50, 75, and 100%). The powders prepared by the two methods were characterized by using X-ray diffraction (XRD), scanning electron microscopy (SEM), and particle size analysis. The as-prepared cylindrical samples were uniaxially cold-pressed at 250 MPa and were sintered at 1450°C under industrial conditions. The sintered samples were characterized using XRD, SEM, and hardness measurements.

In the first method, the density values after sintering were measured for 94WC-6Co to determine physical changes. In the second method, for increasing the hardness and density, materials that were not mechanically alloyed were added to the 94WC-6Co material at different concentrations (25%, 50%, 75%, and 100%); the maximum density was obtained without the additive after 24 h mechanical alloying. Changes in the hardness and density were monitored.

Mechanical alloying was carried out at 1200 rpm using an 8000D™ Spex grinder. A WC grinder vessel and balls (6.35 mm) were used. The ball-to-powder ratio was fixed at 10 : 1. The amount of impurities added to the powder mixture from the grinder (grinder vessel and balls) was determined by weighing the powder mixture before and after alloying. In order to prevent contamination of the powder mixture during mechanical alloying, the powders were shifted to a vacuumed room (Plas-Labs™) before grinding. Ar gas was used for creating an inert atmosphere.

The mechanically alloyed powder mixtures were uniaxially pressed using a hydraulic press with 12.7 mm diameter molds at 250 MPa. All the prepared samples were sintered under vacuum at 1450°C.

The particle size distribution of all the powders was measured by using a Malvern Instruments Mastersizer 2000 apparatus. Particle size measurements were conducted in pure water supplied from a commercial Elga® device.

The powders are agitated in a Bandelin SONOREX™ ultrasonic bath to separate the agglomerated particles.

The phase analysis of the sintered samples prior to mechanical alloying was carried out by using an X-ray diffractometer (Bruker™ D8 Advance 40 kV and 40 mA) using CuKα radiation. The crystalline phases of all the green materials were identified by X-ray diffractometry (XRD) using a Bruker D8 Advanced Series diffractometer with CuKα (1.54060 Å) radiation in the 2θ range of 10–90°, with increments of 0.02° at a rate of 2°/min. International Center for Diffraction Data® (ICDD) powder diffraction files were utilized for the crystalline phase identification of the green materials. As seen in Figure 1, with the increase in the mechanical alloying time for the 94WC-6Co material, the concentrations of the Co_6W_6C and Co_3W_3C phases decreased. It is desirable to avoid the formation of these phases under the adopted reaction environment because they affect the hardness and density of the final material.

The sintered samples were placed on hot bakelite (Struers™ ThermoFast Black resin with Struers LaboPress-1 bakelite). The samples were then polished by Struers Tegrapol-15 automatic polishing device for microstructural observations and hardness measurements.

Electron backscatter diffraction (EBSD) analysis was performed for investigating the effects of mechanical alloying treatment on the material surface. EBSD patterns are generated within a small interaction volume at the surface of a sample, with a penetration depth of less than 50–100 nm. When considering how to prepare a sample for EBSD analysis, it is important to recognize that, in order to obtain high quality patterns, the surface deformation that is typically introduced during standard metallographic preparation is minimized.

On the other hand, microhardness measurements were performed after the polishing. The measurements were conducted by Vicker's test using a Shimadzu™ microhardness test device, under a load of 1000 g for a period of 15 s. For microhardness measurements, three indentations were made and measured at different points on each specimen. The average

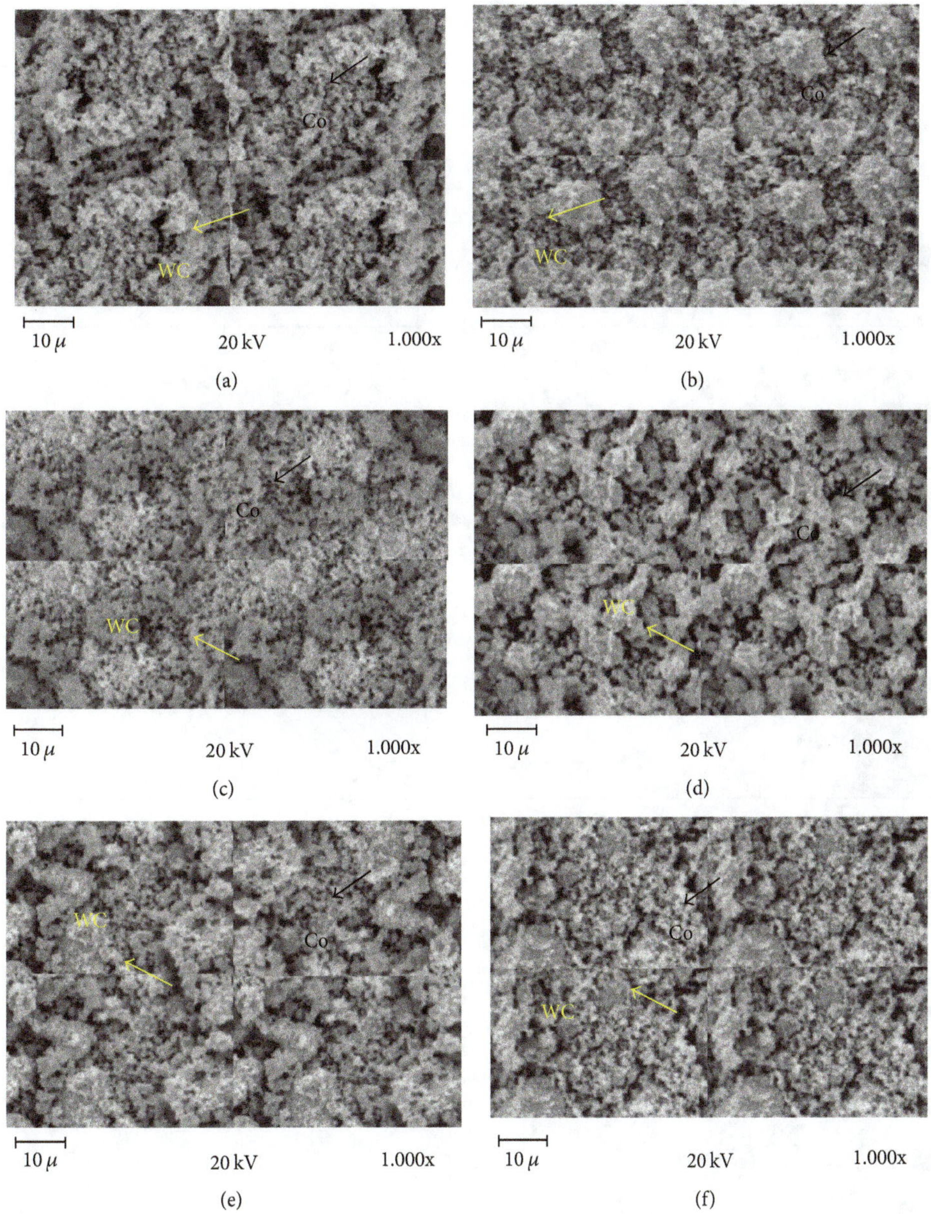

FIGURE 1: SEM images of the mechanically alloyed samples for (a) 20 min, (b) 1 h, (c) 3 h, (d) 6 h, (e) 12 h, and (f) 24 h.

value for each specimen was then determined. Microstructural characterization of all the precursors, mechanically alloyed powders, and sintered samples was carried out using a JEOL™ JSM-T330 electron microscope.

Scanning electron micrographs were obtained using a JEOL JSM-T330 electron microscope (SEM). Samples were prepared by placing a drop of the latex dispersion onto a brass surface, in a dust-free box at room temperature. The samples were then coated with gold, and scans were recorded at angles of 0° and 45° to the vertical direction.

3. Results and Discussion

3.1. Characterization of Powders. The powder prepared with the composition of 94WC-6Co was mechanically alloyed for different durations. The effect of mechanical alloying time on the sintered samples was investigated by observing the formation of powders and new phases and particle size distribution in the sample. The SEM images of the 94WC-6Co powders mechanically alloyed for different durations are shown in Figure 1. In the case of the powder sample subjected to mechanical alloying for 20 min, the particle size of WC decreased, and the Co particles were distributed homogeneously (Figure 1(a)). Therefore, a duration of 20 min was found to be optimum for creating homogeneous powders. The SEM images reveal that the powder mechanically alloyed for 1 h (Figure 1(b)) contained agglomerated particles, while the particles were distributed in the powder that was subjected to mechanical alloying for 3 h (Figure 1(c)). The

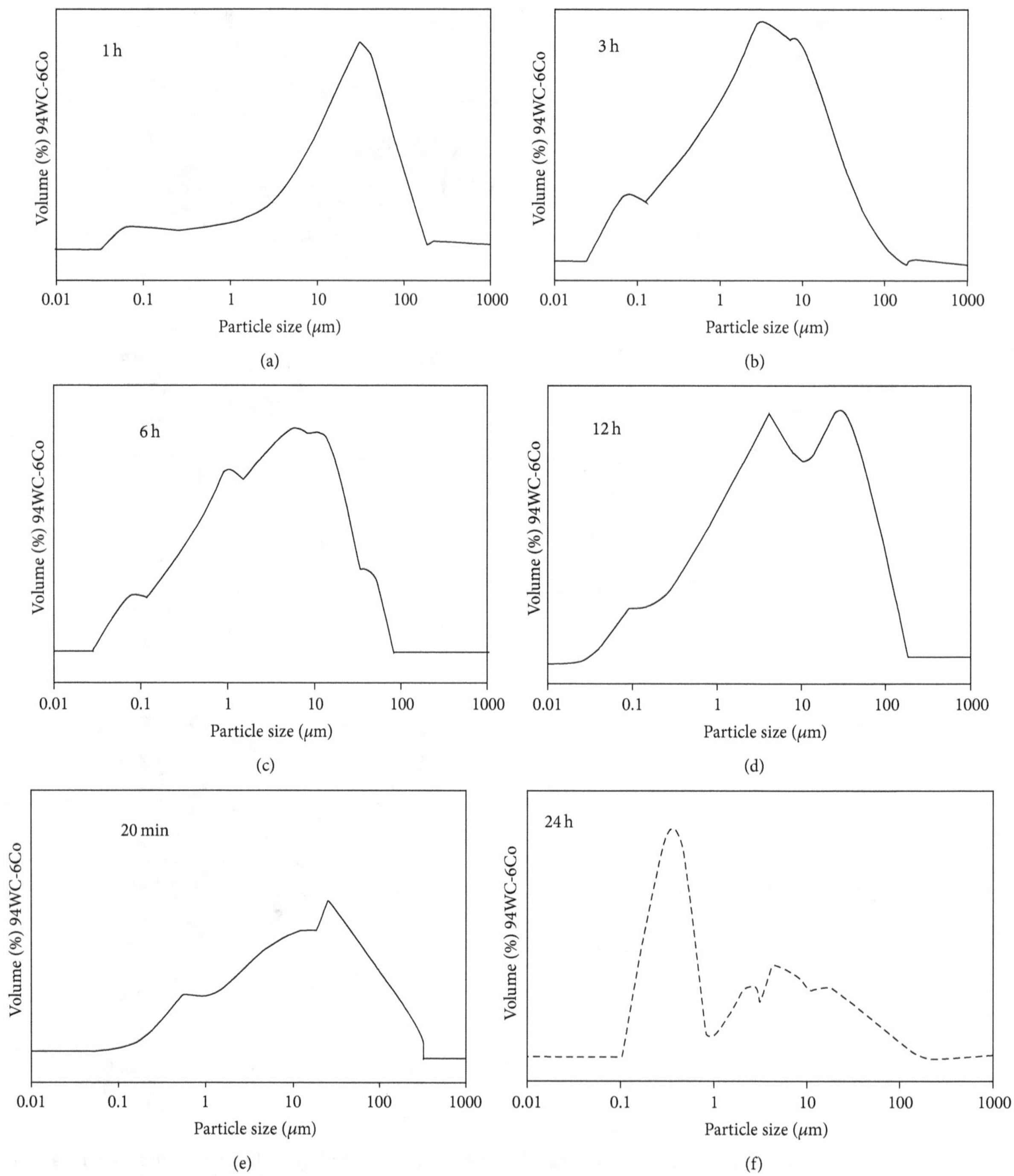

FIGURE 2: Particle size distribution of the powders subjected to mechanical alloying for (a) 1 h, (b) 3 h, (c) 6 h, (d) 12 h, (e) 24 h, and (f) 20 min. Comparison of the particle size distribution of the powders mechanically alloyed for 20 min, 12 h, and 24 h.

powder mechanically alloyed for 6 h (Figure 1(d)) showed the highest agglomeration. The powders subjected to mechanical alloying for 12 (Figure 1(e)) and 24 h showed similar particle distribution.

These results are also representative of the metallographic nature of the samples. The surfaces of the samples are seen in the figure.

Figure 1 also shows the microstructural formation of the WC and Co particles. The green arrow in the SEM images indicates the WC and Co particles. The larger particles comprise WC, and smaller agglomerated particles comprise the Co particles.

The particle size distributions of the mechanically alloyed powders are given in Figure 2. The average particle sizes are 13 μm, 6.3 μm, and 2.9 μm after mechanical alloying for 20 min, 12 h, and 24 h, respectively. According to these results, the average particle size of the powders decreased with increasing alloying time.

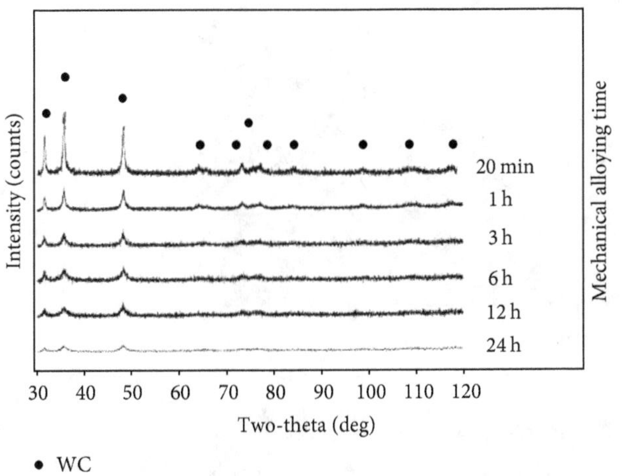

FIGURE 3: XRD diagrams of the powders mechanically alloyed for different durations.

None of the mechanically alloyed particles showed the formation of a new phase. The XRD patterns of the powders mechanically alloyed for different durations of time (20 min, 1 h, 3 h, 6 h, 12 h, and 24 h) are shown in Figure 3. The peak heights decreased and the peak widths increased with an increase in the mechanical alloying time, indicating that the crystal size of the powders decreased with an increase in the mechanical alloying time. As seen in Figure 3, when the mechanical alloying time increases (24 min, 1 h, 3 h, 3 h, 6 h, 12 h, and 24 h), the peak heights increased and peak widths decreased. For example, peak height for 24 min alloying is smaller than that for 1 h and the sum of the peak widths corresponding to 24 min alloying is larger than that for 1 h alloying.

The SEM images of the powders obtained by adding the 94WC-6Co powder mixture mechanically alloyed for 24 hours to the 94WC-6Co powder mixture that was not mechanically alloyed at different concentrations (25%, 50%, 75%, and 100%) are shown in Figure 4. The powder prepared by mechanical alloying for 24 h contained fine particles, although a slight degree of agglomeration was observed. In other words, according to the figure, as the concentration of the powder increases, the particle size decreases and an agglomerated structure is formed because of the increase in the number of fine particles.

An increase in the mechanical alloying time decreases the crystal size. Prolonged mechanical alloying with the effects of heat can lead to a decrease in the crystal size and increase in the surface area, so that the material properties can be controlled better.

By using SEM images of the WC particles in the second method, the particle effects on the samples were investigated. Mechanically alloyed 94WC-6Co particles at different concentrations (25%, 50%, 75%, and 100%) are shown in Figure 4. Because mechanical alloying for 24 h is sufficient for decreasing the particle size, all the powder mixtures prepared by using the second method have fine particles that result in an agglomerated structure. According to Figure 4, with an increase in the number of fine particles, the powder concentration increases, particle size decreases, and an agglomerated structure is formed. The particle sizes obtained for different mechanical alloying times are shown in Figure 2. According to these data, with increasing alloying time, the particle size distribution decreases.

3.2. Characterization of Sintered Samples

3.2.1. Effect of Mechanical Alloying Time on Sintered Samples. In this method, the effect of the mechanical alloying time on the sintering behavior of the 94WC-6Co powder mixtures was investigated. Figure 5 shows the effect of mechanical alloying on the density of the powders. With an increase in the mechanical alloying time, the size distribution of the powders decreased and their surface area increased. Hence, their green density decreased with an increase in the mechanical alloying time. The powders mechanically alloyed for 6, 12, and 24 h showed final condensation.

Figure 5 shows the effect of mechanical alloying time on the density values. With an increase in the alloying time, the particle size and density of the powder decrease but the surface area increases. The density of the samples sintered for 6, 12, and 24 h indicated full condensation. Besides, according to Figure 5, the green density reaches the maximum value (100.46%) and then decreases suddenly, which leads to abrupt changes in the microstructure, such as cracking.

The green densities of the pressed samples were determined by measuring the volumetric sizes. The Archimedes method was used for measuring the densities of the sintered samples. In this method, the samples were weighed first in air and then in liquid; the difference between the two weights was multiplied by the liquid density. In this study, to prevent oxidation, ethyl alcohol was used as the liquid.

The SEM images of the sintered 94WC-6Co samples mechanically alloyed for different durations are shown in Figure 6. The sample mechanically alloyed for 20 min contained fine particles. The mechanically alloyed samples had bigger particles as compared to the samples that were not mechanically alloyed (Figure 6(a)), and had a conventional hard metal structure [20]. The WC particles in the sample mechanically alloyed for 12 h grew in a direction different from those in the sample mechanically alloyed for 24 h (Figures 6(b) and 6(c)).

Because of the prolonged mechanical alloying (12 h, 24 h) with heating, it is reasonable to expect the WC-Co particles to move in different directions. The motion of the particles in different directions (Figures 6(c) and 6(d)) during mechanical alloying provided the desired level of hardness to the sample. In other words, movement in these directions increases the material hardness.

Note that the directions between the particles are evident for these three alloying times (20 min, 12 h, and 24 h); hence, the corresponding SEM images have been considered.

The XRD patterns of the samples mechanically alloyed for different durations are shown in Figure 7. In the case of the sample mechanically alloyed for 20 min, the Co_6W_6C and Co_3W_3C phases were not observed. As the mechanical alloying time increased, the formation of the CO_6W_6C phase increased. The Co_6W_6C and Co_3W_3C phases were observed only in the sample mechanically alloyed for 24 h, because of

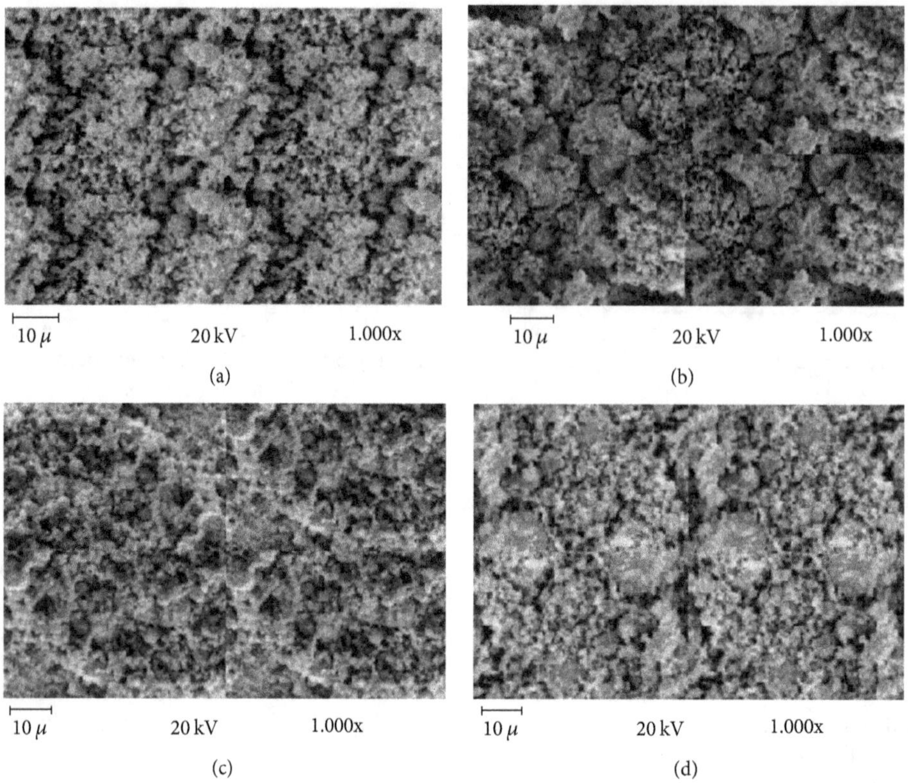

FIGURE 4: SEM images of the powders prepared by mechanical alloying for 24 h at different concentrations: (a) 25%, (b) 50%, (c) 75%, and (d) 100%.

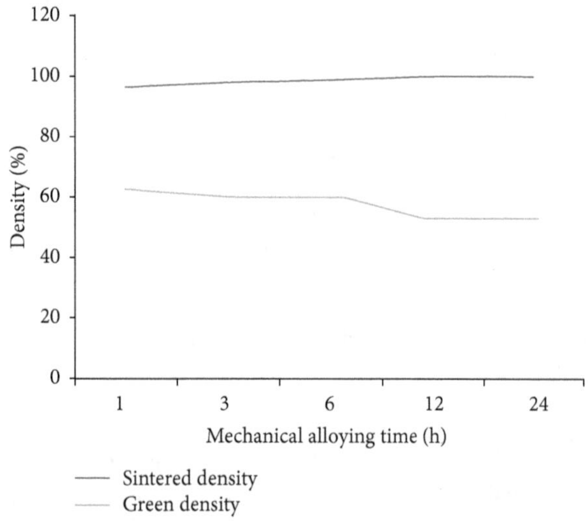

FIGURE 5: Variation in the green and sintered densities with mechanical alloying time.

the oxygen contamination during the mechanical alloying. Oxygen removes carbon from the samples during sintering, thus resulting in the formation of these phases by decreasing the C/W ratio [20]. The XRD patterns and SEM images reveal that the Co_6W_6C and Co_3W_3C phases contained WC particles that grew in the same direction. Therefore, the difference in the carbon concentration in the samples mechanically alloyed for 24 h led to the variation in their microstructures.

The results of the hardness analysis (Figure 8) suggest that the hardness of the samples increased with an increase in the mechanical alloying time.

As per the aim of this study, the density and hardness of the material were determined. From the data in Figures 5 and 8, the expected and desired density and hardness values were obtained. Moreover, because the concentration of agglomerated fine particles during sintering and their densities increased, the material hardness is expected to increase. As seen in Figure 3, with an increase in the mechanical alloying time for the 94WC-6 Co material, the concentrations of the Co_6W_6C and Co_3W_3C phases decrease. Because these phases affect the hardness and density of the material, their formation under the given reaction conditions must be suppressed.

According to the sintering data in Figure 5, increasing the mechanical alloying time causes an increase in the density and hardness of the material, as expected. As mentioned above, the density of the material increases with an increase in the alloying temperature and alloying time; this is because if mechanical alloying is continued after cracking of the microstructure, the hardness of the material decreases, so that the high density does not have any positive effect on the material properties.

3.2.2. Effect of Powder Concentration on Sintered Samples. In this method, the effect of powder concentration on the sintered samples subjected to mechanical alloying for 24 h was investigated. From the XRD patterns of the powder samples

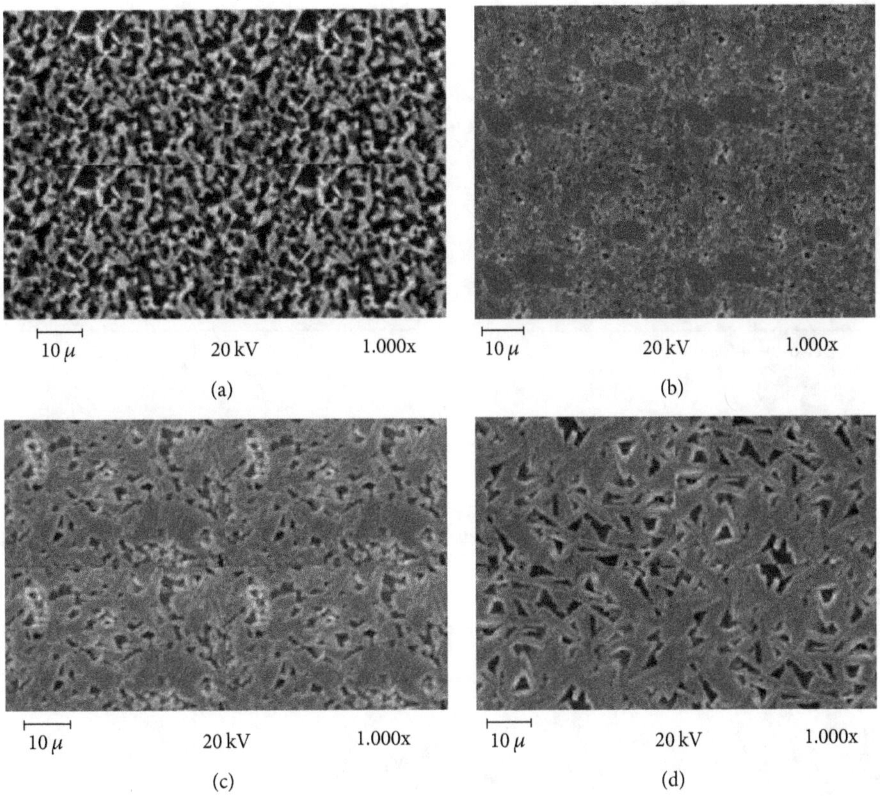

FIGURE 6: (a) SEM images of the unalloyed sample. SEM images of the samples sintered for different mechanical alloying times: (b) 20 min, (c) 12 h, and (d) 24 h.

mechanically alloyed for 24 h, it can be seen that as the crystal size of the powders decreased, the peak height decreased significantly, while the peak width increased. Also, because a higher density was obtained after mechanical alloying for 24 h, this alloying time is adopted for studying the effect of powder concentration on the material characteristics. For sintering processes, the samples were prepared by pressing under 250 MPa and heating at 1450°C. In the first-level experiments, the effect of alloying time on the green and sintered densities of the prepared 94WC-6Co samples was verified. Since the highest density was obtained after alloying for 24 h, this alloying time was adopted in the second-level experiments, wherein different amounts of nonmechanically alloyed 94WC-6Co were added to the 94WC-6Co samples alloyed for 24 h. The green densities and sintering densities were determined after alloying for 1, 3, 6, and 12 h without the use of additives. The maximum density was observed in the case of the sample alloyed for 24 h with 0% additive. The rate of change of density was better controlled for the samples to which different concentrations of 94WC-6Co powders (0, 25, 50, 75, and 100%) were added and alloyed for 24 h. According to XRD results (Figure 9), the width of the peak for the Co_3W_3C phase in the sample containing the 94WC-6Co powder (25%) mechanically alloyed for 24 h decreased as the sample began to crystallize. The Co_3W_3C phase is created only after 100% mechanical alloying of 94WC-6Co for 24 h. The amount of mechanically alloyed powder in this sample was greater than that in all the other samples. When the mechanical alloying time increased, the oxygen contamination also increased. Therefore, the carbide concentration was reduced significantly. The XRD patterns revealed that the Co_6W_6C phase was created before the loss of carbide. The decrease in the carbide concentration allowed the formation of the Co_3W_3C phase. The main materials used initially and the required additives are placed in a mixer and grinder. A vibrating shaft grinder was used. Powders are ground in order to separate the agglomerated particles and minimize the particle size. The grinder type can be chosen as required, and the rotation speed, ball diameter, powder-to-ball ratio, and hard metal content can be optimized. In the final stage of the grinding operation, the particle size used in the industry is 0.5–1.0 μm; it decreases to 100–150 nm when using more specific grinders. Although a wide range of particle sizes can be obtained upon grinding, with this method, the prepared mixture is essentially homogeneous, and particle agglomeration would not occur during sintering. Because the highest hardness (Figure 10) and density were obtained after 24 h of mechanical alloying, we studied the 24 h alloyed samples in the second method (Figure 11).

The XRD patterns of the sintered samples are shown in Figure 9. The peak width of the Co_6W_6C phase in the sample (mechanically alloyed for 24 h) containing 25% (w/w) 94WC-6Co increased with an increase in the mechanical alloying time.

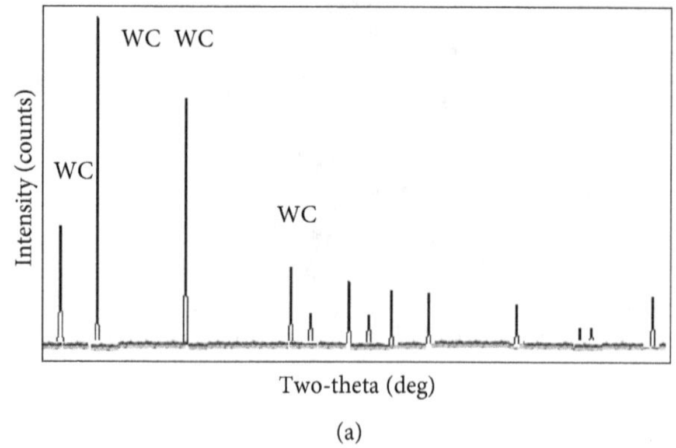

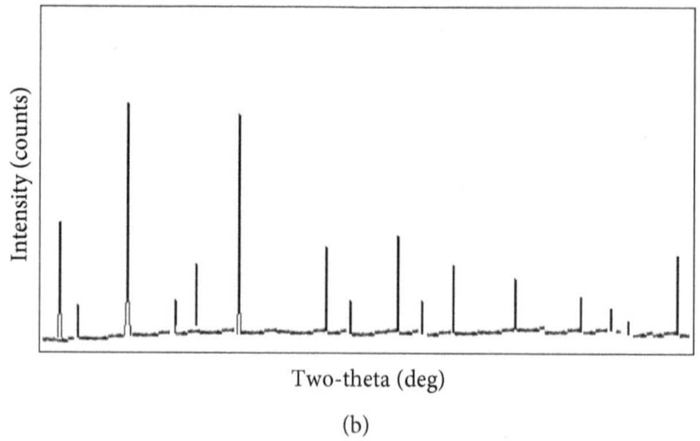

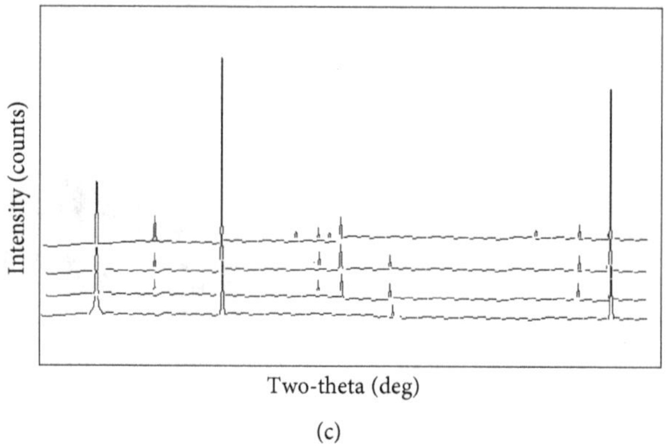

FIGURE 7: XRD patterns of the sintered samples mechanically alloyed for (a) 20 min and (b) 24 h; (c) comparison of the samples mechanically alloyed for 1, 3, 6, and 24 h.

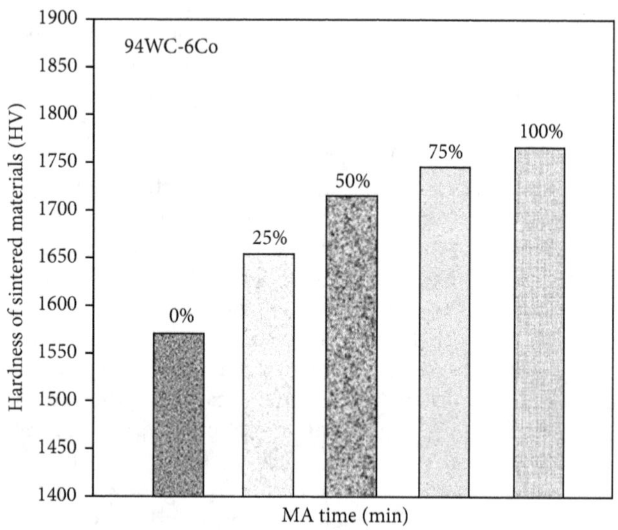

FIGURE 8: Hardness analysis of the sintered samples.

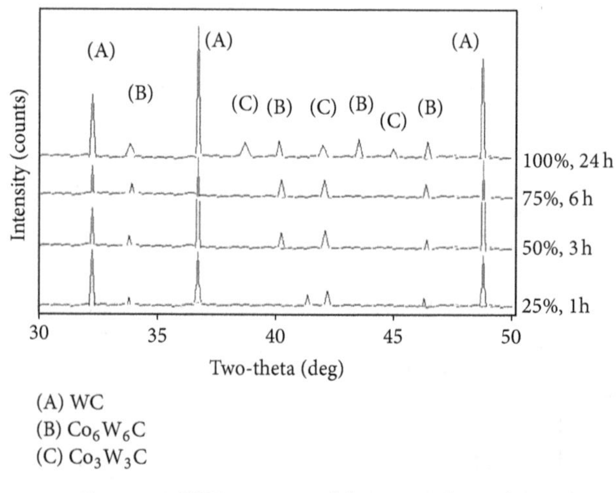

FIGURE 9: XRD patterns of the sintered samples.

The SEM images of the sintered samples with varying concentrations are shown in Figure 12. The sample (mechanically alloyed for 24 h) containing 0% powder was obtained by grinding the precursor powders with the sample mechanically alloyed for 20 min. The SEM image (Figure 12(a)) and XRD pattern of the sample containing 0% powder subjected to mechanical alloying for 24 h reveal the presence of 1–3 μm sized WC particles in the sample. All the observed particles were WC phases. The samples containing 25 and 0% mechanically alloyed powders (Figure 12(b)) had similar structures. With an increase in the concentration of the mechanically alloyed powders, the WC concentration decreased. The XRD patterns reveal that the Co_6W_6C phase was created in the samples containing 50% (Figure 12(c)) and 75% (Figure 12(d))

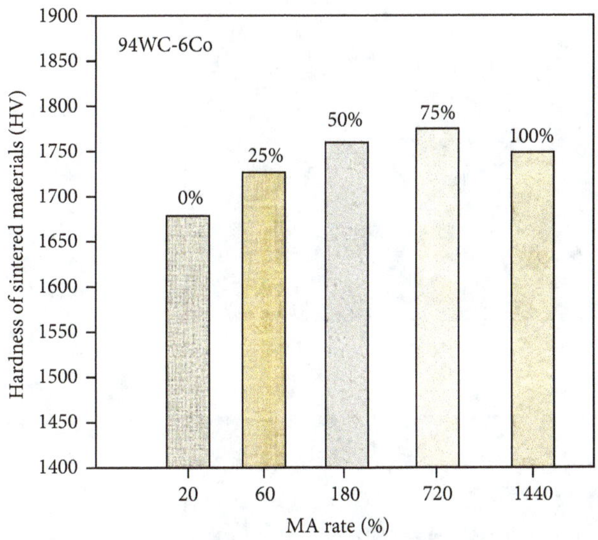

FIGURE 10: Hardness measurements of the sintered samples with varying concentrations of mechanically alloyed powders.

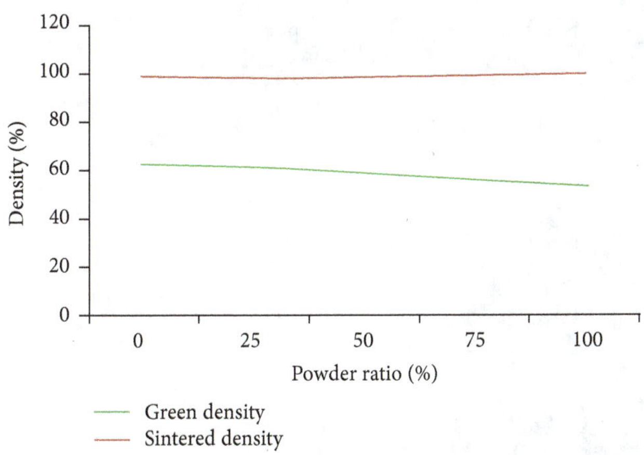

FIGURE 11: Relative density values of the prepared samples at different powder concentrations at a 24 h mechanical alloying time.

mechanically alloyed powder. The C/W ratio that causes the formation of this phase affects the formation of WC. In the sample containing 100% mechanically alloyed powder (Figure 12(e)), the growth of WC particles (light grey) was observed. On the basis of these results, it can be concluded that an increase in the mechanical alloying ratio affects the density positively. It was also found that the carbide concentration affected the microstructure of this system significantly.

In this method, the effect of powder concentration on the sintering behaviors of the samples mechanically alloyed for 24 h was investigated. In the first method (effect of mechanical alloying time on the sintered samples), the XRD patterns of the powder samples mechanically alloyed for 24 h revealed that the peak height decreased significantly and the peak width increased with a decrease in the crystal size of the powder. The 94WC-6Co samples were prepared using the beginning powders at different concentrations and equal composition of the powder mechanically alloyed for 24 h. The powders so prepared were pressed at 250 MPa followed by sintering at 1450°C.

The sample containing 0% powder and mechanically alloyed for 24 h was prepared using the initial powder that was alloyed for 20 min. The sample containing 0% powder contained 1–3 μm sized WC particles. All the observed particles were in the WC phase. The samples with 25 and 0% mechanically alloyed powder had similar microstructures. The samples containing 50 and 75% mechanically alloyed particles experienced deformation with an increase in the concentration of the mechanically alloyed powder and had the Co_6W_6C phase.

The formation of the Co_6W_6C phase, which depended upon the C/W ratio, led to the formation of WC particles. In the sample containing 100% mechanically alloyed powder, the WC particles grew in a detected way. On the basis of the results obtained, we can state that the increasing mechanical alloying ratio affected the density values positively. Moreover, the carbide concentration significantly affected the microstructure of this system. With an increase in the concentration of the mechanically alloyed powder, the hardness of the samples also increased.

4. Conclusions

In this study, an increase in the mechanical alloying time resulted in a decrease in the peak height and an increase in the peak width of the XRD peaks of the samples. The increase in the mechanical alloying time resulted in a decrease in the crystal size and internal strength of the powders, thus causing deformation.

With an increase in the mechanical alloying time (especially after 24 h), the fine particle concentration increased, and agglomeration between the fine particles was observed. The SEM images of the powders revealed that the particle size decreased with an increase in the mechanical alloying time.

The green density of the samples decreased with an increase in the mechanical alloying time, because of the decrease in the surface area of the powders with an increase in the mechanical alloying time. With an increase in the mechanical alloying time, the particle size of the powders decreased, while their surface area increased. The formation of the Co_3W_3C phase was observed in all the sintered samples. When mechanical alloying was carried out for 24 h, the Co_6W_6C phase was also created along with the Co_3W_3C phase.

The sample containing the powder mechanically alloyed for 20 min had fine WC particles. With an increase in the mechanical alloying time, a change in the size of the WC particles was observed. It was also found that when the concentration of the mechanically alloyed powder (24 h) increased, the green density of the samples decreased, while their sintered densities increased. The phase analysis of the samples revealed that the concentration of the Co_6W_6C phase in the sample containing mechanically alloyed powder (24 h mechanical alloying) increased as the ratio of the mechanically alloyed powder increased from 25 to 50–75% and started crystallizing. In the sample containing the powder

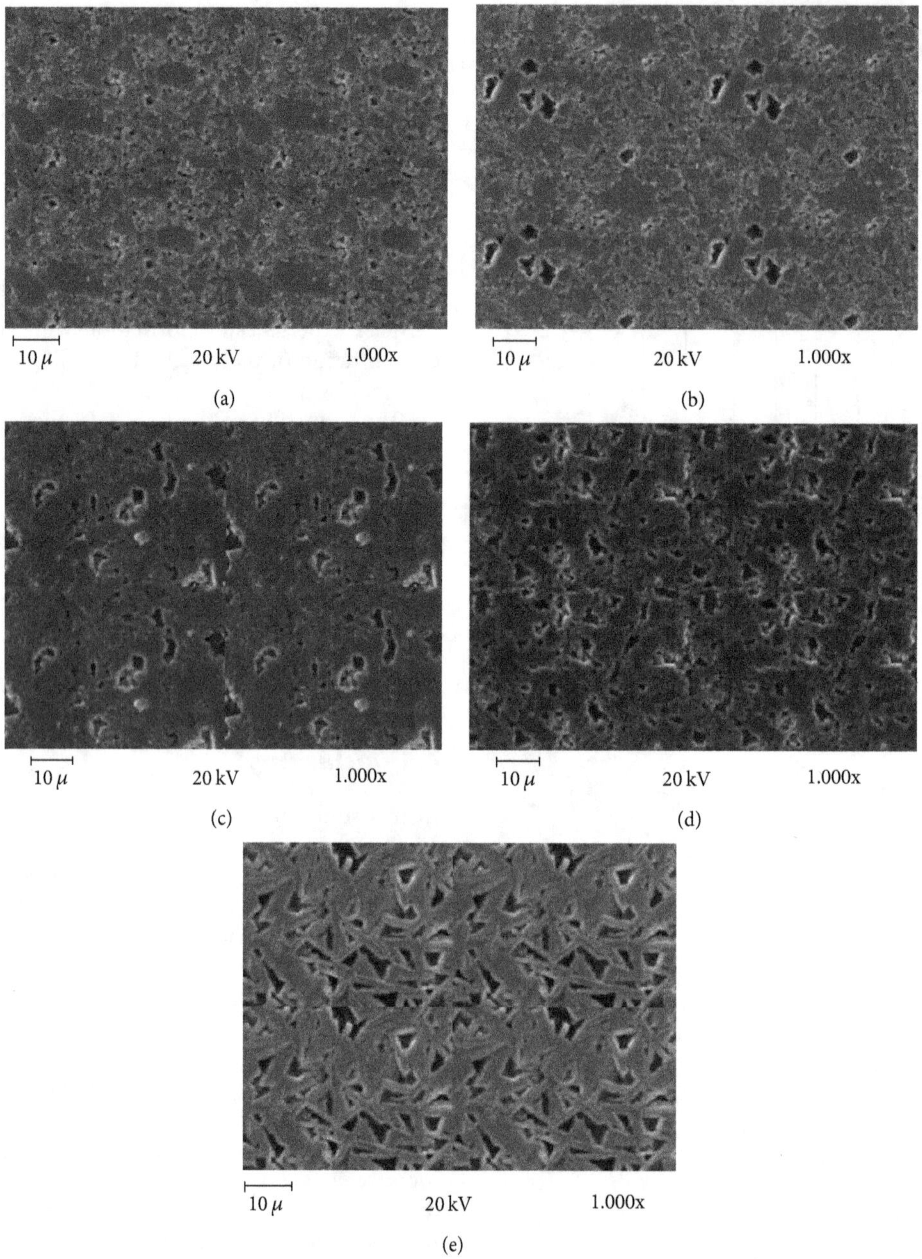

Figure 12: SEM images of the sintered samples with varying concentrations of mechanically alloyed powder: (a) 0, (b) 25, (c) 50, (d) 75, and (e) 100%.

mechanically alloyed for 24 h, both the Co_6W_6C and the Co_3W_3C phases were formed, and their XRD peaks were intense and narrow, indicating that the crystal size of both of these phases increased.

The SEM images of the powder samples that were mechanically alloyed for 24 h showed that WC particles were deformed at 0 and 25% powder concentration. The WC particles in the sample containing 100% mechanically alloyed powder grew in a given direction.

Conflicts of Interest

The authors declare that there are no conflicts of interest regarding the publication of this article.

Acknowledgments

The authors would like to thank Sabahattin Gezerman for his technical support.

References

[1] L. A. Carlsson, F. A. Donald, and R. B. Pipes, "Analysis of composite materials," in *Experimental Characterization of Advanced Composite Materials*, CRC Press, Boca Raton, Fla, USA, 2014.

[2] G. Daniel, "Composite materials: interest and physical properties," in *Composite Materials: Design And Applications*, CRC press, Boca Raton, Fla, USA, 2014.

[3] T. Louis, *Refractory Materials: A Series of Monographs ,Transition Metal Carbides And Nitrides*, University of Minnesota, New York, NY, USA, 1971.

[4] R. M. Raihanuzzaman, Z. Xie, S. J. Hong, and R. Ghomashchi, "Powder refinement, consolidation and mechanical properties of cemented carbides—an overview," *Powder Technology*, vol. 261, pp. 1–13, 2014.

[5] G. Wahl, P. B. Davies, R. F. Bunshah et al., *Ullmann's Encyclopaedia of Industrial Chemistry, Thin Films*, VCH Publishers, Weinheim, Germany, 2014.

[6] J. Yang, M. Odén, M. P. Johansson-Jõesaar, and L. Llanes, "Grinding effects on surface integrity and mechanical strength of WC-Co cemented carbides," *Procedia CIRP*, vol. 13, pp. 257–263, 2014.

[7] O. S. Manakova, A. E. Kudryashov, and E. A. Levashov, "On the application of dispersion-hardened SHS electrode materials based on (Ti, Zr)C carbide using electrospark deposition," *Surface Engineering and Applied Electrochemistry*, vol. 51, no. 5, pp. 413–421, 2015.

[8] X. Zhuang, Y. Zhang, Y. Zhu et al., "The effects of Pd and/or Zr additives on the structures and cyclic stabilities of $Mg_{50}Ni_{50}$-based electrode alloys," *International Journal of Hydrogen Energy*, vol. 40, no. 6, pp. 2768–2774, 2015.

[9] W. H. Zhong, B. Li, R. G. Maguire et al., *Synthesis, Manufacturing and Industry Impacts, Nanoscience and Nanomaterials*, DEStech Publications, Lancaster, Pa, USA, 2012.

[10] I. Konyashin and B. Ries, "Wear damage of cemented carbides with different combinations of WC mean grain size and Co content. Part I: ASTM wear tests," *International Journal of Refractory Metals and Hard Materials*, vol. 46, pp. 12–19, 2014.

[11] B. Yaman and H. Mandal, "Wear performance of spark plasma sintered Co/WC and cBN/Co/WC composites," *International Journal of Refractory Metals and Hard Materials*, vol. 42, pp. 9–16, 2014.

[12] J. Oberste-Berghaus, S. Bouaricha, J. G. Legoux, C. Moreau, and B. Harvey, "Method and apparatus for fine particle liquid suspension feed for thermal spray system and coatings formed therefrom," U.S. Patent 8,629,371, 2014.

[13] A. Jiang, Q. Li, and F. Yang, "Effects of combined addition of Cr_3C_2 and Nbc on microstructure and properties of cemented carbides prepared by WC-Co composite powder," *Materials Science Forum*, vol. 849, pp. 794–800, 2016.

[14] Y. A. Kukushkina, V. V. Sokolov, and M. V. Tomkovich, "Nanoporous carbon prepared by thermochemical treatment of carbide with chlorine," *Russian Journal of Applied Chemistry*, vol. 87, no. 10, pp. 1517–1523, 2014.

[15] I. Konyashin, S. Hlawatschek, B. Ries, F. Lachmann, and M. Vukovic, "Cobalt capping on WC-Co hardmetals. Part I: a mechanism explaining the presence or absence of cobalt layers on hardmetal articles during sintering," *International Journal of Refractory Metals and Hard Materials*, vol. 42, pp. 142–150, 2014.

[16] A. Canakci, T. Varol, H. Cuvalci, F. Erdemir, S. Ozkaya, and E. D. Yalcin, "Synthesis of novel CuSn10-graphite nanocomposite powders by mechanical alloying," *Micro and Nano Letters*, vol. 9, no. 2, pp. 109–112, 2014.

[17] H. G. Joo, D. Q. Zhang, L. X. Gao, T. Liu, and K. Y. Lee, "Corrosion resistance of tungsten carbide-based cermet coatings formed by sinter-brazing," *International Journal of Surface Science and Engineering*, vol. 8, no. 4, pp. 380–391, 2014.

[18] H. Rana, G. Shree, and A. Panwar, "Optimization of abrasive wear characteristics of tungsten carbide and chromium carbide based coating on H-13 steel using detonation gun process by taguchi method," *International Journal of Advanced Materials Manufacturing and Characterization*, vol. 1, no. 1, article 121, 2014.

[19] V. Bounhoure, S. Lay, F. Charlot, A. Antoni-Zdziobek, E. Pauty, and J. M. Missiaen, "Effect of C content on the microstructure evolution during early solid state sintering of WC-Co alloys," *International Journal of Refractory Metals and Hard Materials*, vol. 44, pp. 27–34, 2014.

[20] D. Peştreli, "Effect of mechanical alloying on the sintering behaviour of Wc-co hardmetal system [M.Sc Thesis]," Tech. Rep., Istanbul Technical University, Institute of Science and Technology, Istanbul, Turkey, 2009.

19

Preparation of Dicalcium Phosphate Anhydrous (Monetite) Biological Coating on Titanium by Spray-Drying Method

Xingling Shi,[1,2] Chao Zhao,[3] Lingli Xu,[1] and Qingliang Wang[4]

[1]*School of Materials Science and Engineering, Jiangsu University of Science and Technology, Zhenjiang 212003, China*
[2]*TusPark (Kunshan) Development Co., Ltd., Kunshan 215347, China*
[3]*Hitachi Automotive System (Guangzhou) Ltd., Guangzhou 511300, China*
[4]*School of Materials Science and Engineering, China University of Mining and Technology, Xuzhou 221116, China*

Correspondence should be addressed to Lingli Xu; xulingli311@hotmail.com

Academic Editor: Charles C. Sorrell

Titanium and its alloys have been widely used in the manufacture of endosseous implants due to excellent biocompatibility, low elastic modulus, and good chemical stability. However, the titanium based metals are essentially bioinert materials. In order to improve their bioactivity, biological coatings are usually applied. Recent studies found that, compared with hydroxyapatite coating, dicalcium phosphate anhydrous (DCPA, Monetite) coating maybe more bioactive due to higher solubility and release of Ca and P ions. In this work, DCPA coating was prepared through a novel and simple method. Briefly, high concentration DCPA solution was sprayed onto superhydrophilic titanium and the specimens were dried fast in an air-circulating oven. After repeating the process over 100 times, a compact coating was fabricated. The microstructure, chemical composition, wettability, and *in vitro* bioactivity of the coating were analyzed and evaluated. Results showed that the coating fully covered the substrate and consisted of a large number of uniformly sized DCPA granules that packed together closely. The coating showed good wettability and could keep the property for a long time. After immersion in simulated body fluid for 2 weeks, a large amount of bone-like apatite with low crystallinity was induced implying a good bioactivity.

1. Introduction

Titanium (Ti) and its alloys are important metal materials for the manufacture of load bearing bone implants due to excellent corrosion resistance, biocompatibility, and lower elasticity modulus. However, the Ti based metals are essentially bioinert materials and are insufficient in osseointegration ability. Instead of biological fixation, their combination with human bones mainly depends on mechanical locking. Moreover, a fibrous capsule intervening at the interface of implant and bone is always formed after the implantation, especially when a relatively smooth implant was used [1, 2]. Such kind of fibrous capsule is fragile and its adhesion to both implant and surrounding bone cavity is quite low, and therefore it is easy to peel or form crush induced necrosis under stress and micromovement. In this case, a serious secondary infection is likely to happen and may lead to a premature loss of reconstruction function of the implants. In addition, the fibrous capsule is poorly differentiated and may undergo hyperplasia or transform into tumor tissue under sustained mechanical stimulation. Thus, a fast and strong bonding to the host bone without soft tissue intervention is a prerequisite for the long-term and stable performance of Ti implants.

Bioactive coatings play an important role in improving the osseointegration of metal implant with bone and, among them, the hydroxyapatite (HA) coating is the most widely used [3–5]. However, HA is not osteoinductive and when the implantation site is poor in bone mass and density, such as osteoporosis, even HA coated implants cannot form a rapid biological integration with the host bone. With the increasing desire from patients for early loading and rapid healing, it is urgent to prepare new biological implant coatings with osteoinductivity. By contrast, acidic calcium phosphates, dicalcium phosphate anhydrous ($CaHPO_4$, DCPA, Monetite) and dicalcium phosphate dehydrate ($CaHPO_4{\cdot}H_2O$, DCPD,

Brushite), are more degradable at physiologic pH than HA and will form a Ca^{2+} and PO_4^{3-} supersaturated zone at the interface of implant and bone that may induce osteoblastic differentiation [6]. Among them, DCPA has gained great attention recently as it showed degradation and regeneration properties superior to DCPD. For instance, *in vitro* evaluation showed that DCPA ensured active osteoclast resorption and would also induce higher levels of osteogenic genes expression of bone marrow cells than DCPD [7, 8]. *In vivo* studies are even more promising. Torres and coworkers grafted DCPA and autologous bone granules separately into critical sized bone defects in rabbit calvaria and after 8 weeks they found that the volumes of mineralized tissue formed by two kinds of granules were not significantly different [9, 10]. Oryan et al. compared the effectiveness of calcium phosphates/chitosan-gelatin-platelet gel scaffold on the healing of critical size radial bone defects in rats 8 weeks after implantation. They reported that DCPA filler accelerated bone regeneration and provided a much better balance between substitute resorption and new bone formation compared with β-TCP and HA [11]. DCPA degrades at a mild rate and, unlike β-TCP and DCPD, it will not undergo phase transition into HA in physiologic environment [12]. Such unique properties may bring about its superiority in terms of resorption and remodeling. Therefore, we think DCPA coatings are expected to be a new generation of bioactive coatings used on surface of metal endosseous implants.

Several techniques have been developed for preparation of DCPA coating. Kuroda et al. heated the Ti substrate up to 150°C in a solution with calcium and phosphorus ions by large alternating current (up to 25 A). Water gasified vigorously on Ti surface and DCPA was formed. However, the coating grew in the form of lamellar and brick-like clusters and failed to completely cover the substrate [13]. Ortiz et al. prepared DCPD on metal substrate by electrochemical deposition and chemical deposition separately and then heated the specimens to get DCPA coating. In this procedure, Ca/P ratio was difficult to control as spontaneous grain growth happened during the formation of DCPD coating and therefore the final products might contain calcium pyrophosphate (CPPD) which to some extent weakens the biocompatibility of coatings [14, 15]. DCPA coating can also be prepared by sol-gel method. However, such method not only consumes time as always, but also induces calcium carbonate and other impurities [16]. Obviously, DCPA coatings by these methods can not meet the requirement of biomedical applications. Moreover, the current methods present several drawbacks such as complicated equipment, high safety standard, high current or temperature operation, complicated process, low efficiency, and high energy consumption. Besides, the DCPA granules by these methods usually scatter on substrate surface as clusters and could not cover the substrate completely. Nevertheless, the impurity phase generated during preparation may impair the properties of coating.

In this work, a novel DCPA coating preparation method was proposed based on the dissolution characteristic of dicalcium phosphate. Briefly, high concentration DCPA aqueous solution was prepared and then a DCPA coating was prepared on superhydrophilic Ti substrate by repeated spray-drying treatment. The microstructure, chemical composition, wettability, and *in vitro* apatite inducing ability of the coating were analyzed and evaluated.

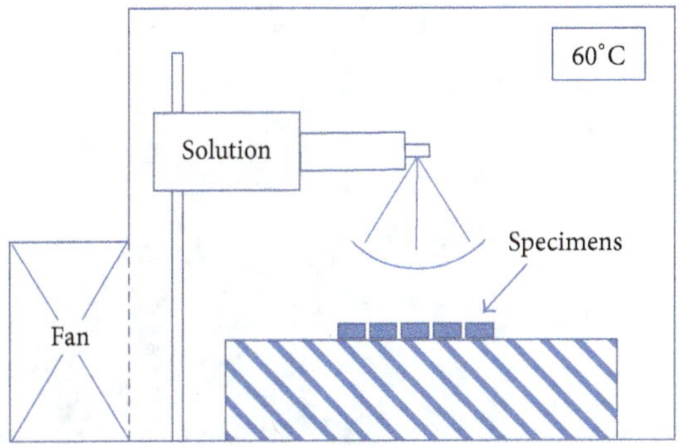

FIGURE 1: Illustration of coating preparation progress.

2. Materials and Methods

2.1. Pretreatment of Specimens. Annealed commercially pure Ti pieces (Grade 2, Baoji Titanium Industry Company Ltd., China), cut into 10 mm × 10 mm × 2.5 mm, were used as substrates. The specimens were wet-polished with 280#, 400#, and 800# SiC abrasive paper successively and then ultrasonically washed in acetone and distilled water for 10 min, respectively. After being dried in air, these specimens were pretreated by acid etching to get superhydrophilic surface. Briefly, the Ti specimens were submerged in a mixture of 98% (wt) H_2SO_4, 37% (wt) HCl, and distilled water with the volume ratio of 5 : 2 : 3 and etched for 30 min. After that, the specimens were sufficiently washed with tap water, neutralized by diluted $NaHCO_3$ solution, and then ultrasonically washed for 10 min in distilled water. The pretreated specimens were placed in distilled water before use.

2.2. Solution Preparation and Coating Fabrication. High concentration DCPA aqueous solutions, 11.5 g/L (unsaturated), were prepared by dissolving analytically pure DCPA powder (Sinopharm Chemical Reagent Co. Ltd., China) into diluted HCl solution of 0.15 mol/L with the assistance of ultrasonicator. Pretreated Ti specimens were taken out of distilled water, placed into an air-circulating oven set at 60°C. Immediately after drying, the specimens were repeatedly sprayed with DCPA solution for certain times by a programmable automatic electric sprayer (TS-T, Shanghai Kezhe Biochemical Technology Co., Ltd., China). The spraying distance was 3 cm, the spray radius was 4 cm, and the spray volume was 0.1 ml per time. A interval of 2 min was set between two successive sprays to ensure complete drying of the specimen surface. The experiment setup is illustrated in Figure 1. After the final spray and drying, specimen were washed thoroughly with distilled water and then were finally dried in the oven for 30 min.

2.3. Characterizations. Product phases and crystallinity were investigated by X-ray diffraction (XRD, D/MAX-3B, Rigaku

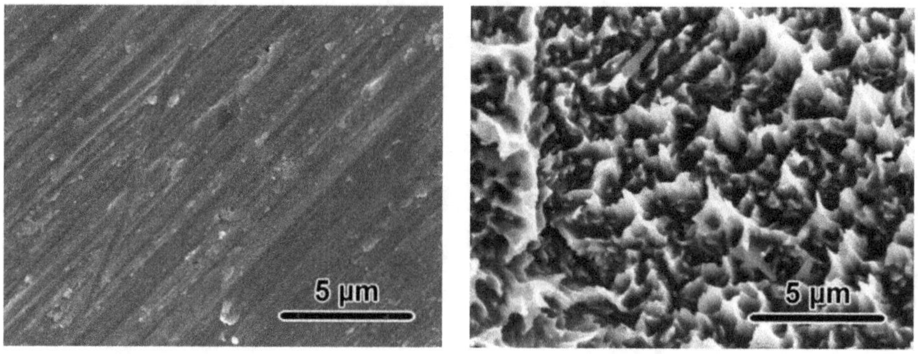

(a) (b)

FIGURE 2: Surface morphology of specimens before and after pretreatment: (a) sandpaper polished; (b) acid etched specimens.

(a) (b)

FIGURE 3: Contact angles of specimens before and after pretreatment: (a) sandpaper polished; (b) acid etched specimens.

Co., USA). The surface morphologies of the specimens were observed with scanning electronic microscope (SEM, S-3000, Hitachi Co., Japan). Surface roughness of specimens was analyzed by laser scanning confocal microscopy (VKX100, Keyence Co., Japan). The contact angle of distilled water on DCPA coating was measured by the static contact angle measuring instrument (JC2000A, Shanghai Zhongchen Digital Technic Apparatus Co., Ltd., China). Bioactivity of DCPA coating was evaluated through the simulated body fluid (SBF) immersion test. The main ion concentrations of the SBF were nearly equal to that of human plasma and the pH was adjusted to 7.4 by trihydroxymethyl aminomethane (Tris) and 1.0 mol/L HCl solution [17]. DCPA coated specimens were immersed in 1.0 mol/L NaOH solution at 60°C for 1 h and after thorough rinse they were immersed in SBF for 14 days in an incubator set at 37°C. After that, specimens were taken out, rinsed with distilled water, and dried in air at room temperature. And then the immersed specimens were analyzed by SEM and XRD.

3. Results and Discussions

3.1. Pretreatment. Surface morphologies of specimens before and after pretreatment were shown in Figure 2. After acid etching, texture of scratches generated during the wet abrasion had disappeared and submicron sized etch pits were found distributing all over the surface. The surface roughness was found increased, from 0.26 μm to 1.61 μm, after acid etching treatment. Surface wettability of specimens was obviously improved as shown in Figure 3. The contact angle on polished surface was about 37°, whereas the water drop spread on acid etched surface fast and the contact angle became 0°. It meant that acid etching had imposed superhydrophilicity to the Ti specimens. However, such excellent wettability gradually degraded when specimens were kept in ambient air. As shown in Figure 4, the contact angle (CA) increased with the increase of storage time. After 14 days storage it was over 100°. In order to preserve the superhydrophilicity, specimens were submersed in distilled water in a sealed glass bottle. As it could be seen, the CA was still 0° after 14 days immersion in distilled water.

3.2. Characterizations of DCPA Coating

3.2.1. Surface Morphology. Figure 5 showed the SEM observation results of acid etched Ti specimens after spray-drying treatment with different repeating times. At the beginning, when the number of repetitions was low, round shaped clumps were formed on the surface, as shown in Figure 5(a). They were separated, occupying one or more etch pits, but closely attached to the specimens surface. However, the acid etched surface was still visible, because of the limited

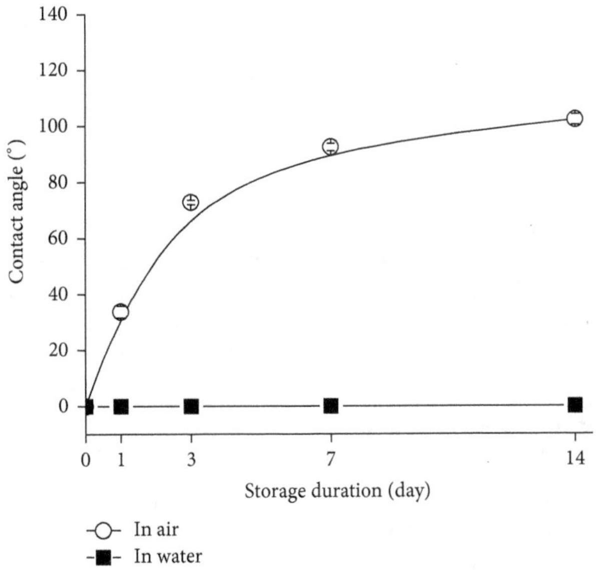

FIGURE 4: The changes of contact angles with increasing storage duration.

FIGURE 5: Surface morphology changes with increasing spray-drying repetition number: (a) 20 times; (b) 60 times; (c) 100 times; (d) 140 times.

numbers of these particles. As the spray-drying operation repeated, more round clumps were formed, and they started to pile up and contact each other, as shown in Figure 5(b). When the repeating times increase to 100, the whole specimens surface was covered with numerous cubic granules of about 1.5 μm in diameter and the acid etched pits disappeared. These granules stacked up tightly with each other and had formed a compact layer, Figure 5(c). The later XRD result confirmed that these are DCPA particles. However, further increase in repeating times, up to 140, induced a less compact coating with particles showing a hairy surface, as shown in Figure 5(d).

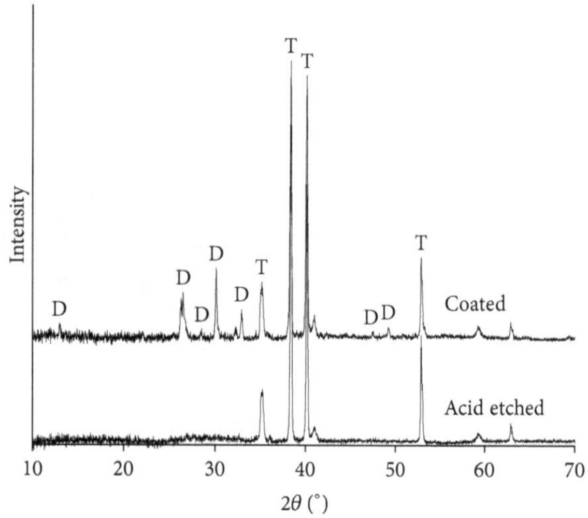

FIGURE 6: XRD patterns of specimens after 100 times of spray-drying treatment.

3.2.2. Coating Phase Analysis. Figure 6 showed the XRD patterns of acid etched Ti after 100 times of spray-drying treatment. It could be seen that several diffraction peaks appeared besides those from pure Ti substrate and all these new peaks could be ascribed to DCPA (PDF #01-077-0128). Among them, the three most intense ones, located at 2θ = 26.4°, 30.2°, and 32.8°, could be assigned to the (220), (112), and (102) planes, respectively. The diffraction results have excluded existence of by-products in the coating as peaks from other crystalized Ca-P compound and amorphous phases were not found. Besides, all the diffraction peaks were sharp in shape which indicated well crystallized structures and an atomic ratio that was close to the ideal stoichiometry. According to calculations using Scherrer formula, the crystal in the cubic granules was 40.59 nm in average. Thermodynamically, the product of Ca-P precipitation from solution could be predicted based on the diagram of Ca-P compounds solubility versus pH. When the pH was less than 4.0, DCPA crystals could preferentially be precipitated from the liquid if the ions were supersaturated [6, 18]. Thus, in this method, low solution pH was the controlling factor for the pure DCPA product.

It could be inferred that, after each spraying, the former precipitated granules on substrate would be dissolved and precipitated again during the following drying stage. With the increase of repeating times and the accumulation of solute, a compact DCPA coating would be formed gradually. Compared with that by Kuroda et al., the DCPA coating prepared by this spray-drying method could cover the acid etched surface completely and the granules in the coating were uniform in size, contacted with each other more closely according to SEM observation [13, 19]. The superhydrophilic surface facilitated even and fast evaporation of water once a thin liquid layer was formed by spraying and was thought to be an important precondition for the homogenous precipitation of DCPA granules. However, as water evaporated quickly, once a compact coating had been formed, the following sprays would only induce partial dissolution of these granules and thus the top surface of the coating became hairy as the spray-drying repeating times reached 140.

3.2.3. Wettability. The wettability of DCPA coating was evaluated by measuring the contact angle, as shown in Figure 7. Same as acid etched specimens, the freshly prepared DCPA coated ones also showed superhydrophilicity. Moreover, after being stored in ambient air for 4 weeks, it still kept an excellent hydrophilicity with the contact angle only rising about 15°. These results showed that, the DCPA coating not only inherited superhydrophilicity of acid etched surface but also preserved a good wettability after a long storing duration.

Currently, surface of Ti implants designed for osseointegration are usually roughened by sandblast and/or acid etching. Submicron scaled pits generated on the Ti surface mimic the microstructure of natural bone and facilitate osteogenic differentiation of stem cells. Besides, such roughening could impose superhydrophilicity to surface of implant which greatly improves the bioactivity of Ti. For instance, the protein in the body fluid could be rapidly adsorbed to the implant surface at the implantation moment and show conformation favorable for cell attachment, proliferation, and osteogenic differentiation [20]. However, a fresh acid etched surface also possesses high surface energy, and therefore carbon containing molecules in the ambient air or packages will easily be adsorbed onto its surface and keep accumulating as storage time increases. Such contamination impairs the excellent wettability and degrades osteoconductivity of rough implants and has become a universal problem that challenges both the producer and surgeons [21–24]. This study demonstrated that hydrophilicity of the bone implants with rough surface could be preserved by slightly soluble bioactive coatings and hence the bioactivity degradation will be overcome.

3.2.4. In Vivo Apatite Induction. The surface morphology observation and XRD results of DCPA coated specimens (spray-drying 100 times) after immersion in SBF for 2 weeks were shown in Figures 8 and 9. As could be seen from the SEM image, the whole surface of specimens was covered by spherical crystal clusters after immersion. It was the typical surface morphology of bioactive materials after SBF immersion test [25, 26]. In the XRD pattern peaks ascribed to apatite could be observed. Meanwhile, it should be noticed that apatite was of low crystallinity and signals of amorphous phase were found at the diffraction angles from 20° to 40°. In addition, the diffraction peaks of DCPA remained clearly and their relative intensities versus those of Ti substrate did not seem to decrease. It indicated that the DCPA coating did not undergo obvious degradation during the immersion. The SBF immersion experiment tallies with *in vivo* animal experiment very well and is reproducible and easy to be operated, so it is widely used to evaluate the bioactivity of materials. The spontaneous deposition of apatite from SBF within 2 weeks indicated that the DCPA coating fabricated by this method

FIGURE 7: Contact angles of DCPA coatings: (a) freshly prepared; (b) after 4 weeks storage in air.

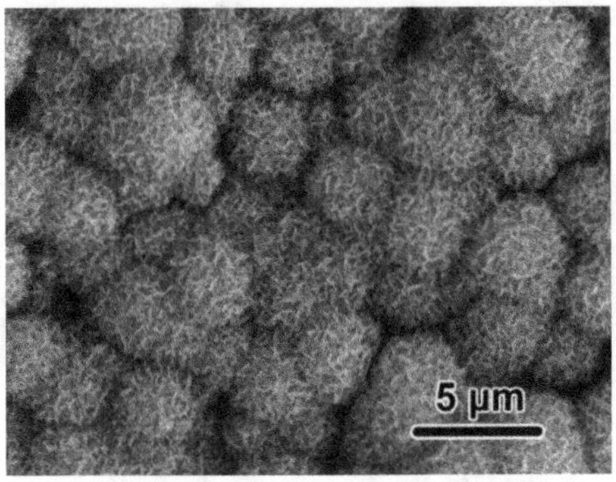

FIGURE 8: Surface morphology of coated specimens after being immersed in SBF for 2 weeks.

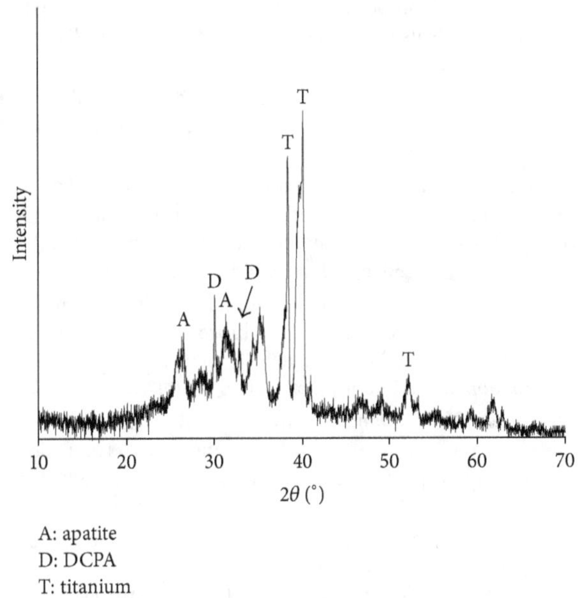

FIGURE 9: Phase analysis of coated specimens after being immersed in SBF for 2 weeks.

has good bioactivity and it is hopeful of inducing new bone to grow rapidly on the surface of implants and increasing the initial stability of implants.

4. Conclusion

DCPA coating was fabricated on titanium substrate through a novel and simple spray-drying method. The coating was compact, fully covered substrate and was consist of uniform sized DCPA granules that packed together closely. The coating showed good wettability and could keep the property for a long time. After immersion in simulated body fluid for 2 weeks, a large amount of bone-like apatite with low crystallinity was induced on the coating surface, which implied good bioactivity. This method is feasible and efficient, and it is expected to be used for other substrates as long as the surface showed good hydrophilicity.

Conflicts of Interest

The authors declare that they have no conflicts of interest.

Acknowledgments

The authors gratefully acknowledge the financial support from National Natural Science Foundation of China (51275514), Natural Science Foundation of Jiangsu Province (BK20160566), China Postdoctoral Science Foundation (2016M601754), and Jiangsu Planned Projects for Postdoctoral Research Funds (1421601085B).

References

[1] L. Le Guéhennec, A. Soueidan, P. Layrolle, and Y. Amouriq, "Surface treatments of titanium dental implants for rapid osseointegration," *Dental Materials*, vol. 23, no. 7, pp. 844–854, 2007.

[2] B. Elliott and T. Goswami, "Implant material properties and their role in micromotion and failure in total hip arthroplasty," *International Journal of Mechanics and Materials in Design*, vol. 8, no. 1, pp. 1–7, 2012.

[3] X. Shi, L. Xu, T. B. Le et al., "Partial oxidation of TiN coating by hydrothermal treatment and ozone treatment to improve its osteoconductivity," *Materials Science and Engineering C: Materials for Biological Applications*, vol. 59, pp. 542–548, 2016.

[4] S. R. Paital and N. B. Dahotre, "Calcium phosphate coatings for bio-implant applications: materials, performance factors, and methodologies," *Materials Science and Engineering: R: Reports*, vol. 66, no. 1-3, pp. 1–70, 2009.

[5] A. Haider, S. Haider, S. S. Han, and I.-K. Kang, "Recent advances in the synthesis, functionalization and biomedical applications of hydroxyapatite: a review," *RSC Advances*, vol. 7, no. 13, pp. 7442–7458, 2017.

[6] K. Ishikawa, S. Matsuya, Y. Miyamoto, and K. Kawate, "9.05—bioceramics," in *Comprehensive Structural Integrity*, 2003.

[7] M. Montazerolghaem, M. Karlsson Ott, H. Engqvist, H. Melhus, and A. J. Rasmusson, "Resorption of monetite calcium phosphate cement by mouse bone marrow derived osteoclasts," *Materials Science and Engineering C: Materials for Biological Applications*, vol. 52, pp. 212–218, 2015.

[8] F. Tamimi, D. L. Nihouannen, H. Eimar, Z. Sheikh, S. Komarova, and J. Barralet, "The effect of autoclaving on the physical and biological properties of dicalcium phosphate dihydrate bioceramics: Brushite vs. monetite," *Acta Biomaterialia*, vol. 8, no. 8, pp. 3161–3169, 2012.

[9] T. Jesús, T. Iskandar, C. A. Jatsue et al., "Monetite granules versus particulate autologous bone in bone regeneration," *Annals of Anatomy*, vol. 200, pp. 126–133, 2015.

[10] F. Tamimi, J. Torres, U. Gbureck et al., "Craniofacial vertical bone augmentation: a comparison between 3D printed monolithic monetite blocks and autologous onlay grafts in the rabbit," *Biomaterials*, vol. 30, no. 31, pp. 6318–6326, 2009.

[11] A. Oryan, S. Alidadi, and A. Bigham-Sadegh, "Dicalcium phosphate anhydrous: an appropriate bioceramic in regeneration of critical-sized radial bone defects in rats," *Calcified Tissue International*, vol. 101, no. 5, pp. 530–544, 2017.

[12] Z. Sheikh, Y. L. Zhang, L. Grover, G. E. Merle, F. Tamimi, and J. Barralet, "In vitro degradation and in vivo resorption of dicalcium phosphate cement based grafts," *Acta Biomaterialia*, vol. 26, pp. 338–346, 2015.

[13] K. Kuroda, Y. Miyashita, R. Ichino, M. Okido, and O. Takai, "Preparation of calcium phosphate coatings on titanium using the thermal substrate method and their in vitro evaluation," *Materials Transactions*, vol. 43, no. 12, pp. 3015–3019, 2002.

[14] I. Y. Ortiz, A. Raybolt dos Santos, A. M. Costa et al., "In vitro assessment of zinc apatite coatings on titanium surfaces," *Ceramics International*, vol. 42, no. 14, pp. 15502–15510, 2016.

[15] K. Li, B. Wang, B. Yan, and W. Lu, "Microstructure in vitro corrosion and cytotoxicity of Ca-P coatings on ZK60 magnesium alloy prepared by simple chemical conversion and heat treatment," *Journal of Biomaterials Applications*, vol. 28, no. 3, pp. 375–384, 2013.

[16] H. Eshtiagh-Hosseini, M. R. Houssaindokht, M. Chahkandhi, and A. Youssefi, "Preparation of anhydrous dicalcium phosphate, DCPA, through sol-gel process, identification and phase transformation evaluation," *Journal of Non-Crystalline Solids*, vol. 354, no. 32, pp. 3854–3857, 2008.

[17] L. Xu, X. Shi, C. Ouyang, and W. Liu, "In vitro apatite formation, protein adsorption and initial osteoblast responses on titanium surface enriched with magnesium," *Rare Metal Materials and Engineering*, vol. 46, no. 6, pp. 1512–1517, 2017.

[18] H. Oonishi and H. Oomamiuda, "Chapter 9 degradation/resorption in bioactive ceramics in orthopaedics," in *Handbook of Biomaterial Properties*, Springer, New York, NY, USA, 1998.

[19] A. V. Zavgorodniy, R. S. Mason, R. Z. LeGeros, and R. Rohanizadeh, "Adhesion of a chemically deposited monetite coating to a Ti substrate," *Surface and Coatings Technology*, vol. 206, no. 21, pp. 4433–4438, 2012.

[20] X. Shi, L. Xu, M. L. Munar, and K. Ishikawa, "Hydrothermal treatment for TiN as abrasion resistant dental implant coating and its fibroblast response," *Materials Science and Engineering C: Materials for Biological Applications*, vol. 49, pp. 1–6, 2015.

[21] W. Att, N. Hori, F. Iwasa, M. Yamada, T. Ueno, and T. Ogawa, "The effect of UV-photofunctionalization on the time-related bioactivity of titanium and chromium-cobalt alloys," *Biomaterials*, vol. 30, no. 26, pp. 4268–4276, 2009.

[22] H. Minamikawa, W. Att, T. Ikeda, M. Hirota, and T. Ogawa, "Long-term progressive degradation of the biological capability of titanium," *Materials* , vol. 9, no. 2, p. 102, 2016.

[23] X. Shi, L. Xu, K. B. Violin, and S. Lu, "Improved osseointegration of long-term stored SLA implant by hydrothermal sterilization," *Journal of the Mechanical Behavior of Biomedical Materials*, vol. 53, pp. 312–319, 2016.

[24] X. Shi, L. Xu, Q. Wang, Sunarso, and L. Xu, "Hydrothermal sterilization improves initial osteoblast responses on sandpaper-polished titanium," *Materials*, vol. 10, no. 7, article 812, 2017.

[25] T. Kokubo and H. Takadama, "How useful is SBF in predicting in vivo bone bioactivity?" *Biomaterials*, vol. 27, no. 15, pp. 2907–2915, 2006.

[26] L. Xu, X. Shi, Z. Zhu, Z. Wang, and Y. Jin, "Improved in vitro osteoconductivity of pure titanium modified with magnesium by hydrothermal treatment," *Rare Metal Materials and Engineering*, vol. 49, no. 9, pp. 2613–2617, 2017.

20

High Temperature Corrosion of Nickel in $NaVO_3$-V_2O_5 Melts

J. Porcayo-Calderon,[1] J. J. Ramos-Hernandez,[2] Jan Mayén,[3] E. Porcayo-Palafox,[1] G. K. Pedraza-Basulto,[4] J. G. Gonzalez-Rodriguez,[1] and L. Martinez-Gomez[2,5]

[1]*CIICAp, Universidad Autónoma del Estado de Morelos, Avenida Universidad 1001, 62209 Cuernavaca, MOR, Mexico*
[2]*Instituto de Ciencias Físicas, Universidad Nacional Autónoma de México, Avenida Universidad s/n, 62210 Cuernavaca, MOR, Mexico*
[3]*CENIDET, Tecnológico Nacional de México, Prolongación Palmira s/n Esq. Apatzingán, Col. Palmira, 62490 Cuernavaca, MOR, Mexico*
[4]*Facultad de Ingeniería, UNACAR, 24180 Ciudad del Carmen, CAM, Mexico*
[5]*Corrosion y Protección (CyP), Buffon 46, 11590 México City, Mexico*

Correspondence should be addressed to J. Porcayo-Calderon; jporcayoc@gmail.com

Academic Editor: Patrice Berthod

Many alloys used at high temperature in industrial processes are Ni-based and many others contain it in appreciable quantities, so it is of interest to evaluate the performance of pure nickel in order to determine the behavior of its alloys once the elements responsible for their protection have been depleted due to accelerated corrosion processes in the presence of vanadium-rich molten salts. Due to this, this work presents the study of Ni behavior in $NaVO_3$-V_2O_5 mixtures at different temperatures. The behavior of pure nickel was determined by both electrochemical and mass loss measurements. The results show that the aggressiveness of the vanadium salts is increased by increasing both the V_2O_5 content and temperature. V_2O_5 addition considerably increases the current densities of the anodic and cathodic reactions. The corrosion process of Ni is modified due to the presence of its corrosion products, and its presence increases the activation energy by at least one order of magnitude. Although nickel shows a high reactivity in vanadium-rich salts, its reaction products are highly stable and protect it from the corrosive medium because the corrosion reactions trap the vanadium and block the migration of nickel ions.

1. Introduction

Many alloys for engineering applications at high temperature are Fe- or Ni-based. Their excellent mechanical properties and corrosion resistance make them useful for manufacturing components of industrial equipment such as industrial gas turbines and other energy conversion systems [1–6]. Corrosion resistance of alloys depends mainly on their chemical composition. In high temperature applications it is essential that the alloys be capable of developing protective oxides such as Cr_2O_3, Al_2O_3, or SiO_2. Chemical and physical characteristics of this type of oxides, such as good adhesion, low gas permeability, high thermodynamic, and low diffusion coefficients, provide excellent protection to alloys in gaseous environments mainly [7–9].

However, the protection afforded by this type of oxide can be limited when molten phases are involved in the corrosion process. In these cases, the accelerated degradation of the materials may occur due to the so-called molten salt corrosion, hot corrosion, or high temperature corrosion. This type of corrosion is characterized by the accelerated degradation of the alloy when covered by a thin layer of molten salt at high temperature in an oxidant environment. In these cases, the integrity of any alloy will be compromised if its protective oxides do not show low solubility in the melt [1, 10, 11].

The solubility of the protective oxides is greater in the presence of vanadium-rich fused salts (sulfate-vanadate type) than in pure Na_2SO_4; this is because the vanadate ions increase the acidic fluxing of any oxide [10]. On the other hand in Na_2SO_4-V_2O_5 mixtures the maximum solubility occurs in the interval 10 to 30% Na_2SO_4 [12]. In many industrial systems where the use of heavy fossil fuels is common, the species causing the catastrophic corrosion of the materials are complex compounds of vanadium, β and γ sodium vanadates ($Na_2O{\cdot}V_2O_4{\cdot}5V_2O_5$ and $5Na_2O{\cdot}V_2O_4{\cdot}11V_2O_5$, resp.) being the most corrosive vanadium species [6, 12–14].

When accelerated corrosion decreases the concentration of the element responsible for forming the protective oxide, corrosion will proceed catastrophically if the rest of the elements of the alloy are unable to form stable corrosion products. For example, in most alloys the element responsible for forming a protective oxide is Cr; however, in the presence of acidic vanadium salts it is not capable of forming a stable protective oxide [15].

Therefore, because many Ni-based superalloys show excellent corrosion resistance due to their ability to develop a protective oxide, Cr_2O_3, this study evaluates the performance of pure nickel in the presence of $NaVO_3$-V_2O_5 mixtures from 600 to 900°C. This is a model system (few reaction products, NiO and vanadates) in order to determine if this element would be able to provide some protection to the alloy once the element responsible for protection has been depleted.

2. Experimental Procedure

2.1. Corrosive Salts. The corrosive medium employed consisted of different $NaVO_3 : V_2O_5$ ratios, from pure $NaVO_3$ to pure V_2O_5, and two intermediates ratios that promote the formation of the highly corrosive complex vanadates, namely, sodium β-vanadyl-vanadate ($Na_2O{\cdot}V_2O_4{\cdot}5V_2O_5$) and sodium γ-vanadyl-vanadate ($5Na_2O{\cdot}V_2O_4{\cdot}11V_2O_5$). In previous studies these complex vanadates have been synthesized from the following ratios: $2NaVO_3 : 5V_2O_5$ and $10NaVO_3 : 7V_2O_5$, respectively [15], where $NaVO_3$ is the vanadium salt with the highest basicity, V_2O_5 is the compound with the highest acidity, and $2NaVO_3 : 5V_2O_5$ and $10NaVO_3 : 7V_2O_5$ ratios are compounds with intermediate acidity and high oxygen adsorption capacity. Since the $NaVO_3$ can be synthesized from the $Na_2O : V_2O_5$ ratio, it is considered that by increasing the molar ratio of V_2O_5 the acidity of the resulting compound increases. For simplicity, hereafter the above-mentioned compounds will be identified according to the following nomenclature: NV is $NaVO_3$, NV3 is sodium γ-vanadyl-vanadate, NV6 is sodium β-vanadyl-vanadate, and V is V_2O_5, where N is Na_2O. Its known melting points are 630°C, 627°C, 535°C, and 691°C, respectively [14].

2.2. Electrochemical Measurements. Electrochemical evaluations were performed using a three-electrode arrangement. As pseudo-reference electrode and counter electrode a platinum wire was used. Nickel samples were spot-welded with a Ni20Cr wire in order to get electrical connection. Ni20Cr wire was isolated with a ceramic tube and sealed with refractory cement. The test temperatures selected were 700, 800, and 900°C, which are above the melting point of the salts. The test temperature was measured constantly, using a type K thermocouple and was controlled to 2°C with respect to the test temperature. Before starting the electrochemical measurements, the electrode array is immersed in the molten salt and the open-circuit potential of the working electrode is measured until steady state (±5 mV) is reached; in all cases this was achieved in about 30 minutes. In all experiments, the atmosphere above the melt was static air. Potentiodynamic polarization curves and linear polarization resistance (LPR) measurements were carried out. Potentiodynamic polarization curves were obtained polarizing the working electrode from −300 to 300 mV at scanning rate of 1 mV/s, and the electrochemical parameters (Tafel slopes, E_{corr} and I_{corr}) were obtained from the active regions of the corresponding anodic and cathodic zones. LPR measurements were obtained applying a potential of ±10 mV, and the variation in current intensity associated with that polarization was measured during 24 hr. An ACM Instrument potentiostat, model GillAC, controlled by a personal computer was used for the electrochemical experiments.

2.3. Mass Loss Corrosion Testing. Before mass loss tests, the specimens were cleaned with acetone and dried, subsequently weighed by means of an analytical digital scale with a precision of 0.00001 g, and then packed in the mixture of salts in porcelain crucibles with 500 mg/cm^2 of vanadium salts. The corrosion tests were carried out in electric furnaces in static air during 100 hours at temperatures of 700, 800, and 900°C. Four specimens were exposed to each test condition. After corrosion tests the corrosion rate was measured as mass loss. Three specimens of each condition test were descaled and chemically cleaned according to ASTM G1 standard, and the other specimen was mounted in thermosetting resin in cross section and polished to analyze the subsurface corrosive attack using a Karl Zeiss DSM-950 scanning electronic microscope (SEM) with an X-ray energy dispersive (EDS) analyzer. In short, the specimen descaling was realized mechanically with a cutter in order to remove the coarse and soft layers of products of corrosion; subsequently the heavily adhered corrosion products were removed by chemical cleaning in a solution of 15% hydrochloric acid (v/v) at room temperature with immersion times of 3 minutes, and cleaning by light brushing until the complete removal of the corrosion products (constant weight of the metal).

In all cases high purity Ni (99.9%) was used. Ni specimens were obtained from rectangular bars and cut in samples with dimensions of 10 × 5 × 5 mm. In order to obtain a homogeneous and defined reaction area, specimens were polished with silicon carbide sandpaper to grade 600. Subsequently, the specimens were washed with distilled water and acetone before performing the corrosion tests.

3. Results and Discussion

3.1. Polarization Curves. Figures 1–3 show the potentiodynamic polarization curves for Ni in the different vanadium salts at 700, 800, and 900°C, respectively. It is observed that, in the presence of NV3, NV6, and V, at all test temperatures, the Ni polarization curves were shifted to higher current densities and nobler potentials than those observed in NV. No significant variation in E_{corr} values was observed in the presence of NV3, NV6, and V, at all test temperatures. In all cases, by increasing the vanadium content in the salt an increase in the current densities of the anodic and cathodic

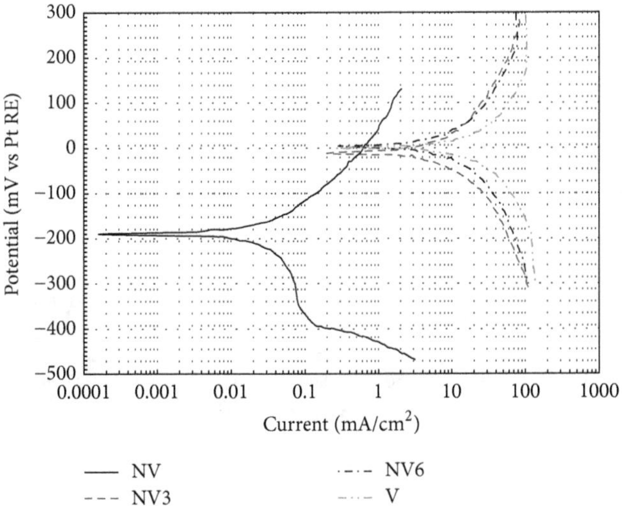

FIGURE 1: Potentiodynamic polarization curves for Ni in NV, NV3, NV6, and V at 700°C.

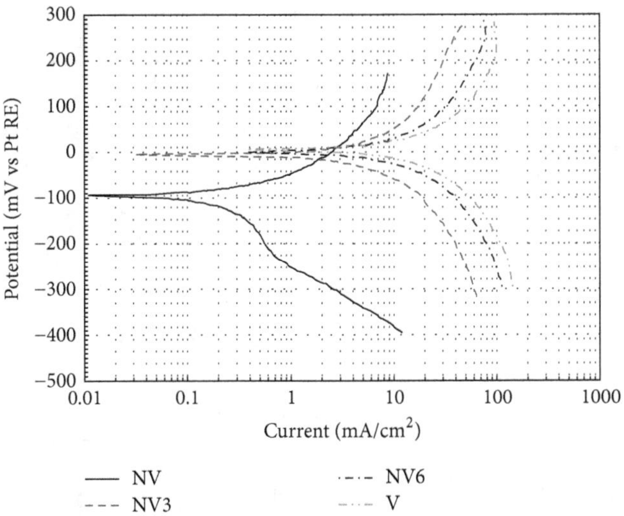

FIGURE 2: Potentiodynamic polarization curves for Ni in NV, NV3, NV6, and V at 800°C.

branches was observed. This difference is most noticeable as the temperature increases.

The observed behavior is associated with the conductive nature of the vanadium salts, because in the molten state the vanadium salts act as electronic conductors or ionic conductors [16]. The conductive nature of the vanadium salts depends on the Na_2O concentration, V^{5+}/V^{4+} ratio, and oxygen partial pressure, such that by increasing the Na_2O concentration the ionic conduction increases, and by increasing the V^{5+}/V^{4+} ratio and the partial pressure of oxygen, the electronic conduction increases. Because vanadium can coexist in different oxidation states (the most stable oxidation states are +5, +4, and +3), the aggressiveness of the vanadium salts increases by increasing the V^{5+} proportion (higher concentration of V_2O_5) and the partial pressure of oxygen (higher oxidation state of vanadium is favored); on the other hand, its aggressiveness is reduced by

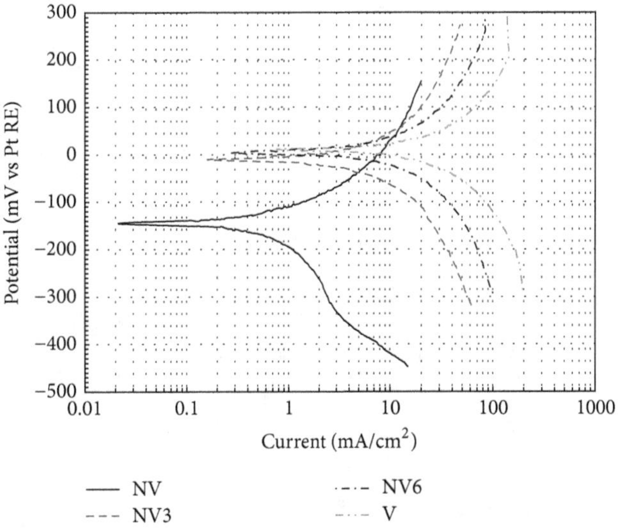

FIGURE 3: Potentiodynamic polarization curves for Ni NV, NV3, NV6, and V at 900°C.

increasing the concentration of Na_2O, due to an increase in the proportion of V^{+4} and V^{+3} (increase in melt basicity) [11, 17, 18]. Therefore, the presence of these multivalent ions enhanced the charge transport through the melt either by counterdiffusion of multivalent cations or electron hopping [11].

From the current corrosion densities calculated from Figures 1–3, the effect of the NV : V ratio and temperature on the nickel corrosion rate was determined. Figure 4 shows the effect of NV : V ratio and temperature on Ni corrosion rate. Although the corrosiveness of the salts NV3 and NV6 shows little difference, it is clear that the aggressiveness of the melt increases by increasing its acidity (higher proportion of V^{+5}) and temperature, and the contrary happens by increasing its basicity (higher concentration of Na_2O). Similar trends have been observed in studies using chromium [15]. In particular, the NV3 and NV6 compounds have little difference in their V content, and they have been considered as the only vanadium species with the highest oxygen absorbing capacity [12, 19]. The increase in the aggressiveness of the molten salts by increasing the V concentration is congruent with previous reports indicating that the V_2O_5 catalyzes the oxidation reaction of Ni [20].

3.2. LPR Measurements. Figures 5–7 show the I_{corr} variation (from LPR measurements) with time for Ni in the different vanadium salts from 700 to 900°C. The observed trends are similar to all test temperatures; namely, there is a strong variation in the I_{corr} values in the first two hours of test and subsequently a pseudo-stationary state is reached. This behavior shows the high reactivity of Ni in the vanadium salts and that the corrosion products formed are highly protective. In all cases, the vanadium salts corrosiveness increases with both the melt acidity and temperature, and again, it is observed that the NV3 and NV6 species show little difference in their degree of aggressiveness. Figure 8 shows the appearance of the corrosion front of Ni evaluated in NV3

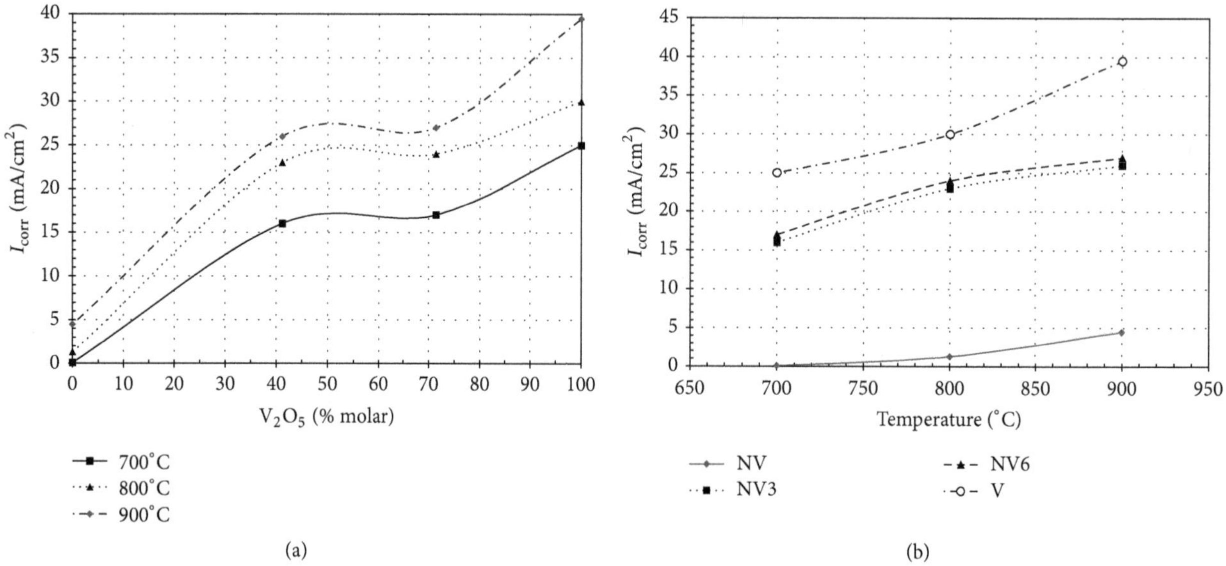

FIGURE 4: Effect of NV : V ratio (a) and temperature (b), on Ni corrosion rate.

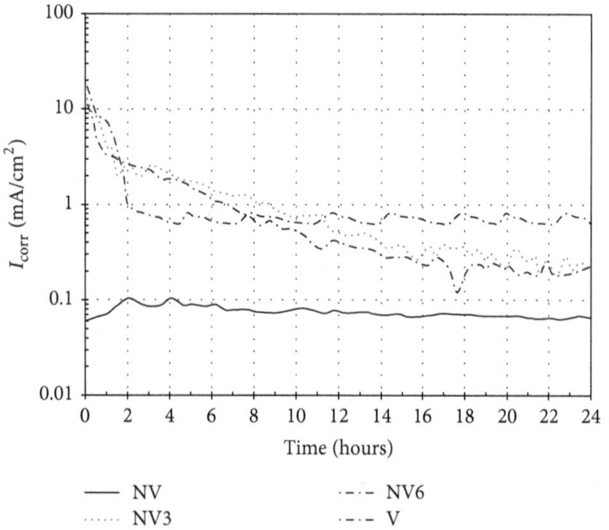

FIGURE 5: I_{corr} variation versus time for Ni in V-rich molten salts at 700°C.

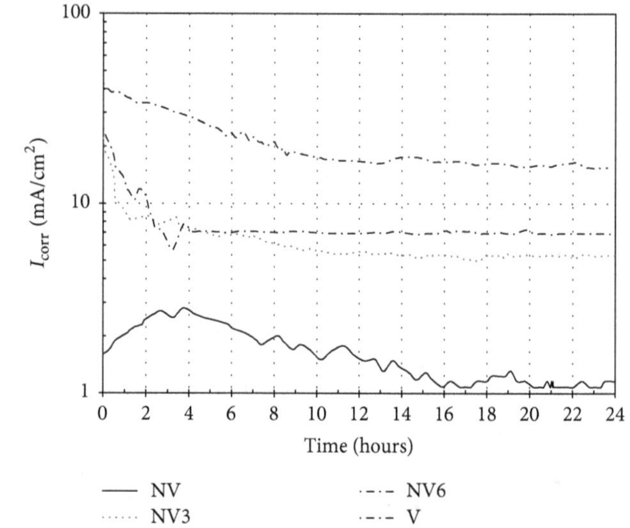

FIGURE 6: I_{corr} variation versus time for Ni in V-rich molten salts at 800°C.

at 800°C. A similar behavior was observed at all temperatures and with the different vanadium salts.

In almost all cases a thin film rich in oxygen and nickel (NiO) onto Ni surface and a thick and dense layer of corrosion products rich in Ni-V-O onto the thin film were observed. Corrosion products showed a defined form of crystals growing either from the Ni surface or from the NiO film towards the melt. On the other hand, the presence of Ni was not observed in the vanadium-rich zone. All this is congruent with the tendency of the LPR values; that is to say, after the initial attack (first hours of immersion) the corrosion products formed a protective barrier that limited the access of the vanadium salts to the Ni surface.

In order to evaluate if there was a change in the corrosion process kinetics between the information obtained from the polarization curves and LPR measurements, the activation energies were determined based on the I_{corr} values obtained from both tests. In the case of the I_{corr} values obtained from the LPR tests, the average of the last 4 hours of test was considered. The dependence between corrosion rate and temperature can be obtained from the Arrhenius equation:

$$I_{corr} = k \exp\left(-\frac{\Delta E}{RT}\right),$$
$$R \ln\left(I_{corr}\right) = R \ln k - \frac{1}{T} \Delta E, \tag{1}$$

where ΔE (J mol^{-1}) is the activation energy, k is a constant, and R is the gas constant (8.314472 J/K-mol). Figure 9 shows the $R\ln(I_{corr})$ versus $1/T$ relationship of the

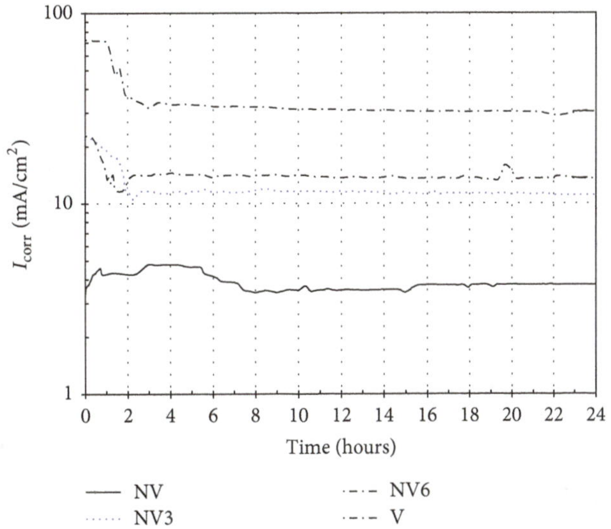

FIGURE 7: I_{corr} variation versus time for Ni in V-rich molten salts at 900°C.

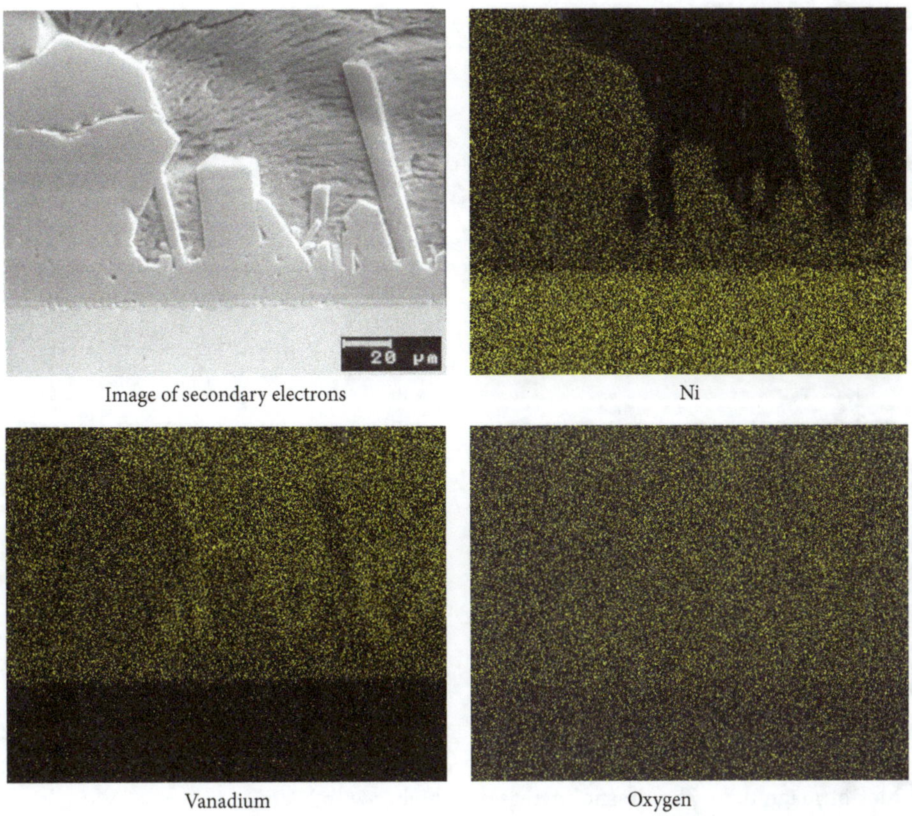

FIGURE 8: Cross sectional aspect of Ni and element mapping, after LPR measurements in NV3 at 800°C for 24 hours.

I_{corr} values obtained from the potentiodynamic polarization curves and the LPR measurements. In the first case the information derived from the potentiodynamic polarization curves corresponds to the interaction of the bare Ni surface with the molten salts, and in the second case due to the corrosion process of Ni with corrosion products adhered to its surface (NiO and nickel vanadates). In the case of a bare Ni surface, the activation energies determined were 193, 23, 22, and 21.5 kJ mol^{-1} for NV, NV3, NV6, and V, respectively. These values show that the Ni corrosion rate increases with the vanadium content of the molten salts and that the corrosiveness of the NV3, NV6, and V is at least one magnitude order greater than that of the NV. In contrast, the activation energy values after 20 hours of corrosion varied between 181 and 202 kJ mol^{-1} for the different vanadium salts. The corrosiveness of the NV did not change significantly,

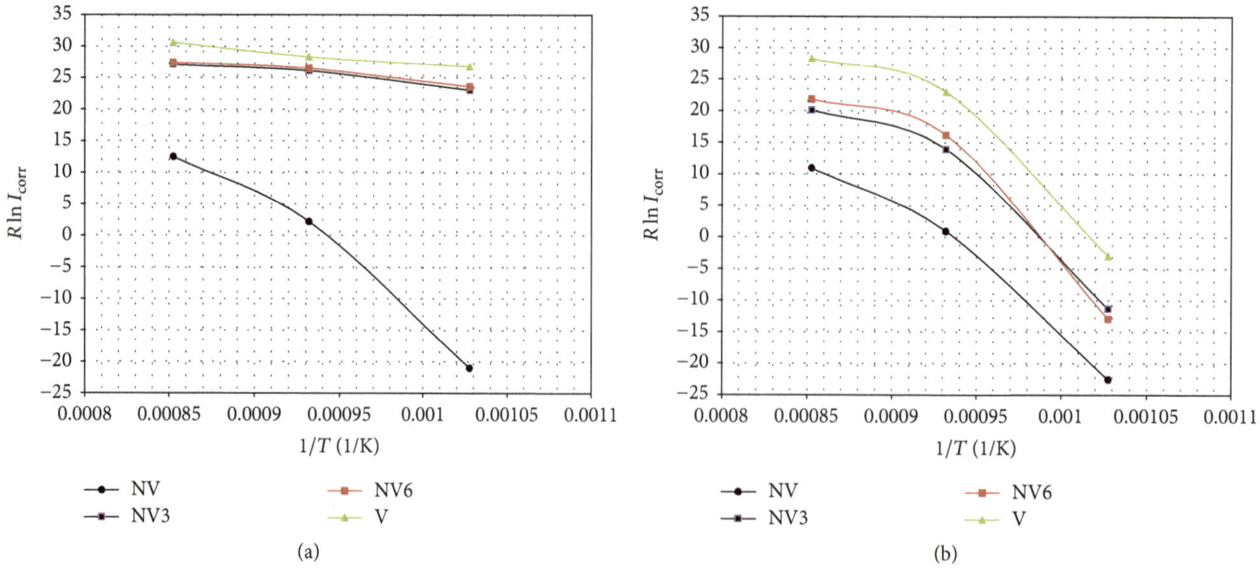

FIGURE 9: $R\ln(I_{corr})$-$1/T$ relationship for Ni evaluated in vanadium salts. (a) Data from the potentiodynamic polarization curves. (b) Data from LPR measurements.

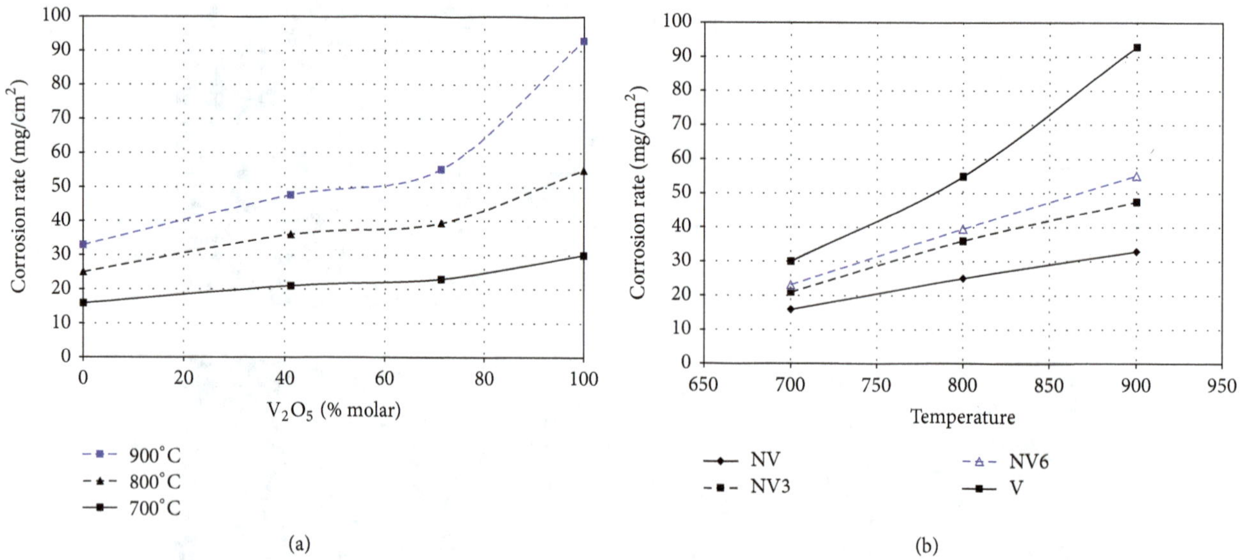

FIGURE 10: (a) Effect of the NV : V ratio and (b) temperature on Ni corrosion rate after 100 hours.

but it is clear that the reactivity of the other vanadium salts decreased by one order of magnitude. The drastic increase in the activation energy values shows that the corrosion products adhering to the Ni surface provide high protection.

3.3. Mass Loss Tests. Figure 10 shows the effect of the different vanadium salts on the corrosion rate of Ni, expressed in terms of weight loss (mg/cm^2) at different temperatures after 100 hours of immersion. From the figures it is observed that, at higher exposure times, the Ni behavior in the different vanadium salts shows the same tendency as observed in the previous tests; namely, the aggressiveness of the molten salts increases with the vanadium content and temperature. Corrosiveness of NV3 and NV6 was similar, being slightly higher with NV6.

Figure 11 shows the cross sectional images of the Ni samples after being immersed in vanadium salts at 700, 800, and 900°C for 100 hours. In all cases similar characteristics are observed. At all temperatures onto Ni surface a nickel oxide layer was observed (in NV3, NV6, and V at 900°C it is not very evident), and on it a thick layer of corrosion products rich in V, Ni, and O. It is clear that nickel was rapidly corroded by the vanadium salts; however the thick layer of corrosion products formed can act as a barrier against the penetration of the corrosive agent.

Regardless of the molten salt, the X-ray diffraction analysis of the corrosion products showed that these were

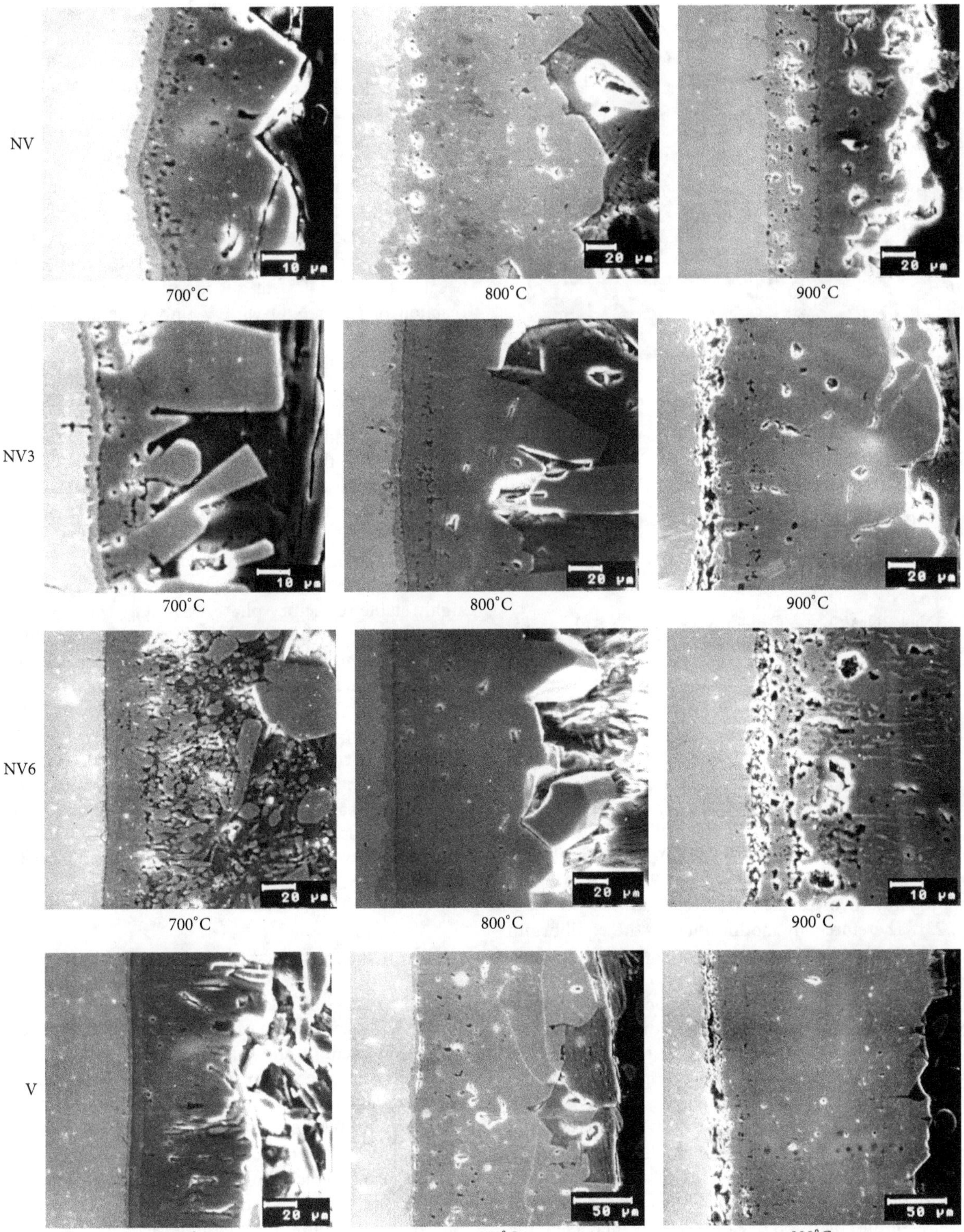

FIGURE 11: Micrographs of the cross sections of Ni exposed to NV3 at 600, 700, 800, and 900°C, during 100 hours.

mainly composed of nickel oxide (NiO) and nickel vanadate ($Ni_3V_2O_8$); similar observations have been made in other studies [21, 22]. This indicates the high corrosiveness of the vanadium salts because they increase the acidic solubility of the metallic oxides [10]. According to the chemical composition of the vanadium salts used, the Ni degradation occurred according to the following reactions:

$$2Ni + O_2 \longrightarrow 2NiO \quad (2)$$

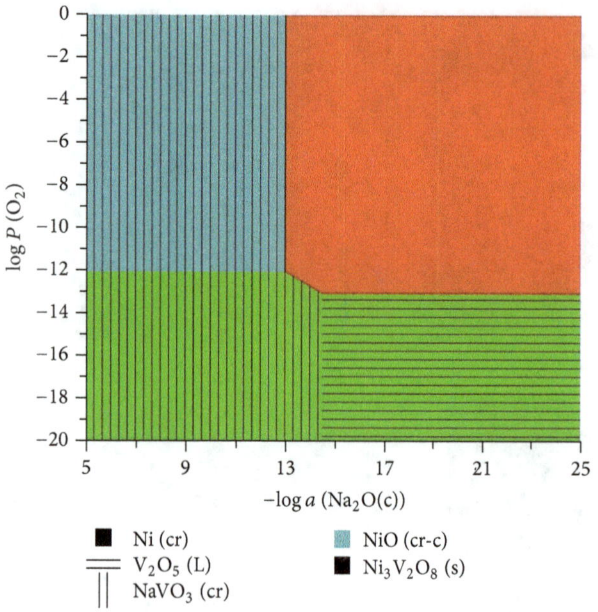

FIGURE 12: Thermodynamic stability diagrams for Ni-Na-V-O at 900°C.

$$3Ni + 2NaVO_3 + \frac{1}{2}O_2 \longrightarrow Ni_3V_2O_8 + Na_2O \quad (3)$$

$$3Ni + V_2O_5 + 2O_2 \longrightarrow Ni_3V_2O_8 + \frac{1}{2}O_2 \quad (4)$$

$$3NiO + 2NaVO_3 \longrightarrow Ni_3V_2O_8 + Na_2O \quad (5)$$

$$3NiO + V_2O_5 \longrightarrow Ni_3V_2O_8 \quad (6)$$

In the corrosion process by vanadium salts in molten state, the melt acts as n-type semiconductors where it is possible that oxygen diffuses as O^{-2} or VO_3^-, and the transfer of charge probably involves the redox reaction $V^{+5} \leftrightarrow V^{+4}$ [16, 20, 23]. Therefore, any modification to the equilibrium of these processes will affect the corrosiveness of the melt. The performance of a material in a highly corrosive medium depends on the chemical stability of both the metallic elements and their compounds such as oxides and vanadates [22]. Therefore, the observed decrease in the Ni corrosion rate as time elapses (LPR measurements) may be due to the high melting point of the nickel vanadate (1220°C), which is considered a refractory compound [24]. The presence of this compound may have limited the mobility of aggressive species towards the metal surface.

Because the Na_2O reduces the acidity of the molten salts, the increase in the corrosion rate observed by increasing the V concentration is consistent with the reactions mentioned above, since at higher V content the proportion of Na_2O formed is reduced. Based on the above reactions, in addition to the equilibrium, $2NaVO_3 = V_2O_5 + Na_2O$, the phase stability diagram shown in Figure 12 was constructed, according to the procedure indicated elsewhere [25].

According to the diagram, it is observed that, in a wide range of partial oxygen pressures equivalent to the melt-atmosphere interface (≈0.21 atm) and to the melt-metal interface (≈1×10^{-4} to 1×10^{-12} atm), if acid vanadium salts (NV3, NV6, and V) are present the only reaction product will be $Ni_3V_2O_8$ and, on the contrary, in low acidity salts (NV) the additional presence of NiO is possible. This is congruent with the micrographs shown in Figure 11. In these it is observed that only in the presence of NV at 900°C a NiO thick layer is evident onto Ni surface. According to other studies the low aggressiveness in NV melts is due to the fact that during its reaction with Ni or NiO, in addition to forming the metallic vanadate, it also generates Na_2O as a byproduct; this further reduces the acidity of the melt and affects the charge transport because the $V^{5+} : V^{4+}$ ratio is modified [11]. In contrast, the reaction of Ni or NiO with V only generates the metallic vanadate without any modification in the acidity of the melt. In the case of the NV3 and NV6 species the decrease in their corrosiveness, compared to that of the V, is due to the presence of NV which reduces the acidity of the mixture. However, its aggressiveness is much greater than that shown by NV, because these complex vanadates in molten state also have the ability to absorb a large amount of oxygen which provides a rapid transport of oxygen to the metal surface [12, 19].

The high initial reactivity of Ni in the presence of vanadium salts has a beneficial effect because it forms a highly stable refractory phase, $Ni_3V_2O_8$, which reduces the corrosiveness of the melt by trapping the vanadium and blocking the migration of nickel ions [3, 23, 26], and the stability of the refractory phase still prevails in the presence of Na_2SO_4 [2]. The results obtained are congruent with other studies where NiO has been proposed as a corrosion inhibitor in vanadium-rich molten salts [2, 3, 24], or Ni-derivatives to remove vanadium compounds [26]. Notwithstanding the above, in the case of nickel-based alloys with a multielement chemical composition the results may be different due to synergistic dissolution reactions between the different oxides developed [11, 27–29].

4. Conclusions

From the results of this work, it can be concluded that Ni shows a high reactivity in vanadium molten salts. An increase in both the acidity and temperature of the melt causes a noticeable increase in the current densities of the anodic and cathodic reactions due to an increase in the transport of charge by the presence of vanadium in multivalent state (V^{5+}, V^{+4}). Corrosion products form a dense and protective barrier onto Ni surface which limits the access of the vanadium salts since no presence of Ni into vanadium melt was observed. The corrosion products increase the activation energy of the corrosive process by at least one order of magnitude. The initial reactivity of the nickel forms a highly stable refractory phase that reduces the corrosiveness of the melt by trapping the vanadium and blocking the migration of nickel ions. The corrosiveness of the vanadium salts increased in the order NV < NV3 < NV6 < V. Low corrosiveness of the NV is due to the fact that during the corrosion process the Na_2O is released as a reaction byproduct, and the melt basicity is increased affecting the charge transport and decreasing the

$V^{5+}:V^{4+}$ ratio. On the other hand, the high corrosiveness of V is due to the fact that its corrosion reactions only form metallic vanadates which do not affect the melt acidity.

Conflicts of Interest

The authors declare that there are no conflicts of interest regarding the publication of this paper.

Acknowledgments

Financial support from Consejo Nacional de Ciencia y Tecnología (CONACYT, México) (Project 159898) is gratefully acknowledged.

References

[1] X. Montero and M. C. Galetz, "Inhibitors and coatings against vanadate-containing oil ash corrosion of boilers," *Surface and Coatings Technology*, vol. 304, pp. 211–221, 2016.

[2] E. Rocca, L. Aranda, and M. Molière, "Chemistry of ash-deposits on gas turbines hot parts: Reactivity of nickel, zinc and iron oxides in (Na, V, S) molten salts," *Materials Science Forum*, vol. 595-598, pp. 169–176, 2008.

[3] E. Rocca, P. Steinmetz, and M. Moliere, "Revisiting the inhibition of vanadium-induced hot corrosion in gas turbines," *Journal of Engineering for Gas Turbines and Power*, vol. 125, no. 3, pp. 664–669, 2003.

[4] C. Cuevas-Arteaga, J. Uruchurtu-Chavarín, J. González, G. Izquierdo-Montalvo, J. Porcayo-Calderón, and U. Cano-Castillo, "Corrosion evaluation of Alloy 800 in sulfate/vanadate molten salts," *Corrosion*, vol. 60, no. 6, pp. 548–560, 2004.

[5] H. Singh, D. Puri, and S. Prakash, "An overview of Na_2SO_4 and/or V_2O_5 induced hot corrosion of Fe-and Ni-based superalloys," *Reviews on Advanced Materials Science*, vol. 16, pp. 27–50, 2007.

[6] J. G. Gonzalez-Rodriguez, S. Haro, A. Martinez-Villafañe, V. M. Salinas-Bravo, and J. Porcayo-Calderon, "Corrosion performance of heat resistant alloys in Na_2SO_4-V_2O_5 molten salts," *Materials Science and Engineering A*, vol. 435-436, pp. 258–265, 2006.

[7] W. J. Quadakkers and M. J. Bennet, "Oxidation induced lifetime limits of thin walled, iron based, alumina forming, oxide dispersion strengthened alloy components," *Materials Science and Technology*, vol. 10, no. 2, pp. 126–131, 1994.

[8] I. Wolf, H. J. Grabke, and P. Schmidt, "Carbon transport through oxide scales on Fe-Cr alloys," *Oxidation of Metals*, vol. 29, no. 3-4, pp. 289–306, 1988.

[9] H. W. Grünling and R. Bauer, "The role of silicon in corrosion-resistant high temperature coatings," *Thin Solids Films*, vol. 95, no. 1, pp. 3–20, 1982.

[10] Y. S. Zhang and R. A. Rapp, "Solubilities of CeO_2, HfO_2 and Y_2O_3 in fused Na_2SO_4-30 mol%$NaVO_3$ and CeO_2 in Pure Na_2SO_4 at 900∘C," *Corrosion*, vol. 43, no. 6, pp. 348–352, 1987.

[11] Y.-S. Hwang and R. A. Rapp, "Thermochemistry and solubilities of oxides in sodium sulfate-vanadate solutions," *Corrosion*, vol. 45, no. 11, pp. 933–937, 1989.

[12] G. W. Cunningham and A. d. S. Brasunas, "The effects of contamination by vanadium and sodium compounds on the air-corrosion of stainless steel," *Corrosion*, vol. 12, pp. 389–405, 1956.

[13] A. Wong, R. I. Marchan, and L. Martinez, "Molten salt corrosion of heat resistant alloys," in *Proceedings of the CORROSION 95*, Paper 465, NACE International, Houston, TX, USA, 1995.

[14] A. Wong-Moreno, Y. Mujica, and L. Martinez, "High temperature corrosion enhanced by fuel oil ash seposits," in *Proceedings of the CORROSION 94*, paper 185, NACE International, Houston, TX, USA, 1994.

[15] J. Porcayo-Calderon, V. M. Salinas Bravo, R. A. Rodríguez-Díaz, and L. Martinez-Gomez, "Effect of the $NaVO_3$-V_2O_5 ratio on the high temperature corrosion of Chromium," *International Journal of Electrochemical Science*, vol. 10, pp. 4928–4945, 2015.

[16] R. B. Dooley and J. R. Wilson, "The Corrosion of a 50Cr-50Ni alloy in liquid vanadate systems in the temperature range 750–950∘C," *Journal of Engineering for Power*, vol. 97, no. 3, pp. 422–428, 1975.

[17] R. Mittelstadt and K. Schwerdtfeger, "The dependence of the oxidation state of vanadium on the oxygen pressure in melts of VOx, Na_2O-VOx, and CaO-SiO_2-VOx," *Metallurgical Transactions B*, vol. 21, no. 1, pp. 111–120, 1990.

[18] I. B. Singh, "Corrosion and sulphate ion reduction studies on Ni and Pt surfaces in with and without V_2O_5 in $(Li,Na,K)_2SO_4$ melt," *Corrosion Science*, vol. 45, no. 10, pp. 2285–2292, 2003.

[19] N. D. Phillips and C. L. Wagoner, "Use of differential thermal analysis in exploring minimum temperature limits of oil ash corrosion," *Corrosion*, vol. 17, no. 8, pp. 396t–400t, 1961.

[20] E. Chassagneux and G. Thomas, "Studies on V_2O_5 induced hot corrosion," *Materials Chemistry and Physics*, vol. 17, no. 3, pp. 273–284, 1987.

[21] S.-I. Muroi, M. Someno, and M. Kobayashi, "Corrosion products of pure metals in V_20_5 Melt," *Tetsu-To-Hagane*, vol. 58, no. 6, pp. 773–786, 1972.

[22] O. Sotelo-Mazón, J. Porcayo-Calderon, C. Cuevas-Arteaga, J. J. Ramos-Hernandez, J. A. Ascencio-Gutierrez, and L. Martinez-Gomez, "EIS Evaluation of Fe, Cr, and Ni in $NaVO_3$ at 700∘C," *Journal of Spectroscopy*, vol. 2014, 10 pages, 2014.

[23] E. Chassagneux, G. Thomas, and M. Soustelle, "Corrosion of Nickel by V_2O_5: Critical phenomenon due to O_2 pressure," *Materials Science and Engineering*, vol. 87, pp. 379–382, 1987.

[24] E. Rocca, L. Aranda, M. Moliere, and P. Steinmetz, "Nickel oxide as a new inhibitor of vanadium-induced hot corrosion of superalloys—comparison to MgO-based inhibitor," *Journal of Materials Chemistry*, vol. 12, no. 12, pp. 3766–3772, 2002.

[25] J. J. Ramos-Hernandez, J. Porcayo-Calderon, V. M. Salinas-Bravo, C. D. Arrieta-Gonzalez, J. G. Gonzalez-Rodriguez, and L. Martinez-Gomez, "Phase stability diagrams for high temperature corrosion processes," *Mathematical Problems in Engineering*, vol. 2013, Article ID 542061, 7 pages, 2013.

[26] N. Di Salvia, M. P. Malavasi, G. Di Salvia, and A. Vaccari, "Formation of Ni and Mg vanadates during the flameless oxy-combustion of heavy fuels," *Fuel Processing Technology*, vol. 138, pp. 534–539, 2015.

[27] R. A. Rapp, "Hot corrosion of materials," *Pure and Applied Chemistry*, vol. 62, pp. 113–122, 1990.

[28] R. A. Rapp, "Hot corrosion of materials: A fluxing mechanism?" *Corrosion Science*, vol. 44, no. 2, pp. 209–221, 2002.

[29] Y. Longa-Nava, Y. S. Zhang, M. Takemoto, and R. A. Rapp, "Hot Corrosion of Nickel-Chromium and Nickel-Chromium-Aluminum Thermal-Spray Coatings by Sodium Sulfate-Sodium Metavanadate Salt," *Corrosion*, vol. 52, no. 9, pp. 680–689, 1996.

21

Experimental Study on the Tensile Property of a Novel Oriented Linear Porous Metal

Bin Liu,[1] Xin-gang Wang,[2] Yong Tang,[1] Wei Yuan,[1] Chao-bin Fang,[1] and Zhen-ping Wan[1]

[1]*School of Mechanical and Automotive Engineering, South China University of Technology, Guangzhou 510640, China*
[2]*China Academy of Space Technology (Xi'an), Xi'an 710000, China*

Correspondence should be addressed to Yong Tang; ytang@scut.edu.cn

Academic Editor: Hossein Moayedi

A novel oriented linear porous metal (OLPM) with high porosity was fabricated through solid-sintering method with cutting copper fibers. In this study, tensile experiments were conducted to investigate the fracture process of the OLPM and the effects of porosity and sintering parameters on the tensile properties. The typical tensile stress-strain plot of this material was obtained based on a large number of tensile test results. The plot can be divided into three stages, namely, initial linear elastic stage, plastic deformation stage, and tensile fracture stage. Based on the results, the fracture mechanism of the OLPM was further studied. Moreover, the porosity and sintering parameters were also varied to investigate their influence on the tensile properties. Tensile strength and plastic deformation were found to decrease with increasing porosity ranging from 70% to 90%. A higher sintering temperature produced a higher tensile strength for the OLPM sintered in the temperature range of 700°C–900°C, but the strength decreased at 1000°C. In addition, the extension of holding time could also slightly affect the tensile strength. Finally, the tensile properties of the OLPM are significantly higher than those of commercial porous metal and porous metal fiber sintered sheet.

1. Introduction

Porous metals with unique structural and functional properties have received extensive attention because of the interesting combinations of their outstanding thermal, acoustic, electrical, and mechanical properties [1, 2]. Porous metal fiber sintered sheet (PMFSS), which is a new type of porous metal, has a three-dimensional reticulated structure characterized by high porosity and large specific surface area, and this material exhibits great potential for various applications in industry. To date, PMFSS is being widely used in defense and military, petrochemical industry, metallurgical machinery, environmental protection for its excellent performance in filtration and separation [3], flow field and gas diffusion [4], energy absorption [5], biomedical device [6], catalytic reaction [7, 8], heat transfer [9], and other applications. With the expansion of their application fields and the increasing complexity of their working environment, some special requirements for their properties of porous metals are becoming increasingly strict. In this study, a novel oriented linear porous metal (OLPM) with oriented linear porous structure and large porosity feature was produced, and this material has shown good potential in flow distribution and pressure drop, and it has been used in methanol steam reforming and presented a good performance [10, 11].

To broaden the application of this material, the relationship between its mechanical properties and its special structure should be elucidated.

In the past decades, many research groups have given significant attention to the fabrication, characterization, and application of porous metals [12, 13]. Thus far, many studies have developed new fabrication methods and improved the mechanical properties of porous metals. Markaki and Clyne [14, 15] produced novel porous sheets using liquid phase sintering method of short stainless steel fibers with 100 μm diameter. They found that the porous sheets had a porosity varying from 75% to 95% and a tensile strength below 1 MPa. Liu et al. [16, 17] developed a porous steel wire mesh material with open porosities prepared by the metallurgical route. The effect of the forming pressure and sintering parameters on the porous structure and mechanical behavior were further discussed. Moreover, Xi et al. [18, 19] fabricated a PMFSS using the vacuum sintering of stainless steel fibers with a diameter of 100 μm, and they introduced several methods to study the compressive and shear properties of PMFSS. Recently, Jin et al. [20] proposed a two-dimensional micromechanics random beam model to investigate the elastic-plastic behavior of PMFSS. In our previous studies [21–23], the tensile, compressive, and shear properties of the PMFSS have been investigated through experimental testing and microstructure observation method. These results provide an effective guideline in improving the design and manufacture methods of PMFSS for different engineering applications.

However, only a few studies have been conducted in the field of oriented linear porous metal. Huang et al. [24] fabricated a kind of porous copper with long cylindrical pores through directional solidification, and the compressive properties were investigated through experimental methods. Liu et al. [25] fabricated lotus-type porous copper bars with 45 mm diameter and 120 mm length through unidirectional solidification and the compressive deformation process was investigated, and the deformation mechanism was analyzed and discussed.

The above reported studies show that detailed research on the tensile properties of oriented linear porous metal using cutting metal fibers is lacking in the current literatures. This study investigated the tensile fracture process of this material produced by the solid-sintering of copper fibers. After the microstructures are examined, the tensile properties of the oriented linear porous metal are further analyzed and discussed by varying the porosity and sintering parameters. Finally, the tensile property of OLPM was compared with the commercial porous metal and the PMFSS.

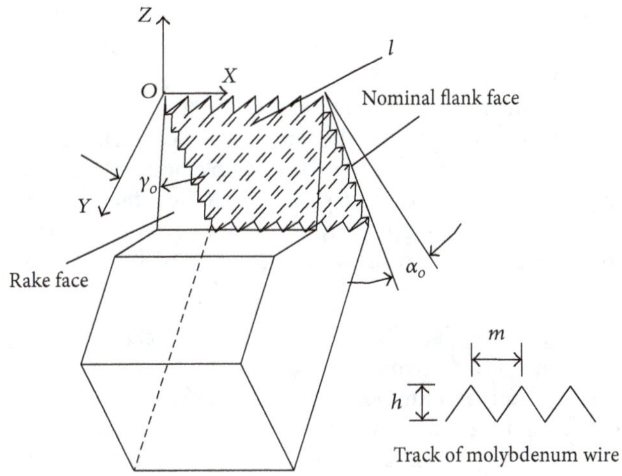

FIGURE 1: Schematic drawing of the multitooth tool.

FIGURE 2: In situ photo of copper fiber manufacturing.

2. Experimental Procedure

2.1. Manufacturing Process of OLPM. As reported previously [10], the preparation of oriented linear porous metal mainly includes three processes, namely, the fabrication of cutting fibers, the mold-pressing of parallel fibers, and the sintering with high temperature in a protective gas atmosphere.

First, long copper fibers were fabricated by a multitooth tool on a horizontal lathe. Figure 1 shows the schematic drawing of the multitooth tool; Figure 2 shows the fiber manufacturing process. The tool was installed on the turret with a setting angle of 45 degrees, and it has multiple tiny teeth fabricated through wire electrical discharge machining for achieving the simultaneous fabrication of multiple fibers. The distance and height of the tiny teeth were 0.3 mm and 0.2 mm, respectively. The diameter of the cutting fibers was approximately 100 μm. The cutting fibers were collected by twining the fibers in a slab parallel to one another.

Second, the copper fibers were cut into several segments with fixed length of 120 mm. Thereafter, multiple segments of copper fibers were uniformly arranged in parallel and put into the packing chamber of a mold pressing equipment, in which multiple stainless steel plates patterned with hollow chamber were first stacked and then tightened with four screws and two end plates. Through this means, the semifinished OLPM with the same shape as the predetermined packing chamber was obtained.

Later, the mold pressing equipment was put into a box-type furnace, which provided the hydrogen gas atmosphere with a constant pressure of 0.3 MPa. The sintering temperature was set in the range of 700°C to 1000°C and was controlled by a programmable temperature controller. Stage heating method was used to optimize the heating rate. The heating rate was kept at 300°C/h when the temperature was below 800°C, whereas it reduced to 200°C/h as the temperature rose above 800°C. The holding times were set as 30 min, 60 min, 90 min, or 120 min. When the sintering process was completed, the furnace was cooled to 150°C, and then the mold pressing equipment was taken out of the furnace and continued to cool down to room temperature. Lastly, the mold pressing equipment was disassembled and hence the OLPM with uniform thickness was formed.

The average porosity of the OLPM can be calculated using the quality-volume method formulated by the following because the obtained porous metal has a regular geometric shape:

$$\theta\,(\%) = \left(1 - \frac{M}{\rho V}\right) \times 100, \quad (1)$$

where θ is the porosity of the porous metal, V is the volume of the porous metal (cm^3), M is the mass of the porous metal (g), and ρ is the density of red copper (g/cm^3).

The dimension of the OLPM sample is 120 mm $*$ 15 mm $*$ 2 mm. Before being tested, the sample was cleaned ultrasonically with acetone for 5 min and then with deionized water for 5 min; finally, it was left to dry in air at ambient temperature. The microscopic morphology of the OLPM was observed using a scanning electron microscope (JSM-6380LA, Japan).

2.2. Tensile Test of OLPM. The tensile test was carried out on a PC-controlled electronics universal testing machine (RGL-20A) with a maximum load of 20 kN, performing both test control and data acquisition. Before test, the sample was clapped in the machine to make sure that the test part was 60 mm in length. To avoid the deformation of the three-dimensional network structure in the clamping process, rubber pads were placed between clamps. All the samples were pulled to failure at a constant tensile rate of 1.5 mm/min; and the tensile direction was the same with the direction of metal fibers. During each test, the load and displacement were recorded continuously. The strain is defined as the ratio between the sample elongation and the initial gauge length. The tensile stress is calculated by dividing the pulling force by the initial section of the sample. The maximum stress in the stress-strain plots was determined as the tensile strength of the OLPM. In order to obtain an accurate value of tensile strength, three specimens were tested in each case, and the average was taken.

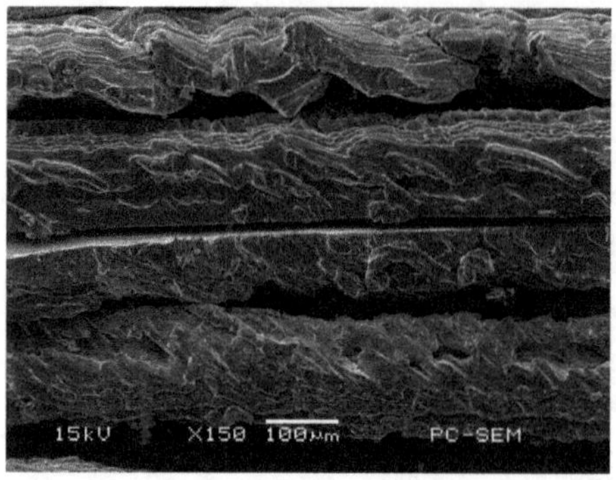

FIGURE 3: SEM image of the OLPM.

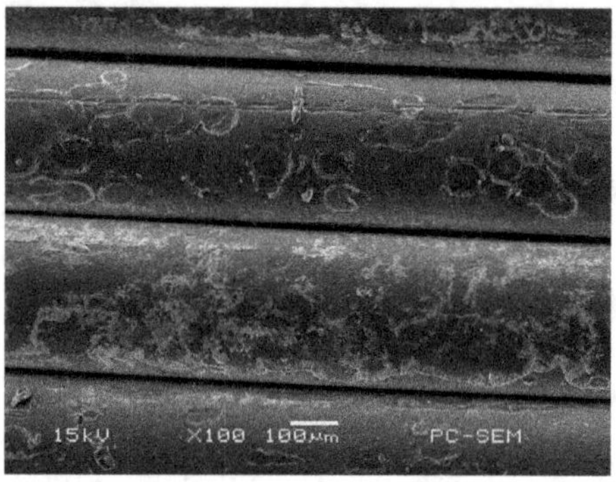

FIGURE 4: SEM image of the CPM.

3. Results and Discussion

3.1. Microstructure of OLPM. The SEM image of the OLPM sintered at 800°C for 30 min is shown in Figure 3. It can be seen from Figure 3 that there are many sintering joints and pore structures in the OLPM. The contact points among metal fibers transferred into sintering joints and formed the pore structure and the three-dimensional network structure of the OLPM during the sintering process. The pore structure formed among the parallel fibers and sintering joints showed different appearance, like long strip pore structure and point pore structure, but they are all along with the direction of the metal fiber. We can get a wide range of porosity (60%–98%) by changing the quantity of metal fibers and the molding pressure.

By further investigation of Figure 3, we can see that there are many microstructures on the surface of metal fibers. These rough surface morphologies are the unique features of the OLPM compared with other commercial porous metals shown in Figure 4. The surface of the metal fiber in the Biało et al. research is also very smooth [26]. These metal fibers are fabricated by melt-extraction [27] or bundle drawing method [28].

What is more, the rough surface can favor the sintering joints and improve the mechanical properties of the oriented linear porous metal. The rough surface increases the friction between fibers and provides an additional occlusal force that makes the fibers bond tightly, thus forming a relatively strong structure. More importantly, the rough surface can provide higher specific surface area and can promote adhesive strength of the primer to the metallic support when the OLPM is used as catalyst support [10, 11].

3.2. Tensile Fracture Process of OLPM. Figure 5 shows the representative tensile stress-strain curve of an OLPM sample with 70% porosity sintered at 700°C for 30 min. It can been seen from this figure that the tensile behavior of the OLPM sample can be divided into three stages: (a) initial linear elastic stage, (b) plastic deformation stage, and (c) fracture stage, which correspond to OA, AB, and BC in the curve, respectively.

The figure shows that the engineering stress increased linearly with the increasing strain in the initial stage, namely, the OA part of the plot. This process only lasts for a short time, and the elastic deformation of this sample is about 2%. The relationship between stress and strain fits well with Hooke's law of elasticity. That is to say, in the initial elastic process, the oriented linear porous metal elongated in the tensile direction, and only a slight deformation occurs on the three-dimensional reticulation.

After the elastic process, the sample reaches the plastic deformation process quickly. In this stage, deformation begins to occur in some metal fibers first, where the local stress concentration occurs, fibers reach the yield limitation first, and the plastic deformation starts here. The local plastic deformation results in stress release and redistribution, which leads to a greater plastic deformation of more fibers with

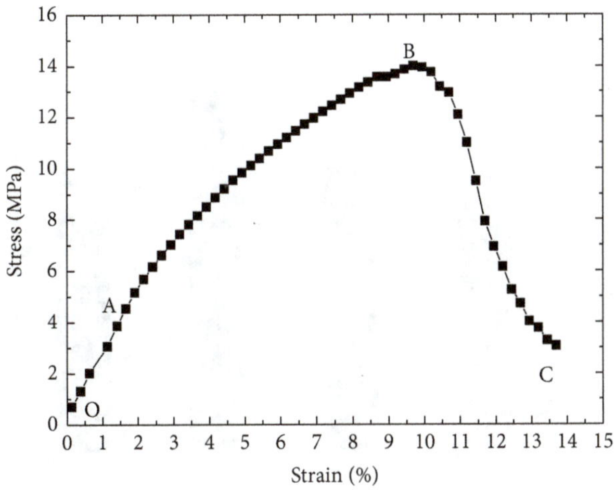

FIGURE 5: Typical tensile stress-strain curve of an oriented linear porous metal sample.

FIGURE 6: Fracture image of the OLPM in the thickness direction.

FIGURE 7: Appearance of the fractured OLPM.

tension continuously increasing. As the plastic deformation capacity of the three-dimensional reticulated skeleton is exhausted, the stress reached the ultimate tensile strength.

With the fracture of the metal fiber gradually evolving into the sintering joint and the entire three-dimensional reticulated structure, the three-dimensional structure becomes prone to fracture. And the porous metal reaches the fracture stage; the engineering stress decreased sharply with the increasing strain in this stage; and this phenomenon lasts until the structure cannot undertake any load. And finally, the sample fails completely.

3.3. Tensile Fracture Mechanism of OLPM. To get a better understanding of the fracture of the OLPM, the facture mechanism of this new material was investigated in this study. The fracture process and the fractured samples were observed and analyzed.

The fracture image of the OLPM in the thickness direction was shown in Figure 6. It is clearly shown in Figure 6 that the fracture region becomes thicker and looser than any other part of the material because the fracture process first occurs in the surface of the material, where the restriction of the fibers is relatively smaller than that in the center of the material and where the deformation is larger than that in the center.

In addition, an obvious necking phenomenon is observed in the width direction along with the tensile load during the fracture process, as shown in Figure 7. This figure indicates that the oriented linear porous metal is a typical plastic material. Besides that, there is no obvious fracture trace in the fracture region of the OLPM. These phenomena are quite different from our previous studies with PMFSS [21], which is also a porous material made from metal fiber. This phenomenon can be caused by the following reasons. First, the metal fibers in the OLPM are parallel to one another, and the sintering joints are also found along with the direction of the metal fiber; that is, all the metal fibers in the OLPM are almost in the same condition, and this condition is good for the material to maintain the plastic property of the metal fiber itself. Figure 8 shows the fracture morphology of metal fibers in the OLPM. It clearly shows that the metal fiber also undergoes a necking deformation under the tensile strength and maintains the original performance itself. Second, unlike the PMFSS, no fiber-to-fiber cross sintering joints and no pore structure perpendicular to the fiber direction restrict the elongation and deformation in the load direction. So the material can extend in the tensile direction. This structure is helpful to strengthen the mechanical properties and the ductility of the OLPM.

3.4. Influence of Porosity. Porosity is one of the most important factors of porous metals, and it greatly influences the mechanical properties of porous metals [29, 30]. To analyze the influence of porosity on the tensile property of the OLPM, a single factor experiment was conducted. Three kinds of OLPM sintered at 800°C for 30 min with porosities of 70%, 80%, and 90% were used to investigate the effect of porosity on the tensile property.

Figure 9 shows the tensile stress-strain plots of the OLPM with different porosities sintered at 800°C for 30 min. The tensile strength of the OLPM decreases greatly with the increase of porosity. As shown in Figure 8, the tensile strength of the sample with 70% porosity is 16.68 MPa, whereas those

FIGURE 8: SEM image of fractured metal fibers in OLPM.

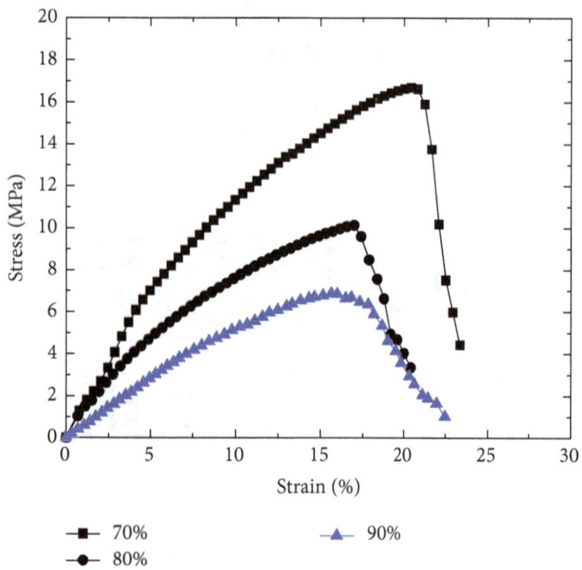

FIGURE 9: Tensile stress-strain curves of the oriented linear porous metal with different porosities.

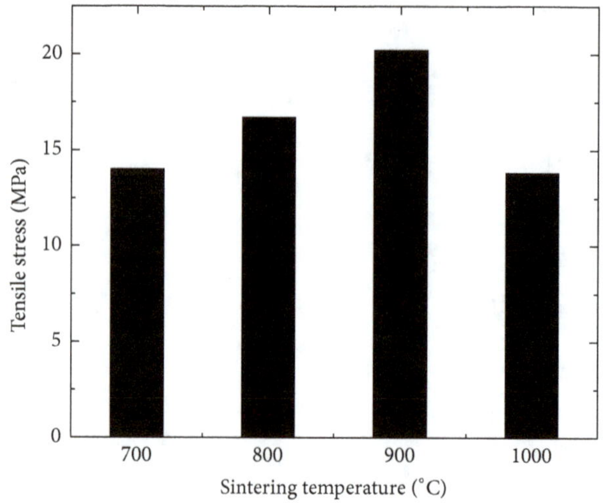

FIGURE 10: Tensile strength of the oriented linear porous metal with different sintering temperatures.

of the samples with the porosities of 80% and 90% are only 10.12 MPa and 7.37 MPa, respectively. This result fits well with the result of Qiao et al. [19]. This result is mainly attributed to the following two aspects. On the one hand, the OLPM sample with lower porosity has more fibers and sintering joints in its cross-section to withstand the tensile force. Conversely, the sample with higher porosity has less fibers and contact regions to form sintering joints. On the other hand, in order to obtain high porosity, sparse fibers are distributed in the mold and low compaction force is applied, which hinder the formation of solid-sintering joints and lead to the sintering joints weaker than those of low porosity. And the stress concentration first occurs at them. These two aspects both result in low tensile strength in high porosity.

In addition, Figure 9 also illustrates that the sample with lower porosity shows a longer plastic stage. The total deformation of OLPM with 70% porosity is 20.85%, while the ductility of OLPMs with 80% and 90% porosity is 16.97% and 15.66%, respectively. That is because higher porosity means less metal fibers in the OLPM, and the faster the fracture process transfer into the three-dimensional structure of the OLPM, the smaller the deformation of the whole process.

After the plastic deformation stage, the fracture rates of the samples with different porosities are not identical. The stress-strain curve of the sample with lower porosity drops more sharply, which means that the sample fractures faster.

3.5. Influence of Sintering Parameters. As important manufacturing parameters, sintering temperature and sintering time greatly influence the forming process of the oriented linear porous metal.

3.5.1. Sintering Temperature. Four samples with 70% porosity sintered at 700°C, 800°C, 900°C, and 1000°C for 30 min were selected to investigate the influence of sintering temperature on the tensile strength of the oriented linear porous metal. Figure 10 shows the tensile strength plot of the OLPM samples sintered at different temperatures.

Obviously, the tensile strength of the OLPM increases with sintering temperatures between 700°C and 900°C, but it decreases at 1000°C. This mainly because the formation of sintering joints is controlled by atom diffusion, which is a random thermal motion. High sintering temperature increases the energy of copper atoms [31, 32]. On the one hand, it increases the number of activated copper atoms and the opportunity of copper fibers forming the sintering joints. On the other hand, it is beneficial for the migration of copper atoms, thus accelerating the sintering process [33]. Therefore, the number of sintering joints increases and sintering joints grow larger. These phenomena will lead to the improvement of the tensile property of the OLPM. Therefore, in the range of 700°C to 900°C, the tensile strength increases with sintering temperature. However, during the sintering process, the metal fiber undergoes a heat treatment process, the increasing sintering temperature enlarges the grain size of the copper fiber when the temperature is higher than the recrystallization temperature, and the microstructure on the surface of the metal fiber will disappear at high temperature

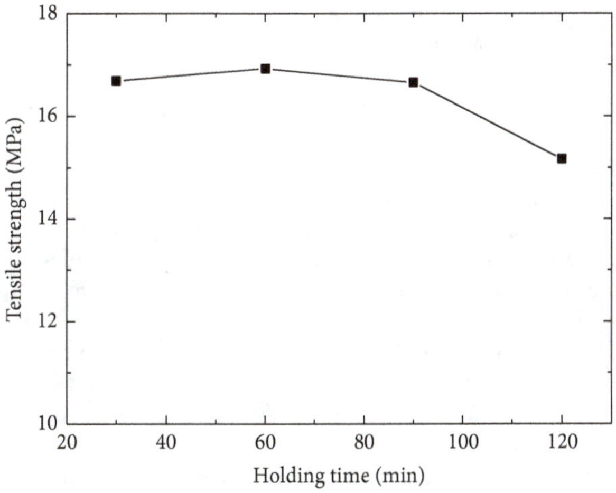

FIGURE 11: Tensile strength of the oriented linear porous metal with different holding times.

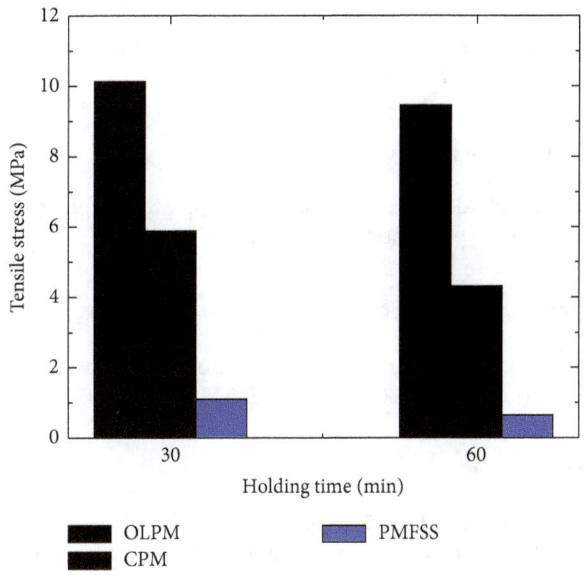

FIGURE 12: Tensile strength of three different porous metals.

[34]. These phenomena will all lead to the decrease of tensile property at 1000°C. This situation provides us with a guide that a reasonable sintering temperature to fulfill different requirements could be chosen.

3.5.2. Holding Time. The tensile property of the porous metal with the porosity of 70% sintered at 800°C for 30 min, 60 min, 90 min, and 120 min is shown in Figure 11. From Figure 11, we can conclude that holding time has slight effect on the tensile property of OLPM. When the sintering time was extended from 30 min to 60 min, the tensile strength increased from 16.68 MPa to 16.92 MPa. However, the sintering time of 60 min leads to the maximum tensile strength at the sintering temperature of 800°C, and the tensile strength decreases to 16.65 MPa and 15.16 MPa when the sintering time increases to 90 min and 120 min, respectively. The difference between different holding time is 11%.

The results are caused by the following two aspects. On the one hand, sintering joints formed between fibers become larger with the increase of sintering time, leading to the improvement of the tensile properties of the porous metal. Meanwhile, the increase of sintering time caused the grain of the copper fibers to become coarse, leading to the decrease in the mechanical property of the fiber itself and then to the decrease of the tensile properties of the material. The combination of the two aspects leads to the result. In addition, the lengthening of the sintering time increases the manufacturing cost of the oriented linear porous metal. Therefore, improving the tensile strength of the porous metal by extending the sintering time is not logical.

3.6. Comparing with Commercial Porous Metal and PMFSS. The tensile strength of the oriented linear porous metal was compared with the commercial porous metal and the porous metal fiber sintered sheet [21]. Figure 12 shows the tensile strengths of three kinds of materials sintered at 800°C for 30 min and 60 min with 80% porosity. The figure shows that the tensile strength of the OLPM is significantly higher than those of the other two kinds of materials. This result is because that the surface of the metal fiber in CPM is smooth, as shown in Figure 4. It is relatively difficult for the metal fiber to form sintering joints between the smooth surfaces [35], so the tensile strength is relatively lower than that of the oriented linear porous metal. The metal fibers in the PMFSS occlude one another, and not all the metal fibers can resist the tension force, so the tensile strength is significantly lower than that of the oriented linear porous metal.

These results obtained provide us with an important technical guideline for different potential applications of the OLPM, especially for the conditions with the tensile loading.

4. Conclusion

(1) A novel oriented linear porous metal with a three-dimensional network structure was produced through the solid-state sintering method with cutting copper fibers. With a special long strip pore structure and numerous microstructures on the surface of the metal fiber, this material exhibits unique properties during the tensile fracture process. And the tensile process of the OLPM included the following three stages: the elastic stage at low strain, the plastic deformation stage, and the damage and fracture of the overall network structure.

(2) The tensile strength depends heavily on the porosity. The tensile strength of the OLPM decreased rapidly with the increasing porosity, and the extent of the deformation increases with the decreasing porosity.

(3) The sintering parameters also play important role in the tensile properties of the OLPM. The tensile strength increases with sintering temperature from 700°C to 900°C and then decreases at 1000°C. The extension of holding time slightly affected the tensile strength of the OLPM, and 900°C is the optimal sintering temperature of this material.

(4) The tensile strength of the OLPM is much higher than the PMFSS and the CPM. The results obtained in this work

provide an important technical guideline for different potential applications of the OLPM, especially for the conditions with the tensile loading.

Conflict of Interests

The authors declare that there is no conflict of interests regarding the publication of this paper.

Acknowledgment

This work was supported by the National Natural Science Foundation of China (Grants nos. 51275180 and 51475172).

References

[1] L.-P. Lefebvre, J. Banhart, and D. C. Dunand, "Porous metals and metallic foams: current status and recent developments," *Advanced Engineering Materials*, vol. 10, no. 9, pp. 775–787, 2008.

[2] G. J. Davies and S. Zhen, "Metallic foams: their production, properties and applications," *Journal of Materials Science*, vol. 18, no. 7, pp. 1899–1911, 1983.

[3] A. Nabovati, E. W. Llewellin, and A. C. M. Sousa, "A general model for the permeability of fibrous porous media based on fluid flow simulations using the lattice Boltzmann method," *Composites Part A: Applied Science and Manufacturing*, vol. 40, no. 6-7, pp. 860–869, 2009.

[4] J.-K. Kuo and C.-K. Chen, "A novel Nylon-6–S316L fiber compound material for injection molded PEM fuel cell bipolar plates," *Journal of Power Sources*, vol. 162, no. 1, pp. 207–214, 2006.

[5] Z. Bo and C. Tianning, "Calculation of sound absorption characteristics of porous sintered fiber metal," *Applied Acoustics*, vol. 70, no. 2, pp. 337–346, 2009.

[6] G. Ryan, A. Pandit, and D. P. Apatsidis, "Fabrication methods of porous metals for use in orthopaedic applications," *Biomaterials*, vol. 27, no. 13, pp. 2651–2670, 2006.

[7] W. Yuan, Y. Tang, X. Yang, and Z. Wan, "Toward using porous metal-fiber sintered plate as anodic methanol barrier in a passive direct methanol fuel cell," *International Journal of Hydrogen Energy*, vol. 37, no. 18, pp. 13510–13521, 2012.

[8] Y. Tang, W. Zhou, M. Q. Pan, H. Q. Chen, W. Y. Liu, and H. Yu, "Porous copper fiber sintered felts: an innovative catalyst support of methanol steam reformer for hydrogen production," *International Journal of Hydrogen Energy*, vol. 33, no. 12, pp. 2950–2956, 2008.

[9] A. S. Abduljalil, Z. B. Yu, and A. J. Jaworski, "Selection and experimental evaluation of low-cost porous materials for regenerator applications in thermoacoustic engines," *Materials and Design*, vol. 32, no. 1, pp. 217–228, 2011.

[10] M. Q. Pan, Y. Tang, X. L. Wei, and J. Xiang, "Oriented linear cutting fiber sintered felt as an innovative catalyst support for methanol steam reforming," *International Journal of Hydrogen Energy*, vol. 36, no. 12, pp. 7066–7073, 2011.

[11] M. Q. Pan, X. L. Wei, and Y. Tang, "Factors influencing methanol steam reforming inside the oriented linear copper fiber sintered felt," *International Journal of Hydrogen Energy*, vol. 37, no. 15, pp. 11157–11166, 2012.

[12] J. Banhart, "Manufacture, characterisation and application of cellular metals and metal foams," *Progress in Materials Science*, vol. 46, no. 6, pp. 559–632, 2001.

[13] H. Nakajima, "Fabrication, properties and application of porous metals with directional pores," *Progress in Materials Science*, vol. 52, no. 7, pp. 1091–1173, 2007.

[14] A. E. Markaki, V. Gergely, A. Cockburn, and T. W. Clyne, "Production of a highly porous material by liquid phase sintering of short ferritic stainless steel fibres and a preliminary study of its mechanical behaviour," *Composites Science and Technology*, vol. 63, no. 16, pp. 2345–2351, 2003.

[15] T. W. Clyne, A. E. Markaki, and J. C. Tan, "Mechanical and magnetic properties of metal fibre networks, with and without a polymeric matrix," *Composites Science and Technology*, vol. 65, no. 15, pp. 2492–2499, 2005.

[16] P. Liu, G. He, and L. H. Wu, "Fabrication of sintered steel wire mesh and its compressive properties," *Materials Science and Engineering A*, vol. 489, no. 1-2, pp. 21–28, 2008.

[17] P. Liu, G. He, and L. Wu, "Uniaxial tensile stress—strain behavior of entangled steel wire material," *Materials Science and Engineering A*, vol. 509, no. 1-2, pp. 69–75, 2009.

[18] Z. P. Xi, H. P. Tang, J. Y. Wang, Y. J. Wu, L. F. Dong, and B. J. Yang, "Study on the mechanical properties of porous metals," *Rare Metal Materials and Engineering*, vol. 36, pp. 555–558, 2007.

[19] J. C. Qiao, Z. P. Xi, H. P. Tang, J. Y. Wang, and J. L. Zhu, "Compressive behavior of porous metal fibers," *Rare Metal Materials and Engineering*, vol. 37, no. 12, pp. 2173–2176, 2008.

[20] M. Z. Jin, C. Q. Chen, and T. J. Lu, "The mechanical behavior of porous metal fiber sintered sheets," *Journal of the Mechanics and Physics of Solids*, vol. 61, no. 1, pp. 161–174, 2013.

[21] W. Zhou, Y. Tang, M. Q. Pan, X. L. Wei, and J. H. Xiang, "Experimental investigation on uniaxial tensile properties of high-porosity metal fiber sintered sheet," *Materials Science and Engineering A*, vol. 525, no. 1-2, pp. 133–137, 2009.

[22] W. Zhou, Y. Tang, B. Liu et al., "Compressive properties of porous metal fiber sintered sheet produced by solid-state sintering process," *Materials & Design*, vol. 35, pp. 414–418, 2012.

[23] Z. Wan, B. Liu, W. Zhou, Y. Tang, K. S. Hui, and K. N. Hui, "Experimental study on shear properties of porous metal fiber sintered sheet," *Materials Science and Engineering A*, vol. 544, pp. 33–37, 2012.

[24] F. Huang, T.-W. Yang, Z.-J. Li, Z.-H. Li, Q.-L. Jin, and R. Zhou, "Anisotropic compressive properties of ordered porous copper," *The Chinese Journal of Nonferrous Metals*, vol. 21, no. 3, pp. 604–610, 2011.

[25] X.-H. Liu, D. Yao, X.-F. Liu, and J.-X. Xie, "Deformation behaviors and constructive relation of lotus-type porous copper under compressive direction perpendicular to pores," *The Chinese Journal of Nonferrous Metals*, vol. 19, no. 7, pp. 1237–1244, 2009.

[26] D. Biało, L. Paszkowski, W. Wiśniewski et al., "Properties of high porosity structures made of metal fibers," in *Recent Advances in Mechatronics*, pp. 470–474, Springer, Berlin, Germany, 2007.

[27] N. I. Baik, Y. Choi, and K. Y. Kim, "Fabrication of stainless steel and aluminum fibers by PDME method," *Journal of Materials Processing Technology*, vol. 168, no. 1, pp. 62–67, 2005.

[28] R. De Bruyne, "Bundle-drawn metal fibers," *Advanced Materials and Processes*, vol. 147, no. 6, pp. 33–34, 1995.

[29] P. S. Liu, "Mechanical relations for porous metal foams under several typical loads of shearing, torsion and bending," *Materials Science and Engineering A*, vol. 527, no. 29-30, pp. 7961–7966, 2010.

[30] P. S. Liu, T. F. Li, and C. Fu, “Relationship between electrical resistivity and porosity for porous metals,” *Materials Science and Engineering A*, vol. 268, no. 1-2, pp. 208–215, 1999.

[31] R. M. German, *Sintering Theory and Practice*, Wiley, New York, NY, USA, 1996.

[32] P. Wang, *The Powder Metallurgy*, Metallurgical Industry Press, Beijing, China, 1997.

[33] J. Pan and A. C. F. Cocks, “A numerical technique for the analysis of coupled surface and grain-boundary diffusion,” *Acta Metallurgica et Materialia*, vol. 43, no. 4, pp. 1395–1406, 1995.

[34] X. L. Wei, M. Q. Pan, D. H. Dehuai, Q. Wang, Y. Tang, and W. Zhou, “Effect of heating temperature on microstructure and mechanical properties of continuous copper fiber,” in *Proceedings of the International Conference on Industrial Mechatronics and Automation (ICIMA ’09)*, pp. 475–478, IEEE, Chengdu, China, May 2009.

[35] D. Gouvêa and R. H. R. Castro, “Sintering: the role of interface energies,” *Applied Surface Science*, vol. 217, no. 1-4, pp. 194–201, 2003.

22

On Best Corrected Mixture Problems in Metallurgy

F. Dubeau and M. E. Ntigura Habingabwa

Département de Mathématiques, Université de Sherbrooke, 2500 Boulevard de l'Université, Sherbrooke, QC, Canada J1K 2R1

Correspondence should be addressed to F. Dubeau; francois.dubeau@usherbrooke.ca

Academic Editor: Risto Lahdelma

We consider three simple mixture problems occurring in metallurgy. The first problem considered is the classic minimum cost mixture problem. For the second problem, we consider finding a correction to a given mixture, a premix, without considering the cost of this premix. We only consider the cost and the weight of the quantity used as a correction. We show that the minimum cost correction does not correspond to the minimum weight correction, and we built the Pareto curve that gives all intermediate solutions between these two extreme solutions. Finally, the third problem is the correction problem for a nonfree premix. The correction is done to obtain a minimum cost corrected mixture.

1. Introduction

Using materials, such as scrap metal for recycling, in order to form blends with desired specifications on some basic chemical elements is a well-known mixture problem met in foundries. People working in these fields bring to our attention the three different but related problems that we are going to consider in this paper. The first problem, called the *basic problem*, is to make a minimum cost mixture while checking the constraints on the specifications. It is a classic mixture problem solved by linear programming. For the second problem, called the *correction problem*, we start with a given mixture, called here a *premix*, which does not satisfy all the constraints on the specifications. Without considering the cost of this premix, we look for a *correction*, which is itself a mixture of available materials, that we will add to the premix in such a way that the resulting *corrected mixture*, which is the premix and the correction added together, satisfies all the constraints on the specifications. The correction can be done at minimum cost or at minimum weight. This problem is an example of a bicriteria linear programming problem, and we are going to identify its Pareto set which gives the link between minimum cost solution and minimum weight solution. Finally, the third problem is similar to the preceding correction problem, but now we take into account the cost of the given premix. We look for a minimum unit cost corrected mixture. This is an example of a linear-fractional programming problem. It is important to point out that the questions raised in this paper and the data used in the examples are inspired from a real situation.

Mixture problems are important applications in Operations Research [1]. One such application is the diet problem which received a lot of attention [2–7]. A variant of this problem is the pooling problem [8]. Also other approaches could be used to analyze similar mixture problems; for example, the fuzzy approach could be used to introduce uncertainties on the data [9].

2. Basic Problem

2.1. Problem Description. For the basic problem, there are N materials that can be used to produce a mixture. Let $\pi_n \geq 0$ be the weight of the nth material in the final mixture ($n = 1, \ldots, N$). The total weight of the mixture is

$$z_{\text{weight}}(\pi) = \sum_{n=1}^{N} \pi_n. \tag{1}$$

The total cost of the mixture of weight z_{weight} is then

$$z_{\text{cost}}(\pi) = \sum_{n=1}^{N} c_n \pi_n, \tag{2}$$

where $c_n \geq 0$ is the unit costs of the nth material (for $n = 1, \ldots, N$).

The specifications of the mixture and of the materials are the amount (proportion % of the content) of known basic chemical elements (aluminium, carbon, chromium, copper, iron, etc.) contained in a unit weight of the material. Typically, there may be between 10 and 20 basic chemical elements that characterize a chemical mixture. Each nth material is characterized by M specifications noted by $s_{m,n}$ ($m = 1, \ldots, M$).

The specifications of the mixture are determined by a linear combination of the specifications of the N elements of the mixture. If we denote by spec_m, the mth specification of the mixture, we have

$$\text{spec}_m z_{\text{weight}}(\pi) = \sum_{n=1}^{N} s_{m,n} \pi_n. \tag{3}$$

The specifications of the mixture are subjected to two types of constraints.

Type 1. Each specification must be in a well-defined range; that is to say,

$$\text{spec}_m^{\min} \leq \text{spec}_m \leq \text{spec}_m^{\max} \tag{4}$$

for $m = 1, \ldots, M$, where $\text{spec}_m^{\min} \geq 0$ and $\text{spec}_m^{\max} \leq 1$ are, respectively, the minimum and maximum value allowed for the mth specification of the mixture.

Type 2. Specifications can be connected or related by J linear relations of the form

$$\sum_{m=1}^{M} \alpha_{j,m} \text{spec}_m \begin{Bmatrix} \leq \\ = \\ \geq \end{Bmatrix} \beta_j \tag{5}$$

for $j = 1, \ldots, J$. For example, in Figure 1, we illustrate a situation where there are 4 links between the specifications of two chemical elements denoted as A and B. The pair of specifications $(\text{spec}_A, \text{spec}_B)$ must be within the region bounded by the 4 given straight lines.

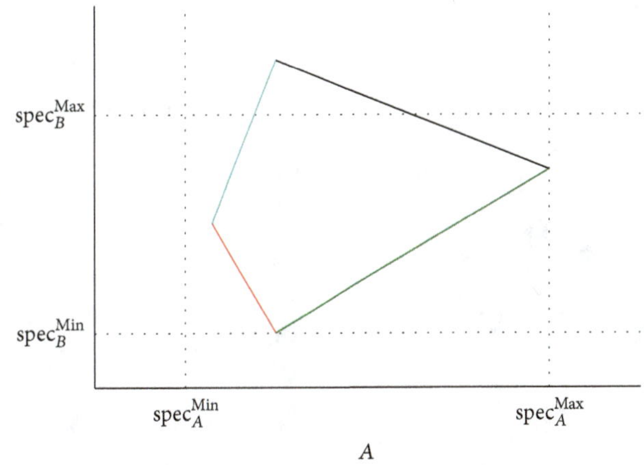

FIGURE 1: Related specifications.

2.2. Problem Formulation. The basic problem is to look for a minimum cost mixture of unit weight; that is to say, $z_{\text{weight}}(\pi) = 1$, satisfying constraints of Types 1 and 2. The decision variables are π_n for $n = 1, \ldots, N$; they will then be the proportions of each ingredient in the unit weight mixture. The mathematical formulation is therefore

$$\min \quad z_{\text{cost}}(\pi) = \sum_{n=1}^{N} c_n \pi_n,$$

subjet to (i) $\sum_{n=1}^{N} \pi_n = 1$;

(ii) for $m = 1, \ldots, M$:

$$\sum_{n=1}^{N} s_{m,n} \pi_n \begin{cases} \leq \text{spec}_m^{\max}, \\ \geq \text{spec}_m^{\min}; \end{cases}$$

(iii) for $j = 1, \ldots, J$:

$$\sum_{n=1}^{N} \left[\sum_{m=1}^{M} \alpha_{j,m} s_{m,n} \right] \pi_n \begin{Bmatrix} \leq \\ = \\ \geq \end{Bmatrix} \beta_j;$$

(iv) $\pi_n \geq 0$ for $n = 1, \ldots, N$.

$$(P)$$

Let $\mathcal{S}$ be the notation for the set of feasible solutions to this problem.

2.3. Numerical Example. In our numerical example coming from a real context, we have $N = 8$ materials and the mixture has specifications on $M = 11$ chemical elements. The minimum and maximum values for the specifications of the mixture are listed in Table 1. We have 2 specifications on 11 basic chemical elements; then there are 22 constraints concerning the concentration of the basic chemical elements in the mixture. Specifications and costs of the 8 materials are given in Table 2.

Constraints of the first type are completely determined using the data in Tables 1 and 2. Constraints of the second type require some clarification. We consider 4 constraints of Type 2 connecting carbon ($m = 2$) and chromium ($m = 3$) specifications. Constraints are described geometrically in Figure 2. They indicate that the pair $(\text{spec}_2, \text{spec}_3)$ should be inside the parallelogram enclosed by 4 straight lines. With the data provided, the constraints are

$$\begin{aligned}
&\text{line no. 1: } \text{spec}_3 - \frac{50}{9}\text{spec}_2 \leq 9.2500 \\
&\text{line no. 2: } \text{spec}_3 - \frac{50}{9}\text{spec}_2 \geq 7.6944 \\
&\text{line no. 3: } \text{spec}_3 + 10\text{spec}_2 \leq 54.05 \\
&\text{line no. 4: } \text{spec}_3 + 10\text{spec}_2 \geq 51.25.
\end{aligned} \tag{6}$$

The coefficients of these constraints are given in Table 3.

TABLE 1: Minimum and maximum specifications in % of weight in the mixture.

	Specifications (%) for $M = 11$ chemical element of the mixture			
m	Chemical element		$\text{spec}_m^{\min}$	$\text{spec}_m^{\max}$
1	Al	Aluminium	0	0.2
2	C	Carbon	2.7	2.98
3	Cr	Chromium	23.25	25.25
4	Cu	Copper	0	0.2
5	Fe	Iron	65	85
6	Mn	Manganese	0.3	0.6
7	Mo	Molybdenum	0	0.3
8	Ni	Nickel	0	0.2
9	P	Phosphorus	0	0.06
10	S	Sulphur	0	0.06
11	Si	Silicone	0.3	0.8

TABLE 2: Characteristics of available materials.

	$s_{m,n}$ specifications (%) of $N = 8$ available materials							
	n							
m	1	2	3	4	5	6	7	8
1	0	1	0	0	0	0.0096	0.0340	0
2	0.2744	0	80	7.2096	1.7184	6.5088	0.0768	0
3	0.1666	0	0	48.2880	51.9552	0	0	0
4	0.1666	0	0	0	0	0	0	0
5	96.1772	0	0	40.6650	42.1304	15.2928	21.7728	69.1200
6	0.8232	0	0	0	0	73.2768	0.1632	0
7	0.0490	0	0	0	0	0	0	0
8	0.1862	0	0	0	0	0	0	0
9	0.0392	0	0	0.0192	0.0192	0.2688	0.0096	0
10	0	0	0	0.0384	0.0192	0.0096	0	0
11	0.1176	0	0	2.6592	0.0672	0.5952	17.1616	0.0900
	Material unit cost c_n							
	0.1450	1.3500	0.1100	0.5050	0.6750	0.5760	0.3715	0.7250

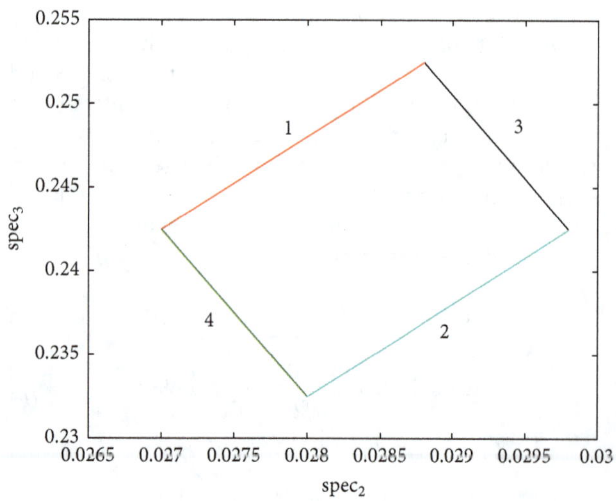

FIGURE 2: Coupled specifications of carbon and chromium.

For the problem (P), the constraints (i) and (iv) ensure that the set of feasible solutions is bounded if it is not empty. This set is actually not empty and we get the optimal solution of Table 4, with (minimum) unit cost equal to 0.344984\$. If we take away the materials $n = 5$ and $n = 6$, set $\pi_5 = 0$ and $\pi_6 = 0$ in the model, this new problem has no solution, and the feasible set is empty. It means that it is impossible to make a mixture which satisfies all the constraints with the 6 remaining materials.

3. Correction to a Fixed Quantity of "Free" Premix

3.1. Problem Description. For the second problem we assume that we have in stock a given mixture already done called a *premix*, with its own specifications, but which does not satisfy some constraints for a new mixture. The problem we consider is to add to this premix a *correction*, which is a combination

TABLE 3: Coefficients to Type 2 constraints.

j	n = 1	2	3	4	5	6	7	8	β_j
				$J = 4$ constraints of Type 2					
1	−1.3578	0	−444.4444	8.2347	42.4085	−36.1600	−0.4267	0	≤9.2500
2									≥7.6944
3	2.9106	0	800	120.3840	69.1392	65.0880	0.7680	0	≤54.0500
4									≥51.2500

TABLE 4: Optimal solution of the basic mixture problem.

π_1	π_2	π_3	π_4	π_5	π_6	π_7	π_8
0.530478	0	0.004486	0.272517	0.192517	0	0	0

of the N available materials, in such a way that the *corrected mixture* obtained by adding the correction to the premix satisfies all the new constraints on the specifications. We are going to obtain this correction without considering the cost of the premix.

Let us set

$$\begin{aligned} &\pi_0^a \geq 0 \quad \text{for the weight of the premix in the corrected mixture,} \\ &\pi_n^a \geq 0 \quad \text{for the weight of the } n\text{th material in the corrected mixture } (n = 1, \ldots, N). \end{aligned} \tag{7}$$

As we want to correct a given weight of the premix, we set $\pi_0^a = 1$, and the decision vector will be $\pi^a = (\pi_1^a, \ldots, \pi_N^a)$.

The weight of the correction to add to the premix to obtain an acceptable mixture is given by

$$z_{\text{weight}}(\pi^a) = \sum_{n=1}^{N} \pi_n^a. \tag{8}$$

If we consider a minimum weight correction, we have to minimize $z_{\text{weight}}(\pi^a)$. The total weight of the corrected mixture is given by the following formula:

$$z_{\text{weight}}^{\text{total}}(\pi^a) = \sum_{n=0}^{N} \pi_n^a = 1 + z_{\text{weight}}(\pi^a). \tag{9}$$

The cost of the correction is

$$z_{\text{cost}}(\pi^a) = \sum_{n=1}^{N} c_n \pi_n^a. \tag{10}$$

If we consider a minimum cost correction, we have to minimize $z_{\text{cost}}(\pi^a)$.

The M specifications of the premix are denoted by $s_{m,0}$ $(m = 1, \ldots, M)$, and each material is also characterized by the M specifications previously described and denoted by $s_{m,n}$ $(m = 1, \ldots, M)$. The specifications spec_m of the resulting corrected mixture are then defined by

$$\text{spec}_m z_{\text{weight}}^{\text{total}}(\pi^a) = \sum_{n=0}^{N} s_{m,n}\pi_n^a = s_{m,n} + \sum_{n=1}^{N} s_{m,n}\pi_n^a, \tag{11}$$

and the constraints of Types 1 and 2 are directly applicable.

3.2. Problem Formulation. Since the problem we consider is to correct a unit weight of a premix ($\pi_0^a = 1$) by adding to that premix a minimum cost or a minimum weight correction, the mathematical formulation is

$$\begin{aligned}
&\min \quad z_{\text{cost}}(\pi^a) = \sum_{n=1}^{N} c_n \pi_n^a, \\
&\min \quad z_{\text{weight}}(\pi^a) = \sum_{n=1}^{N} \pi_n^a, \\
&\text{subjet to} \quad \text{(i)} \ \sum_{n=1}^{N} \pi_n^a = \pi_{\text{tot}}^{\text{a}}; \\
&\text{(ii) for } m = 1, \ldots, M: \\
&\sum_{n=0}^{N} s_{m,n}\pi_n^a - \text{spec}_m^{\max} \pi_{\text{tot}}^a \leq \text{spec}_m^{\max}, \\
&\sum_{n=0}^{N} s_{m,n}\pi_n^a - \text{spec}_m^{\min} \pi_{\text{tot}}^a \geq \text{spec}_m^{\min}, \\
&\text{(iii) for } j = 1, \ldots, J: \\
&\sum_{n=0}^{N} \left[\sum_{m=1}^{M} \alpha_{j,m} s_{m,n} \right] \pi_n^a - \beta_j \pi_{\text{tot}}^a \begin{Bmatrix} \leq \\ = \\ \geq \end{Bmatrix} \beta_j; \\
&\text{(iv)} \ \pi_0^a = 1, \\
&\pi_n^a \geq 0 \quad \text{for } n = 1, \ldots, N.
\end{aligned} \tag{P^a}$$

Table 5: Premix specifications in % of weight.

m	Chemical elements		$\text{spec}_{m,0}$ specifications (%) of the premix $\text{spec}_m^{\min}$	$s_{m,0}$	$\text{spec}_m^{\max}$
1	Al	Aluminium	0	0.15	0.2
2	C	Carbon	2.7	2.98	2.98
3	Cr	Chrome	23.25	25.25	25.25
4	Cu	Copper	0	0.19	0.2
5	Fe	Iron	65	80	85
6	Mn	Manganese	0.3	0.55	0.6
7	Mo	Molybdenum	0	0.25	0.3
8	Ni	Nickel	0	0.17	0.2
9	P	Phosphorus	0	0.05	0.06
10	S	Sulphur	0	0.05	0.06
11	Si	Silicone	0.3	0.70	0.8

In this model, π_{tot}^a is an intermediate variable. We denote by $\mathcal{S}^a$ the set of feasible solutions to this problem.

Let us point out that the solution to this problem given by $\pi_0^a = 1$ and $\pi^a = (\pi_1^a, \ldots, \pi_N^a)$ is not proportions. The proportions of premix and materials in the optimal corrected mixture are

$$\tilde{\pi}_i^a = \frac{\pi_i^a}{z_{\text{weight}}^{\text{total}}(\pi^a)} = \frac{\pi_i^a}{1 + z_{\text{weight}}(\pi^a)} \quad (12)$$

for $i = 0, \ldots, N$.

3.3. The Pareto Set. Problem (P^a) is a problem with two criteria [10]: $z_{\text{cost}}(\pi^a) = c\pi^a$ and $z_{\text{weight}}(\pi^a) = u\pi^a$, where $c = (c_1, \ldots, c_N)$ and $u = (1, \ldots, 1)$. The feasible set in the criteria space $\mathbb{R}^2$ is defined by

$$\begin{aligned} \mathcal{S}_c^a &= \left\{z^a(\pi^a) \in \mathbb{R}^2 \mid z^a(\pi^a) = C\pi^a \text{ for } \pi^a \in \mathcal{S}^a\right\} \\ &= C\mathcal{S}^a, \end{aligned} \quad (13)$$

where

$$z^a(\pi^a) = \begin{pmatrix} z_{\text{cost}}(\pi^a) \\ z_{\text{weight}}(\pi^a) \end{pmatrix} = \begin{pmatrix} \sum_{n=1}^{N} c_n \pi_n^a \\ \sum_{n=1}^{N} \pi_n^a \end{pmatrix} = C\pi^a \quad (14)$$

$$\text{with } C = \begin{pmatrix} c_1 & \cdots & c_N \\ 1 & \cdots & 1 \end{pmatrix}.$$

A feasible solution $\pi^a \in \mathcal{S}^a$ is an *efficient feasible solution* if and only if there is no other feasible solution $\overline{\pi}^a \in \mathcal{S}^a$ such that

(i) $z_i(\overline{\pi}^a) \le z_i(\pi^a)$ for every $i \in \{\text{cost}, \text{weight}\}$;

(ii) $z_j(\overline{\pi}^a) < z_j(\pi^a)$ for at least one $j \in \{\text{cost}, \text{weight}\}$, that is to say that no other feasible solution improves both criteria or at least one criterion without damaging the other criterion.

The set of efficient solution is the *efficient set* or the *Pareto set* denoted by $\mathcal{E}^a$. The efficient set or Pareto set in the criteria space is $\mathcal{E}_c^a = C\mathcal{E}^a$. By considering the weighted sum of the two criteria using parameter $\lambda \in (0, 1)$ and the problem

$$\begin{aligned} \min \quad & z_\lambda(\pi^a) \\ & = (1 - \lambda) z_{\text{cost}}(\pi^a) + \lambda z_{\text{weight}}(\pi^a) \\ \text{subject to} \quad & \pi^a \in \mathcal{S}^a, \end{aligned} \quad (P_\lambda)$$

we can use the following relation for the Pareto set $\mathcal{E}^a$ [10]:

$$\mathcal{E}^a = \bigcup_{\lambda \in (0,1)} \arg\min z_\lambda(\pi^a). \quad (15)$$

The efficient set $\mathcal{E}^a$ being a union of a finite number of faces of $\mathcal{S}^a$ follows that $\mathcal{E}_c^a$ is of the same form since it is the image of $\mathcal{E}^a$ by a linear transformation. As a subset of $\mathbb{R}^2$, $\mathcal{E}_c^a$ is formed by a finite number of pairwise segments connected by their endpoints, also called *efficient vertices* or *efficient extreme points*. It is a polygonal line. To each vertex of that polygonal line is associated an interval of values of λ, noted $[\underline{\lambda}, \overline{\lambda}]$. The union of all those intervals is $[0, 1]$. Likewise, a unique value of λ is assigned to each segment which corresponds to the common bound of the intervals of the two endpoints of the segment. See [11] for a complete description of $\mathcal{E}_c^a$.

3.4. Numerical Example. Table 5 shows the specifications of the premix. Note that these specifications satisfy the constraints of Type 1 but not the constraints of Type 2. Type 2 constraint coefficients are given in Table 6. The efficient vertices in the criteria space are given in Table 7, and the corresponding optimal solutions in the decision space appear in Table 8. In these tables, the minimum cost solution is on the first line, while the minimum weight solution is on the last line. The other solutions correspond to intermediate vertices on the Pareto curve $\mathcal{E}_c^a$ given in Figure 3. This curve gives

TABLE 6: Coefficients of Type 2 constraints.

	$J = 4$ constraint of Type 2									
	n									
j	0	1	2	3	4	5	6	7	8	β_j
1	8.6944	−1.3578	0	−444.4444	8.2347	42.4085	−36.160	−0.4267	0	≤9.2500
2										≥7.6944
3	55.0500	2.9106	0	800	120.3840	69.1392	65.088	0.7680	0	≤54.0500
4										≥51.2500

TABLE 7: Efficient vertices in the criteria space $\mathbb{R}^2$.

$\underline{\lambda}$	$\overline{\lambda}$	z^a_{cost}	z^a_{weight}
0	0.8402832	0.0028353	0.0195544
0.8402832	0.9094854	0.0042987	0.0192762
0.9094854	0.9791369	0.0049710	0.0192093
0.9791369	1	0.0066641	0.0191732

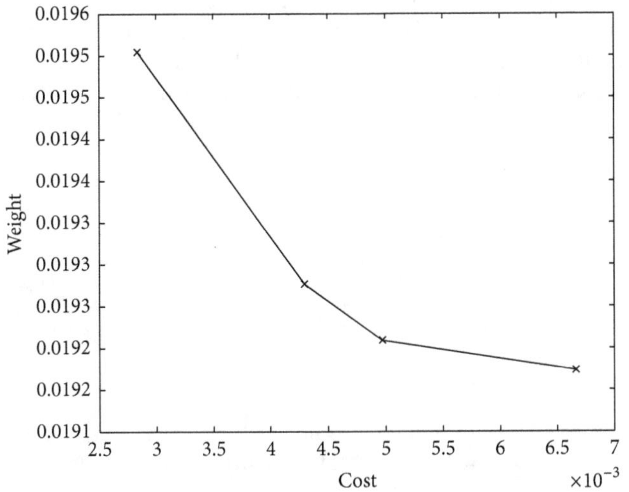

FIGURE 3: Pareto curve $\mathscr{E}^a_c$ in the criteria space $\mathbb{R}^2$.

us the best pairs $(z_{\text{cost}}, z_{\text{weight}})$, going from the minimum cost efficient solution to the minimum weight efficient solution. For any point $(z_{\text{cost}}, z_{\text{weight}}) \in \mathscr{E}^a_c$, z_{cost} is the minimum cost for all feasible correction π^a such that $z_{\text{weight}}(\pi^a) = z_{\text{weight}}$, and, conversely, z_{weight} is the minimum weight for all feasible correction π^a such that $z_{\text{cost}}(\pi^a) = z_{\text{cost}}$.

The solution of this problem shows that the correction at minimum cost effectively costs 0.0028353\$ and weights 0.0195544 weight units. The unit cost of the correction is then

$$\frac{0.0028353\$}{0.0195544 \text{ weight unit}} = 0.1449962\frac{\$}{\text{weight unit}}. \quad (16)$$

The corrected mixture formed thus weighs 1.0195543 weight units. On the other hand, the correction at minimum weight requires an addition of 0.0191732 weight unit, which costs 0.0066641\$. In this case, the unit cost of the correction is

$$\frac{0.0066641\$}{0.0191732 \text{ weight unit}} = 0.3475736\frac{\$}{\text{weight unit}}. \quad (17)$$

The corrected mixture formed thus weighs 1.0191732 weight units.

The proportions of premix and materials in the optimal corrected mixture are reported in Table 9 for the efficient vertices.

Note that the correction obtained does not use materials $n = 5$ and $n = 6$; they could be removed. Recall that the effect of removing these two materials eliminates the possibility of finding a feasible mixture with only the available materials, without premix.

3.5. A Theoretical Result. We conclude this section by establishing a relation between the feasible sets $\mathscr{S}$ and $\mathscr{S}^a$. They are both closed subsets of $\mathbb{R}^N$, and they can be empty or nonempty. The set $\mathscr{S}$ can be nonempty while $\mathscr{S}^a$ is empty. When $\mathscr{S}$ is nonempty, it is bounded because

$$u^t\pi = \sum_{n=1}^{N} \pi_n = 1,$$
$$\pi_n \geq 0 \quad (18)$$
$$\text{for } (n = 1, \ldots, N).$$

When $\mathscr{S}^a$ is nonempty, it can be bounded (hence compact) or unbounded (thus not compact). The following result links these two situations. All possible situations are reported in Table 10.

Theorem 1. *If $\mathscr{S}^a$ is nonempty, we have that*

(i) *$\mathscr{S}^a$ is bounded (thus compact) if and only if $\mathscr{S}$ is empty;*

(ii) *$\mathscr{S}^a$ is unbounded (thus not compact) if and only if $\mathscr{S}$ is nonempty.*

Proof. If $\mathscr{S}^a$ is nonempty, let us take $0 \neq \pi \in \mathscr{S}$ and also take $\pi^a \in \mathscr{S}^a$. For $t \geq 0$, we set $\pi^a(t) = \pi^a + t\pi$. It is easily verified that $\pi^a(t) \in \mathscr{S}^a$ and thus $\mathscr{S}^a$ is unbounded. Conversely, if $\mathscr{S}^a$ is unbounded, there are $\pi^a(t) \in \mathscr{S}^a$ and $\pi \in \mathbb{R}^N$ such that $u^t\pi = 1$ and $\pi^a(t) = \pi^a + t\pi \in \mathscr{S}^a$ (for all $t \geq 0$). Then we check that $\pi \in \mathscr{S}$ by substituting $\pi^a(t)$ in the constraint of (P^a), dividing by t and passing to the limit as t goes to infinity. □

TABLE 8: Efficient vertices in the decision space $\mathbb{R}^8$.

π_0^a	π_1^a	π_2^a	π_3^a	π_4^a	π_5^a	π_6^a	π_7^a	π_8^a
1	0.0195543	0	0	0	0	0	0	0
1	0.0126373	0	0	0	0	0	0.0066389	0
1	0.0113950	0	0	0	0	0	0.0066381	0.0011760
1	0.0124767	0	0	0	0	0	0	0.0066965

TABLE 9: Proportions of premix and materials for efficient vertices.

$\tilde{\pi}_0^a$	$\tilde{\pi}_1^a$	$\tilde{\pi}_2^a$	$\tilde{\pi}_3^a$	$\tilde{\pi}_4^a$	$\tilde{\pi}_5^a$	$\tilde{\pi}_6^a$	$\tilde{\pi}_7^a$	$\tilde{\pi}_8^a$
0.9808207	0.0191793	0	0	0	0	0	0	0
0.9810883	0.0123983	0	0	0	0	0	0.0065133	0
0.9811530	0.0111802	0	0	0	0	0	0.0065130	0.0011538
0.9811875	0.0122420	0	0	0	0	0	0	0.0065705

TABLE 10: Relationship between the feasible sets $\mathcal{S}$ and $\mathcal{S}^a$.

$\mathcal{S}$	$\mathcal{S}^a$
Empty	Empty
empty	Nonempty, closed and bounded (compact)
Nonempty, closed and bounded (compact)	empty
Nonempty, closed and bounded (compact)	Nonempty, closed and unbounded

4. Correction to a Fixed Quantity of "Nonfree" Premix

4.1. Problem Description. It would have been more realistic to consider that the premix has a possible nonzero cost in the preceding problem. So we reconsider the preceding problem assuming a possible nonzero unit cost c_0 for the premix. We consider one criterion which is the unit cost of the corrected mixture.

Let us set

$$\pi_0^f \geq 0 \quad \text{for the weight of the premix in the corrected mixture,}$$
$$\pi_n^f \geq 0 \quad \text{for the weight of the } n\text{th material in the corrected mixture } (n = 1, \ldots, N). \tag{19}$$

As we want to correct a given weight of the premix, we set $\pi_0^f = 1$, and the decision vector will be $\pi^f = (\pi_1^f, \ldots, \pi_N^f)$.

For the total weight, we have

$$z_{\text{weight}}^{\text{total}}\left(\pi^f\right) = \sum_{n=0}^{N} \pi_n^f = 1 + \sum_{n=1}^{N} \pi_n^f = 1 + z_{\text{weight}}\left(\pi^f\right) \tag{20}$$

because we have assumed that $\pi_0^f = 1$, and the total cost is

$$z_{\text{cost}}^{\text{total}}\left(\pi^f\right) = \sum_{n=0}^{N} c_n \pi_n^f = c_0 + \sum_{n=1}^{N} c_n \pi_n^f = c_0 + z_{\text{cost}}\left(\pi^f\right) \tag{21}$$

which take into account the fact that the premix might have a nonzero cost, so $c_0 \geq 0$.

For this problem, the criteria will be the unit cost of the corrected mixture, premix, and correction added together, given by the ratio

$$z^f\left(\pi^f\right) = \frac{z_{\text{cost}}^{\text{total}}\left(\pi^f\right)}{z_{\text{weight}}^{\text{total}}\left(\pi^f\right)} = \frac{c_0 + \sum_{n=1}^{N} c_n \pi_n^f}{1 + \sum_{n=1}^{N} \pi_n^f} = \frac{c_0 + z_{\text{cost}}\left(\pi^f\right)}{1 + z_{\text{weight}}\left(\pi^f\right)}. \tag{22}$$

4.2. Problem Formulation. The decision variables will be again π_n^f for $n = 1, \ldots, N$, and therefore the formulation is

$$\min \quad z^f\left(\pi^f\right) = \frac{c_0 + \sum_{n=1}^{N} c_n \pi_n^f}{1 + \sum_{n=1}^{N} \pi_n^f} \qquad \text{subject to} \quad \pi^f \in \mathcal{S}^f. \tag{P^f}$$

Let $\mathcal{S}^f$ be the set of feasible solution to this problem. Obviously, we have $\mathcal{S}^f = \mathcal{S}^a$ because the constraints of this problem are the constraints of the preceding problem (P^a).

4.3. Linear-Fractional Program. The problem (P^f) is an example of a linear-fractional program. Under the assumption that the feasible region $\mathcal{S}^f$ is nonempty and bounded, the Charnes-Cooper transformation [12]

$$\eta_0 = \frac{1}{1 + \sum_{i=1}^{N} \pi_i^f}, \qquad \eta_n = \frac{\pi_i^f}{1 + \sum_{i=1}^{N} \pi_i^f} \quad \text{for } n = 1, \ldots, N, \tag{23}$$

translates the linear-fractional program above to the equivalent linear program:

$$
(\widetilde{P}^f)\quad
\begin{aligned}
&\min \quad \tilde{z}^f(\eta) = \sum_{n=0}^{N} c_n \eta_n \\
&\text{subject to} \quad \text{(i)} \ \sum_{n=0}^{N} \eta_n = 1; \\
&\text{(ii) for } m = 1, \dots, M: \\
&\sum_{n=0}^{N} s_{m,n}\eta_n \le \text{spec}_m^{\max}, \\
&\sum_{n=0}^{N} s_{m,n}\eta_n \ge \text{spec}_m^{\min}; \\
&\text{(iii) for } j = 1, \dots, J: \\
&\sum_{n=0}^{N} \left[\sum_{m=1}^{M} \alpha_{j,m} s_{m,n} \right] \eta_n \left\{ \begin{matrix} \le \\ = \\ \ge \end{matrix} \right\} \beta_j; \\
&\text{(iv) } \eta_n \ge 0 \quad \text{for } n = 0, \dots, N.
\end{aligned}
$$

Let $\widetilde{\mathcal{S}}^f$ be the set of feasible solutions $\eta = (\eta_0, \dots, \eta_N)$ of $(\widetilde{P}^f)$. Let us observe that $(\widetilde{P}^f)$ is similar to the first problem (P), and the premix can now be considered as one of the available materials. The fact that we assume $\mathcal{S}^a$ nonempty and bounded, Theorem 1 implies that (P) has no solution; hence, $(\widetilde{P}^f)$ has no solution $\eta = (\eta_0, \dots, \eta_n) \in \widetilde{\mathcal{S}}^f$ with $\eta_0 = 0$. It follows that any solution of $(\widetilde{P}^f)$, if it exists, is such that $\eta_0 > 0$ and that the corresponding solution of (P^f) is given by

$$\pi_n^f = \frac{\eta_n}{\eta_0} \quad \text{for } n = 0, \dots, N, \tag{24}$$

and hence $\pi_0^f = 1$. In case the feasible region $\mathcal{S}^a$ is nonempty and unbounded, $(\widetilde{P}^f)$ might have a solution $\eta = (\eta_0, \dots, \eta_n) \in \widetilde{\mathcal{S}}^f$ with $\eta_0 > 0$; then the solution of (P^f) is again given by the preceding formula. But it might happen that $(\widetilde{P}^f)$ has a solution $\eta = (\eta_0, \dots, \eta_n) \in \widetilde{\mathcal{S}}^f$ with $\eta_0 = 0$; then the resulting mixture is not a corrected mixture. It means that it is better to use original materials $(n = 1, \dots, N)$ to produce the desired mixture because it is cheaper than trying to use the premix. Hence,

$$\pi_n = \eta_n \quad \text{for } n = 1, \dots, N, \tag{25}$$

is a solution of the original problem (P). In this case, let us define π^f by the formula $\pi_0^f = 0$, and $\pi_n^f = \eta_n$ for $n = 0, \dots, N$, as before.

4.4. A Parametric Analysis. We could be interested by the variation of the criteria, the unit cost of the resulting (corrected or not) mixture, with respect to the cost c_0. This information is obtained from a parametric analysis of the criteria with respect to the parameter c_0. We split the criteria of $(\widetilde{P}^f)$ in two parts:

$$\tilde{z}^f(\eta) = \sum_{n=1}^{N} c_n \eta_n + c_0 \eta_0 = \tilde{z}_1^f(\eta) + c_0 \tilde{z}_2^f(\eta) \tag{26}$$

and we consider the following bicriteria problem:

$$
(\widetilde{P}^{fs})\quad
\begin{aligned}
&\min \quad \tilde{z}_1^f(\eta) = \sum_{n=1}^{N} c_n \eta_n \\
&\min \quad \tilde{z}_2^f(\eta) = \eta_0 \\
&\text{subject to} \quad \eta \in \widetilde{\mathcal{S}}^f.
\end{aligned}
$$

The Pareto curve of this problem, obtained by considering the weighted sum, is

$$\tilde{z}_\lambda^f(\eta) = (1-\lambda)\tilde{z}_1^f(\eta) + \lambda \tilde{z}_2^f(\eta) \tag{27}$$

and the correspondence

$$\lambda = \frac{c_0}{1 + c_0} \quad \text{or } c_0 = \frac{\lambda}{1-\lambda}, \tag{28}$$

allows us to find a piecewise linear continuous increasing concave representation for the optimal minimal unit cost $z_*^f(c_0) = \min z^f(\pi^f)$ in terms of c_0.

Since

$$
\begin{aligned}
z_{\text{cost}}(\pi^f) &= \sum_{n=1}^{N} c_n \pi_n^f \\
&= \begin{cases} \dfrac{1}{\eta_0}\displaystyle\sum_{n=1}^{N} c_n \eta_n = \dfrac{1}{\eta_0}\tilde{z}_1^f(\eta) & \text{if } \eta_0 > 0 \ (\text{or } \pi_0^f = 1), \\ \displaystyle\sum_{n=1}^{N} c_n \eta_n = \tilde{z}_1^f(\eta) & \text{if } \eta_0 = 0 \ (\text{or } \pi_0^f = 0), \end{cases} \\
z_{\text{weight}}(\pi^f) &= \sum_{n=1}^{N} \pi_n^f \\
&= \begin{cases} \dfrac{1}{\eta_0}\displaystyle\sum_{n=1}^{N} \eta_n = \dfrac{1}{\eta_0}(1-\eta_0) & \text{if } \eta_0 > 0 \ (\text{or } \pi_0^f = 1), \\ \displaystyle\sum_{n=1}^{N} \eta_n = 1 & \text{if } \eta_0 = 0 \ (\text{or } \pi_0^f = 0), \end{cases}
\end{aligned}
\tag{29}
$$

we have

$$z_*^f(c_0) = \frac{c_0 \pi_0^f + z_{\text{cost}}(\pi^f)}{\pi_0^f + z_{\text{weight}}(\pi^f)}, \tag{30}$$

that is, valid not only for $\pi_0^f = 1$ but also for $\pi_0^f = 0$ and where π^f is the optimal solution.

TABLE 11: Efficient vertices of $(\widetilde{P}^{fs})$ and (P^f) in the criteria space $\mathbb{R}^2$.

(a)

$\underline{\lambda}$	$\overline{\lambda}$	$\widetilde{z}_1^f(\eta)$	$\widetilde{z}_2^f(\eta)$
0	0.124887	0.002781	0.980820
0.124887	0.127242	0.005732	0.960140
0.127242	0.249832	0.011545	0.920268
0.249832	0.343079	0.024633	0.880968
0.343079	0.604078	0.045238	0.841514
0.604078	0.710795	0.045834	0.841124
0.710795	0.874033	0.056158	0.836911
0.874033	1	0.056666	0.836838

(b)

$\underline{c}_0$	$\overline{c}_0$	$z_{\text{cost}}^f(\pi^f)$	$z_{\text{weignt}}^f(\pi^f)$
0	0.142710	0.002835	0.019554
0.142710	0.145794	0.005970	0.041513
0.145794	0.333035	0.012545	0.086638
0.333035	0.522255	0.027961	0.135113
0.522255	1.525752	0.053758	0.188333
1.525752	2.450600	0.054491	0.188885
2.450600	6.938604	0.067101	0.194869
6.938604	$+\infty$	0.067714	0.194974

TABLE 12: Optimal solutions of $(\widetilde{P}^{fs})$ and (P^f) in the decision space $\mathbb{R}^7$.

(a)

η_0	η_1	η_2	η_3	η_4	η_5	η_6	η_7	η_8
0.980820	0.019179	0	0	0	—	—	0	0
0.960140	0.038506	0	0.001352	0	—	—	0	0
0.920268	0.079282	0	0.000448	0	—	—	0	0
0.880968	0.098547	0	0	0.020483	—	—	0	0
0.841514	0.096656	0	0	0.061829	—	—	0	0
0.841124	0.094791	0	0	0.062039	—	—	0.002044	0
0.836911	0.081505	0	0	0.064288	—	—	0.001878	0.015415
0.836838	0.081852	0	0	0.064325	—	—	0	0.016983

(b)

π_0^f	π_1^f	π_2^f	π_3^f	π_4^f	π_5^f	π_6^f	π_7^f	π_8^f
1	0.019554	0	0	0	—	—	0	0
1	0.040105	0	0.001408	0	—	—	0	0
1	0.086151	0	0.000486	0	—	—	0	0
1	0.111862	0	0	0.023251	—	—	0	0
1	0.114859	0	0	0.073473	—	—	0	0
1	0.112696	0	0	0.073758	—	—	0.002430	0
1	0.097389	0	0	0.076816	—	—	0.002244	0.018420
1	0.097811	0	0	0.076867	—	—	0	0.020295

4.5. Numerical Example. We consider two situations with the preceding data.

Case 1. We use all the materials, without $n = 5$ and $n = 6$. In this case, $\mathcal{S}$ is empty and $\mathcal{S}^a$ is nonempty and bounded; hence η_0 will always be strictly positive. Tables 11 and 12 present results for $(\widetilde{P}^{fs})$ and the corresponding values for (P^f).

Case 2. We consider all the $N = 8$ materials; then $\mathcal{S}$ is nonempty and $\mathcal{S}^a$ is nonempty and not bounded; hence it

Table 13: Efficient vertices of $(\widetilde{P}^{fs})$ and (P^f) in the criteria space $\mathbb{R}^2$.

(a)

$\underline{\lambda}$	$\overline{\lambda}$	$\widetilde{z}_1^f(\eta)$	$\widetilde{z}_2^f(\eta)$
0	0.124887	0.002781	0.980820
0.124887	0.127242	0.005732	0.960140
0.127242	0.249832	0.011545	0.920267
0.249832	0.261550	0.024633	0.880970
0.261550	0.270235	0.153784	0.516323
0.270235	1	0.344983	0

(b)

$\underline{c}_0$	$\overline{c}_0$	$z_{\text{cost}}^f(\pi^f)$	$z_{\text{weignt}}^f(\pi^f)$
0	0.142710	0.002835	0.019554
0.142710	0.145794	0.005970	0.041513
0.145794	0.333035	0.012545	0.086640
0.333035	0.354188	0.027961	0.135112
0.354188	0.370304	0.297848	0.936768
0.370304	$+\infty$	0.344984	1

Table 14: Optimal solutions of $(\widetilde{P}^{fs})$ and (P^f) in the decision space $\mathbb{R}^9$.

(a)

η_0	η_1	η_2	η_3	η_4	η_5	η_6	η_7	η_8
0.980820	0.019179	0	0	0	0	0	0	0
0.960140	0.038506	0	0.001352	0	0	0	0	0
0.920267	0.079284	0	0.000448	0	0	0	0	0
0.880970	0.098547	0	0	0.020482	0	0	0	0
0.516323	0.277318	0	0	0.151270	0.055086	0	0	0
0	0.530479	0	0.004485	0.272517	0.192517	0	0	0

(b)

π_0^f	π_1^f	π_2^f	π_3^f	π_4^f	π_5^f	π_6^f	π_7^f	π_8^f
1	0.019554	0	0	0	0	0	0	0
1	0.040104	0	0.001408	0	0	0	0	0
1	0.086153	0	0.000486	0	0	0	0	0
1	0.111862	0	0	0.023249	0	0	0	0
1	0.537102	0	0	0.292976	0.106690	0	0	0
0	0.530479	0	0.004485	0.272517	0.192517	0	0	0

might happen that η_0 be 0 for large values of c_0. Tables 13 and 14 present results for $(\widetilde{P}^{fs})$ and the corresponding values for (P^f).

Let us observe that the results for $c_0 = 0$ correspond to the results obtained in Section 3.

5. Conclusion

Three simple mixture problems occurring in metallurgy were analyzed. We first considered the basic mixture problem. We then considered a correction problem to a "free" premix. We have shown that, in the case of a free premix, the minimal cost correction does not generally correspond to the minimum weight correction. We built the Pareto curve that gives all intermediate values between these two extremes. The same problem, with a "nonfree" premix, leads to a unique minimal unit cost solution.

Competing Interests

The authors declare that there is no conflict of interests regarding the publication of this paper.

Acknowledgments

This work has been financially supported by an individual discovery grant from NSERC (National Sciences and Engineering Research Council of Canada) for the first author.

References

[1] G. B. Dantzig, *Linear Programming and Extensions*, Princeton University Press, Princeton, NJ, USA, 1963.

[2] G. B. Dantzig, "The diet problem," *Interfaces*, vol. 20, no. 4, pp. 43–47, 1990.

[3] F. Dubeau, P.-O. Julien, and C. Pomar, "Étude ce cas: un porcin dans un environnement sain," *INFOR*, vol. 46, pp. 199–217, 2008.

[4] S. G. Garille and S. I. Gass, "Stigler's diet problem revisited," *Operations Research*, vol. 49, no. 1, pp. 1–13, 2001.

[5] L. M. Lancaster, "The evolution of the diet model in managing food systems," *Interfaces*, vol. 22, no. 5, pp. 59–68, 1992.

[6] L. M. Lancaster, "The history of the application of mathematical programming to menu planning," *European Journal of Operational Research*, vol. 57, no. 3, pp. 339–347, 1992.

[7] G. J. Stigler, "The cost of subsistence," *American Journal of Agricultural Economics*, vol. 27, no. 2, pp. 303–314, 1945.

[8] C. Audet, J. Brimberg, P. Hansen, S. Le Digabel, and N. Mladenović, "Pooling problem: alternate formulations and solution methods," *Management Science*, vol. 50, no. 6, pp. 761–776, 2004.

[9] A. Rong and R. Lahdelma, "Fuzzy chance constrained linear programming model for optimizing the scrap charge in steel production," *European Journal of Operational Research*, vol. 186, no. 3, pp. 953–964, 2008.

[10] R. E. Steuer, *Multiple Criteria Optimization*, Wiley Series in Probability and Mathematical Statistics: Applied Probability and Statistics, John Wiley & Sons, New York, NY, USA, 1986.

[11] F. Dubeau and A. Kadri, "Computation and visualization of the Pareto set in the criterion space for the bicriteria linear programming problem," *International Journal of Mathematics and Computation*, vol. 15, no. 2, pp. 1–15, 2012.

[12] A. Charnes and W. W. Cooper, "Programming with linear fractional functionals," *Naval Research Logistics Quarterly*, vol. 9, pp. 181–186, 1962.

23

Physical Metallurgy of Modern Creep-Resistant Steel for Steam Power Plants: Microstructure and Phase Transformations

V. C. Igwemezie,[1,2] C. C. Ugwuegbu,[2] and U. Mark[2]

[1]*Renewable Energy Marine Structures Centre for Doctoral Training (REMS CDT), Cranfield University, Cranfield, Bedfordshire MK43 0AL, UK*
[2]*Department of Materials & Metallurgical Engineering, Federal University of Technology Owerri, Ihiagwa, Nigeria*

Correspondence should be addressed to V. C. Igwemezie; victor.igwemezie@cranfield.ac.uk

Academic Editor: Sunghak Lee

The fact that the microstructure of steel depends on its composition and the heat treatment given to it has been heavily exploited in the design of steel for power plant applications. To obtain a steel that can function at the higher temperature where power plants operate without failure for extended life, heat treatment is needed to produce fine and highly stable dispersion of carbides, nitrides, and intermetallic compounds in the microstructure of the material. A significant contribution also comes from solid solution strengthening by substitutional solutes. We review here various types of phases, microstructures, functions, and interacting effects of the various alloying elements in the design of steel for modern power plant application.

1. Introduction

The critical role that steel has played in the construction of our society is largely due to its superior adaptation to the wide ranges of mechanical requirements. The strength and toughness of steel can vary with changes in its microstructure, that is, the size and shape of grains and the detailed arrangement of the atoms [1].

There are behaviours of iron which are unseen from ordinary view but ultimately affect their physical behaviour. The macroscopic shape of steel crystal depends on the environment, processing, and many other factors determined mostly by the fact that a piece of steel is generally polycrystalline and is filled completely with millions of one or more kinds of crystals. The fact that a piece of steel contains varieties of crystals is frequently not apparent to many design engineers, making them treat the metal as a continuum, especially in modelling approaches.

The electrons form the bonds between atoms. The strength of these bonds is a major factor which affects the nature and properties of materials. Another consideration is that microstructure of steel by nature is imperfect. In practice, it is found that equilibrium in engineering steel demands that crystals should be imperfect or away from equilibrium. Steels may not be very useful as they are presently without certain defects. There is always an equilibrium concentration of point defects and more significant are the nonequilibrium, unintended defects introduced in the process of manufacture [15].

In the design of steel for power plant or for high temperature application there are challenging engineering requirements for the materials engineer. The summary of these requirements is as follows [6]:

(1) High creep strength at high temperature

(2) High toughness and resistance to embrittlement

(3) Resistance to steam oxidation and corrosion

(4) Ease of fabrication and weldability

Not all steels are candidates for high temperature application. Study has shown that austenitic steels have higher coefficient of thermal expansion and low conductivity than ferritic steels. In other words, ferrite has much smaller thermal expansion coefficient and high thermal conductivity than austenite, as shown in Table 1.

When considered for elevated temperature service, materials with large thermal expansivities become susceptible to thermal fatigue, especially in thick sections of the kind

Table 1: Comparison of the coefficient of thermal expansion (CTE) and thermal conductivity for ferritic and austenitic steel at 100°C. The compositions (wt%) of the steels are also given [2, 3].

(a)

Steel	CTE/$\mu m^{-1}K^{-1}$	Thermal conductivity/$Wm^{-1}K^{-1}$
Ferritic	9.9	24.9
Austenitic	17.3	16.3

(b)

Steel	AISI type	C	Mn	P	S	Si	Cr	Ni
Ferritic	403	0.15	1.00	0.040	0.03	0.50	11.5–13.0	—
Austenitic	305	0.12	2.00	0.045	0.03	1.00	17–19	10.5–13.0

frequently employed in power generation. This becomes a major problem due to the fact that steam turbines are often turned off and on during their service life. This thermal cycling can cause fatigue in austenitic steels. This has been the main reason why austenitic steels in spite of their superior creep strength are rejected in favour of ferritic steels for the construction of power plant [6]. Martensitic steels have smaller thermal expansion and larger thermal conductivity than austenitic steels and Ni base superalloys and are now offering highest potential to designing thick section steels for power plant application.

In the development of modern steam power plants, the goal is centred around increasing the steam parameters from the current optimum of about 620°C at 28 MPa up to 700°C at 35 MPa in order to improve efficiency and thereby reduce CO_2 emission [19]. Further advanced steam conditions of 700°C and above have been already initiated to gain net efficiency higher than 50% at Thermie AD700 project in Europe and at DOE vision 21 project [13, 20, 21]. In the light of available knowledge so far, these steam conditions cannot be reached with ferritic-martensitic alloys, which are currently the state-of-the-art material for steam power plants [19, 22]. Hence, these projects involve the replacement of 9–12Cr martensitic steels by Ni base superalloys for the highest temperature regimes. To minimize the requirement of expensive Ni base superalloys, as Ni superalloy is about ten times costlier than high temperature steels, 9–12Cr martensitic steels are planned to be used in the next highest temperature components of such very high temperature plants. Therefore, 9–12Cr martensitic steels are still strongly desired to expand the present temperature range up to 650°C and more. It should be noted that research is also going back to austenitic steels to see their possible use for 700°C USC plants in Japan.

2. Phase Transformation in Creep Steel for Power Plant

The general rule for determining if creep will occur in metals or not is [23, 24]

$$T > 0.3\text{–}0.4\,(T_M). \quad (1)$$

Pure iron has a melting temperature, T_M, of about 1536°C and the equation above suggests that pure iron would creep at temperature where traditional steam power plant operates, that is, in temperature range of 450–565°C.

Since service temperature depends on the melting temperature of the steel, a direct approach to solving the creep problem is to use metals with high melting temperatures so that the temperature at which creep occurs is increased. An example of such metals is tungsten with T_M of 3407°C. In this case, creep becomes important above about 1363°C which is far above the aforementioned temperature range for power plants. This apparent advantage comes with considerable consequences. Firstly, W is several times more expensive than steel, which could make affordability of the power plant very unlikely for most countries. This cost implication and the attendant low demand will have a negative impact on the huge investment in the technology. Secondly, the oxide of tungsten (WO_3) sublimes and so forth is not considered as a practical solution. Another approach could be to use nickel-based superalloys. Study has shown that the creep rupture strength of the nickel-based superalloys is very good at high temperatures and they have a sufficiently low expansion coefficient. Unfortunately, they are about 6–10 times more expensive than low alloy steels, depending on the alloy composition and market prices of alloying elements. In addition, nickel-based superalloys are more expensive to weld and machine.

Almost all steels rely on the transformation between *austenite* and *ferrite* for obtaining the desired microstructure. When the high temperature *fcc*-phase austenite in steel decomposes to the less dense *bcc*-phase ferrite, a number of different microstructures and morphologies can form, which depend on the cooling rate, the presence of alloying elements, and the conditions and availability of lower energy nucleation sites for heterogeneous nucleation. The transformation temperature between austenite and ferrite can be greatly affected by the presence of solutes, such as C, Si, Mn, Ni, Mo, Cr, W, Ti, Nb, and V. For example, 0.8 wt% carbon lowers the second solid-state transformation temperature from 910°C to 723°C.

The transformation temperature can also be altered by changing the cooling rate from elevated temperatures. A rapidly cooled steel transforms at a lower temperature and there is shorter amount of time available to accomplish the necessary changes in atomic positions before the material finally cools down to room temperature. The products of such transformation are metastable phases. The advantage of

TABLE 2: Stored energy as function of microstructure, defined relative to microstructure for which stored energy is arbitrarily set to zero [adapted from [4]].

Case	Microstructure	Stored energy, J/mol
Phase mixture in Fe-0.2C-1.5Mn wt% at 27°C		
1	Ferrite, graphite, and cementite	0
2	Ferrite and cementite	70
3	Paraequilibrium ferrite and paraequilibrium cementite	385
4	Bainite and paraequilibrium cementite	785
5	Martensite	1214
Phase mixture in Fe-0.1C-2.2Cr-1Mn wt% at 600°C		
6	Ferrite and paraequilibrium cementite → ferrite and M_2X (e.g., when steel enters service)	63
7	Ferrite and paraequilibrium cementite → ferrite and $M_{23}C_6$ (e.g., after long service time)	0

fast cooling is that the structural scale is reduced, which is advantageous in achieving greater strength and toughness [1].

To obtain a steel that can function at the higher temperature where power plants operate without failure for extended life, pure iron is alloyed with other elements such as C, Cr, Ni, Mn, Mg, W, Ti, V, Mo, and Nb and then heat-treated to produce steel with required creep properties. Creep steels are able to survive for such long periods as 30 years because the operating temperature is only about half of the absolute melting temperature, making the migration of atoms very slow to cause any significant change in the steel microstructure.

In power plants, the longevity of iron alloys relies on the fact that the diffusivities are incredibly small [4, 25]. Notwithstanding this, slow but significant changes are observed to occur over the long service life. Another consequence of the small diffusion coefficients is that the dominant creep mechanism is the climb of dislocations over obstacles with the help of thermal energy. The obstacles are mainly carbide particles which are dispersed throughout the microstructure [4].

Another reason why the changes that are expected over the years must occur at a slow rate is that the steels are generally tempered before they enter service, producing a very stable microstructure and relieving stresses. The stability can be discussed in terms of the free energy stored in the material, as illustrated in Table 2 [4].

The stored energy of the plain carbon steel, (Phase mixture in Fe-0.2C-1.5Mn wt% at 27°C) is described relative to the equilibrium state which is a mixture of ferrite, graphite, and cementite without any defects. The phases in cases 1 and 2 involve the equilibrium partitioning of all elements so as to minimize free energy, that is, minimize the tendency to react further. In cases 3–5, the substitutional solutes are frozen in their position.

It can be seen that bainite and martensite have higher stored energies relative to ferrite because of strain energy contributions and solute trapping. The term *paraequilibrium* means that the phase has the same ratio of iron to substitutional solute atoms as the parent phase from which it formed.

Cases 6 and 7 are for a classical power plant alloy, Fe-0.1C-2.2Cr-1Mo wt% at 600°C. Here, the reduction in stored energy when the microstructure changes from M_2X to $M_{23}C_6$ microstructure is very small. This means that this transformation will occur at a corresponding slow rate. In summary, the reaction rate of a transformation can be controlled by the available *Gibbs free energy*, also referred to as the *driving force*, and by the diffusivity of carbon.

3. The Matrix Microstructure of Creep Steel

The ability of power plant steels to resist creep deformation depends on the presence in the microstructure of fine and highly stable dispersion of carbides and intermetallic compounds which precipitate during tempering or during elevated temperature service [4]. The carbides interfere with the climb and glide of dislocations and retard the coarsening rate of the microstructure as a whole, for example, the size and shape of martensite or bainite plates. In other words, these particles not only interfere with the progress of dislocations but also stabilize the microstructure so that features such as lath boundaries change very slowly during long-term service at elevated temperatures [7]. A significant contribution also comes from solid solution strengthening by substitutional solutes. The solid solution strengthening component is found to become more important after prolonged service at elevated temperature, as the microstructural contribution to strengthening diminishes due to annealing effects.

The heat treatments given to the steels before service are usually so severe that the precipitates are in an overaged condition. The microstructures are therefore relatively stable before entering service, though further microstructural changes are inevitable given that service involves many decades at temperatures where substitutional atoms have significant mobility and this change is sometimes used to monitor remaining life of the steel components.

Majority of steels used in fossil power plants in the whole world are constructed from low alloy steels containing Cr, Mo, and V as the major alloying elements. The total alloy concentration is usually less than 5 wt% [4]. These include

TABLE 3: Chemical compositions (wt%) of some low-alloy creep-resistant steels for power plant construction.

Designation	C	Si	Mn	Ni	Mo	Cr	V
1Cr0.5Mo	0.15	0.25	0.50	—	0.60	0.95	—
0.25CrMoV	0.15	0.25	0.50	0.05	0.50	0.30	0.25
0.5Cr0.5Mo0.25V	0.12	0.25	0.50	—	0.60	0.45	0.25
1CrMoV	0.25	0.25	0.75	0.70	1.00	1.10	0.35
2.25Cr1Mo	0.15	0.25	0.50	0.10	1.00	2.30	0.00

TABLE 4: Heat treatment conditions, matrix phase, and precipitates for steels of varying chromium content [5].

Steel	Water quenching treatment	Tempering/air cooled treatment	Phases present	Precipitates after tempering
2Cr-2W	930°C/hr	730°C/hr	Bainite	$M_{23}C_6$, M_6C, M_2C
5Cr-2W	930°C/hr	750°C/hr	Bainite	$M_{23}C_6$
9Cr-2W	950°C/hr	750°C/hr	Martensite	$M_{23}C_6$
12Cr-2W	1000°C/hr	750°C/hr	Martensite + δ-ferrite (0.16)	$M_{23}C_6$
15Cr-2W	1050°C/hr	750°C/hr	Martensite + δ-ferrite (0.16)	$M_{23}C_6$

steels such as 0.5Cr0.5Mo0.25V, 1CrMoV, and 2.25Cr1Mo, the compositions of which are given in Table 3, along with some other common creep-resistant steels.

These alloys have formed the backbone of the power generation and petrochemical industries for over 50 years for operating temperatures of 565°C or less. 2.25Cr1Mo steel is widely used for superheater tubing in power plant and as filler materials for joining 0.5Cr0.5Mo0.25V steam piping [26].

The primary microstructures of these low alloy creep-resistant steels are essential for leaner steels, allotriomorphic ferrite, and pearlite [6]. In other words, about a decade ago or so, for the classical or common creep steels, the best established steels relied on allotriomorphic ferrite or bainite as the starting microstructure. These microstructures contain very small grains of the order of a quarter of a micron, which provide additional opposition to the movement of dislocations. If diffusion creep was prevalent in these materials, a large grain size would be needed to increase the energy required for diffusion between and through grains [6].

To mitigate oxidation and corrosion in these creep steels, deliberate increase in the concentration of chromium was made. Consequently, δ-ferrite formed, which is detrimental to creep strength. To eliminate this ferrite from the microstructure, balancing additions of other alloying elements were added leading to sufficient increase in the hardenability of the steel. On normalization heat treatment, it is found that the microstructure of this new steel system became martensite instead of bainite for identical cooling conditions.

Abe and Nakazawa (1992) [5] presented a study on the effect of varying chromium content on creep steel. The summary of their study is shown in Table 4.

By varying the chromium content of simple 2 wt% W steel, the microstructure changes from bainitic to martensitic and then to a dual-phase matrix of martensite and δ-ferrite of about 0.16 volume fraction. δ-ferrite is coarse and soft and consequently detrimental to creep strength. The phase can form during the solidification of ferritic steels [27–29]. In other words, higher alloy content steel, such as 9–15 wt% Cr steels, tends to be martensitic on cooling rather than bainitic. This is because, as already mentioned, the larger alloy content causes an increase in the hardenability of the steel to the point where it cannot transform to bainite. This is illustrated by the calculated time-temperature transformation (TTT) diagrams for 2.3 (X), 4.3 (Y), and 9.3 (Z) wt% chromium steels shown in Figure 1.

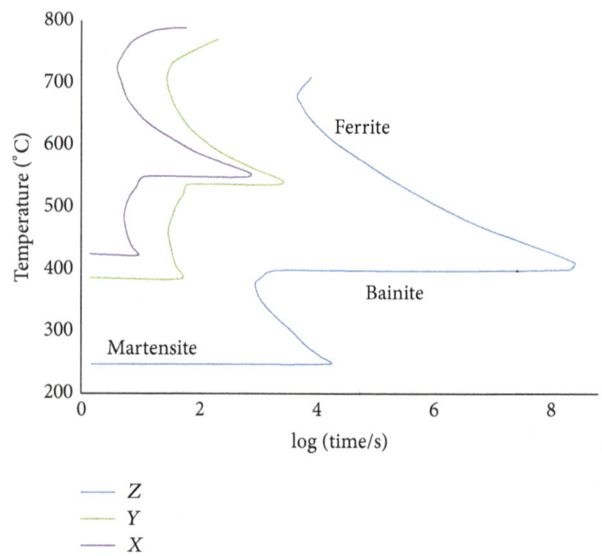

FIGURE 1: Calculated TTT diagrams for steels of composition X (0.15C-0.2Si-0.5Mn-1Mo-2.3Cr wt%), Y (4.3Cr wt%), and Z (9.3Cr wt%) [9–12].

In Figure 1, the upper curve for each case represents diffusion transformation, whereas the lower curve represents bainite. In general, 9–15 wt% Cr steels are fully air hardened, meaning that the microstructure formed on air cooling from normalizing is completely martensitic. Therefore, martensitic matrix is seen as arising by chance. The martensite microstructure is reported to turn out to be the most exciting

Table 5: Nominal chemical compositions (wt%) of ferritic steels.

Steels	C	Si	Mn	Cr	Mo	W	Co	V	Nb	B	N	Others
2Cr												
T22	0.12	0.3	0.45	2.25	1.0	—	—	—	—	—	—	—
HCM2S	0.06	0.20	0.45	2.25	0.1	1.60	—	0.25	0.05	0.003	—	—
9Cr												
T9	0.12	0.60	0.45	9.00	1.0	—	—	—	—	—	—	—
HCM9M	0.07	0.30	0.45	9.00	2.0	—	—	—	—	—	—	—
T91	0.10	0.40	0.45	9.00	1.0	—	—	0.20	0.08	—	0.05	—
NF616	0.07	0.06	0.45	9.00	0.5	1.80	—	0.20	0.05	0.004	0.06	—
Tempaloy F-9	0.06	0.50	0.60	9.00	1.0	—	—	0.25	0.40	0.005	—	—
EM12	0.10	0.40	0.10	9.00	2.0	—	—	0.30	0.40	—	—	—
12Cr												
HT91	0.20	0.40	0.60	12.0	1.0	—	—	0.25	—	—	—	0.5Ni
HT9	0.20	0.40	0.60	12.0	1.0	0.5	—	0.25	—	—	—	0.5Ni
HCM12	0.10	0.30	0.55	12.0	1.0	1.0	—	0.25	0.05	—	0.03	—
HCM12A	0.11	0.10	0.60	12.0	0.4	2.0	—	0.20	0.05	0.003	0.06	1.0Cu
NF12	0.08	0.20	0.50	11.0	0.2	2.6	2.5	0.20	0.07	0.004	0.05	—
SAVE12	0.10	0.30	0.20	11.0	—	3.0	3.0	0.20	0.07	—	0.04	0.07Ta 0.04Nd

ferritic creep-resistant steels. Consequently, modern heat-resistant steels are based on martensite microstructure. These higher alloy content steels for higher temperatures and greater corrosion/oxidation resistance are generally referred to as 9–12 wt% Cr steels [6, 27–29] for the well-established ones.

Martensitic steels usually have to be tempered in order to obtain an optimum combination of strength and toughness. The tempering process usually leads to a decrease in strength due to the precipitation of iron carbides from carbon that was originally in solid solution in the martensite. However, when the steel contains strong carbide forming elements such as Mo, Cr, V, Ti, and W, it becomes possible to recover the strength during prolonged tempering at elevated temperatures. This is called *secondary hardening*; it occurs because substitutional alloying elements (such as Cr) can diffuse during prolonged annealing to precipitate finely dispersed alloy carbides.

The alloy carbides in secondary-hardened steels are found to be more thermodynamically stable than iron carbides and show little tendency to coarsen. However, many of the alloy carbides find it difficult to nucleate, so that the equilibrium carbide is not necessarily the first to precipitate. Instead, there is a sequence of alloy carbides precipitated in an order determined by a combination of kinetic and thermodynamic considerations which leads ultimately towards an equilibrium microstructure.

Many of the precipitates found in power plant steels have crystal structures and compositions which are quite different from those of the ferrite matrix. The precipitate/matrix interfacial energy can therefore be expected to be large, making it difficult for the equilibrium precipitate to nucleate. Consequently, decomposition often starts with the formation of one or more metastable phases which are kinetically favoured. These initial phases eventually dissolve as equilibrium is approached. The phases normally appear and reappear in the course of the microstructure attaining equilibrium. This progression towards equilibrium at service temperatures makes creep steels useful. It is well established that the fracture toughness of many power plant steels deteriorates during service at elevated temperatures for two reasons. Firstly, the carbide particles, particularly those located at the prior austenite boundaries, coarsen and hence provide easier sites for crack or void nucleation. Secondly, the elevated temperatures permit the impurities to diffuse relatively rapidly and saturate the boundaries.

Table 5 shows the composition of a selection of ferritic power plant steels.

One can see that there are three main, well-established groups of ferritic power plant steels as shown in Table 5, usually designated 2, 9, and 12 wt% Cr steels. The 9–12 wt% Cr steels are currently undergoing the most research and development due to their high strength and good corrosion resistance. However, research has been done [8, 17, 30, 31] and currently going on in the development of 15Cr1Mo creep-resistant steels for power plant temperatures up to 700°C.

4. Heat Treatment of Power Plant Steels

In general, the primary microstructures of power plant alloys often consist of δ-ferrite, martensite, bainite, allotriomorphic ferrite, and retained austenite as the major phases obtained following a normalizing heat treatment [32]. The phases formed in these alloy steels are dependent not only on the elements that make up their composition but also on the heat treatments applied to the components before they go into service. Studies have demonstrated how easy it is to change the microstructure of the steels by making little change in the alloy content. The situation can be further complicated by heat treatments applied to the steel, which can have a large effect on the microstructure and the mechanical properties. The heat treatment produces carbides and precipitates that have a direct effect on the properties of the steel. The main intention of the heat treatment is to optimize the mechanical properties of the alloy and to relieve stresses. Generally, two types of heat treatment are applied: a normalization treatment and a tempering treatment.

Normalization involves austenitization at a high temperature in order to dissolve many of the alloying elements into solid solution. The temperature is usually high enough to dissolve stable precipitates already in the steel. The precipitate may be carbides and nitrides so that on cooling from austenitizing temperature they precipitate as a fine dispersion of particles. The austenitization temperature that gives complete dissolution of precipitates varies for different alloy compositions. A too low temperature gives incomplete dissolution of carbide and this leads to loss of creep rupture strength in 12Cr-Mo-V steel. A too high temperature gives undesirable formation of δ-ferrite phase. Recent publications have emphasised that a high Ac_1 temperature (where austenite formation first begins during heating) leads to good creep strength [33–35].

After austenitization, the cooling rate will determine the phases that are formed from austenite and depends on alloy composition. In some low alloy steels slow cooling may form undesirable ferrite. Formation of ferrite, or low bainite if the cooling is too fast, will result in an unsatisfactory distribution of carbides, and this is found to affect creep properties. To control cooling rates modifying composition to change the hardenability or using a suitable cooling medium is necessary.

The normalization is usually followed by tempering heat treatment to obtain the required mechanical properties. This tempering process nucleates precipitates and relieves stress. It is well established that the creep properties improve with subsequent tempering and then eventually degrade after the precipitation of the larger equilibrium alloy carbides. Most creep steels are subjected to very severe tempering (~700°C for several hours) causing general coarsening and the precipitation of ever more stable alloy carbides and intermetallic compounds. The precipitates, phases and composition of a steel, and the heat treatments applied to it are all important and interacting factors affecting creep properties. It is these *solid-state reactions* which ultimately determine the mechanical stability of the steels and, hence, their useful design lives [32].

In practice, two tempering heat treatments are often applied to the alloys after normalization. The first is meant to eliminate any retained austenite in the microstructure as this could influence dimensional stability. An initial temper of 500°C can induce the decomposition of the austenite so machining can be carried out accurately. The second is to produce a "stable" microstructure consisting of a variety of alloy carbides in a ferritic matrix. This tempering produces the required mechanical properties.

Tempering is carried out below the austenite transformation temperature so that there is no transformation to austenite. Temperature may be in excess of 500°C in order to allow the diffusion of substitutional atoms over the dimensions of precipitate growth in a reasonable period of time. 2.25Cr1Mo steel is usually given a stress-relief heat treatment at 700°C for several hours before going into service at a temperature of approximately 565°C [4]. The conditions chosen affect the precipitate size and creep properties. The initial precipitations formed are those for which nucleation is easiest. At longer times, mores table phases may form with the complimentary dissolution of the less stable carbides. Although the nucleation of these latter phases is more difficult, their formation leads to a reduction in the free energy of the system and so is thermodynamically preferable. Typical heat treatment schedule in the design of novel creep steel by Cole (2000) [6] is shown in Table 6.

Table 6: Typical heat treatments schedule for power plant steels where Y0336, Z1092, A, and B are names for some developed creep steels [6].

Steel	Y0336	Z1092	A	B
Normalization				
Temperature, °C	1053	1048	1200	1180
Duration, hr	11.35	11.6	2	2
Cooling rate	AC	AC	AC	AC
Tempering				
Temperature, °C	750	742	800	800
Duration, hr	12.67	12.5	4	4
Cooling rate	AC	AC	AC	AC
Annealing				
Temperature, °C	—	—	740	740
Duration, hr	—	—	4	4
Cooling rate	—	—	AC	AC

AC: air cool.

The heat treatments applied to these steels have to be carefully chosen and controlled to produce a suitable microstructure. The choice has to take into account the composition of the alloy and the phases which will be formed. Normalizing at a too low temperature or tempering at a too high temperature could have a deleterious effect on the creep rupture strength.

Study has shown clearly that carbide particle sizes increase with tempering temperature. This growth is undesirable for creep properties, since a dispersion of fine particles is desired. Sikka et al. (1984) [36] reported that tempering 9Cr-1Mo alloy above a nominal temperature of 760°C produced insignificant effects on creep rupture strength, but tempering below this temperature significantly improved it.

5. Description of Carbides in Power Plant Steels

The evolution of microstructure in steel is a complex and dynamic process involving simultaneous transformations of different types of phases. All the products compete for the limited nucleation sites, resources, and space. A fast cooling condition can preserve the austenite into lower temperature regions, so that finer structures prevail. When many transformations occur at the same time, the fast reaction may consume much of the austenite. The migration of carbon through interstices, the creation rate of nuclei on the austenite grain boundaries, the growth rates of individual phases, and so forth all react and interact with each other [1].

The task in designing power plant steels is usually to model the evolution of these precipitates and dissolution reactions. In dealing with precipitations in creep steels, a simple notation system is used to describe the chemical formula of the precipitate. Here "M" is used to represent the metal content (mixture of metal atoms) and "X" represents the carbon and nitrogen content.

There are varieties of phases that could be found in these power plant steels. The following list shows the well-known precipitates that determine the microstructure of power plant steels and are crucial in the development of creep strain [15].

Types of Precipitates That Can Be Found in Power Plant Steels [15, 20]. They are listed as follows:

Possible phases

Graphite

Epsilon $Fe_{2.4}C$

Cementite Fe_3C

Chi Fe_2C

M_2X M_6C $M_{23}C_6$ M_7C_3 Laves

M_5C_2 Z-phase μ-Phase χ-phase

M_5C_2 is among the latest discovery in 1Cr-0.5Mo steels. μ-phase has also been discovered.

Since these precipitates are often metastable, they do not represent the lowest free energy state. They form because they nucleate easily and they may dissolve as the material approaches equilibrium with time. Therefore, the first alloy carbide to form is generally not the equilibrium carbide, leading to precipitation sequences that produce more stable carbides [6, 15]. Sometimes, carbides which are metastable at first become stable during service when the composition of the steel changes. This is reported to happen when two different kinds of steels are joined together, leading to diffusion fluxes across the junction.

Cementite is iron-rich carbide that forms very rapidly, whereas graphite, despite an equilibrium phase, forms incredibly slowly since it is difficult to nucleate. Cementite containing no alloying element additions is seen to have an approximately hexagonal close-packed (*hcp*) arrangement of metal atoms with localized distortions to accommodate the carbon atoms or simply has an orthorhombic structure. The structure of cementite containing no alloying element additions has the structure Fe_3C.

M_3C denotes predominantly an iron-rich carbide with the same orthorhombic structure as Fe_3C, but here several other alloying elements can be found to partition into this carbide in significant quantities. Other metal elements can substitute for iron in alloy steels. The unit cell dimensions are found to be affected by partial substitution of alloying elements. Mn is found to dissolve in large quantities, as can Cr, with up to one-fifth of the Fe atoms being replaced by Cr (specific to low alloy steels). Ni and Co can also dissolve as they form metastable orthorhombic carbides, although it is found that they usually partition to ferrite. Mo, W, and V have also been found to have limited solubilities in M_3C. In general, M_3C carbides can be referred to using the general formula $(Fe, Cr, Mn, Mo)_3C$ [26].

Tempering causes the cementite to enrich towards its equilibrium composition. It is found that the rate of enrichment is fastest when the particles are of small size and the ferrite is saturated in carbide forming solute atoms [26, 32]. In low Cr steels, it is found that the cementite will enrich slowly and is unlikely to reach its equilibrium composition in the component lifetime but in high Cr and high alloy steels the cementite enriches rapidly and then dissolves as other alloy carbides precipitate. This phenomenon is believed to be the basis of secondary hardening in these steels. The steel softens as the cementite forms and removes carbon from solid solution but is then hardened by the alloy carbides which interact with dislocations, thus giving the so called secondary hardening. It is possible to form more stable carbides, nitrides, and borides than cementite. These are seen to form by the dissolution of a relatively coarse cementite dispersion which is replaced by finer alloy carbide dispersion [32].

It is found that when strong carbide forming elements are present, alloy carbides would form in preference to cementite. However, alloy carbides do not form until the steels are tempered in the range of 500–600°C because the metallic alloying elements cannot diffuse sufficiently rapidly to allow the alloy carbides to nucleate. Carbon, like nitrogen, can diffuse much more rapidly since it moves interstitially through the iron lattice; thus iron carbide tends to be the first precipitates formed in these steels.

In many cases M_2X type precipitates are found to be the first to form after cementite and they have a hexagonal crystal structure (*hcp*). Here, M denotes one or more of these alloying elements, V, Cr, Mn, Fe, Ni, Nb, Mo, Ta, and W; X represents C or N. It has been determined that in complex steels the phase is best regarded as $(Cr, Mo, V)_2(C, N)$. Study found that, in Mo steel containing low Cr content and no nitrogen, such as 2.25Cr-1Mo steels, M_2X is often close to Mo_2C [37]. In 9–12 wt% Cr steels containing nitrogen, the composition of M_2X is found to be closer to Cr_2N.

M_2X is found to exist as fine needle-like particles dispersed throughout the matrix and generally considered to nucleate on dislocations and martensite lath boundaries. Its precipitation is found to often cause secondary hardening, which is particularly seen to be important for low alloy steels, where M_2X plays a major role in the creep strength of the alloy. The precipitation of Mo_2C is usually seen to be the major factor in conferring creep resistance on low alloy ferritic steels. Study found that increasing Ni contents in steel decreases the stability of M_2X phase [26].

Stabilization of a Cr-rich M_2X phase in 12 wt% Cr steels can also cause secondary hardening [38]. Unfortunately, the elements which stabilize M_2X such as Cr and Mo are seen to cause the formation of δ-ferrite, a phase detrimental to creep properties, therefore limiting the quantities which may be added.

In their study, Baker and Nutting (1959) [37] observed that M_7C_3 forms after the formation of M_2X (i.e., M_2C). M_7C_3 is a Cr-rich carbide with the trigonal structure of Cr_7C_3, having a solubility of Fe up to 60%. This phase can form after cementite formation (M_3C) without any M_2X being observed. In Thompson's review, Titchmarsh (1978) [39] is said to have found that the Cr : Fe ratio can be greater than 1 in 2.25Cr1Mo steel. Mn, V, and Mo can also dissolve, with decreasing probabilities, respectively.

It has been found that M_7C_3 tends to nucleate in the vicinity of cementite or at the cementite/ferrite interface of a 2.25Cr-1Mo steel. M_7C_3 seemed to be observed if a sufficient Cr concentration is available. The dissolution of cementite is seen to provide an area sufficiently rich in chromium in which M_7C_3 could precipitate since these will be the most favourable sites [32, 40]. This phase has been found to coarsen rapidly and gave no beneficial contribution to creep rupture strength.

$M_{23}C_6$ is chromium-rich carbide which precipitates after either M_7C_3 or M_2X in 9–12 wt% Cr. It is observed to possess complex *fcc* structure of $Cr_{23}C_6$ [26] and the precipitate may also contain W, Mo, V, and Ni [5, 6, 41]. In steels containing significant amounts of Mo, it is believed that the formula $Fe_{21}Mo_2C_6$ can result and that in steels containing both Cr and Mo its composition can be anywhere within the above. This carbide is often found to precipitate with equilibrium composition and is predominant after tempering [42]. The phase is found to nucleate on the prior austenite grain boundaries or martensite lath boundaries or adjacent to M_7C_3.

Overaging is accompanied by the formation of $M_{23}C_6$ and dissolution of M_2X. It is then often found that M_2X dissolves after long-term exposure at high temperature in 9–12 wt% Cr steels and so does not improve the long-term creep resistance. It was also found that vanadium stabilizes M_7C_3 and so decreases the rate of release into the matrix of carbon and chromium for the growth of $M_{23}C_6$.

Various studies show that the composition of $M_{23}C_6$ varies with time. The carbide is found to enrich in Fe after 5 years in 2.25Cr-1Mo steels with a high Cr content with some low Mo content and after 18 years in service the phase contained much more chromium-rich $M_{23}C_6$. The carbide $M_{23}C_6$ is frequently observed in the form of large or coarse particles and is not thought to contribute directly to creep strength. That is, $M_{23}C_6$ is less effective in hindering creep deformation. Delaying its precipitation would have the effect of stabilizing the finer dispersions of M_2X and MX to longer times with a possible enhancement of creep strength. Study has also shown that the tendency for $M_{23}C_6$ to coarsen should also be greatest in the Cr-rich alloys [6].

However, it has been suggested that $M_{23}C_6$ carbides may stabilize lath boundaries and delay grain growth. In his study, Abe (1999) [13] made a conclusion that $M_{23}C_6$ precipitates are the most effective in 9Cr-W system for pinning lath boundaries and reducing the coarsening rate of the laths. This study shows that coarsening occurs by sideways movement of lath boundaries and the coalescence of two boundaries.

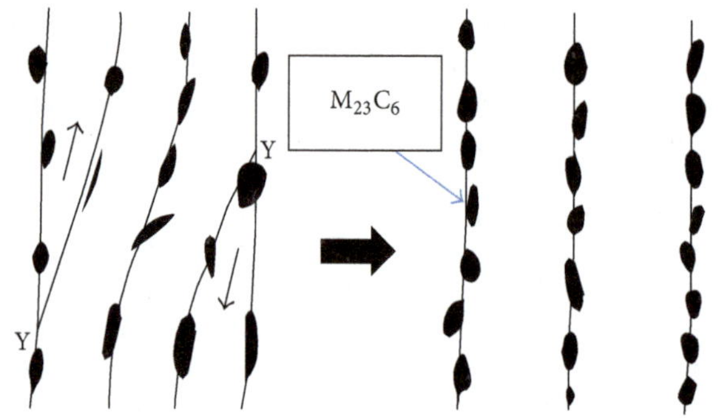

FIGURE 2: Schematic of coarsening of martensite laths by the movement of Y-junction in 9Cr-W steels during creep testing at 550–650°C for up to 15 000 hours, with varying tungsten content [6, 13].

Figure 2 illustrates the movement of Y-junction of triple points of lath boundaries to cause coarsening.

Figure 3 shows an image obtained, using TEM, from a thin foil specimen of 9Cr steel given severe heat treatments.

The microstructure in Figure 3(a) shows tempered martensite which retains the essential lath-like morphology and carbides. The absence of coarsening is said to be due to stabilization of the microstructure by the carbides which nucleated heterogeneously at the lath boundaries [6]. Figure 3(b) shows carbides aligned at the martensite laths boundaries.

9Cr-Mo steels after creep test have shown preferential recovery of the microstructure in the vicinity of prior austenite grain boundaries. This recovery phenomenon is associated with carbides forming and coarsening preferentially at the prior austenite grain boundaries, removing solutes and dissolving smaller precipitates in the surrounding regions. The loss of pining effect by the precipitates allows the recovery of the martensitic microstructure [6]. This process is depicted schematically as in Figure 4 for a 12Cr alloy.

These mechanisms are reported to account for the loss in long-term creep strength often experienced by 9–12Cr wt% steels, caused by the coarsening of the microstructure. This process has been reported to have occurred for modified 9Cr-1Mo steel at 600 and 650°C for 10 000 hours [14]. Figure 5 shows degradation in creep rupture strength for a modified 9Cr-1Mo steel by this process.

For the improvement of long-term creep rupture strength, research has shown that the dissolution of small precipitates and the coarsening of large ones need to be retarded. Recovery of the microstructure also needs to be delayed, especially in the vicinity of prior austenite grain boundaries.

The coarsening of $M_{23}C_6$ carbides in 12CrMo(W)VNbN steel is reported to be accompanied by dissolution of fine MX carbonitrides. Increase in Ni content in 12CrMoV steel results in accelerated microstructure degradation with more rapid coarsening of $M_{23}C_6$, dissolution of MX, and precipitation of coarse M_6X and Fe_2Mo. Increase in nitrogen content results in accelerated microstructure degradation during creep at 650°C with more rapid precipitation of a phase referred to as Z-phase.

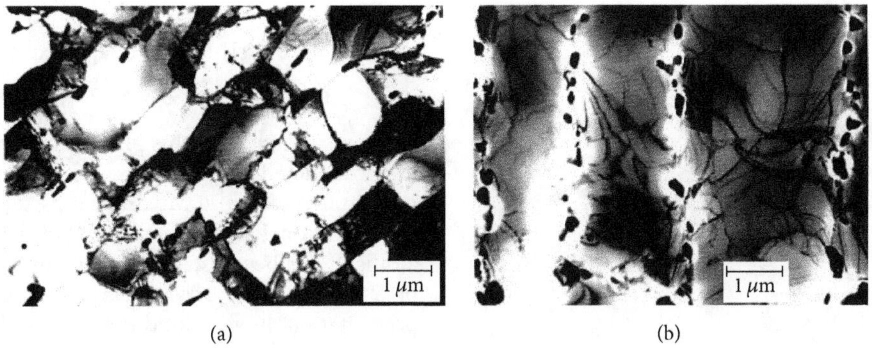

FIGURE 3: TEM micrograph of Z1092 (9Cr steel), showing the general microstructure [6].

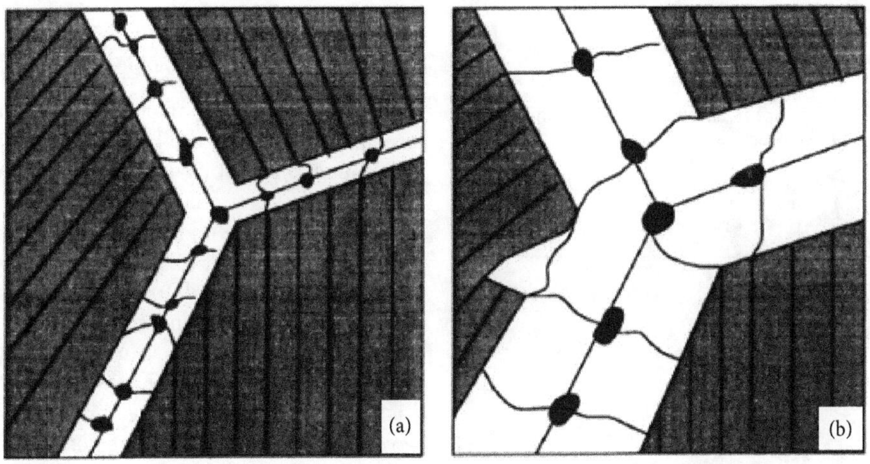

FIGURE 4: Illustration of the degradation mechanism of a 12Cr-0.4Mo-2W-Cu-V-Nb steel. (a) shows the microstructure prior to creep testing and (b) illustrates the microstructure after long-term creep deformation [6].

One of the important observations in the recent literature is that many ferritic steels are strong in creep in the early stages of creep life, but the strength then deteriorates sharply as a consequence of the development of local regions of recovery, where damage is then intensified. The observations reported in [6] confirm that recovery is not usually homogeneous. A possible solution is suggested of the use of a dual heat treatment to obtain more homogeneous dispersion of precipitates and retard localized recovery. The observations also support the design philosophy that the concentration of Ni should be minimized since it causes acceleration of microstructure degradation with more rapid coarsening of $M_{23}C_6$.

In Fe-9Cr-W steels, W content has been found to affect the coarsening of $M_{23}C_6$ during creep testing at 600°C [13], as shown in Figure 6. However, this information is usually interpreted with caution because the addition of tungsten also causes the precipitation of Laves phase (intermetallic compound) [13].

This study found that the coarsening of $M_{23}C_6$ and the martensite laths were reduced by increasing the amount of tungsten. The increase in W also caused the precipitation of Laves phase, which is said to be beneficial in decreasing creep rate and suppressing the recovery of the microstructure.

When steels contain Mo and relatively low level of Cr, M_6C precipitates, which seems to be the equilibrium carbide [35, 43]. Therefore, M_6C is found to be essentially Mo-rich carbide with *fcc* [26]. M_6C is found to nucleate on M_2X and $M_{23}C_6$ interphase boundaries, prior austenite grain boundaries, or martensite lath growing rapidly at the expense of all surrounding carbides, nucleating at existing particles [6]. The transformation of Mo_2C to M_6C is said to occur more rapidly in bainite than in ferrite.

Work reported by Nutting (1999) [44] indicates that this phase may not nucleate from existing carbides, but as the carbides dissolve, solutes are transferred to M_6C by diffusion. The carbide is often coarse and since it causes the dissolution of less stable fine carbides; M_6C *can cause a reduction in creep strength*. It has been shown that the coarsening rate of M_6C is greater than that of $M_{23}C_6$ and so *is a particularly undesirable phase* in creep steel design. This carbide is found to be mainly rich in Mo in 0.5Cr0.5Mo0.25V steel and precipitate in 12CrMoV steels with high Ni content [44].

MX denotes some carbides which may be observed in power plant steels and often has a cubic structure like NaCl [41]. It was observed to be often rich in vanadium, although NbC and TiC have also been found to form. The vanadium carbide structure is basically VC but extends to V_4C_3. There

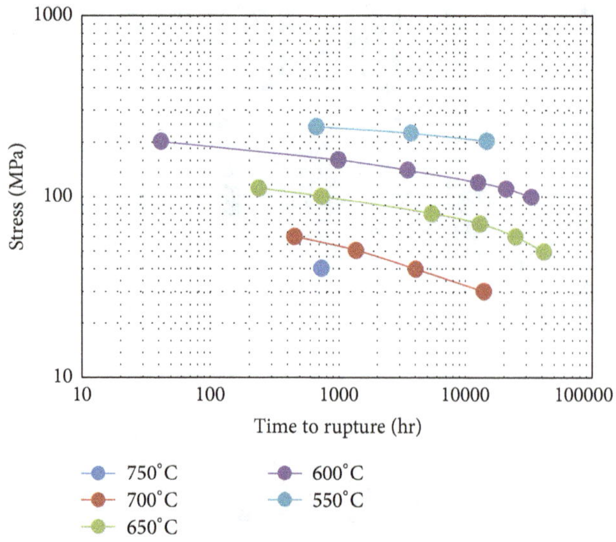

FIGURE 5: Degradation in creep rupture strength for a modified 9Cr-1Mo steel, adapted from [14] or stress versus time to rupture curves for 9Cr-1Mo (or T91).

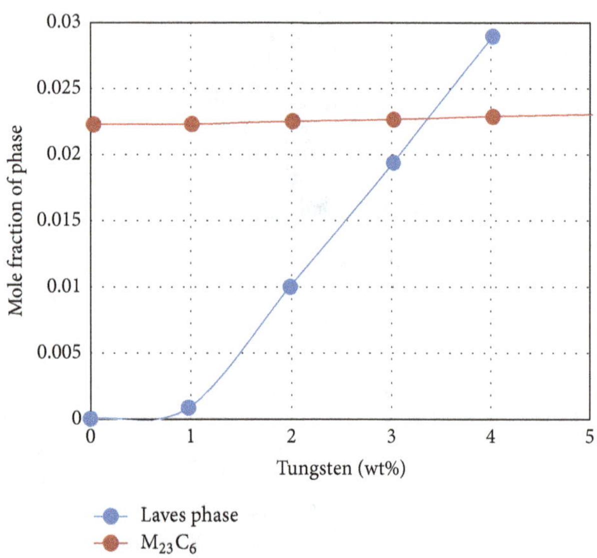

FIGURE 6: Coarsening data from Fe-0.1C-9Cr-0.5Mn-0.3Si wt% steel, containing a variety of tungsten additions, tested in creep at 600°C [13].

is also some solubility of Fe, Cr, and Mo in these MC. The carbides MoC and WC were observed to have hexagonal structures. Other precipitates of this form such as VN were found to be desirable because they form a fine distribution of small and stable particles which are beneficial to creep strength. VN and NbC could be found in a 12 wt% Cr steel containing V and Nb. VN was found as a plate-like phase and NbC in a more spheroidal form.

After long exposure to elevated temperatures *Laves phase* precipitates and it has the general composition Fe_2M, where the alloy content may be W, Mo, or a combination of both. Laves phase is found to have a higher interfacial energy. It has often been observed in 9–12 wt% Cr steels containing W or Mo and in 9Cr1Mo steel of approximate composition:

11 wt% Si, 44 wt% Mo, 17 wt% Cr, and 28 wt% Fe.

It was also found that this phase tends to form preferentially in Cr-depleted regions. These regions were found to be comparatively rich in Mo, indicating that a principal kinetic constraint on Laves phase formation is the diffusion of Mo atoms. It has been established that increasing the Si content of 9Cr-2Mo steel accelerates and increases Laves phase precipitation. It was found that Mn additions were useful in retarding the formation of Laves phase [45]. Precipitation of Laves phase on lath boundaries decreases creep rate and suppressed the recovery of the microstructure. A Si content of 0.3 wt% was found to provide the best balance between recovery of the microstructure and coarsening of the precipitate of some creep steel. Laves phase has been found to initially form on prior austenite grain boundaries followed by precipitation on lath boundaries and finally some precipitation within the laths [45]. It may also benefit from nucleation on existing phases such as $M_{23}C_6$. Study found formation of Laves phase at grain boundaries and martensitic lath interfaces were an important factor in maintaining creep rupture strength. For tungsten containing steels, Laves phase precipitation is found to be beneficial for creep strength than for solid solution strengthening by tungsten. Laves phase containing W is found to be thermally stable but the coarsening rate of the particles after precipitation is high [45]. This gives large Laves phase precipitates after several thousand hours of creep exposure, indicating that the beneficial effects for creep strength would not last into the long-term.

Laves phase is seen to be the equilibrium precipitate at temperatures below ~600°C in the 10CrMoV steel. The variation of the equilibrium volume fraction with temperature is shown in Figure 7. From the graph, Laves phase is not observed beyond about 600°C.

There have been a number of studies of Laves phase precipitation in 9CrMoWV steel (NF616) [46], a commercial alloy designated for service at temperatures above 600°C in power plant. The steel (wt%: Fe-0.106C-8.96Cr-0.47Mo-0.051N-0.069Nb-0.20V-1.83W) contains tungsten which makes it more susceptible to Laves formation, which occurs over a greater range of temperatures than in the 10CrMoV steel. In the study in [47], Laves phase (Fe_2Mo) and progressive coarsening of $(CrFe)_{23}C_6$ carbides with respect to increasing test duration and temperature were observed. The formation and growth of Fe_2Mo are reported to occur at Cr_2N/matrix interface, caused by the depletion of Cr solutes in the vicinity of Cr_2N precipitates [47].

Another phase usually found in 12CrMoVNb steels at temperature greater than 550°C is Z-phase. This phase is a complex nitride phase containing Cr, V, and niobium (Nb) (of the form Cr(Nb, V)N and it is found to form in close association with dissolving NbX particles [47, 48]. As this phase grows, it is seen to gradually replace MX and M_2X nitrogen-rich phases which improve the creep rupture strength [49]. Therefore, Z-phase directly contributes to a reduction in the creep rupture strength of the observed

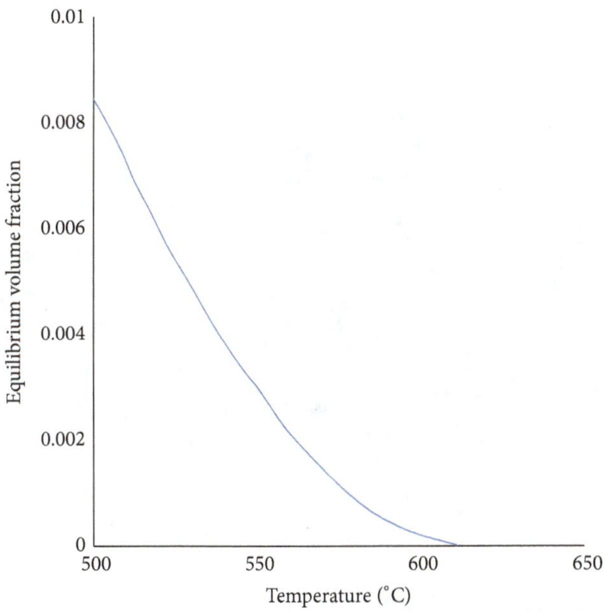

FIGURE 7: Variation of equilibrium volume fraction of Laves phase with temperature in 10CrMoV steel, calculated using MTDATA [7].

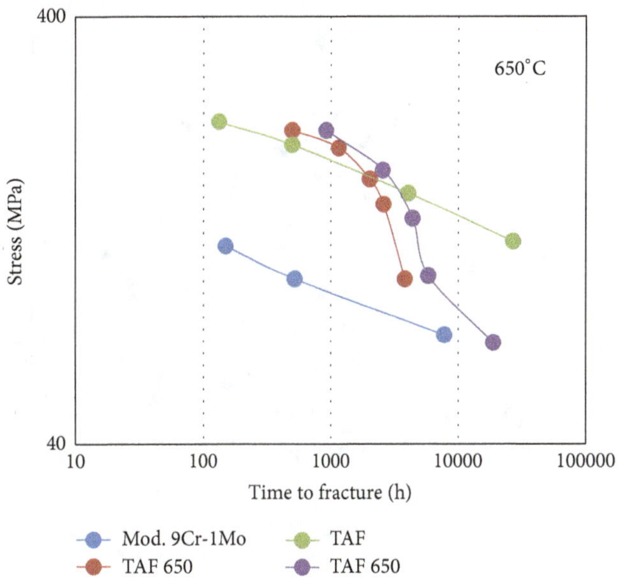

FIGURE 8: Comparison of stress versus time to rupture curves for various martensitic steels at 650°C, adapted from [16].

12CrMoVNb steel. Precipitation of large Z-phase particles, Cr(V, Nb)N, which depletes the matrix of fine MX nitrides, (V, Nb)N, was observed to be the major cause of early breakdown in long-term creep strength of a number of new 9–12%Cr martensitic steels [50].

Study shows that large concentration of niobium (0.9 wt%) enhances the precipitation of Z-phase. In 12Cr steel used by [51], precipitation of Z-phase was observed to be accomplished by Cr atoms diffusing into (V, Nb)N particles and later this structure transforms to cubic or tetragonal Z-phase. It was also reported that Z-phase formed rapidly at prior austenite grain boundaries. Also the formation was fastest at 650°C.

In general, the precipitation of Z-phase, M_6X, and $Fe_2(W, Mo)$ Laves phase during creep causes a loss of creep strength at long times, because they consume existing fine M_2X and MX and or $M_{23}C_6$ precipitates [42, 50]. Retardation of the Z-phase formation is necessary to improve strength of creep-resistant steels [49].

Figure 8 is the sigmoidal behaviour in creep rupture data for four creep steels at 650°C showing the effect of Z-phase formation [16].

The synergetic effect of Z-phase precipitation and tungsten depletion of solid solution due to Fe_2W Laves phase formation is believed to be the reason for the sigmoidal shape of creep rupture strength curve of TAF 650 steel. The loss of creep strength in the modified 9Cr-1Mo (or T91) is reported [16] to be due to preferential recovery of the microstructure in the vicinity of prior austenite boundaries, as shown in Figure 9. The preferential recovery promotes the onset of acceleration creep and hence causes premature rupture. The dissolution of MX and the precipitation of Z-phase are reported to promote the preferential recovery.

FIGURE 9: TEM micrograph after creep rupture testing for 34,141 hr at 600°C and 100 MPa [16].

Generally, it is believed that the loss of creep strength in 9–12Cr steels at 550°C and above is due mainly to the dissolution of fine M_2X and MX carbonitrides during creep exposure, which is driven by the precipitation of coarse Z-phase and or M_6X carbonitrides which are thermodynamically more stable than M_2X and MX [16, 52–54].

A report has been made on the discovery of M_5C_2 in ex-service 1Cr-0.5Mo creep steels [55]. This phase is said to have a monoclinic structure and has not been identified before in creep-resistant Cr-No steels. The precipitates were found to be rod-shaped and appeared to nucleate heterogeneously on M_2C and remain in ferrite regions from which M_2C had disappeared, suggesting that M_5C_2 is more stable than M_2X. The discoveries of M_5C_2 and Z-phase are more recent than the other phases in ferritic power plant steels.

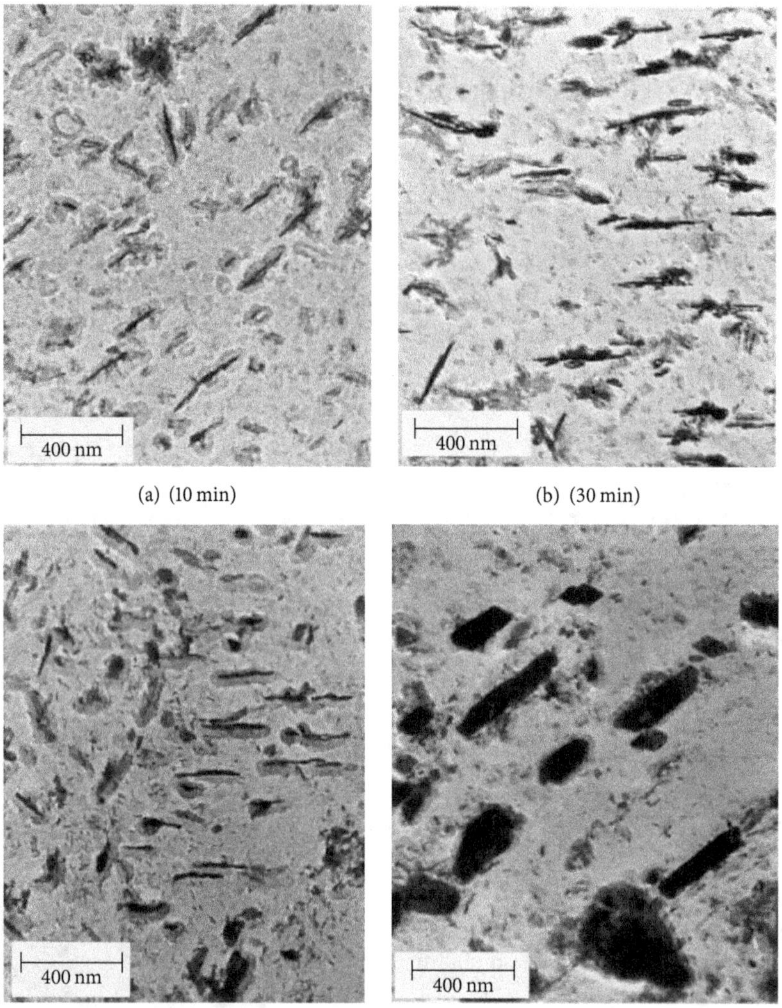

(a) (10 min) (b) (30 min)

(c) (1 hr) (d) (16 hr)

FIGURE 10: This figure shows micrographs which are examples of all the transitions observed at 600°C. [7].

TABLE 7: Heat treatments of 10CrMoV steel for carbide stability study [7].

Temperature, °C	Time, h
600	0.16, 0.5, 1, 4, 16
650	0.16, 0.5, 2, 8
800	0.05, 0.16

A visual picture of some of these carbides can be seen from the study carried out to understand the evolution of microstructure in the early stages of tempering in a bid to establish carbide stability diagram for 10CrMoV steel. The starting material was in quenched condition. The heat treatment given to the test piece was as listed in Table 7.

The result of series of micrographs examined, after a given time, using TEM of samples held at 600°C is as shown in Figure 10.

In Figure 10(a), analysis shows that the small needle-shaped particles are M_2X precipitated along with M_3C. After 30 minutes, the M_2X precipitate intensified, while the M_3C volume fraction remained approximately constant (Figure 10(b)). After 1 hour, there were additional blocky precipitates seen which were identified as $M_{23}C_6$; M_3C was present as well as M_2X (Figure 10(c)). Finally, after 16 hours, the microstructure consisted of relatively large $M_{23}C_6$ particles localized to lath boundaries along with M_2X occurring within the laths (Figure 10(d)).

The sequence observed in this test was summarized as follows:

$$\begin{aligned} M_3C + M_2X &\longrightarrow M_3C + M_2X + M_{23}C_6 \\ &\longrightarrow M_2X + M_{23}C_6 \end{aligned} \quad (2)$$

It was reported that some fine MX phase could be present, although none was said to be found in this particular investigation. The same sequence was reported for test at 650°C, but the precipitation kinetics was faster for all phases at this temperature.

At 800°C, $M_{23}C_6$ was observed after only 3 min, by which time M_3C had dissolved. The approximate carbide stability diagram of this experiment can be found in these publications [26, 45]. At 600°C, the equilibrium precipitate for the 2.25Cr1Mo steel is $M_{23}C_6$, while for the 10CrMoV steel the equilibrium precipitates are $M_{23}C_6$ and M_2X. [7].

A common observation reported here is that once a metastable phase has completed precipitating (or enriching), it will start to dissolve to provide solute for all the more stable phases whose extent of reaction is less. So, in the case of the 10CrMoV steel, M_3C and Laves phase eventually dissolve to provide solute to M_2X and $M_{23}C_6$ which are the stable phases.

For the 10CrMoV steel, the order of events is first the rapid completion of M_3C enrichment as the steel has a very large Cr concentration. This is followed by the completion of *Laves* precipitation; for this particular temperature (600°C), Laves phase is a metastable (brittle) phase. It has been explained that because the precipitation of *Laves* phase is extremely sluggish, at the time it stops precipitating (~30 min), the volume fraction which has formed is very small (~2×10^{-9}). After this time, it dissolves rapidly and so is not experimentally observed. Finally, M_2X and $M_{23}C_6$ form as the equilibrium phases.

This order of precipitation and enrichment is also observed in 2.25Cr1Mo steel but, due to low Cr concentration and higher C content in this steel, it is reported that M_2C precipitation occurs and is finished well before $M_{23}C_6$ precipitation and the volume fractions of M_2C and M_3C are greater. Therefore, precipitation or enrichment of these phases removes a greater proportion of the available solute for the matrix than is removed in the 10CrMoV steel. The reduction in the solute available for $M_{23}C_6$ suppresses its formation to later times [7].

6. Effect of Alloying Elements of Creep Steel on Phase Stability

The large number of alloying elements in creep steel produces a huge variety of microstructure where the matrix can differ as well as the composition. The composition influences the phases' precipitation sequences and distribution.

The elements that make the composition of these steels have specific functions, even though there may be interaction among them. Increasing or decreasing their amount in search of better creep steel hinges on the understanding of their specific roles in the alloy system. These elements are usually

> C, Mn, Ni, Co, Bo, Cu, Cr, Si, W, V, Mo, Ti, Nb, Ta (and Nd), Al, and N.

Elements such as Mn, Mo, Cr, V, W, Ti, and Nb can both form stable carbides and be found in solid solution within the ferrite [26]. The studies on power steels show that Cr, W, and Mo provide solid solution strengthening. The precipitates that commonly form in this alloy system include $M_{23}C_6$, as well as MX (VN and NbC). The drop in the strengthening effect of precipitate hardening is due to the coarsening of these precipitates [6].

The carbon content of the steel determines the amount of carbon in the carbide and the excess remains in solid solution in ferrite. Nitrogen is reported to enter the carbide phase, readily forming carbonitrides with iron and many alloying elements, and separate alloy nitride phases in the presence of Ti or Al [26].

The known characteristics of some of the elements are as follows.

Carbon (C). Carbon stabilizes austenite relative to ferrite. It is essential for the formation of carbides which cause the secondary hardening of power plant steels. It usually has peak value because high levels of carbon lead to unacceptable reductions in mechanical properties such as toughness and may cause cracking after normalization and also after welding. These effects have been studied by [38, 45, 56].

Manganese (Mn). Manganese stabilizes austenite but is often found to have an adverse effect on the creep strength of power plant steel. This phenomenon was attributed to the retention of austenite which will be rich in carbon and nitrogen, so reducing the effects of secondary hardening. Study shows that increasing Mn content may increase the growth rate of M_6X, an undesirable and coarse phase which can remove W from solid solution and cause the dissolution of other more desirable precipitates [57].

Nickel (Ni). Nickel stabilizes austenite and is found to accelerate the precipitation and coarsening of carbide phase and *Laves phase*. This coarsening reduces the creep properties and so is detrimental to creep properties of power plant alloys. Ni concentrations greater than 0.4 wt% have been shown to deteriorate the creep strength of high chromium steel appreciably [54]. This is because Ni increased the rate of dissolution of M_2X as a consequence of precipitation of M_6X phase. It increases the coarsening rate of M_6X and $M_{23}C_6$ [54].

In the search for a creep-resistant steel capable of functioning at temperatures of 650–750°C, in a very recent study, 15Cr1Mo ferritic steel was alloyed with increasing Ni content as shown in Table 8.

The result of that endeavour is shown in Figure 11.

15Cr1Mo steel gave ferritic and martensitic phases upon quenching, with volume fraction of martensite increasing with Ni addition. The precipitates observed were reported to be Laves phase and χ-phase (which are intermetallic compounds), $Cr_{23}C_6$ (chromium carbide), and Z-phase (Cr(V, Nb)N) [8, 31]. Ni was seen to cause preferential precipitation of $Cr_{23}C_6$ at grain boundaries between the ferrite and martensite. Once again Ni is seen to accelerate the precipitation of *Laves phase*. This study seems to show that at 650°C no effect in creep strength is seen by increasing Ni content of 15Cr1Mo steel but there was huge creep strength improvement in comparison with T92 creep steel, whose composition is given as 9Cr-0.5Mo-1.8W-V-Nb, at each temperature. This huge difference is said to be due to increased precipitation of Lave phase and chromium carbide. The coarsening effect of Ni on carbide and Lave phases was not reported.

Cobalt (Co). Cobalt is austenite stabilizer and is found not to coarsen the steel microstructure like Mn and Ni. Study shows that cobalt strengthens the martensitic matrix and

Table 8: 15Cr1Mo steel studied [8].

	C	Cr	Ni	Co	Mo	W	V	Nb	N	B
0.8Ni	0.051	15.16	0.86	1.96	1.03	6.03	0.21	0.06	0.037	0.0031
1.4Ni	0.048	15.13	1.41	1.99	1.03	6.02	0.21	0.058	0.037	0.0031
2.1Ni	0.051	15.07	2.15	2.04	0.98	6.03	0.21	0.054	0.036	0.0031

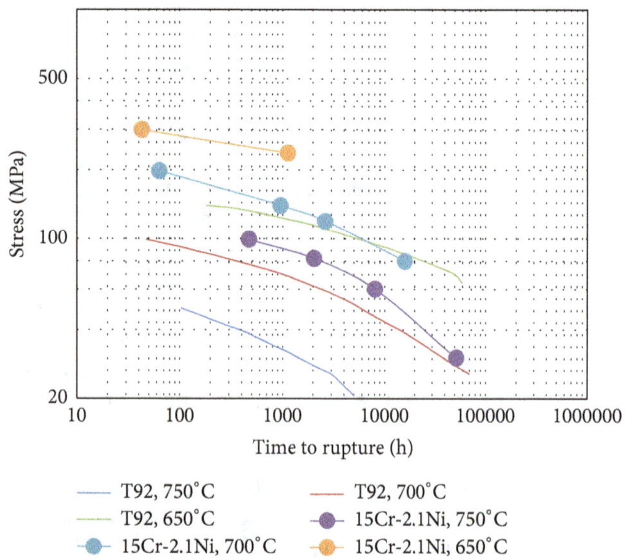

Figure 11: Comparison of stress versus creep rupture strength of 15Cr steels with increasing Ni content and T92 at 650–750°C, adapted from [8]; also see [17].

enhances the stability of fine precipitates such as MX and μ-phase. Cobalt was found to remain in solid solution in 12%Cr steels, even with concentrations of up to 10 wt%. This is beneficial effect for the creep rupture strength of creep steels. Irvine et al. (1960) [38] and Igarashi and Swaragi (1997) [58] have studied this beneficial effect. It is thought that cobalt promotes the formation of a stable oxide layer. However, the effects of cobalt additions on oxidation resistance are not well known. Major disadvantage in the use of Co as alloying element is that it makes steel difficult to recycle. In other words, cobalt-containing steels have a very low scrap value. In recent study, additions of Co and Ni were observed to strongly influence the creep strength and precipitation behaviour of intermetallic compounds (Laves phase, χ-phase, and μ-phase) during creep exposure at high temperature for 15Cr1Mo6W steel [31]. They found that creep strength is increased by precipitation of fine particles and homogenous dispersion of intermetallic particles in the ferritic matrix.

Boron. Study on boron additions shows beneficial effect on high temperature strength of 10Cr steel. It is found that long-term creep rupture strength was improved by B additions up to approximately 0.01 wt%. This increase in rupture strength by boron has been attributed to stabilization of $M_{23}C_6$ at the grain boundaries, retardation of the coarsening of the precipitate, prevention of grain boundary sliding, and thereby dynamic recrystallization. Some research work on this can be found in [14, 57, 59, 60].

However, there appears to be no benefit to creep rupture properties at higher boron percentages. In fact, excessive boron is detrimental to strength due to the formation of undissolved boride phases [45]. Increased levels of boron are also found to lead to problems with weldability and forgeability. Excessive levels of boron can also combine with nitrogen to form quite coarse BN precipitates. This is seen to reduce the amount of finer nitrides in the alloy such as VN, which can *have beneficial effects on creep rupture properties*, such as ductility [61].

Copper. Copper is an austenite stabilizing elements and is often added purely to counteract the undesirable formation of δ-ferrite. Hence, copper is often used as an addition to power plant steels to retard δ-ferrite formation, since it stabilizes the austenite phase. High copper additions, up to 2 wt%, have been found to cause the precipitation of fine Cu particles on martensite lath boundaries which pins them and retarding the recovery of the martensite. It was also reported that Cu additions caused refinement of the prior austenite grain size. One concern with large copper additions is that it may precipitate and become removed from the alloy during heat treatment. Studies in this area have been reported by [18, 53, 62].

Chromium. Chromium stabilizes ferrite and is carbide former. The addition of chromium to ferritic steels has been shown to improve the oxidation resistance. The high Cr steels are extremely corrosion-resistant but the lower Cr steels are susceptible to corrosion [6]. Studies have long established that Cr improves the steam oxidation resistance of ferrous alloys [63, 64]. Large chromium additions of 9-10 wt% provide hardenability, good creep strength, and resistance to corrosion. An increase of 11 wt% Cr was found to markedly increase corrosion resistance at creep steel at 650°C; hence at least 11 wt% Cr has been recommended in the design of a steel for use at 650°C. Disadvantage of high concentrations of Cr is the promotion of undesirable δ-ferrite formation. Ennis et al. (1998) [65] and Abe and Nakazawa (1992) [5] have reported this study.

Silicon. Silicon is a ferrite stabilizer and it is found to influence the rate at which carbide precipitates. It is also found to accelerate the precipitation and coarsening of *Laves* phase in 9Cr steels. Silicon additions, like Cr, promote the formation of undesirable δ ferrite phase [66]. However, silicon can be very important in the formation of protective oxidation layers. It was reported that about 0.8 wt% Si is required to form a protective layer at 650°C for 9 wt% Cr steels, while 0.4 wt% Si

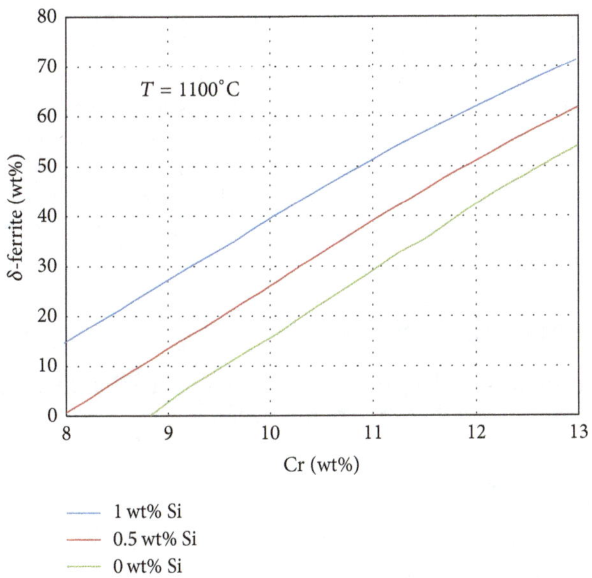

FIGURE 12: δ-ferrite volume fraction as a function of Cr wt% for a steel of composition 0.08C-0/1Si-0.5Mn 8/13Cr-3W-0.2V-0.005B-0.005N wt% [18].

is needed in the case of 12 wt% Cr steels [6]. However, there is optimum silicon content of approximately 0.3 wt% Si to give acceptable corrosion resistance and mechanical strength since it causes acceleration of *Laves* phase precipitation and coarsening. Study has shown that silicon content above 0.4 wt% leads to deterioration in tensile strength, creep rupture strength, and toughness due to δ-ferrite formation. Therefore, the formation of δ-ferrite has been the major problem with silicon additions [18].

Figure 12 shows calculation using Thermo-Calc for a model steel, comparing the amount of δ-ferrite which forms during austenitization with Cr and Si content.

The figure clearly shows that the addition of high Cr or Si content will cause δ-ferrite formation during austenitization around 1100°C. To mitigate against this, austenite stabilizing element is commonly added. These characteristics are balanced with the effect on the mechanical properties when designing ferritic alloys with significant silicon contents. Other reports on Si study can be found in [41, 66, 67].

Tungsten. Tungsten is a ferrite stabilizer and strong carbide former. It has been found to stabilize M_2X though less effective. Tungsten promotes the formation of *Laves* phase. Like Cr and Si, high contents of tungsten may lead to the formation of undesirable δ-ferrite [45].

In general, *tungsten addition has been found to improve the creep rupture strength of ferritic heat-resistant steels.* This effect was attributed to solid solution hardening and precipitation strengthening due to $M_{23}C_6$ and *Laves* phase. $M_{23}C_6$ and *Laves* phase were found to precipitate at grain boundaries and this is believed to be the main factor responsible for the improved creep rupture strength. Study shows that increased W concentrations in a 9Cr alloy reduced the coarsening rate of $M_{23}C_6$ precipitates. In turn, this reduced the coarsening of martensite laths due to the pining effects of $M_{23}C_6$, thus delaying recovery. Report also has it that the coarsening rate of *Laves* phase was very large in these alloys. Reports on effects of tungsten addition to creep steel can be found readily in [29, 46, 68].

Vanadium. Vanadium is a ferrite stabilizer and strong carbide former. Vanadium is a key element due to the stability of its nitride, and this nitride is found to improve the creep rupture strength of steel. Like W, it stabilizes M_2X phase and also MX. Vanadium additions to 10–12 wt% Cr steels have been found to significantly improve the long-term creep strength. The strengthening effect of vanadium may also be enhanced by the removal of available carbon into vanadium carbides, thus leaving Mo in the matrix, where it contributes to solid solution strengthening. Reports on this study can be found in [6, 69–72].

Molybdenum. Molybdenum forms stable carbides and remains in ferrite, contributing to solid solution strengthening. Like W and Va, Mo has been shown to stabilize M_2X phase and a phase given as M_3C_6 phase. Increasing Mo concentration above 1 wt% has been found to promote formation of M_6C, *Laves* phase, and δ-ferrite in 9 wt% Cr steels. Study in this area can be found in [69, 71, 73].

Titanium. Titanium is strong carbide and nitride former and improves the creep rupture strength of ferritic steels. However, increase in Ti may result in a large reduction in rupture ductility due to Ti based inclusions preventing deformation at grain boundaries.

Niobium. Niobium addition is found to give MX carbides and nitrides which are small and stable. These precipitates help to pin grain boundaries and prevent grain growth. The effectiveness of this mechanism relies on dissolving Nb during austenitization; otherwise insoluble NbC remains, causing coalescence of the precipitates, coarsening them, and accelerating recovery. Increased level of Nb may also cause the formation of *undesirable Z-phase*. Further information on this can be found in [69].

Tantalum (and Neodymium). The additions of Ta and Nd to creep steel have been found to be beneficial to the creep rupture strength of weld joints in a 0.1C-11Cr-3W-3Co-V-Nb-N steel. Igarashi and Swaragi (1997) [58] reported that this effect was attributed to the formation of TaN and NdN precipitates. These nitrides were found to be extremely stable and do not coarsen appreciably during welding and the subsequent heat treatment.

Aluminum. The addition of Al is thought to be detrimental to the creep rupture strength, since it leads to the formation of AlN instead of VN or NbN. AlN is a coarse phase and not useful to creep strength, while the others are fine and dispersed and so improve creep property.

Nitrogen. Nitrogen, like carbon, is small enough to occupy interstitial sites in the iron lattice and stabilizes austenite relative to ferrite. It has been found to stabilize MX precipitate in Cr steels, probably of the composition CrN, and this has been one of the reasons suggested why N additions may increase creep strength [38]. Nitrogen addition is found to aid the formation of MX type precipitate, such as VN and NbN, which are fine and generally desirable for creep strength. Potential disadvantage militating against full utilization of nitrogen is said to be the difficulty associated with increasing nitrogen content of ferritic steels above approximately 0.08 wt% without pressurizing the cast in a nitrogen atmosphere. This adds to the cost of designing.

However, study shows that addition of nitrogen to boron containing steels may precipitate BN which can offset the beneficial effects of both B and N. Report on effect of nitrogen on creep steel can also be found in [38, 61, 74].

Phosphorus. The presence of phosphorus in substantial amount is found to lead to segregation to grain boundaries which embrittles the alloy. Study shows that it reduces the volume fraction of beneficial M_2X precipitates and increases creep rate. Phosphorus is said to interact and drive the segregation of Mo at prior austenite grain boundaries and results in loss of ductility in 9Cr1Mo steel. Pilling et al. (1982) [70] and Noble et al. (1990) [75] have some report on the effect of this element on creep steel.

Sulphur. Formation of sulphides, such as MnS, can provide preferential nucleation sites for cavitation. Precipitation occurs on prior austenite grain boundaries when cooling from the austenite. The cavitation is caused by the low interfacial adhesion between MnS and the ferrite matrix [44].

It is usually useful to examine the combined effect of elements that behave in a similar manner rather than individual effects. For example, δ-ferrite promoters are Cr, Si, W, and Mo. For the effect of W and Mo concentrations on 9Cr alloys, it is found that, in the range 0.5 > Mo wt% > 0.8, Mo improved the creep rupture strength [76]. This is just one example of the many interactions between variables which make alloy design extremely difficult.

From above discussions one can see that the elements that make up the creep steel composition are added for a variety of reasons. The carbide formers are Ti, W, Va, Cr, Nb, Mo, Si and the undesirable δ-ferrite promoters tend to be ferrite stabilizers, like Si, Mo, Cr, W, V. and the counteracting elements the austenite stabilizers, that is, C, Mn, Ni, Co, Cu and N.

7. Conclusion

We can see that power plant steels are designed by considering the known effects of each single element and their interactions. This interaction makes the design of this class of steel extremely challenging venture. These elements are added either to stabilize phases which are beneficial to creep resistance or to suppress phases which are detrimental. The nature of steel has created opportunity to design steels that have served man in the high temperature domain, up to 650°C for decades. There are emerging possibilities of alloying with nickel and boron to produce steel that functions at 700°C and above, where the alloys are independent of any carbides as strengthening factors. The current studies on 15Cr steel are keeping the hope alive. The success of this is expected to increase tremendously the efficiency of power plants by 50% and the concomitant reduction in the emissions of gases such as CO_2, NO_x, and SO_x. The immense possibility of altering steel microstructure to create wide ranges of mechanical properties underscores the critical role that steel has played in the construction of our society. It just appears that the potential of steel is limitless.

Competing Interests

The authors declare that there are no competing interests regarding the publication of this paper.

Acknowledgments

The authors are grateful to HKDH Bhadeshia of University of Cambridge for his numerous insightful and inspiring papers and his willingness to be open in information dissemination in phase transformation study. They relied heavily on his research activities. They also wish to commend David Graham Cole whose Ph.D. thesis is a mine of information. His work aided in no small measure in this review.

References

[1] J. Chen, *Modelling of simultaneous transformations in steels [Ph.D. thesis]*, Department of Materials Science and Metallurgy, Fitzwilliam College, University of Cambridge, 2009.

[2] H. K. D. H. Bhadeshia, "Novel aspects of ferritic steels for the generation of electricity," in *Proceedings of the 4th International Symposium on Mechanical Science based on Nanotechnology*, pp. 143–146, Sendai, Japan, 2007.

[3] M. Murugananth and H. K. D. H. Bhadeshia, "Components of the creep strength of welds," in *Mathematical Modelling of Weld Phenomena—VI*, H. Cerjak and H. K. D. H. Bhadeshia, Eds., pp. 243–260, Institute of Materials, 2002.

[4] H. K. D. H. Bhadeshia, A. Strang, and D. J. Gooch, "Ferritic power plant steels: remanent life assessment and approach to equilibrium," *International Materials Reviews*, vol. 43, no. 2, pp. 45–69, 1998.

[5] F. Abe and S. Nakazawa, "Microstructural evolution and creep behaviour of bainitic, martensitic, and martensite-ferrite dual phase Cr-2W steels," *Materials Science and Technology*, vol. 8, no. 12, pp. 1063–1069, 1992.

[6] D. G. Cole, *Designs of heat-resistant steels for small power plant [Ph.D. thesis]*, Materials Science and Metallurgy, Darwin College, University of Cambridge, 2000.

[7] J. D. Robson and H. K. D. H. Bhadeshia, "Modelling precipitation sequences in powerplant steels Part 2—application of kinetic theory," *Materials Science and Technology*, vol. 13, no. 8, pp. 640–644, 1997.

[8] M. Shibuya, Y. Toda, K. Sawada, H. Kushima, and K. Kimura, "Improving the high-temperature creep strength of 15Cr ferritic creep-resistant steels at temperatures of 923–1023 K," *Materials Science and Engineering A*, vol. 652, pp. 1–6, 2016.

[9] H. K. D. H. Bhadeshia, "Estimation of the microstructure and properties of ferritic creep-resistant steels," in *Modelling of Microstructural Evolution in Creep Resistant Materials*, A. Strang

and M. McLean, Eds., pp. 15–38, The Institute of Materials, London, UK, 1999.

[10] H. K. D. H. Bhadeshia, "Thermodynamic analysis of isothermal transformation diagrams," *Metal Science*, vol. 16, no. 3, pp. 159–165, 1982.

[11] F. Brun, T. Yoshida, J. D. Robson, V. Narayan, H. K. D. H. Bhadeshia, and D. J. C. MacKay, "Theoretical design of ferritic creep resistant steels using neural network, kinetic, and thermodynamic models," *Materials Science and Technology*, vol. 15, no. 5, pp. 547–554, 1999.

[12] D. Cole, C. Martin-Moran, A. G. Sheard, H. K. D. H. Bhadeshia, and D. J. C. MacKay, "Modelling creep rupture strength of ferritic steel welds," *Science and Technology of Welding and Joining*, vol. 5, no. 2, pp. 81–89, 2000.

[13] F. Abe, "Coarsening behaviour of martensite lathes in tempered martensitic 9Cr-W steels during creep deformation," in *Proceedings of the 4th International Conference on Recrystallization and Related Phenomena*, pp. 289–294, The Japan Institute of Metals, 1999.

[14] F. Abe, "Research and development of advanced ferritic steel for 650°C USC boilers," in *Proceedings of the at 4th Ultra-Steels Workshop that took place in 2000 at Tsukuba International Congress CenterTitle: Ultra-Steel 2000: Innovative Structural Materials for Infrastructure in 21st Century Sponsor*, pp. 119–129, National Research Institute for Metals, Tsukuba, Japan, 2000.

[15] H. K. D. H. Bhadeshia, "Design of ferritic creep-resistant steels," *ISIJ International*, vol. 41, no. 6, pp. 626–640, 2001.

[16] F. Abe, *High Performance Creep Resistant Steels for 21st Century Power Plants*, National Institute for Materials Science (NIM), Tsukuba, Japan, 1999.

[17] M. Shibuya, Y. Toda, K. Sawada, H. Kushima, and K. Kimura, "Effect of precipitation behavior on creep strength of 15%Cr ferritic steels at high temperature between 923 and 1023 K," *Materials Science and Engineering A*, vol. 592, pp. 1–5, 2014.

[18] F. Abe, M. Igarashi, N. Fujitsuna, and K. Kimura, "Research and development of advanced ferritic steels for 650°C USC boilers," in *Proceedings of the 6th Liege Conference in Materials for Advanced Power Engineering*, vol. 5, part 1, pp. 259–268, 1998.

[19] F. Kauffmann, T. Klein, A. Klenk, and K. Maile, "Creep behavior and in-depth microstructural characterization of dissimilar joints," *Science and Technology of Advanced Materials*, vol. 14, no. 1, Article ID 014203, 2013.

[20] H. K. D. H. Bhadeshia, "Creep-resistant steel-case study," in *Kinetics and Microstructure Modelling Lecture Note for Course MP6*, pp. 1–6, Department of Materials & Metallurgy University of Cambridge, 2013.

[21] R. Viswanathan, J. F. Henry, J. Tanzosh, G. Stanko, J. Shingledecker, and B. Vitalis, "Materials for ultrasupercritical coal fired power plant boilers," in *Proceedings of the 29th International Technical Conference on Coal Utilization and Fuel Systems*, Clearwater, Fla, USA, 2004.

[22] A. Helmrich, Q. Chen, G. Stamatelopoulos, and B. Scarlin, "Materials development for advanced steam boilers," in *Proceedings of the 8th Liège Conference on Materials for Advanced Power Engineering*, J. Lecomte-Beckers, M. Carton, F. Schubert, and P. J. Ennis, Eds., vol. 2, pp. 931–948, September 2006.

[23] W. D. Callister, *Materials Science and Engineering: An Introduction*, John Wiley & Sons, New York, NY, USA, 7th edition, 2007.

[24] J. Rösler, H. Harders, and M. Baeker, *Mechanical Behaviour of Engineering Materials: Metals, Ceramics, Polymers, and Composites*, Springer, Berlin, Germany, 2007.

[25] H. K. D. H. Bhadeshia, *Bainite in Steels: Transformations, Microstructure and Properties*, The Institute of Materials, London, UK, 2nd edition, 2001.

[26] R. C. Thomson, *Carbide composition changes in power plant steels as a method of remanent creep life prediction [Ph.D. thesis]*, Department of Materials Science and Metallurgy, Newnham College, University of Cambridge, 1992.

[27] F. Masuyama and T. Yokoyama, "High temperature materials for power engineering," in *Materials for Advanced Power Engineering*, D. Coutsouradis, J. H. Davidson, J. Ewald et al., Eds., pp. 301–308, Kluwer Academic, Dordrecht, The Netherlands, 1994.

[28] K. Kimura, H. Kushima, F. Abe, K. Yagi, and H. Iries, "Assessment of creep strength properties of 9–12%Cr steels," in *Advances in Turbine Materials, Design and Manufacturing, 4th International Charles Parsons Turbine Conference*, pp. 257–269, The Institute of Materials, London, UK, 1997.

[29] H. K. D. H. Bhadeshia, *Mechanical Properties of Martensite in Heat-resistant Steels*, Proceedings of Ultra-Steel 2000, National Research Institute for Metals, Tsukuba, Japan, 2000.

[30] K. Kimura, Y. Toda, H. Kushima, and K. Sawada, "Creep strength of high chromium steel with ferrite matrix," *International Journal of Pressure Vessels and Piping*, vol. 87, no. 6, pp. 282–288, 2010.

[31] M. Shibuya, Y. Toda, K. Sawada, H. Kushima, and K. Kimura, "Effect of nickel and cobalt addition on the precipitation-strength of 15Cr ferritic steels," *Materials Science and Engineering A*, vol. 528, no. 16-17, pp. 5387–5393, 2011.

[32] J. D. Robson and H. K. D. H. Bhadeshia, "Modelling precipitation sequences in power plant steels part 1—Kinetic theory," *Materials Science and Technology A*, vol. 13, no. 8, pp. 631–639, 1997.

[33] Y. Murata, M. Morinaga, and R. Hashizume, "Development of ferritic steels for steam turbine rotors with the aid of a molecular orbital method," in *Advances in Turbine Materials, Design and Manufacturing, 4th International Charles Parsons Turbine Conference*, A. Strang, W. M. Banks, and R. D. Conroy, Eds., pp. 270–282, The Institute of Materials, London, UK, 1997.

[34] M. Morinaga, R. Hashizume, Y. Murata et al., "Electronic approach to the design of ferritic steels for power plants," in *Materials for Advanced Power Engineering*, D. Coutsouradis, J. H. Davidson, J. Ewald et al., Eds., part 1, pp. 319–328, Kluwer Academic, Dordrecht, The Netherlands, 1994.

[35] C. J. Tillman and D. V. Edmonds, "Alloy carbide precipitation and aging during high-temperature isothermal decomposition of an Fe-4Mo-0.2C alloy steel," *Metals Technology*, vol. 1, no. 10, pp. 456–461, 1974.

[36] V. K. Sikka, M. G. Cowgill, and B. W. Roberts, "Creep properties of modified 9Cr-1Mo steel," in *Proceedings of the Topical Conference on Ferritic Alloys for Use in Nuclear Energy Technologies*, pp. 412–423, TMS-AIME, Warrendale, Pa, USA, 1984.

[37] R. G. Baker and J. Nutting, "The tempering of 2.25Cr1Mo steel after quenching and normalizing," *Journal of the Iron and Steel Institute*, vol. 191, no. 3, pp. 257–268, 1959.

[38] K. J. Irvine and F. B. Pickering, "The temperature characteristics of low-carbon low-alloy steels," *Journal of Iron and Steel Institute*, vol. 194, no. 2, pp. 137–153, 1960.

[39] J. M. Titchmarsh, "The identification of second-phase particles in steels using an analytical transmission electron microscope," in *Proceedings of the 9th International Conference on Electron Microscopy*, J. M. Sturgess, Ed., vol. 1, pp. 618–619, Microscopical Society of Canada, Toronto, Canada, 1978.

[40] J. Beech and D. H. Warrington, "M_7C_3 to $M_{23}C_6$ transformation in chromium containing alloys," *Journal of the Iron and Steel Institute*, vol. 204, no. 5, pp. 460–468, 1966.

[41] J. H. Woodhead and A. G. Quarrell, "Role of carbides in low-alloy creep resisting steels," *Journal of the Iron and Steel Institute*, vol. 203, pp. 605–620, 1965.

[42] F. Abe, "Bainitic and martensitic creep-resistant steels," *Current Opinion in Solid State and Materials Science*, vol. 8, no. 3-4, pp. 305–311, 2004.

[43] D. V. Edmonds and R. W. K. Honeycombe, "Structure and properties of an isothermally transformed Fe-4Mo-0.2C alloy," *Journal of the Iron and Steel Institute*, vol. 211, pp. 209–216, 1973.

[44] J. Nutting, "The structural stability of low alloy steels for power generation application," in *Advanced Heat Resistant Steel for Power Generation*, R. Viswanathan and J. Nutting, Eds., pp. 12–30, Institute of Materials, Minerals and Mining, 1999.

[45] J. D. Robson, *Modelling of carbide and Laves phase precipitation in 9–12 wt% chromium steels [Ph.D. thesis]*, Materials Science and Metallurgy, Clare College, University of Cambridge, 1996.

[46] M. Ohgami, H. Minura, and H. Naoi, "Creep rupture properties and microstructures of a new ferritic W containing steel," in *Proceedings of the 5th International Conference on Creep of Materials*, pp. 69–73, Lake Buena Vista, Fla, USA, 1992.

[47] B. K. Choudhary, "Tertiary creep behaviour of 9Cr-1Mo ferritic steel," *Materials Science and Engineering A*, vol. 585, pp. 1–9, 2013.

[48] A. Strang and V. Vodárek, "Z phase formation in martensitic 12CrMoVNb steel," *Materials Science and Technology*, vol. 12, no. 7, pp. 552–556, 1996.

[49] K. Sawada, K. Suzuki, H. Kushima, M. Tabuchi, and K. Kimura, "Effect of tempering temperature on Z-phase formation and creep strength in 9Cr-1Mo–V–Nb–N steel," *Materials Science and Engineering A*, vol. 480, no. 1-2, pp. 558–563, 2008.

[50] H. K. Danielsen and J. Hald, "On the nucleation and dissolution process of Z-phase Cr(V,Nb)N in martensitic 12%Cr steels," *Materials Science and Engineering A*, vol. 505, no. 1-2, pp. 169–177, 2009.

[51] L. Cipolla, H. K. Danielsen, D. Venditti, P. E. Di Nunzio, J. Hald, and M. A. J. Somers, "Conversion of MX nitrides to Z-phase in a martensitic 12% Cr steel," *Acta Materialia*, vol. 58, no. 2, pp. 669–679, 2010.

[52] J. Pilling and N. Ridley, "Tempering of 2.25 Pct Cr-1 Pct Mo low carbon steels," *Metallurgical Transactions A*, vol. 13, no. 4, pp. 557–563, 1982.

[53] A. Iseda, Y. Sawaragi, S. Kato, and F. Masuyama, "Development of a new 0.1C-11Cr-2W-0.4Mo-1Cu steel for large diameter and thick wall pipe for boilers," in *Creep: Characterization, Damage and Life Assessments, Proceedings of the 5th International Conference on Creep of Materials, Lake Buena Vista, Florida, USA*, ASM International, 1992.

[54] V. Vodárek and A. Strang, "Effect of nickel on the precipitation processes in 12CrMoV steels during creep at 550°C," *Scripta Materialia*, vol. 38, no. 1, pp. 101–106, 1997.

[55] S. D. Mann, D. G. McCulloch, and B. C. Muddle, "Identification of M_5C_2 carbides in ex- service 1Cr-0.5Mo steels," *Metallurgical and Materials Transactions A*, vol. 26, article 509, 1995.

[56] T. Fujita and K. Asakura, "Effects of carbon on creep rupture strength and toughness of high Cr-Mo heat-resisting steels containing V and Nb," *Transactions of the Iron and Steel Institute of Japan*, vol. 26, no. 12, pp. 1073–1079, 1986.

[57] K. Miyata, M. Igarashi, and Y. Sawaragi, "Effect of trace elements on creep properties of 0.06C-2.25Cr-1.6W-0.1Mo-0.25V-0.05Nb steel," *ISIJ International*, vol. 39, no. 9, pp. 947–954, 1999.

[58] M. Igarashi and Y. Swaragi, "Development of 0.1C-11Cr-3W-3Co-V-Nb-Ta-Nd-N ferritic steel for USC boilers," in *Proceedings of the International Conference on Power Engineering*, pp. 107–112, Japan Society of Mechanical Engineers (JSME), Tokyo, Japan, 1997.

[59] I. M. Park and T. Fujita, "Long-term creep rupture properties and microstructure of 12% Cr heat resisting steels," *Transactions of the Iron and Steel Institute of Japan*, vol. 22, no. 11, pp. 830–837, 1982.

[60] M. Hättestrand and H.-O. Andrén, "Boron distribution in 9–12% chromium steels," *Materials Science and Engineering A*, vol. 270, no. 1, pp. 33–37, 1999.

[61] F. Abe, "Effect of boron on microstructure and creep strength ofadvanced ferritic power plant steels," *Procedia Engineering*, vol. 10, pp. 94–99, 2011.

[62] T. Tsuchiyama, Y. Futamura, and S. Takaki, "Strengthening of heat resistant martensitic steel by Cu addition," in *Creep and Fracture of Engineering Materials and Structures*, T. Sakuma and K. Yagi, Eds., vol. 171–174 of *Key Engineering Materials*, pp. 411–418, 2000.

[63] H. L. Solberg, G. A. Hawkins, and A. A. Potter, "Corrosion of unstressed steel specimens by high temperature steam," *Transactions of the American Society of Mechanical Engineers*, vol. 64, pp. 303–316, 1942.

[64] I. A. Rohrig, R. M. Van Duzer, and C. H. Fellows, "High temperature steam corrosion studies at Detroit," *Transactions of the American Society of Mechanical Engineers*, vol. 66, pp. 277–290, 1944.

[65] P. J. Ennis, Y. Wouters, and W. J. Quadakkers, "The effects of oxidation on the service life of 9–12 wt% Cr steels," in *Advanced Heat Resistant Steels for Power Generation, San Sabastian, Spain*, R. Viswanathan and J. Nutting, Eds., pp. 457–467, IoM Communications, 1998.

[66] Y. Kato, M. Ito, Y. Kato, and O. Furukim, "Effect of Si on precipitation behaviour of Nb-laves phase and amount of Nb in Solid solution at elevated temperature in high purity 17%Cr-0.5%Nb Steels," *Materials Transactions*, vol. 51, no. 9, pp. 1531–1535, 1986.

[67] N. Fujitsuna, M. Igarashi, and G. R. Booker, "Acceleration of Fe3W precipitation and its effect on creep deformation behaviour of 8.5Cr-2W-VNb steels with Si," in *Creep and Fracture of Engineering Materials and Structures*, T. Sakuma and K. Yagi, Eds., vol. 171–174 of *Key Engineering Materials*, pp. 469–476, 2000.

[68] J. Vanaja, K. Laha, M. D. Mathew, T. Jayakumar, and E. Rajendra Kumar, "Effects of tungsten and tantalum on creep deformation and rupture properties of reduced activation ferritic-martensitic," *Procedia Engineering*, vol. 55, pp. 271–276, 2013.

[69] T. Fujita, K. Asakura, T. Sawada, T. Takamatsu, and Y. Otoguro, "Creep rupture strength and microstructure of low C-10Cr-2Mo heat-resisting steels with V and Nb," *Metallurgical Transactions A*, vol. 12, no. 6, pp. 1071–1079, 1981.

[70] J. Pilling and N. Ridley, "Tempering of 2.25 Pct Cr-1 Pct Mo Low Carbon Steels," *Metallurgical Transactions A*, vol. 13, no. 4, pp. 557–563, 1982.

[71] D. J. Gooch, "Creep fracture of 12CrMoV steel," *Metal science*, vol. 16, no. 2, pp. 79–89, 1982.

[72] T. Fujita and N. Takahashi, "The effects of V and Nb on the long period creep rupture strength of 12 wt% Cr heat-resisting steel

containing Mo and B," *Transactions of the Iron and Steel Institute of Japan*, vol. 18, pp. 269–278, 1979.

[73] V. Foldyna, A. Jakobova, R. Riman, and A. Gemperle, "Effect of structural factors on the creep properties of modified chromium steels," *Steel Research*, vol. 62, no. 10, pp. 453–458, 1991.

[74] F. Masuyama and M. Ohgami, "Creep behaviour of high nitrogen ferritic steel," in *Proceedings of the 7th International Symposium on Aspects of High Temperature Deformation and Fracture in Crystalline Materials*, Y. Hosoi, H. Yoshinaga, H. Oikawa, and K. Maruyama, Eds., pp. 325–332, Japan Institute of Metals, Sendai, Japan, 1993.

[75] F. W. Noble, B. A. Senior, and B. L. Eyre, "The effect of phosphorus on the ductility of 9Cr-1Mo steels," *Acta Metallurgica et Materialia*, vol. 38, no. 5, pp. 709–717, 1990.

[76] H. Naoi, M. Ohgami, Y. Hasegawa et al., "Mechanical properties of 12Cr-W-Co ferritic steels with high creep rupture strength," in *Materials for Advanced Power Engineering', Part 1*, D. Coutsouradis, J. J. Davidson, J. Ewald, and T. Fujita, Eds., vol. 1, pp. 425–434, Kluwer Academic, Dordrecht, The Netherlands, 1994.

Permissions

All chapters in this book were first published in AMSE, by Hindawi Publishing Corporation; hereby published with permission under the Creative Commons Attribution License or equivalent. Every chapter published in this book has been scrutinized by our experts. Their significance has been extensively debated. The topics covered herein carry significant findings which will fuel the growth of the discipline. They may even be implemented as practical applications or may be referred to as a beginning point for another development.

The contributors of this book come from diverse backgrounds, making this book a truly international effort. This book will bring forth new frontiers with its revolutionizing research information and detailed analysis of the nascent developments around the world.

We would like to thank all the contributing authors for lending their expertise to make the book truly unique. They have played a crucial role in the development of this book. Without their invaluable contributions this book wouldn't have been possible. They have made vital efforts to compile up to date information on the varied aspects of this subject to make this book a valuable addition to the collection of many professionals and students.

This book was conceptualized with the vision of imparting up-to-date information and advanced data in this field. To ensure the same, a matchless editorial board was set up. Every individual on the board went through rigorous rounds of assessment to prove their worth. After which they invested a large part of their time researching and compiling the most relevant data for our readers.

The editorial board has been involved in producing this book since its inception. They have spent rigorous hours researching and exploring the diverse topics which have resulted in the successful publishing of this book. They have passed on their knowledge of decades through this book. To expedite this challenging task, the publisher supported the team at every step. A small team of assistant editors was also appointed to further simplify the editing procedure and attain best results for the readers.

Apart from the editorial board, the designing team has also invested a significant amount of their time in understanding the subject and creating the most relevant covers. They scrutinized every image to scout for the most suitable representation of the subject and create an appropriate cover for the book.

The publishing team has been an ardent support to the editorial, designing and production team. Their endless efforts to recruit the best for this project, has resulted in the accomplishment of this book. They are a veteran in the field of academics and their pool of knowledge is as vast as their experience in printing. Their expertise and guidance has proved useful at every step. Their uncompromising quality standards have made this book an exceptional effort. Their encouragement from time to time has been an inspiration for everyone.

The publisher and the editorial board hope that this book will prove to be a valuable piece of knowledge for researchers, students, practitioners and scholars across the globe.

List of Contributors

Agnes Yung Weng Lee, Soh Fong Lim, S. N. David Chua, Khairuddin Sanaullah, Rubiyah Baini and Mohammad Omar Abdullah
Faculty of Engineering, Universiti Malaysia Sarawak, 94300 Kota Samarahan, Malaysia

Xiaojun Wang, Yulong Zhuo, Shuqiang Deng, Yongxin Li, Wen Zhong and Kui Zhao
Jiangxi Key Laboratory of Mining Engineering, Jiangxi University of Science and Technology, Ganzhou, Jiangxi 341000, China
School of Resources and Environment Engineering, Jiangxi University of Science and Technology, Ganzhou, Jiangxi 341000, China

Ji-Yeon Shim and Bong-Yong Kang
Carbon & Light Materials Application R&D Group, KITECH, Jeonju-si, Republic of Korea

Kalyan Kumar Singh and Saurabh Singh
Department of Mechanical Engineering, Indian School of Mines, Dhanbad 826004, India

Anil Kumar Shrivastava
Department of Mechanical Engineering, Galgotias College of Engineering & Technology, Greater Noida 201306, India
Jan Koplík, Jaromír Pořízka, Lukáš Kalina, Jiří Másilko and Matěj Březina
Faculty of Chemistry, Materials Research Centre, Brno University of Technology, Purkyňova 118, 612 00 Brno, Czech Republic

Gyuhe Kim, Taechang Ahn and Hui Yun Hwang
Department of Mechanical Design Engineering andong National University, 1375 Gyeongdong-ro andong-si, Gyeongsangbuk-do 760-749, Republic of Korea

Behzad Fotovvati, Steven F. Wayne, Gladius Lewis and Ebrahim Asadi
Department of Mechanical Engineering, University of Memphis, Memphis, TN 38152, USA

Yingda Qian, Yuanlan Liang, Xuguang Luo, Kaiyan He, Lingyu Wan, Qingyi Yang and Zhe Chuan Feng
College of Physical Science and Technology, Guangxi University, Nanning 530004, China
Laboratory of Optoelectronic Materials and Detection Technology, Guangxi Key Laboratory for the Relativistic Astrophysics, Guangxi University, Nanning 530004, China

Wenhong Sun
College of Physical Science and Technology, Guangxi University, Nanning 530004, China

Hao-Hsiung Lin
Graduate Institute of Electronics and Department of Electrical Engineering, National Taiwan University, Taipei 106-17, Taiwan

Devki N. Talwar
Department of Physics, Indiana University of Pennsylvania, Indiana, PA 15705-1087, USA
Department of Physics, University of North Florida, Jacksonville, FL 32224, USA

Ting-Shan Chan
National Synchrotron Radiation Research Center, Hsinchu 300-76, Taiwan

Ian Ferguson
Department of Electrical and Computer Engineering, Missouri University of Science and Technology, Rolla, MO 65409, USA

Yanni Wei
Department of Materials Science and Engineering, Xi'an University of Technology, 5 South Jinhua Road, Xi'an 710048, China

Fu Sun
Shaanxi Zhituo Solid-State Additive Manufacturing Technology Company, 70 West of Chaoyang Street, Weinan 714000, China

J. J. Ramos-Hernandez
Facultad de Química, Universidad Nacional Autónoma de Mexico, Edificio B, Cd. Universitaria, 04510 Ciudad de Mexico, Mexico

J. Porcayo-Calderon and J. G. Gonzalez-Rodriguez
CIICAp, Universidad Autónoma del Estado de Morelos, Avenida Universidad 1001, 62209 Cuernavaca, MOR, Mexico

Jan Mayén
CIATEQ A.C., Eje 126 No. 225, Zona Industrial, 78395 San Luis Potosí, SLP, Mexico

G. Lara-Rodriguez
Instituto de Investigaciones en Materiales, Universidad Nacional Autónoma de Mexico, Cd. Universitaria, 04510 Ciudad de Mexico, Mexico

V. M. Salinas-Bravo
Instituto Nacional de Electricidad y Energías Limpias, Avenida Reforma 113, Colonia Palmira, 62490 Cuernavaca, MOR, Mexico

L. Martinez-Gomez
Instituto de Ciencias Físicas, Universidad Nacional Autónoma de Mexico, Avenida Universidad s/n, 62210 Cuernavaca, MOR, Mexico
Corrosion y Protección (CyP), Buffon 46, 11590 Mexico City, Mexico

Er-Hui Zhao
School of Mechanical Engineering, Beijing Institute of Technology, Beijing, China

Biao Ma and He-Yan Li
School of Mechanical Engineering, Beijing Institute of Technology, Beijing, China
Collaborative Innovation Center of Electric Vehicles in Beijing, Beijing, China

Mingshun Yang, Zimeng Yao, Yan Li, Pengyang Li and Lang Bai
Faculty of Mechanical and Precision Instrument Engineering, Xi'an University of Technology, Xi'an 710048, China

Fengkui Cui
School of Mechatronics Engineering, Henan University of Science and Technology, Luoyang, China

Alexei Pustovalov and Vladimir An
Institute of High Technology Physics, National Research Tomsk Polytechnic University, 30 Lenin Ave., Tomsk 634050, Russia

Jin-Chun Kim
School of Materials Science and Engineering, University of Ulsan, 93 Daehak-ro, Nam-Ku, Ulsan 44610, Republic of Korea

Tomáš Ibehej, Jakub Hromádka and Rudolf Hrach
Department of Surface and Plasma Science, Faculty of Mathematics and Physics, Charles University, V Holesovickach 2, 180 00 Prague 8, Czech Republic

Oleg Gorshkov, Dmitry Filatov, Maria Shenina and Alexander Bobrov
Research and Educational Center for Physics of Solid State Nanostructures, Lobachevsky State University of Nizhny Novgorod, 23 Gagarin Ave., Nizhny Novgorod 603950, Russia

Ivan Antonov, Alexander Kasatkin and Maria Koryazhkina
Research Institute for Physics and Technology, Lobachevsky State University of Nizhny Novgorod, 23 Gagarin Ave., Nizhny Novgorod 603950, Russia

Irina Korotaeva and Mikhail Kudryashov
Department of Physics, Lobachevsky State University of Nizhny Novgorod, 23 Gagarin Ave., Nizhny Novgorod 603950, Russia

Bo Zheng
MacDermid Enthone, MacDermid Performance Solutions, 26 Tuas West Road, Singapore 638382
Singapore Institute of Manufacturing Technology, 2 Fusionopolis Way, No. 08-04, Innovis, Singapore 138634

Lai Peng Wong
MacDermid Enthone, MacDermid Performance Solutions, 26 Tuas West Road, Singapore 638382

Linda Y. L. Wu
School of Materials Science and Engineering, Nanyang Technological University, 50 Nanyang Avenue, Singapore 639798

Zhong Chen
Singapore Institute of Manufacturing Technology, 2 Fusionopolis Way, No. 08-04, Innovis, Singapore 138634

Hong Zhou, Sijia Sun and Hao Ding
Beijing Key Laboratory of Materials Utilization of Nonmetallic Minerals and Solid Wastes, National Laboratory of Mineral Materials, School of Materials Science and Technology, China University of Geosciences, Beijing 100083, China

Ahmet Ozan Gezerman and Burcu Didem Çorbacıoğlu
Department of Chemical Engineering, Faculty of Chemical and Metallurgical Engineering, Yıldız Technical University, Istanbul, Turkey

Xingling Shi
School of Materials Science and Engineering, Jiangsu University of Science and Technology, Zhenjiang 212003, China
TusPark (Kunshan) Development Co., Ltd., Kunshan 215347, China

Chao Zhao
Hitachi Automotive System (Guangzhou) Ltd., Guangzhou 511300, China

Lingli Xu
School of Materials Science and Engineering, Jiangsu University of Science and Technology, Zhenjiang 212003, China

Qingliang Wang
School of Materials Science and Engineering, China University of Mining and Technology, Xuzhou 221116, China

J. Porcayo-Calderon, E. Porcayo-Palafox and J. G. Gonzalez-Rodriguez
CIICAp, Universidad Autónoma del Estado de Morelos, Avenida Universidad 1001, 62209 Cuernavaca, MOR, Mexico

J. J. Ramos-Hernandez
Instituto de Ciencias Físicas, Universidad Nacional Autónoma de México, Avenida Universidad s/n, 62210 Cuernavaca, MOR, Mexico

Jan Mayén
CENIDET, Tecnológico Nacional de México, Prolongación Palmira s/n Esq. Apatzingán, Col. Palmira, 62490 Cuernavaca, MOR, Mexico

G. K. Pedraza-Basulto
Facultad de Ingeniería, UNACAR, 24180 Ciudad del Carmen, CAM, Mexico

L. Martinez-Gomez
Instituto de Ciencias Físicas, Universidad Nacional Autónoma de México, Avenida Universidad s/n, 62210 Cuernavaca, MOR, Mexico
Corrosion y Protección (CyP), Buffon 46, 11590 México City, Mexico

Bin Liu, Yong Tang, Wei Yuan, Chao-bin Fang and Zhen-ping Wan
School of Mechanical and Automotive Engineering, South China University of Technology, Guangzhou 510640, China

Xin-gang Wang
China Academy of Space Technology (Xi'an), Xi'an 710000, China

Index

www.ingramcontent.com/pod-product-compliance
Lightning Source LLC
LaVergne TN
LVHW081018210926
848485LV00004B/897